ROBERT L. CHILDRESS

School of Business Administration
University of Southern California

Fundamentals of Finite Mathematics

PRENTICE-HALL, INC., Englewood Cliffs, New Jersey

Library of Congress Cataloging in Publication Data

CHILDRESS, ROBERT L.
   Fundamentals of finite mathematics.

   Includes index.
   1. Mathematics—1961–   I. Title.
QA39.2.C45   510   75-37725
ISBN   0-13-339325-9

© 1976
by Prentice-Hall, Inc.
Englewood Cliffs, New Jersey

Printed in the United States of America

10   9   8   7   6   5   4   3   2   1

Prentice-Hall International, Inc., *London*
Prentice-Hall of Australia Pty. Limited, *Sydney*
Prentice-Hall of Canada, Ltd., *Toronto*
Prentice-Hall of India Private Limited, *New Delhi*
Prentice-Hall of Japan, Inc., *Tokyo*
Prentice-Hall of Southeast Asia Pte. Ltd., *Singapore*

1900455

To Shirley

Contents

Tables   483

Selected Answers to Problems   519

Index   541

Preface

During the past two decades the importance of applied mathematical training for students of business, economics, government, and the sciences has been well-established. Courses of instruction have gradually evolved to provide this training. The sequence of courses often begins with finite mathematics and extends through advanced topics in operations research. Needless to say, many students who follow this entire sequence are well prepared to apply their mathematical training in a number of different professions.

As we well know, however, not all students have the inclination to take the entire mathematical sequence. For this reason, the introductory courses in the sequence are especially important. An introductory course, such as finite mathematics, must be taught with the objectives of providing useful mathematical techniques as well as providing fundamental concepts that are preliminary to more advanced studies. In other words, the course must be applied yet theoretical. To some it must serve as a beginning and to others as an end. It must stimulate the sophisticated student, but not at the expense of the average student. Most important, the course must provide the excitement and realism that will encourage students to enthusiastically continue their studies of mathematics.

These objectives cannot be accomplished by a text alone. If so, teaching would merely involve assigning the appropriate material and measuring the student's progress. Neither can they be easily met by a teacher without the aid of a text. Most students require the detailed explanation, the example, and the application. The teacher together with a suitable text can, however, accomplish such objectives with a good number of students. *Fundamentals of Finite Mathematics* has been designed with this in mind. Throughout the text, emphasis has been placed on explaining the subject matter in a conceptually rigorous yet easily readable and understandable fashion. Detailed examples and numerous applications have been included to reinforce the conceptual material as well as provide the student with some insight into the usefulness of the mathematical techniques. The only prerequisite for understanding these concepts, applications, and examples is a good course in algebra.

The text begins with a fairly standard presentation of sets and an abbreviated section on logic. The next two chapters introduce the concepts of functions and systems of equations. In so doing, the student is afforded the opportunity to review fundamental algebraic operations such as the solution procedures for simultaneous equations. Special care is taken in these chapters to illustrate the importance of functions and systems of equations in decision making. Examples include cost and revenue functions, breakeven analysis, predicting equations, and an important section on the exponential function and its use in determining growth rates. Chap. 4 provides a thorough discussion and numerous applications of the rules of counting. This is followed in Chap. 5 by a discussion of probability and in Chap. 6 by a discussion of probability functions. Measures of central tendency and variability are introduced along with discrete and, to a limited extent, continuous probability functions. The hypergeometric, binomial, and normal distributions are included. Chap. 7 continues the discussion of probability by introducing several important topics in probability and statistics. These include Chebyshev's inequality, the concept of sampling, an introduction to the central limit theorem, and an elementary presentation of confidence intervals and tests of hypothesis. Chap. 8 presents an introductory approach to decision theory and an exposure to game theory.

Chap. 9 introduces matrices and matrix algebra. Examples are included to show the parallel between the row operations of matrix algebra and the ordinary algebraic operations used in Chap. 2 for solving simultaneous equations. The student is asked to apply these concepts to Markov chains in Chap. 10. The concept of a Markov chain is introduced by using an example involving brand loyalty taken from the field of marketing. Chap. 11 provides an introduction to linear programming and presents the graphical solution to the linear programming problem. For those instructors who are so inclined, the chapter offers examples of realistic applications of linear programming that cannot be found in other texts written at this level. Chap. 12 discusses the simplex algorithm. Chap. 13 presents the transportation and assignment problems. Chap. 14 provides a discussion of mathematics of finance. A discussion of logarithms and rules of exponents is included as an Appendix.

Most instructors will have no more than one semester available for the material in this text. This, of course, means that certain subjects must be emphasized at the expense of others. Since opinions differ on the relative importance of the various topics, no universal outline is suggested. Chaps. 7,

10, and 12 require more time for thorough coverage than some of the other chapters and can be omitted in a basic course. Instructors wishing to emphasize probability will definitely want to cover Chaps. 4 through 8. If the emphasis is placed on operations research, Chaps. 9 through 13 become of primary importance. Chaps. 1 and 2 are prerequisites to the remaining chapters in the text.

The textual material is supplemented by 400 problems and 300 examples. The solution procedure for all of the examples is given in detail. Selected answers to the problems are included in the Appendix. For student reference, the text includes amount, present value, and annuity tables along with tables for common and natural logarithms, exponential functions, squares and square roots, the binomial distribution, and the standard normal distribution.

I am grateful to the numerous students at the University of Southern California who have contributed helpful suggestions for improving the text. My thanks also go to my department chairman, Dr. George Schick, for creating the kind of atmosphere that makes writing a textbook possible. I want to thank Professor Pat Wheatley of California Polytechnic University and Roland Sink of Pasadena City College for their suggestions on the table of contents. I am again grateful to Miss Chris Kutschinski who worked tirelessly on typing the manuscript.

ROBERT L. CHILDRESS
Los Angeles, California

To the Student

Students often approach a course in mathematics with a certain degree of fear or perhaps hesitation regarding their ability to successfully master the course material. Perhaps in the past these fears have been well-founded. After all, it does take a special kind of training to understand a complicated proof or to answer a question that seems more like a riddle than a problem which one might someday be faced with in a business situation. Fortunately, during the past decade a revolution has occurred in the teaching of mathematics. Whether this revolution has been brought about by students clamoring for applicability in their courses or by instructors tiring of teaching abstract concepts without applications does not matter. The important point is that the *revolution of applicability* in college mathematics has come.

Both student and teacher share in the benefits of this new approach in teaching mathematics. Students find it easier to understand concepts that are illustrated by concrete, practical examples. Moreover, they often find that the concepts and examples can be directly applied both in other courses and in their work. Instructors benefit from the fact that far more students understand the importance and applicability of the subject matter. This, of course, makes the job of the instructor much more satisfying and rewarding.

To illustrate what we mean by applicability in college mathematics, consider the following problem taken from Chap. 13 of this text. A firm has three factories and four warehouses. The cost of shipping from each factory to each warehouse depends on the distance the product must be shipped, freight rates, etc. These costs are shown in the table on the following page.

| Factory (F_i) | Warehouse (W_j) | | | | Factory Capacity |
	W_1	W_2	W_3	W_4	
F_1	$0.30	$0.25	$0.40	$0.20	100
F_2	0.29	0.26	0.35	0.40	250
F_3	0.31	0.33	0.37	0.30	150
Warehouse Requirement	100	150	200	50	500

The cost of shipping one unit from factory 1 to warehouse 1 is $0.30. Similarly, the cost of shipping 1 unit from factory 2 to warehouse 3 is $0.35. The capacities of the three factories during the current period are given in the last column of the table and the requirements at each warehouse are given in the bottom row of the table. The objective of the analyst is to develop a shipping schedule that results in minimum total shipping costs. Although it is not obvious at this time, the reader will find that problems like this are common in all forms of business and that they are simple to solve. Once the reader understands the few simple steps needed to solve this kind of problem, he will be able to *apply* this mathematical technique to many realistic business problems.

What, then, do we mean by the revolution of applicability in college mathematics? Simply, we mean that the subjects taught in a college mathematics course are directly applicable in the administration of a business, social, or governmental enterprise. The reader, perhaps to his amazement, will find that the topics discussed in this text and by his instructor will be directly applicable to his work.

What Is Finite Mathematics?

The term *finite mathematics* is commonly used to describe a group of important mathematical topics that are not based on calculus. Although this group of topics varies in both level and degree among alternative textbooks, most instructors would include sets, functions, probability, linear algebra, and linear programming as important topics in finite mathematics. These topics, along with selected applications based on these topics, are included in this text. Instead of offering a brief summary of each of these topics at this time, we shall introduce the topics in an appropriate sequence in the text.

Tips on Reading This Text

The chapters that follow introduce important topics in finite mathematics. In reading these chapters the student should first concentrate on understanding the concepts introduced in each chapter. Once the concepts are understood, the symbolism and problem formulations can be mastered by careful study of the examples. Similarly, the discussion in each chapter along with the examples should enable the reader to understand the solution procedure. Finally, the examples in each of the chapters along with the problems at the conclusion of each chapter should be worked. These examples and problems illustrate important applications of the topics introduced in the chapters. So that the student can check his work, selected answers to odd-numbered problems are included at the end of the text.

To illustrate these tips for studying finite mathematics, consider a simple problem in breakeven analysis. An individual is considering the purchase of a new Beachcraft Bonanza which will be rented by the hour for business and pleasure flying. The Bonanza is a high-performance, single-engine aircraft that rents for $36 per flying hour. The fixed cost of owning this aircraft is $500 per month. These costs include depreciation, insurance, hanger rental, and certain maintenance fees. The variable cost, which includes gasoline, reserve for engine and propeller overhaul, etc., is $20 per hour. The individual wants to determine the number of rental hours per month required to breakeven.

As in most mathematical problems, the concept illustrated above is simple. Breakeven is merely the number of hours that the aircraft must be flown so that revenue and cost are equal. To determine this number of hours, we recognize that revenue is $36 per hour and costs are $500 per month plus $20 per hour. We formulate the problem by letting x represent flying hours. Revenue is thus

$$R = 36x$$

and costs are

$$C = 500 + 20x$$

Since breakeven occurs when revenue and costs are equal, we equate revenue and costs to obtain

$$R = C$$

$$36x = 500 + 20x$$

To solve for the breakeven number of hours, we subtract $20x$ from both sides of the equation and obtain

$$36x - 20x = 500 + 20x - 20x$$

$$16x = 500$$

$$x = 31.25 \text{ hours}$$

The aircraft must fly 31.25 hours per month to breakeven.

Although this example is simple, it illustrates the important tips for studying mentioned above. The concept discussed in the example is break-even. The problem formulation involves equating total revenue with total costs. The solution is obtained by solving an algebraic equation for breakeven sales (e.g., hours of flight time per month). A sufficient number of examples and problems concerning breakeven are given in the text to enable the reader to master the concept and to apply the method in an actual business situation.

As the student begins this course in finite mathematics, certain points should be kept in mind. First, the topics discussed in this text and by the instructor are important in the administration of a modern enterprise. Consequently, the reader should make every attempt to master and under-stand the important concepts discussed in his class. The discussion, examples, and problems given in this text and by the instructor should enable the student to learn to formulate and solve certain classes of mathematical problems. Most important, the student should learn how to apply the mathematical techniques discussed in the course to the administration of an enterprise.

Chapter One

Sets and Logic

The concept of sets and the algebra of sets is enjoying increasing popularity and usage in business and the sciences. One reason for this popularity is that an understanding of the basic concepts of sets and set algebra provides a form of language through which the business specialist or practitioner can communicate important concepts and ideas to his associates. Specialists in electrical engineering, for example, use set algebra in the design of circuits. At the opposite extreme, specialists in written communications use sets in the analysis of statements and preparation of reports. Students of business administration use sets in the study of probability, statistics, programming, optimization, etc. This chapter provides an introduction to sets and set algebra along with selected applications. The reader can expect additional applications later in this text as well as in his studies of marketing, finance, economics, statistics, and the sciences.

1.1 The Concept of a Set

A *set* is defined as a collection or aggregate of objects. To illustrate this definition, the members of the reader's finite mathematics course are a set of students. Similarly, the members of the American Economic Association are a set of individuals, the faculty of the school of business is a set of instructors, and the *Encyclopaedia Britannica* is a set of books.

From these examples the student can see that the concept of a set is relatively straightforward. There are, however, certain requirements for the collection or aggregate of objects that constitute the set. These requirements are:

1. The collection or aggregate of objects must be well-defined, i.e., we must be able to determine unequivocally whether or not any object belongs to the set.
2. The objects of a set must be distinct, i.e., we must be able to distinguish between the objects.
3. The order of the objects within the set must be immaterial, i.e., a, b, c is the same set as c, b, a.

On the basis of these requirements for a set, suppose we are asked to verify that the students in a finite mathematics class are a set. To determine if the students are a set, we ask three questions. First, can we determine if a student is registered for the course? Second, is it possible to distinguish between the students? Third, is the order of individuals in the class immaterial, i.e., is the class the same set whether arranged alphabetically by student or by social security number? If the answer to all three questions is yes, we conclude that the class is a set.

Example. Verify that the letters in the word Mississippi satisfy the requirement for a set. The letters in this word are m, i, s, and p. These letters are well-defined, distinct, and the order of the letters is immaterial. Therefore, the letters are a set.

Example. A coin is tossed three times. Denoting a head by H and a tail by T, determine the set that represents all possible outcomes of the three tosses.

The possible outcomes are $\{HHH, HHT, HTH, THH, HTT, THT, TTH, TTT\}$, where HHH represents a head on the first toss, a head on the second toss, and a head on the third toss. HHT represents a head on the first toss, a head on the second toss, and a tail on the third toss, etc.

Example. The set of digits is defined as the numbers $\{0, 1, 2, 3, 4, 5, 6, 7, 8, 9\}$. Determine which of the following numbers are members of this set: 3, 7.2, IV, 137, $\frac{4}{3}$.

The digit 3 is the only member of the set. The reader should be able to verify that the remaining numbers do not belong to the set of digits.

1.2 Specifying Sets and Membership in Sets

It is customary to designate a set by a capital letter. For instance, the set of digits defined in the above example could be designated as D.

The objects that belong to the set are termed the *elements* of the set or *members* of the set. The elements or the members of the set are designated by one of two methods: (1) the roster method or (2) the descriptive method. The roster method involves listing within braces all members of the set. The descriptive method (also called the defining property method) involves describing the membership in a manner such that one can determine if an object belongs in the set.

To illustrate the specification of sets and membership, consider again the set of digits. The set of digits is designated by the capital letter D. The elements in the set (or alternatively the members of the set) are shown either by listing all the elements in the set within braces or by describing within braces the membership. If the roster method is used, the set of digits would appear as

$$D = \{0, 1, 2, 3, 4, 5, 6, 7, 8, 9\}$$

This is read as "D is equal to that set of elements 0, 1, 2, 3, 4, 5, 6, 7, 8, 9." If the descriptive method of specifying the set is used, the set would be

$$D = \{x \mid x = 0, 1, 2, 3, \ldots, 9\}$$

This is read as "D is equal to that set of elements x such that x equals 0, 1, 2, 3, ..., 9." In interpreting the symbolism used in set notation, it is useful to think of the left brace as shorthand for "that set of elements" and the vertical line as shorthand for "such that." Commas are used to separate the elements, and the three periods mean "continuing in the established pattern." The 9 in the above set is interpreted as the final number of the set. The right brace designates set completion.

Example. The positive integers or "natural numbers" are the numbers 1, 2, 3, 4, 5, ... Show the set of natural numbers.

Assume that the set of natural numbers is represented by N. We cannot, of course, list all of the members of the set. We therefore use the descriptive method of specifying set membership and write

$$N = \{x \mid x = 1, 2, 3, 4, 5, \ldots\}$$

Example. Develop the set notation for the English alphabet.

We can use either the roster method or the descriptive method of specifying set membership. Representing the set of letters in the English alphabet by A, the set is

$$A = \{a, b, c, d, e, f, g, h, i, j, k, l, m, n, o, p, q, r, s, t, u, v, w, x, y, z\}$$

The descriptive method would conserve some space,

$$A = \{a, b, c, \ldots, y, z\}$$

Either method is acceptable. When one is using the descriptive method, however, it is important to remember that the description must be sufficient for one to determine the membership of the set. The elements must be well-defined, distinct, and order must be immaterial.

Example. The possible convention sites for the Western Farm Equipment Association are Los Angeles, San Francisco, Phoenix, and Las Vegas. The set of convention sites is

$$S = \{\text{Los Angeles, San Francisco, Phoenix, Las Vegas}\}$$

1.2.1   SET MEMBERSHIP

The Greek letter $\in$ (epsilon) is customarily used to indicate that an object belongs to a set. If A again represents the set of letters in the English alphabet, then $a \in A$ means that a is an element of the alphabet. The symbol $\notin$ (epsilon with a slashed line) represents nonmembership. We could thus write $\alpha \notin A$, meaning that alpha is not a member of the English alphabet. Similarly, referring to the convention site set S, we can write Los Angeles $\in S$ and San Diego $\notin S$.

1.2.2   FINITE AND INFINITE SETS

A set is termed *finite* or *infinite*, depending on the number of elements in the set. The set A defined above is finite since it has 26 members, the letters in the English alphabet. The set D is also finite, since it has only the ten digits. The set N of positive integers or natural numbers is infinite, since the process of counting continues infinitely.

Example. Rational numbers are defined as that set of numbers a/b, where a represents all integers, both positive and negative including 0, and b represents all positive and negative integers, excluding 0. Develop the set R of rational numbers and specify whether the set is finite or infinite.

Letting R represent the set of rational numbers, we have

$$R = \{a/b \mid a = \text{all integers including 0}, b = \text{all integers excluding 0}\}$$

The set of rational numbers is infinite. Examples of rational numbers include any number that can be expressed as the ratio of two positive or negative whole numbers, such as $\frac{3}{2}$, $-\frac{6}{2}$, 7, etc.

Example. The individuals who are members of the American Economic Association comprise a set. Assuming that a list of the membership is available, we could write the set as

$$S = \{\text{all members of the American Economic Association}\}$$

This set is finite, although quite large. It is a set because the elements are well-defined, distinct, and of inconsequential order.

1.3 Set Equality and Subsets

Two sets P and Q are said to be equal, written $P = Q$, if every element in P is in Q and every element in Q is in P. Set equality thus requires all elements of the first set to be in a second set and all elements in the second set to be in the first set. As an example, consider the set

$$P = \{0, 1, 2, 3, 4, 5\} \quad \text{and} \quad Q = \{2, 0, 1, 3, 5, 4\}$$

These sets are equal, since every element in P is in Q and every element in Q is in P.

The student will often have occasion to consider only certain elements of a set. These elements form a *subset* of the original set. As an example, assume that S represents the stockholders of company XYZ. Those stockholders who are employees of the company represent a subset of S. Thus, if Mr. Jones is a stockholder of XYZ and is employed by XYZ, he is a member of the subset of stockholders who are employed by the company.

Subset can be defined as follows: A set R is a subset of another set S if every element in R is in S. For example, if $S = \{0, 1, 2, 3, 4\}$ and $R = \{0, 1, 2\}$, then every element in R is in S, and R is a subset of S. The symbol for subset is $\subseteq$. R is a subset of S is written $R \subseteq S$.

Example. Let A represent the letter in the English alphabet and C represent the letters in the word "corporation." Verify that C is a subset of A.

Since $A = \{a, b, c, d, e, f, g, h, i, j, k, l, m, n, o, p, q, r, s, t, u, v, w, x, y, z\}$ and $C = \{c, o, r, p, a, t, i, n\}$, we see that $C \subseteq A$.

Example. Let D represent the set of digits and I represent the set of all integers including 0. Since $D = \{0, 1, 2, 3, 4, 5, 6, 7, 8, 9\}$ and $I = \{\ldots, -5, -4, -3, -2, -1, 0, 1, 2, 3, 4, 5, \ldots\}$ the finite set D is a subset of the infinite set I, i.e., $D \subseteq I$.

In the preceding example D was defined as the set of all digits including 0. We previously defined N as the set of all positive integers excluding 0. Assume that one is interested in determining if D is a subset of N. By inspection of the sets, we see that D includes the element 0, whereas N does not. Consequently, D is not a subset of N. This is written as $D \nsubseteq N$. The notation for subset $\subseteq$ and not subset $\nsubseteq$ parallels the notation for member $\in$ and not member $\notin$.

1.3.1 PROPER SUBSETS

The term *subset* is often differentiated from that of *proper subset*. A proper subset is designated by the symbol $\subset$. A proper subset P is a subset of another set U, written $P \subset U$, if all elements in P are in U but all elements in U are not in P. This simply means that for P to be a proper subset of U, then U must have all elements that are in P plus at least one element that is not in P. As an example, if

$$S = \{0, 1, 2, 3, 4\}$$

and

$$R = \{0, 1, 2\}$$

then R is a proper subset of S, i.e., $R \subset S$.

Example. Verify that the set C, the letters in corporation, is a proper subset of A, the letters in the English alphabet. Since every letter in C is in A, but every letter in A is not in C, we conclude that $C \subset A$.

1.3.2 UNIVERSAL SET

In discussing sets and subsets, the term *universal set* is often encountered. The term "universal set" is applied to the set that contains all the elements the analyst will wish to consider. If, for example, we are interested in categorizing the stockholders of the XYZ Company, the universal set would be all stockholders of the XYZ Company. The various categories of stockholders would then be subsets of the universal set. Similarly, if the analyst were interested in certain combinations of letters, the universal set would be defined as A, the letters of the English alphabet. It would then be possible to specify various subsets of the universal set A, such as C, the letters in the word corporation.

The universal set contains all elements under consideration. In contrast, the *null* set is defined as a set that has no elements or members. To illustrate, assume that we define a universal set as consisting of three students who are to take an exam. We define subsets as consisting of the students who score an A on the exam. If we refer to the students as S_1, S_2, and S_3, the students who scored an A could be $\{S_1, S_2, S_3\}$, $\{S_1, S_2\}$, $\{S_1, S_3\}$, $\{S_2, S_3\}$, $\{S_1\}$, $\{S_2\}$, $\{S_3\}$, $\{\ \}$. Note that there are eight possible outcomes and these are shown as subsets. One of the subsets is the null set $\{\ \}$. The null set, also referred to as the *empty* set, is designated by the Greek letter ϕ (phi). In this example of eight possible subsets, one of the possibilities is the universal set $\{S_1, S_2, S_3\}$ and another is the null set ϕ. This illustrates that both the universal set and the null sets are included as subsets of the universal set. This concept is also illustrated by the following examples.

Example. A stock market analyst is concerned with the price movement of the stock of the three large automobile manufacturers. Using the ticker symbols of Chrysler, Ford, and General Motors, determine the universal set representing upward movements in price of the stocks and specify all possible subsets.

The universal set representing upward price movement in all three stocks is $U = \{C, F, GM\}$. The possible subsets are $\{C, F, GM\}$, $\{C, F\}$, $\{C, GM\}$, $\{F, GM\}$, $\{C\}$, $\{F\}$, $\{GM\}$, $\{\ \}$.

We again note that there are eight possible subsets. One of the subsets is the universal set $U = \{C, F, GM\}$ and another is the null set $\phi = \{\ \}$.

Example. A businessman, vacationing at Santa Anita Racetrack, places a bet on the first three races. If we denote winning race 1 as R_1, winning race 2 as R_2, etc., the universal set is $U = \{R_1, R_2, R_3\}$. The possible subsets are $\{R_1, R_2, R_3\}$, $\{R_1, R_2\}$, $\{R_1, R_3\}$, $\{R_2, R_3\}$, $\{R_1\}$, $\{R_2\}$, $\{R_3\}$, $\{\ \}$.

1.3.3. COUNTING SUBSETS

The number of possible subsets of the universal set can be calculated through the use of a straightforward formula. The number of possible subsets is given by the formula

$$N = 2^n \tag{1.1}$$

where n represents the number of elements in the universal set and N is the number of possible subsets. If the universal set contains three members, there are eight possible subsets. Similarly, for a universal set with four members there are $N = 2^4 = 16$ possible subsets.

1.4 Set Algebra

Set algebra consists of certain operations on sets whereby the sets are combined to produce other sets. As an example, consider a group of students who are enrolled in finite mathematics and another group who are enrolled in introductory finance. From these two sets we can specify, using the algebra of sets, a third set containing as members those individuals who are enrolled in both courses. These operations are most easily illustrated through use of the Venn diagram.

1.4.1 VENN DIAGRAM

The *Venn diagram*, named after the English logician John Venn (1834–83), consists of a rectangle that conceptually represents the universal set. Subsets

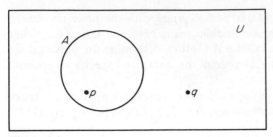

Figure 1.1.

of the universal set are represented by circles drawn within the rectangle or universal set. In Fig. 1.1 the universal set U is represented by the rectangle and the subset A by the circle. The Venn diagram shows that p is a member of A and that q is not a member of A. Both p and q are members of the universal set.

1.4.2 COMPLEMENTATION

The first set operation we consider is that of *complementation*. Let P be any subset of a universal set U. The complement of P, denoted by P' (read "P complement"), is the subset of elements of U that are not members of P. The complement of P is indicated by the shaded portion of the Venn diagram in Fig. 1.2.

   Example. For the universal set $D = \{0, 1, 2, 3, 4, 5, 6, 7, 8, 9\}$ with the subset $P = \{0, 1, 3, 5, 7, 9\}$, determine P'.
   The complement of P contains all elements in D that are not members of P. Thus, $P' = \{2, 4, 6, 8\}$.

1.4.3 INTERSECTION

A second set operation is *intersection*. Again, let P and Q be any subsets of a universal set U. The intersection of P and Q, denoted by $P \cap Q$ (read "P

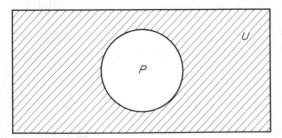

Figure 1.2.

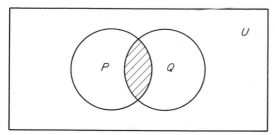

Figure 1.3.

intersect Q"), is the subset of elements of U that are members of both P and Q. $P \cap Q$ is shown by the shaded area of Fig. 1.3.

Example. For the universal set $D = \{0, 1, 2, 3, 4, 5, 6, 7, 8, 9\}$ with subsets $P = \{0, 1, 3, 5, 7, 9\}$ and $Q = \{0, 2, 3, 5, 9\}$, determine the intersection of P and Q.

The intersection of P and Q is the subset that contains the elements in D that are simultaneously in P and Q. Thus,

$$P \cap Q = \{0, 3, 5, 9\}$$

With a little thought the reader can also recognize that

$$P \cap U = P$$

and

$$P \cap P' = \phi$$

Sets such as $P \cap P' = \phi$ which have no common members are termed *disjoint* or *mutually exclusive*.

1.4.4   UNION

A third set operation is *union*. If we again let P and Q be any subsets of a universal set U, then the union of P and Q, denoted by $P \cup Q$ (read "P union Q"), is the set of elements of U that are members of either P or Q. $P \cup Q$ is shown by the shaded portion of the Venn diagram in Fig. 1.4.

Example. For the universal set $D = \{0, 1, 2, 3, 4, 5, 6, 7, 8, 9\}$ with subsets $P = \{0, 1, 3, 5, 7, 9\}$ and $Q = \{0, 2, 3, 5, 9\}$, determine the union of P and Q.

The union of P and Q is the subset that contains the elements in D that are in P or in Q. Thus,

$$P \cup Q = \{0, 1, 2, 3, 5, 7, 9\}$$

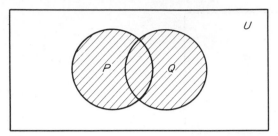

Figure 1.4.

From the definitions given it also follows that

$$P \cup U = U$$

$$P \cup P' = U$$

1.4.5 OTHER SET OPERATIONS

Two additional set operations are sometimes included in the algebra of sets. The two operations are *difference* and *exclusive union*. Since subsets formed by either of these set operations can also be formed by use of complementation, intersection, and union, these operations are often excluded in discussing set algebra.

Let P and Q be any subsets of the universal set U. The difference of P and Q, denoted by $P - Q$ (read "P minus Q"), is the subset that consists of those elements that are members of P but are not members of Q. This subset is shown in the Venn diagram in Fig. 1.5. The difference of $P - Q$ can also be expressed as $P \cap Q'$.

The exclusive union of P and Q, denoted by $P \cup Q$ (read "P exclusive union Q"), is the set of elements of U that are members of P or of Q but not of both. The subset is shown in the Venn diagram in Fig. 1.6. The subset can also be expressed as $(P \cap Q)' \cap (P \cup Q)$ or $(Q' \cap P) \cup (P' \cap Q)$.

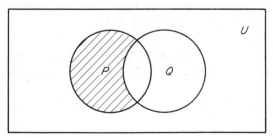

Figure 1.5.

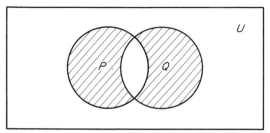

Figure 1.6.

1.4.6   COMBINING SET OPERATIONS

Part of the utility of the algebra of sets occurs because of the ability to combine two or more sets into new sets through the use of the set operations. As an example, consider a national firm with district offices. The accounting system of this firm is computerized, and the various accounts are stored in a data bank accessible to the computer. The collection department manager at the national office is concerned with accounts that are 60 days past due. Accounts over 60 days past due are subsets of the set of accounts receivable. In analyzing the accounts due in both the western and southwestern districts, the manager would request the union of the two subsets of past due accounts. If he were interested in customers who simultaneously had accounts past due in both districts, the manager would request the intersection of the subsets.

The use of set algebra to form sets can be illustrated by Fig. 1.7. This figure consists of a Venn diagram, representing the universal set, and two subsets P and Q. Areas in the Venn diagram are shown by the letters a, b, c, and d. The set that consists of areas a, b, c, and d is the universal set U, while the set P consists of a and b, the set Q consists of b and c, etc. Using this notation, we can use the algebra of sets to construct all possible subsets. The student should carefully study Fig. 1.7 and the construction of subsets using the set operations describing the areas a, b, and c presented in Table 1.1 before proceeding.

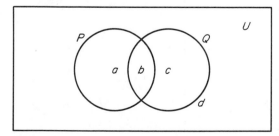

Figure 1.7.

Table 1.1.

Area	*Set*
a, b, c, d	U
a, b	P
b, c	Q
a, d	Q′
c, d	P′
b	$P \cap Q$
a, b, c	$P \cup Q$
d	$(P \cup Q)'$
a	$P \cap Q'$ or $P - Q$
c	$P' \cap Q$ or $Q - P$
a, c	$(P \cap Q') \cup (P' \cap Q)$, or
	$(P \cap Q)' \cap (P \cup Q)$, or
	$(P \cup Q)$, or
	$(P - Q) \cup (Q - P)$
a, c, d	$(P \cap Q)'$

The construction of subsets using the algebra of sets and Venn diagrams is further illustrated by the following examples.

Example. *Business Month* is developing a profile of its subscribers. Of the information requests returned by the subscribers to the business magazine, the following data were obtained: 30 percent were in a service industry, 40 percent were self-employed, 20 percent sold through retail channels. Of those who were self-employed, 40 percent were in the service industry, 20 percent were in a retail business, and 10 percent were in a retail service business. The response also indicated that 50 percent of the retail businesses were service-oriented. From these data, develop a Venn diagram that shows the reader profile as subsets of the response set.

The circles shown in the Venn diagram in Fig. 1.8 represent the service

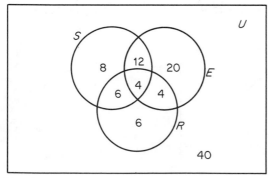

Figure 1.8.

industry S, self-employed E, and retail business R. From the data we know that $S = 30$, $E = 40$, and $R = 20$. Since 40 percent of the self-employed are in the service industry, it follows that $E \cap S = 16$. Similarly, $E \cap R = 8$ and $E \cap R \cap S = 4$. We also know from the data that $R \cap S = 10$.

The example can be expanded to give the reader additional practice in the use of set operations to specify subsets. This is done by developing set notation for statements describing subsets of the *Business Month* Venn diagram. The statements and subsets are given in Table 1.2 and refer to Fig. 1.8. This table and figure also deserve careful study.

Table 1.2.

Statement	*Set*	*Number*
1. Self-employed in retail service	$E \cap R \cap S$	4
2. Self-employed in retail nonservice	$E \cap R \cap S'$	4
3. Self-employed in nonretail nonservice	$E \cap R' \cap S'$	20
4. Self-employed in nonretail service	$E \cap R' \cap S$	12
5. Non-self-employed in nonretail service	$E' \cap R' \cap S$	8
6. Non-self-employed in retail service	$E' \cap R \cap S$	6
7. Non-self-employed in retail nonservice	$E' \cap R \cap S'$	6
8. Non-self-employed, nonservice, nonretail	$E' \cap R' \cap S'$	40
9. Service or self-employed	$S \cup E$	54
10. Nonretail service or nonretail self-employed	$(R' \cap S) \cup (R' \cap E)$	40
11. Non-self-employed in service	$E' \cap S$	14

Example. The Venn diagram shown in Fig. 1.9 is subdivided into areas a_1, a_2, a_3, a_4, a_5, and a_6. The areas are described by the subsets given as follows:

Area	*Set*
a_1, a_2, a_3, a_4	$P \cup Q$
a_5	$R \cap (P \cup Q)'$
a_3, a_4	$(P \cap R) \cup (Q \cap R)$
a_1, a_2, a_5	$(R' \cap P) \cup (R' \cap Q) \cup ((P \cup Q)' \cap R)$

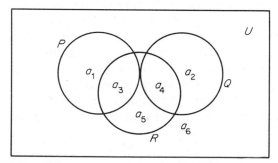

Figure 1.9.

1.4.7 LAWS OF THE ALGEBRA OF SETS[1]

The algebra of sets is composed of certain laws. In some cases, these laws are similar to the algebra of numbers. We shall state these laws in the form of postulates, thus relieving the reader of the burden of studying mathematical proofs of the laws. The less obvious postulates are illustrated with Venn diagrams.

$$\text{Postulate 1: } P \cup Q = Q \cup P$$

$$\text{Postulate 2: } P \cap Q = Q \cap P$$

Postulates 1 and 2 are termed the *commutative law*. These postulates state that the order in which we combine the union or intersection of two sets is immaterial. This corresponds to the commutative law in ordinary algebra, which states that $p + q = q + p$ and that $p \cdot q = q \cdot p$.

$$\text{Postulate 3: } P \cup (Q \cup R) = (P \cup Q) \cup R$$

$$\text{Postulate 4: } P \cap (Q \cap R) = (P \cap Q) \cap R$$

Postulates 3 and 4 are termed the *associative law*. These postulates state that the selection of two of three sets for grouping in a union or intersection is immaterial. Thus, the order in which the sets are combined is immaterial. These postulates correspond to the associative law in ordinary algebra, which enables us to state that $p + (q + r) = (p + q) + r$ and that $p \cdot (q \cdot r) = (p \cdot q) \cdot r$.

$$\text{Postulate 5: } P \cup (Q \cap R) = (P \cup Q) \cap (P \cup R)$$

$$\text{Postulate 6: } P \cap (Q \cup R) = (P \cap Q) \cup (P \cap R)$$

Postulates 5 and 6 are called the *distributive law*. Postulate 5 has no analogous postulate in ordinary algebra. Postulate 6 corresponds to the distributive law in ordinary algebra that enables us to state that $p(q + r) = p \cdot r + p \cdot r$.

Since Postulates 5 and 6 are not as obvious as Postulates 1 through 4, we shall verify one of them through Venn diagrams. To verify Postulate 5, we must show that the area represented by $P \cup (Q \cap R)$ is the same as that represented by $(P \cup Q) \cap (P \cup R)$. The area representing the set $P \cup (Q \cap R)$ is shown in Fig. 1.10(a). In Fig. 1.10(a), the intersection of Q and R is shown by the shading from lower left to upper right, and P is shown by shading from upper left to lower right. The union of P with $Q \cap R$, represented by both the shaded and crosshatched area in the figure, gives $P \cup (Q \cap R)$.

---

[1] This section can be omitted without loss of continuity.

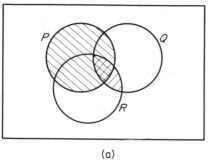 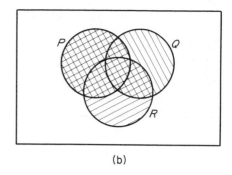

(a)   (b)

Figure 1.10.

The area representing the set $(P \cup Q) \cap (P \cup R)$ is shown in Fig. 1.10(b). In Fig. 1.10(b), the union of P and Q is shown by shading from upper left to lower right, and the union of P and R is shown by shading from lower left to upper right. The intersection of the two shaded areas gives the cross-hatched area described as the set $(P \cup Q) \cap (P \cup R)$. Since the shaded and crosshatched area in Fig. 1.10(a) equals the crosshatched area in Fig. 1.10(b), we conclude that Postulate 5 is true. Postulate 6 is verified in the same manner as Postulate 5.

Postulate 7: $P \cap P = P$

Postulate 8: $P \cup P = P$

Postulate 9: $P \cup \phi = P$

Postulates 7 through 9 follow directly from the operations of union and intersection. They can be verified by the Venn diagram in Fig. 1.11.

Postulate 10: $P \cap U = P$

Postulate 10: $P \cup P' = U$

Postulate 12: $P \cap P' = \phi$

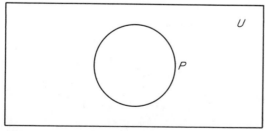

Figure 1.11.

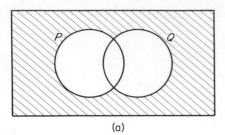 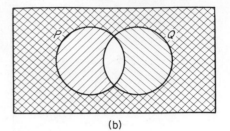

(a)   (b)

Figure 1.12.

Postulates 10 through 12 are based upon the definition of the null set, the universal set, and the complement. These postulates are obvious and can also easily be verified by the Venn diagram shown in Fig. 1.11.

$$\text{Postulate 13: } (P \cup Q)' = P' \cap Q'$$

$$\text{Postulate 14: } (P \cap Q)' = P' \cup Q'$$

Postulates 13 and 14 are termed De Morgan's law. Postulate 13 states that the complement of the union of two sets is equal to the intersection of the complement of each set. Postulate 14 states that the complement of the intersection of two sets is equal to the union of the complement of each set. We shall verify Postulate 13 through the use of Venn diagrams. The student is asked to verify Postulate 14.

Figure 1.12(a) shows the set $(P \cup Q)'$ as the area shaded from upper left to lower right. In Fig. 1.12(b), P' is shown by shading from upper left to lower right and Q' is shown by shading from lower left to upper right. The intersection of P' with Q' is shown by the crosshatched area. Since the shaded area in Fig. 1.12(a) equals the crosshatched area in Fig. 1.12(b), it follows that $(P \cup Q)' = P' \cap Q'$.

The laws of set algebra are used analagously with the laws of ordinary algebra. Just as the laws of ordinary algebra can be used to simplify algebraic expressions, the laws of set algebra can be used to simplify sets. This is illustrated by the following examples.

Example. Simplify the set $(A \cup B) \cup (A \cap B)$.

1. Let $P = (A \cup B)$:	$P \cup (A \cap B)$
2. Postulate 5:	$(P \cup A) \cap (P \cup B)$
3. Substitute for P:	$((A \cup B) \cup A) \cap ((A \cup B) \cup B)$
4. Postulate 1:	$(A \cup A \cup B) \cap (A \cup B \cup B)$
5. Postulate 7:	$(A \cup B) \cap (A \cup B)$
6. Postulate 7:	$A \cup B$

In this example, it is important for the student to use parentheses to enclose $P = (A \cup B)$. This is further illustrated by the following example.

Example. On page 11 and again on page 12 we stated that $(P \cap Q)' \cap (P \cup Q)$ was equal to $(Q' \cap P) \cup (P' \cap Q)$. Using the laws of set algebra, verify this relationship.

Given:	$(P \cap Q)' \cap (P \cup Q)$
1. Postulate 14:	$(P' \cup Q') \cap (P \cup Q)$
2. Let $R = (P' \cup Q')$	$R \cap (P \cup Q)$
3. Postulate 6:	$(R \cap P) \cup (R \cap Q)$
4. $(P' \cup Q') = R$	$((P' \cup Q') \cap P) \cup ((P' \cup Q') \cap Q)$
5. Postulate 7:	$(P \cap P') \cup (P \cap Q') \cup (P' \cap Q) \cup$ $(Q \cap Q')$
6. Postulate 12:	$(P \cap Q') \cup (P' \cap Q)$

Thus,

$$(P \cap Q)' \cap (P \cup Q) = (P \cap Q') \cup (P' \cap Q)$$

Example. Simplify the expression $P \cup (P' \cap Q)$.

Given:	$P \cup (P' \cap Q)$
1. Postulate 5:	$(P \cup P') \cap (P \cup Q)$
2. Postulate 11:	$U \cap (P \cup Q)$
3. Postulate 10:	$P \cup Q$

The expression simplifies to $P \cup Q$.

Example. Show that the set $(P \cup Q') \cap (P \cap Q)'$ equals Q'.

Given:	$(P \cup Q') \cap (P \cap Q)'$
1. Postulate 14:	$(P \cup Q') \cap (P' \cup Q')$
2. Postulate 5:	$Q' \cup (P \cap P')$
3. Postulate 12:	Q'

The primary difficulty in understanding this example is in converting $(P \cup Q') \cap (P' \cup Q')$ to $Q' \cup (P \cap P')$. This step involves rewriting (Postulate 1) the expression as $(Q' \cup P) \cap (Q' \cup P')$. Postulate 5, the distributive law, is applied to write the expression as $Q' \cup (P \cap P')$. Since $P \cap P'$ is equal to ϕ, the expression reduces to Q'.

1.5   Cartesian Product

Suppose that one is asked to list the possible outcomes of two tosses of a coin. Since either a head or a tail occurs on a single toss, the possible outcomes are described by the set $O = \{(H, H), (H, T), (T, H), (T, T)\}$. There are four elements in this set and, corresponding to the requirements given for a set, the order of the elements is immaterial. Are each of the elements, however, distinct? To answer this question we must ask if the element (H, T) differs from the element (T, H). The answer is that these elements do differ, since the order of the occurrence is important. If, for instance, an individual bet $1 that a head would occur on the first toss of the coin and $1 that a tail would occur on the second toss, he would be $2 richer if element (H, T) occurred and $2 poorer if (T, H) occurred.

The elements $(H, H), (H, T), (T, H)$, and (T, T) are examples of *ordered pairs*. One of the two components of each ordered pair is designated as the first element of the pair, and the other, which need not be different from the first, is designated as the second element. If the first element is designated as a and the second element is designated as b, we have the ordered pair (a, b). This ordered pair differs from the ordered pair (b, a), and both ordered pairs differ from the set $\{a, b\}$.

Ordered pairs are formed by the *Cartesian product* of two sets. If A and B are two sets, the Cartesian product of the sets, designated by $A \times B$, is the set containing all possible ordered pairs (a, b) such that $a \in A$ and $b \in B$. If the set A contains the elements a_1, a_2, a_3 and the set B contains the elements b_1 and b_2, the Cartesian product $A \times B$ is the set $A \times B = \{(a_1, b_1), (a_1, b_2), (a_2, b_1), (a_2, b_2), (a_3, b_1), (a_3, b_2)\}$. All possible ordered pairs are included in the set.

The concept of the Cartesian product is quite useful in many decision problems. The student of probability and statistics will often be asked to consider the possible outcomes of an experiment. As an example, consider again the problem of determining all possible outcomes of two tosses of a coin. If we define the outcome of the first toss as the set $O_1 = \{H, T\}$ and the outcome of the second toss as the set $O_2 = \{H, T\}$, then the Cartesian product $O_1 \times O_2$ gives all possible outcomes of the two tosses. As we have seen, these outcomes are $O_1 \times O_2 = \{(H, H), (H, T), (T, H), (T, T)\}$.

The Cartesian product of two sets can be determined quite easily with the aid of a box diagram. Figure 1.13 shows the box diagram for the Cartesian product of O_1 and O_2. The method of constructing the diagram involves listing the elements of O_1 to the left of the box and O_2 above the box. The blanks in the box are then filled in with the ordered pairs. The Cartesian product $O_1 \times O_2$ consists of the elements in the box, i.e., $(H, H), (H, T), (T, H), (T, T)$.

$$O_2$$

	H	T
H	H, H	H, T
T	T, H	T, T

O_1 (label to the left of the rows)

Figure 1.13.

The Cartesian product can be expanded to combine more than two sets. This means that the concepts discussed for ordered pairs can, for example, be applied to *ordered triplets*. To illustrate, assume that we are asked to list all possible outcomes of three tosses of a coin. Denoting $O_i = \{H, T\}$, where O_i represents the possible outcomes on the ith toss (i.e., $i = 1$ represents the first toss, $i = 2$ represents the second toss, etc.), the possible outcomes would be given by the Cartesian product $O_1 \times O_2 \times O_3$. This Cartesian product is determined by finding $O_1 \times O_2$ and then $(O_1 \times O_2) \times O_3$. From Fig. 1.13, we know that $O_1 \times O_2 = \{(H, H), (H, T), (T, H), (T, T)\}$. $(O_1 \times O_2) \times O_3$ is shown in Fig. 1.14. The Cartesian product of $O_1 \times O_2 \times O_3$ is $\{(H, H, H), (H, T, H), (T, H, H), (T, T, H), (H, H, T), (H, T, T), (T, H, T), (T, T, T)\}$.

The box diagrams of Figs. 1.13 and 1.14 provide a straightforward method of determining of the Cartesian products of sets. With some practice, the student can apply this concept without difficulty. Before illustrating the concept with examples, however, let us carry the Cartesian product one additional step. Assume that we are asked to list all possible outcomes of 4 tosses of a coin. There are 16 such outcomes. Most individuals would be extremely hard pressed to think of all 16. If we use the concept of the Cartesian product, the task becomes routine. The possible outcomes are given by the set $O_1 \times O_2 \times O_3 \times O_4$. The box diagram can be expressed in terms of $(O_1 \times O_2) \times (O_3 \times O_4)$ or by any other grouping of the parentheses. Figure 1.15 shows the Cartesian product of $(O_1 \times O_2) \times (O_3 \times O_4)$.

The elements in the box diagram such as (H, H, H, T) contain four members. Mathematicians call such an element an ordered "4-tuple." Using this

$$O_1 \times O_2$$

	H, H	H, T	T, H	T, T
H	H, H, H	H, H, T	H, T, H	H, T, T
T	T, H, H	T, H, T	T, T, H	T, T, T

O_2 (label to the left of the rows)

Figure 1.14.

$$O_3 \times O_4$$

	H, H	H, T	T, H	T, T
H, H	H, H, H, H	H, H, H, T	H, H, T, H	H, H, T, T
H, T	H, T, H, H	H, T, H, T	H, T, T, H	H, T, T, T
T, H	T, H, H, H	T, H, H, T	T, H, T, H	T, H, T, T
T, T,	T, T, H, H	T, T, H, T	T, T, T, H	T, T, T, T

$O_1 \times O_2$ (row label at left)

Figure 1.15.

term, an ordered pair could be referred to as an ordered 2-tuple, an ordered triplet as an ordered 3-tuple, etc. In general, then, an element containing n members is referred to as an ordered *n-tuple*.

Example. A retailer specializes in three products: color television, black and white television, and stereos. He offers a service contract with the sale of each of the products, which the customer may or may not elect to purchase. Determine the possible combination of sales options.

Let the products be represented by the set $P = \{C, B, S\}$ and the sales contract by $R = \{E, E'\}$. The Cartesian product of $P \times R$ gives the combination of sales options. The box diagram in Fig. 1.16 shows the elements of $P \times R$.

P

	C	B	S
E	E, C	E, B	E, S
E′	E′, C	E′, B	E′, S

R (row label at left)

Figure 1.16.

Example. A builder has three basic floor plans: single story, two story, and trilevel. Each of these plans can have either a shake roof or a wood shingle roof. In addition, the plans are available with or without fireplaces. Determine the number of combinations of plans and show these plans in a box diagram.

Let the basic floor plans be represented by the set $F = \{1, 2, 3\}$, the roofing material by the set $R = \{S, W\}$, and the fireplace option by the set $O = \{f, f'\}$. The Cartesian product of R and F is shown in Fig. 1.17. The combina-

$$F$$

	1	2	3
R S	S, 1	S, 2	S, 3
W	W, 1	W, 2	W, 3

Figure 1.17.

tion of plans is represented by the set $\{F \times R \times O\}$ and is shown in Fig. 1.18. There are twelve combinations of plans.

Example. An advertising agency is placing ads for three products. The media available for advertising are radio, television, and newspaper. The

$$F \times R$$

O	f	f, S, 1	f, S, 2	f, S, 3	f, W, 1	f, W, 2	f, W, 3
	f'	f', S, 1	f', S, 2	f', S, 3	f', W, 1	f', W, 2	f', W, 3

Figure 1.18.

ads will be written by either the agency or the sponsor. Develop the possible combinations with the aid of a box diagram (see Fig. 1.19).

$$M$$

		R	T	N
	1	1, R	1, T	1, N
P	2	2, R	2, T	2, N
	3	3, R	3, T	3, N

Figure 1.19.

Let the products be represented by the set $P = \{1, 2, 3\}$, the media by the set $M = \{R, T, N\}$, and the source of the advertisement by the set $S = \{A, A'\}$. The box diagram for $P \times M \times S$ is shown in Fig. 1.20.

$$M \times P$$

S		1, R	1, T	1, N	2, R	2, T	2, N	3, R	3, T	3, N
	A	A, 1, R	A, 1, T	A, 1, N	A, 2, R	A, 2, T	A, 2, N	A, 3, R	A, 3, T	A, 3, N
	A'	A', 1, R	A', 1, T	A', 1, N	A', 2, R	A', 2, T	A', 2, N	A', 3, R	A', 3, T	A', 3, N

Figure 1.20.

The total number of members formed by the Cartesian product of two sets is given by the product of the number of elements in each set. Thus, if set A contains five elements and set B contains four elements, then the Cartesian product $A \times B$ contains $5(4) = 20$ elements. The rule applies to the Cartesian product of more than two sets. For the three sets A, B, and C, containing N_1, N_2, and N_3 elements respectively, the Cartesian product of the sets $A \times B \times C$ contains $N_1 \cdot N_2 \cdot N_3$ elements.

Example. Set A has ten elements, set B has six elements, set C has twelve elements, and set D has three elements. Determine the number of elements in the Cartesian product of $A \times B \times C \times D$.

The number of elements is given by product of the number of elements in each set. Thus, the Cartesian product contains $10 \cdot 6 \cdot 12 \cdot 3$, or 2160 elements.

1.6   Applications to Logic[2]

The algebra of sets can be applied to problems of logic. To illustrate, assume that we receive the following guidelines concerning the allocation of expenditures in a firm:

> *Advertising expenditures are to be directed toward men who are college graduates or are over 30 years old, but not to college graduates under 30 years old.*

To simplify this unnecessarily complicated directive, let A represent men who are college graduates and B represent men who are over 30 years old. The set $A \cup B$ then represents men who are college graduates or over 30 years old. Similarly, the set $A \cap B'$ represents college graduates under 30 years old. Since advertising expenditures are not to be directed toward college graduates who are under 30 years old, the set describing the allocation of advertising expenditures is

$$(A \cup B) \cap (A \cap B')'$$

This set can be simplified by using either the laws of set algebra or a Venn diagram. To simplify the set using the laws of set algebra, we perform the following steps:

---

[2] This section can be omitted without loss of continuity.

Given: $(A \cup B) \cap (A \cap B')'$
  1. Postulate 14: $(A \cup B) \cap (A' \cup B)$
  2. Postulate 5: $B \cup (A \cap A')$
  3. Postulate 12: $B \cup \phi$
  4. Postulate 9: B

Interestingly, the analysis shows that the advertising guidelines can be reduced to the simple statement, "Advertising expenditures should be directed to men who are over 30 years old." The reader should verify that the same result can be shown through use of the Venn diagram.

Example. Simplify the following edict made by the president of Amalgamated Industries:

> *Due to recent cutbacks in the major divisions of our firm, employees who are over 60 years of age or have over 30 years of service with the company are eligible for early retirement. This excludes, however, individuals who have 30 years of service but are under 60.*

To simplify this policy statement, let A represent the set of individuals over 60 years old and B represent the set of individuals who have over 30 years of service with the company. Those individuals who are over 60 years of age or have over 30 years with the company are described by the set $A \cup B$. Individuals under 60 years of age who have over 30 years of service are given by the set $A' \cap B$. Those eligible for early retirement are thus described by the set $(A \cup B) \cap (A' \cap B)'$. This set is simplified as shown in the preceding example to give $(A \cup B) \cap (A' \cap B)' = A$. The analysis shows that regardless of the number of years of service, only individuals over 60 years of age are eligible for early retirement.

Example. The president of Amalgamated Industries was told that his policy on early retirement specifically included all individuals over 60 years of age, regardless of the length of service to the company. Upon hearing this interpretation, he revised the policy as follows:

> *Due to recent cutbacks in the major divisions of our firm, employees who are over 60 years of age and have over 30 years of service with the company are eligible for early retirement. This specifically excludes individuals who are over 60 but have less than 30 years of service with the company.*

Simplify this policy statement by using a Venn diagram.

The Venn diagram describing the president's policy is shown in Fig. 1.21.

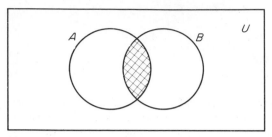

Figure 1.21.

Sets A and B are defined as before. Individuals who are over 60 years of age and have over 30 years of service are described by the set $A \cap B$. Those who are over 60 years of age but have less than 30 years of service are described by the set $A \cap B'$. The individuals described by the president as eligible for early retirement are described by the intersection of these sets, $(A \cap B) \cap (A \cap B')'$. This set can be reduced by using set algebra to the set $A \cap B$, the set shown by the crosshatched area in the Venn diagram.

These examples have used the operations of intersection, union, and complementation together with the laws of set algebra to clarify certain complicated statements. It can be seen that these operators are employed in the logical analysis of statements in much the same fashion as they were in the earlier description of sets and set membership.

1.6.1 STATEMENTS AND CONNECTIVES

One of the important concepts in logic is that of the *statement*. A statement is a simple declaration. For instance, "The sun is shining" and "I received an A on the last examination" are simple statements. The distinguishing characteristic of a statement is that the statement makes an assertion. The assertion made by the statement can be either true or false, but not both. It should be emphasized that interrogations (such as "Is the sun shining?" and imperatives (such as "Please turn in your homework.") do not assert anything and, consequently, are not statements.

The set operators are used to combine simple statements to form *compound* statements. The operations of complementation, intersection, and inclusive and exclusive union that were discussed in Sec. 1.4 have analogous counterparts in logic. These counterparts, termed *connectives* in logic, are negation, conjunction, and inclusive and exclusive disjunction.

The *negation* of a statement A is the statement "not A." To illustrate, let

$$A = \text{The sun is shining}$$

The negation of A is

$$A' = \text{The sun is not shining}$$

The *conjunction* of two statements A and B is the statement "A and B." This is denoted by $A \cap B$. For instance, if

$$A = \text{The sun is shining}$$

and

$$B = \text{I will go swimming}$$

then

$$A \cap B = \text{The sun is shining and I will go swimming}$$

Similarly,

$$A' \cap B' = \text{The sun is not shining and I will not go swimming}$$

The *inclusive disjunction* of two statements A and B is the statement A or B or both and is denoted by $A \cup B$. The *exclusive disjunction* of the two statements is A or B but not both and is denoted by $A \cup B$. Defining the statement A as

$$A = \text{I will major in engineering}$$

and the statement B as

$$B = \text{I will major in business}$$

then the inclusive disjunction of the two statements is

$$A \cup B = \text{I will major in engineering or business or both}$$

The exclusive disjunction of the two statements is

$$A \cup B = \text{I will major in engineering or business but not both}$$

Example. Define the statements A and B as $A =$ "The Dow-Jones stock market averages advanced" and $B =$ "Corporate profits were reported higher during the past quarter." Use the connectives (i.e., operators) negation, conjunction, and inclusive and exclusive disjunction to describe the following compound statements:

1. The Dow-Jones average advanced, and corporate profits were reported higher during the quarter.

Answer: $A \cap B$.

2. The Dow-Jones average fell, but corporate profits were reported higher during the quarter.

Answer: $A' \cap B$.

3. Although not reflected by the Dow-Jones average that continued to advance, corporate profits were reported lower during the past quarter.

Answer: $A \cap B'$.

Example. Define the statements A and B as A = "The labor contract is inflationary" and B = "The price increase is inflationary." Use the logical connectives to describe the following compound statements.

1. The labor contract is inflationary, but the price increase is not.

Answer: $A \cap B'$.

2. Either the labor contract or the price increase is inflationary.

Answer: $A \cup B$.

3. Neither the labor contract nor the price increase is inflationary.

Answer: $(A \cup B)'$.

4. Taken together, the labor contract and the price increase are inflationary.

Answer: $A \cap B$.

5. The labor contract could be considered inflationary or the price increase could be considered inflationary, but certainly both would not be considered inflationary.

Answer: $A \cup B$.

1.6.2   LOGICAL ARGUMENTS

The principles of set algebra can easily be extended to the analysis of logical arguments. A *logical argument* consists of a series of statements or *premises* followed by a conclusion. The argument is termed *logically true* if the conclusion is the logical consequence of the premises. Conversely, the argument is *logically false* if the conclusion does not follow from the premises.

The statements or premises upon which the argument is based are termed *factual*. The factual statements may themselves be either true or false. Consequently, an argument that is logically true need not be based on correct facts.

To illustrate the concept of a logical argument, consider the following statement:

Automobile insurance premiums in states such as Massachusetts that have passed "no fault" insurance laws are lower than the premiums in states such as California that have not passed the "no fault" insurance laws. Consequently, insurance premiums in Massachusetts are lower than those in California.

To examine the validity of the argument, let

> A = Automobile insurance premiums in states that have passed "no fault" laws are lower than the premiums in states that have not passed the laws.
> B = Massachusetts has a "no fault" insurance law.
> C = California does not have a "no fault" insurance law.
> D = Massachusetts insurance premiums are lower than those in California.

Statements A, B, and C are premises. Statement D is the conclusion based upon the three premises. To determine if D is logically true, consider the Venn diagram in Fig. 1.22. The figure shows that B is a subset of A and that C is not a subset of A. Since the insurance premiums in all states that are members of A are lower than those in all states that are not members of A, it logically follows that the insurance premiums in Massachusetts are lower than those in California. The argument, therefore, is logically true.

Example. Determine the validity of the following argument:

> *Rapid expansion in an economy is often accompanied by inflation. Germany and Japan have experienced rapid economic growth during the past decade. Therefore, Germany and Japan have experienced inflation.*

Let

> A = Rapid expansion is often accompanied by inflation.
> B = Germany and Japan have experienced rapid economic growth.
> C = Germany and Japan have experienced inflation.

To determine the validity of the argument, consider the Venn diagram in Fig. 1.23. The diagram illustrates that the argument is logically true only if C is

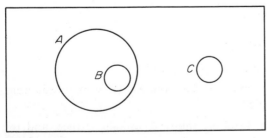

Figure 1.22.

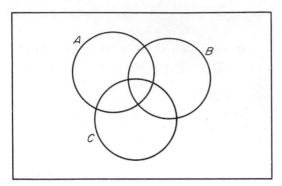

Figure 1.23.

a member of the conjunction of A and B. Since there is no factual statement to the effect that C is a member of the conjunction of A and B, we conclude that the argument is logically false.

PROBLEMS

1. In a recent editorial, a journalist referred to the Republican voters in an upcoming election. Would it be possible to define this group of voters as a set?

2. In a recent opinion poll, 100 individuals were questioned concerning a legislative proposal. Both the opinion and the sex of the respondent were noted. Give the subsets that a political analyst might consider important.

3. An economist referred to the set of all family units in the United States whose combined family income is less than $6000. Is the use of the term "set" appropriate in this instance?

4. Let set $A = \{5, 7, 9, 10, 12, 14\}$. List elements in the following sets that are also in A.
   (a) $\{x \mid x = \text{all odd numbers}\}$
   (b) $\{x \mid 2x + 6 = 20\}$
   (c) $\{x \mid x - 3 = \text{an odd integer}\}$

5. List all possible subsets of the set $S = \{a, b, c\}$.

6. A student can take one or more of five courses. How many different course selections are possible?

7. A restaurant offers pickles, ketchup, mustard, onions, lettuce, and tomatoes on its hamburgers. How many different hamburgers are possible?

8. For the sets $U = \{0, 1, 2, 3, 4, 5, 6, 7, 8, 9\}$, $R = \{0, 2, 4, 6, 8\}$, $S = \{2, 3, 4, 5, 6\}$, and $T = \{5, 6, 7, 8, 9\}$, determine the membership of the following
   (a) $R \cap S$
   (b) $R \cap S \cap T'$
   (c) $R \cap S' \cap T'$
   (d) $R \cup S'$
   (e) $(R \cap S \cap T)'$

9. Simplify the following expressions:
   (a) $(P \cup Q)' \cup (P \cap Q')$
   (b) $(P \cap Q) \cap (P \cup Q)'$
   (c) $(P \cap Q)' \cup P$

10. Kawar Travel Agency handles the winter travel arrangements for all ski clubs in the Los Angeles, San Diego, and Phoenix areas that are affiliated with the United States Ski Association. The billing system for the agency is computerized. From time to time, the agency has need for the information stored in the computer and must call for it by set nomenclature. Following are examples of the type of information which is available:

    A = ski clubs in Los Angeles
    B = ski clubs in San Diego
    C = ski clubs in Phoenix
    D = ski clubs whose billings
        are more than 45 days old

    Develop set nomenclature for the following:
    (a) All the ski clubs in California with chapters both in Los Angeles and San Diego that have billings with the agency.
    (b) The billings in Phoenix or San Diego that are more than 45 days old.
    (c) The billings in Phoenix or Los Angeles less than 45 days old.

11. A highway construction crew is made up of men who can operate heavy equipment as follows: bulldozer, 33; crane, 22; cement mixer, 35; mixer and dozer, 14; mixer and crane, 10; dozer and crane, 10; mixer, dozer, and crane, 4. How many men are in the construction crew?

12. An analysis of the membership at a local country club shows that 60 percent of the members are men, 40 percent of the members score below 85, and 50 percent of the individuals belonging to the club are over 30 years old. It was also determined that 50 percent of the men are over 30 years old, 60 percent of the men score below 85, 40 percent of the men over 30 years old score below 85, and that 10 percent of the members who score below 85 are women over 30 years old.
    (a) What percentage of the membership are men over 30 years old that score below 85?
    (b) What percentage of the membership are men that score below 85?

13. The Johnson Corporation has 500 employees, of which 200 got a raise, 100 got a promotion, and 80 got both.

    (a) How many employees got a promotion with no raise?

    (b) How many employees got neither a raise nor a promotion?

14. A student surveyed 525 people. He determined that 350 read the newspaper for news, 215 listened to radio, and 140 watched TV. Additionally, 75 read the newspaper and watched television, 40 listened to radio and watched TV, and 100 read the newspaper and listened to radio. If 25 used all three sources of news, how many people utilized none of the three?

15. Define the sets F, C, and S, respectively, as stockholders in Ford, Chrysler, and U.S. Steel. If there are 500 stockholders in Ford, 800 stockholders in Chrysler, 700 stockholders in U.S. Steel, 200 stockholders in both Ford and Chrysler, 100 stockholders in both Ford and U.S. Steel, 100 stockholders in both Chrysler and U.S. Steel, and 50 stockholders in all three, determine numerically the number of stockholders in the following sets:

    (a) $F \cup C$

    (b) $(F \cap C) \cup (F \cap S)$

    (c) $(C \cap S) \cap F'$

    (d) $(F \cap C)' \cap (F \cap S)'$

16. A color television set and a black and white television set are being distributed by a company. Both sets offer the option of a remote control unit. Determine the Cartesian product of the two sets.

17. A firm is about to market a new line of hair spray. The product can be called Hair Mist (m), Hair Set (s), or Hair Hold (h). The product can be placed in either the cosmetics section (c) or in the personal care section (p) of a store and can be distributed either through wholesalers (w) or directly to the retailer (r). Determine the possible number of combinations and construct a box diagram to show these combinations.

18. A firm must raise additional capital. Two methods have been suggested, stock or bonds. With either method, the firm has the choice of a public or private offering of the securities. Depending on market conditions, the offering could be undersubscribed or oversubscribed. Determine all possibilities for the offering. How many possibilities must be considered?

19. An automobile manufacturer offers 5 different models of cars. Each model has 9 separate option packages and is available in 14 colors. If a car dealer were to maintain a complete inventory of cars, what is the minimum number of cars the dealer must stock?

20. Consider the following two statements:

    A = "The illegal sale of drugs is rising."

    B = "The Department of H. E. W. is concerned with rising drug traffic."

Describe in words the statements represented by the following logical operations:

(a) $A' \cap B$

(b) $A \cap B$

(c) $(A \cup B)'$

(d) $A \cup B$

(e) B'

21. A railroad company executive released the following directive:

> *All passenger service to cities where revenue is less than $5000 per month or the number of passengers is fewer than 500 per month will be discontinued. This does not include cities that have fewer than 500 passengers per month which produce revenue of $5000 per month, or more.*

Where will passenger service be discontinued?

SUGGESTED REFERENCES

CANGELOSI, VINCENT E., *Compound Statements and Mathematical Logic* (Columbus, Ohio: Charles E. Merrill Books, Inc., 1967).

FREUND, JOHN E., *College Mathematics with Business Applications* (Englewood Cliffs, N.J.: Prentice-Hall, Inc., 1975), 1.

HANNA, SAMUEL C., and JOHN C. SABER, *Sets and Logic* (Homewood, Ill.: Richard D. Irwin, Inc., 1971).

KEMENY, JOHN G., et al., *Finite Mathematical Structures* (Englewood Cliffs, N.J.: Prentice-Hall, Inc., 1958), 1, 2.

KEMENY, JOHN G., et al., *Finite Mathematics with Business Applications*, 2nd ed. (Englewood Cliffs, N.J.: Prentice-Hall, Inc., 1972), 1, 2.

LIPSCHUTZ, SEYMOUR, *Set Theory and Related Topics*, Schaum's Outline Series (New York, N.Y.: McGraw-Hill Book Company, Inc., 1964).

THEODORE, CHRIS A., *Applied Mathematics: An Introduction* (Homewood, Ill.: Richard D. Irwin, Inc., 1975), 1–4.

Chapter Two

Functions, Relations, And Systems Of Linear Equations

For mathematics to be of use in business or the sciences, it is important that the relationships that exist between variables be formally defined. For instance, the businessman knows that profits are related to the number of units of product sold and the cost of the product. If it is possible to state these relationships mathematically, the breakeven point can be calculated and profit can be forecast for any specific number of units of product sold. In this chapter, we shall discuss the properties of functions, relations, and systems of linear equations. Examples including breakeven analysis and allocation of resources are used to illustrate these concepts.

2.1 Linear Functions

A *function* is a mathematical relationship in which the values of a single dependent variable are determined from the values of one or more independent variables. Functions that have more than one independent variable are termed *multivariate functions*. Functions that have a single independent variable are termed *univariate functions*. This section introduces one of the most elementary univariate functions, the linear function. Multivariate and nonlinear univariate functions are discussed in Chap. 3.

The functional form of the *linear function* is

$$f(x) = a + bx \qquad (2.1)$$

where $f(x)$ is the dependent variable, x is the independent variable, and a and b are *parameters* of the function. The parameter a is the value of the dependent variable when x is zero and b is the coefficient of the independent variable.

The linear function has one dependent and one independent variable. The symbol $f(x)$, read "f of x", represents values of the dependent variable, and x represents values of the independent variable; $f(x)$ varies according to the rule of the function as x varies. For the linear function, the *rule of the function* states that b is to be multiplied by x and this product added to a. This sum determines the value of the dependent variable $f(x)$. Since the value of the dependent variable depends on the value of the independent variable, $f(x)$ is termed the dependent variable and x is termed the independent variable. The term *variable* refers to a quantity that is allowed to assume different numerical values.

The properties of a linear function can be illustrated by an example. The linear function $f(x) = -4 + 2x$ is graphed in Fig. 2.1.

For each value of the independent variable x, there is one and only one value of the dependent variable $f(x)$. The value of the dependent variable is calculated by the rule of the function. The rule of the function in the example in Fig. 2.1 states that $f(x)$ is equal to -4 plus $2x$.

Both $f(x)$ and x are termed variables, since they are both permitted to take on different numerical values. The table in Fig. 2.1 lists six possible values of x and $f(x)$. For this linear function, however, there are an infinite number of possible values of the variables. For example, between $x = 0$ and $x = 1$ there are an infinite number of possible values of x. Similarly, there are an infinite number of possible values of x between $x = 1$ and $x = 2$. From the infinite number of possible values, we have selected the six values in the table to illustrate the function.

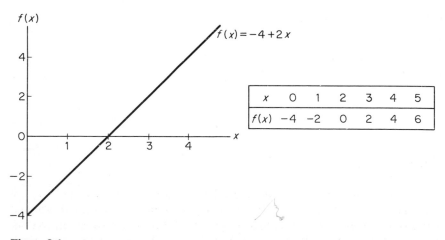

Figure 2.1.

2.1.1 DOMAIN AND RANGE

The permissible values of the independent variable are termed the *domain* of the function. If, for instance, the analyst is interested in all positive values of the independent variable (including zero), he would state that the domain of the function consists of all positive numbers. The function graphed in Fig. 2.1 could thus be expressed as

$$f(x) = -4 + 2x \qquad \text{for } x \geq 0$$

where $x \geq 0$ is read "*x* greater than or equal to zero." The domain of this function is shown by $x \geq 0$ to consist of all positive numbers.

In many situations, the analyst will be concerned with only selected values of the independent variable. If he were concerned with only positive integer values of x, then the domain of the function would be values of the independent variable that are positive integers. As an example,

$$f(x) = -4 + 2x \qquad \text{for } x = 0, 1, 2, 3, \ldots$$

This function consists of the set of ordered pairs $(x, f(x))$,

$$\{(0, -4), (1, -2), (2, 0), (3, 2), (4, 4), \ldots\}$$

The function is graphed in Fig. 2.2.

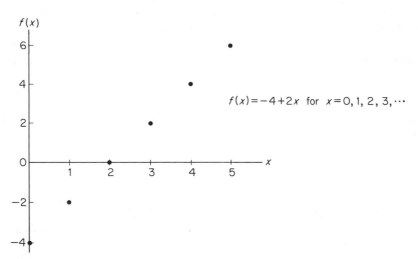

Figure 2.2.

The permissible values of the dependent variable are termed the *range* of the function. The range of the function is those numbers that the dependent variable assumes as the independent variable takes on all values in the domain. In the example illustrated in Fig. 2.1, the domain of the function is all values of x along the continuous line between $x = 0$ and $x = \infty$. The range for this function contains all the numbers along the continuum from $f(x) = -4$ to $f(x) = \infty$. Similarly, if the domain of the function consists of positive integer values, the corresponding range consists of the values calculated by $f(x) = -4 + 2x$ for $x = 0, 1, 2, 3, \ldots$. The concepts of function, domain, and range are further illustrated by the following examples.

Example. Harvey West, an analyst for Pacific Soft-Drink Company, is attempting to develop a cost function for a diet cola. From the accounting department he has learned that fixed costs are $10,000 and variable costs are $0.03 per bottle. Present capacity limitations are 50,000 bottles per month, or 600,000 bottles per year. Mr. West wishes to develop the cost function and to specify the domain and range of the function.

We shall represent the number of bottles of the diet cola by x and the total cost by $f(x)$. Since total cost is composed of fixed and variable costs, the cost function on a yearly basis is

$$f(x) = 10,000 + 0.03x$$

The domain of the function is $x = 0$ to $x = 600,000$. The corresponding range is $f(x) = \$10,000$ to $f(x) = \$28,000$.

Example. Bill Short owns a small auto repair shop. Mr. Short acts as supervisor and employs three auto repairmen. His shop is operated on a 40-hour per week basis, and his employees are guaranteed 40 hours of pay each week. Weekly labor costs consist of a fixed component and a sum that varies with the amount of overtime during the week. The weekly salaries of Mr. Short and the three repairmen total $750. The overtime rate is $5.00 per hour for the employees. Mr. Short does not draw overtime pay. The maximum amount of overtime is 60 hours per week. Determine the cost function and the domain and range of the function.

The cost function consists of the sum of the weekly salaries and the overtime pay. If we represent the overtime hours per week by x and the total weekly salary by $f(x)$, then the cost function is

$$f(x) = 750 + 5x$$

The domain of the function is $x = 0$ to $x = 60$, and the corresponding range of the function is $f(x) = \$750$ to $f(x) = \$1050$.

2.1.2 ESTABLISHING LINEAR FUNCTIONS

It is important to be able to construct a graph of a linear function and to be able to establish a linear function from a set of data. We begin this section by reviewing the procedure for constructing the graph. We then show how one can determine the parameters of the function given the appropriate set of data.

To illustrate the procedure for graphing a linear function, consider the function

$$f(x) = 8 - 2x \qquad \text{for } -2 \leq x \leq 8$$

A linear function can be graphed from two data points $(x_1, f(x_1))$ and $(x_2, f(x_2))$.[1] To determine two data points, we select two values of x and calculate the corresponding values of $f(x)$. Selected values of x and $f(x)$ for our function are given in the table in Fig. 2.3. Any two of these values are plotted on the coordinate axes and the points are connected by a straight line. It is often convenient to use the *intercept* as one of the data points. The intercept is given by the parameter a and is equal to the value of $f(x)$ when x equals 0, i.e., the point at which the function intercepts the vertical axis. From the graph of the function in Fig. 2.3, we see that the intercept is $f(0) = 8$. Connecting this point and any other data point by a straight line gives the graph of the function.

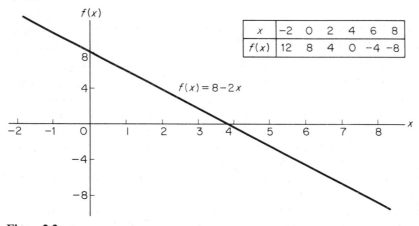

x	-2	0	2	4	6	8
$f(x)$	12	8	4	0	-4	-8

$f(x) = 8 - 2x$

Figure 2.3.

The *slope* of a linear function is defined as the change in the value of the dependent variable divided by the change in the value of the independent variable. Expressed algebraically, the slope of a linear function is

$$\text{slope} = \frac{f(x_2) - f(x_1)}{x_2 - x_1} \qquad (2.2)$$

It can be shown that the slope of a linear function is equal to the coefficient of the independent variable, b. Since

$$f(x_2) = a + bx_2$$

---

[1] This also applies for a linear equation.

and

$$f(x_1) = a + bx_1$$

the slope from Eq. (2.2) is

$$\text{slope} = \frac{f(x_2) - f(x_1)}{x_2 - x_1}$$

Substituting $f(x_2) = a + bx_2$ and $f(x_1) = a + bx_1$ into Eq. (2.2) gives

$$\text{slope} = \frac{a + bx_2 - a - bx_1}{x_2 - x_1}$$

or

$$\text{slope} = \frac{b(x_2 - x_1)}{x_2 - x_1} = b$$

The slope of the linear function $f(x) = a + bx$ is thus equal to the coefficient of the independent variable, b.

Example. Graph the linear function $f(x) = -3 + 1.5x$ for $0 \le x \le 8$. Give the intercept and slope of the function.

A linear function can be graphed from two data points. One obvious data point is the ordered pair $(0, -3)$. Since there are an infinite number of values of x in the domain $0 \le x \le 8$, there are an infinite number of ordered pairs $(x, f(x))$ that could be used for a second data point. One of these ordered pairs is the data point $(4, 3)$. The graph of the function is shown in Fig. 2.4.

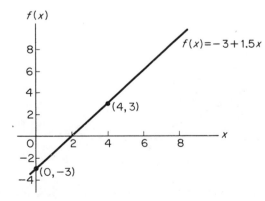

Figure 2.4.

The intercept of the function is $a = f(0) = -3$. The slope of the function is given by the coefficient of the independent variable, b, and is $b = 1.5$.

Example. Use Eq. (2.2) to show that the slope of the linear function in the preceding example is 1.5.

To determine the slope of the function, we arbitrarily select two values of the independent variable, say $x = 4$ and $x = 5$. The values of the dependent variable are

$$f(4) = -3 + 1.5(4) = 3.0$$

and

$$f(5) = -3 + 1.5(5) = 4.5$$

Dividing the change in the value of the dependent variable by the change in the value of the independent variable gives

$$\text{slope} = \frac{f(5) - f(4)}{5 - 4} = \frac{4.5 - 3.0}{1} = 1.5$$

The slope of the function is $b = 1.5$.

Example. Graph the linear function $f(x) = 6 - x$ for $-2 \le x \le 6$. Give the intercept and slope of the function.

To graph the linear function we need only to determine two data points. One logical choice is the ordered pair (0, 6), i.e., the intercept of the function. We arbitrarily selected the ordered pair (4, 2) as the second data point. This pair was determined by evaluating the function for $x = 4$. The intercept of the function is $a = 6$ and the slope is $b = -1$. The graph of the function is shown in Fig. 2.5.

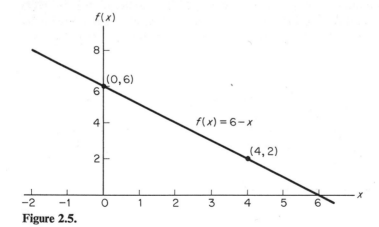

Figure 2.5.

We stated earlier that it is often necessary to determine the function from a set of data points. In these instances, the analyst must determine the appropriate type of function and the parameters of the function. As an illustration, consider the data points plotted in Fig. 2.6. Let us assume that the analyst is reasonably confident that the data points can be described by a linear function. Since the general form of the linear function is

$$f(x) = a + bx$$

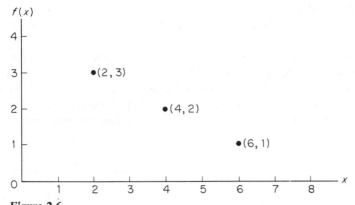

Figure 2.6.

the function can be established by determining the values of the parameters a and b.

The procedure for establishing functions through data points involves substituting the data points into the general form of the function and solving the resulting equations simultaneously for the parameters of the function. Two data points are required for a linear function. From the data points in Figure 2.6, we arbitrarily select the points (2, 3) and (6, 1). Substituting the data point (2, 3) into the general form of the linear function, $f(x) = a + bx$, gives the equation

$$3 = a + b(2)$$

Substituting the data point (6, 1) into the general form of the linear function gives a second equation,

$$1 = a + b(6)$$

These two equations are solved simultaneously by the method of substitution. Substituting $a = 3 - b(2)$ from the first equation into the second equation gives

$$1 = 3 - b(2) + b(6)$$

or

$$4b = -2$$

and

$$b = -0.5$$

From the first equation we obtain

$$a = 3 - b(2)$$

Since $b = -0.5$, substituting for b gives

$$a = 3 - (-0.5)(2) = 4$$

The linear function is thus

$$f(x) = 4 - 0.5x$$

Example. Verify that an alternative selection of data points in Fig. 2.6 results in the same function, $f(x) = 4 - 0.5x$.

The function $f(x) = 4 - 0.5x$ was established using the data points $(2, 3)$ and $(6, 1)$. The same function can be established using any combination of two of the three data points. To illustrate, we shall establish the linear function using the data points $(4, 2)$ and $(6, 1)$.

The parameters of the linear function are determined by substituting the two data points $(4, 2)$ and $(6, 1)$ into the general form of the function, $f(x) = a + bx$. This substitution gives

$$2 = a + b(4)$$

$$1 = a + b(6)$$

Substituting $a = 2 - b(4)$ from the first equation into the second gives

$$1 = 2 - b(4) + b(6)$$

or

$$2b = -1$$

and

$$b = -0.5$$

From the first equation we obtain

$$a = 2 - b(4)$$

$$a = 2 - (-0.5)(4) = 4$$

The linear function is again $f(x) = 4 - 0.5x$.

Example. The Holmgren Electronics Company manufactures an electronic module used in television sets. On the basis of market studies, they have found that the price they are able to obtain for the module varies with the number of modules manufactured. Specifically, the relationship between the price and the number of units manufactured is as follows:

Price per Module	Units Manufactured (Thousands)
$1.00	3.0
1.25	2.5
1.50	2.0
1.75	1.5
2.00	1.0

In an effort to determine the relationship between price p and the number of units manufactured q, Anna Holmgren made a plot of the data given above. This plot is shown in Fig. 2.7.

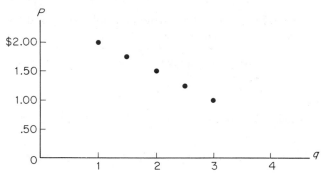

Figure 2.7.

From the plot of the data, Ms. Holmgren noted that price appeared to be linearly related to the number of units manufactured. Determine this relationship.

Since price is a linear function of quantity, we can substitute any two data points from the demand schedule into the function

$$p = a + bq$$

and solve the resulting two equations for a and b. Substituting the data points $(q = 3.0, \ p = \$1.00)$ and $(q = 1.0, \ p = \$2.00)$ into the linear function gives

$$1 = a + b(3)$$

and

$$2 = a + b(1)$$

Solving the two equations simultaneously by the method of substitution gives $a = 2.50$ and $b = -0.50$. The linear demand function is thus

$$p = 2.50 - 0.50q$$

Example. Joe Findley is a stockholder in Electronic Products, Inc. In studying the annual report to stockholders, Joe notes that Electronic Products' computer is being depreciated on a straight-line basis. This means that the depreciation charge is constant each year. If the book value of the computer was $1 million in year 2 and $700,000 in year 5, determine the year in which the computer is fully depreciated, i.e., in which the book value will equal the scrap value. The scrap value of the computer is $200,000.

The function that describes the depreciation and book value can be determined by substituting the data points (2, $1,000,000) and (5, $700,000) into the linear function

$$B(t) = a + bt$$

where $B(t)$ is the book value in year t, a is the original book value, and b is the yearly straight-line decpreciation. The two equations are

$$1,000,000 = a + b(2)$$
$$700,000 = a + b(5)$$

These equations are solved simultaneously to give $a = \$1,200,000$, $b = -\$100,000$ and

$$B(t) = 1,200,000 - 100,000t$$

The year in which the computer is fully depreciated is determined by equating the function with $200,000 and solving for t.

$$200,000 = 1,200,000 - 100,000t$$

The solution is $t = 10$ years. The depreciable life of the computer is 10 years.

2.1.3   DISTANCE BETWEEN TWO POINTS

In some instances it is useful to be able to determine the straight-line "distance" between two points on a linear function. To show what we mean by the distance between two points, consider the function plotted in Fig. 2.8. The data points (8, 5) and (16, 7) are connected by a heavily shaded line. The distance between the two points is the length of the heavily shaded line.

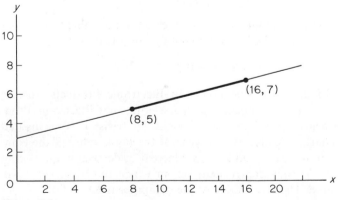

Figure 2.8.

The length of this line is determined from the *Pythagorean theorem.*[2] As the reader will perhaps remember from geometry, the Pythagorean theorem enables one to determine the distance between the data points (x_1, y_1) and (x_2, y_2). The formula is

$$d = \sqrt{(y_2 - y_1)^2 + (x_2 - x_1)^2} \tag{2.3}$$

In Fig. 2.8 we see that $x_1 = 8$, $x_2 = 16$, $y_1 = 5$, and $y_2 = 7$. The distance between the two data points in the figure is

$$d = \sqrt{(7 - 5)^2 + (16 - 8)^2}$$
$$d = \sqrt{4 + 64} = \sqrt{68}$$

From Table A.8, Squares and Square Roots, in the Appendix at the end of this book, the distance is

$$d = 8.2462$$

Example. Determine the distance between the data points (3, 6) and (10, 16).

$$d = \sqrt{(16 - 6)^2 + (10 - 3)^2}$$
$$d = \sqrt{100 + 49} = \sqrt{149}$$
$$d = 12.2066$$

Example. The state highway commission has established two alternative routes between points A and C. The direct route, although shorter, passes over some locally hilly terrain. A circuitous route through point B bypasses the hilly terrain. Highway engineers estimate that the cost per mile of highway construction for the direct route would be 10 percent greater than the cost per mile for the circuitous route. The two alternative routes are shown on the map in Fig. 2.9. If the sole criterion is to minimize the cost of highway construction, which route should be selected?

The coordinates of the points A, B, and C are (20, 10), (70, 60) and (120, 50). The distance from A to B is

$$d_{AB} = \sqrt{(60 - 10)^2 + (70 - 20)^2} = \sqrt{5000} = 70.71 \text{ miles}$$

The distance from B to C is

$$d_{BC} = \sqrt{(60 - 50)^2 + (120 - 70)^2} = \sqrt{2600} = 50.99 \text{ miles}$$

---

[2] The Pythagorean theorem states that the square of the hypotenuse of a right triangle is equal to the sum of the squares of the two sides of the triangle.

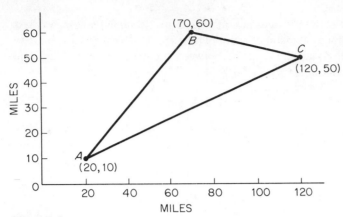

Figure 2.9.

The total distance via the circuitous route is 121.70 miles.
The distance via the direct route is

$$d_{AC} = \sqrt{(50 - 10)^2 + (120 - 20)^2} = \sqrt{11{,}600} = 107.70 \text{ miles}$$

The circuitous route is $(121.70 - 107.70)(100)/107.70 = 13$ percent longer than the direct route. Based on the criterion of minimizing the cost of highway construction, the direct route should be selected.

2.2 Functions and Set Terminology

In Sec. 2.1, a function was defined as a mathematical relationship in which the values of a single dependent variable are determined from the values of one or more independent variables. The language of sets can be used to offer a more general definition of a function. This definition is as follows:

> *Let A and B be two (nonempty) sets and let x represent an element from A and y an element from B. A rule, method, or procedure that associates each element x in A with a unique, or single, y in B is termed a function. A function is also called a mapping from A to B.*

To illustrate this definition of a function, suppose that we have two sets A and B. Let A be the set

$$A = \{x_1, x_2, x_3\}$$

and B be the set

$$B = \{y_1, y_2, y_3\}$$

Suppose further that for each element x in set A there is associated a uniquely determined element y in set B. A rule that associates each x in set A with a unique y in set B is termed a function. One possible function from A to B is the set of ordered pairs

$$\{(x_1, y_1), (x_2, y_2), (x_3, y_3)\}$$

Another possible function is the set of ordered pairs

$$\{(x_1, y_3), (x_2, y_2), (x_3, y_1)\}$$

Additional functions could be defined from set A to set B. The only requirement would be that each value of x in A be associated with a unique, or single, value of y in B.

The concept of a function is illustrated in Fig. 2.10. This figure shows a mapping of the elements x in set A to the elements y in set B. The important concept illustrated by the figure is that for each element x in set A there is one and only one associated element y in set B. Once again, a rule that associates each element in A with a unique element in B is termed a function. The elements in A are the *domain of the function* and those in B are the *range of the function*. Expressed in set notation, the function in Fig. 2.10 is

$$\{(x_1, y_1), (x_2, y_2), (x_3, y_3)\}$$

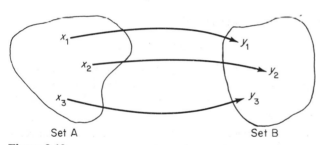

Set A Set B

Figure 2.10.

Fig. 2.11 shows a mathematical relationship that is not a function. Notice in this figure that the element x_1 in set A is associated with both y_1 and y_2 in set B. The requirement for a function is that each element in the domain be uniquely associated with a unique element in the range. Since x_1 is associated

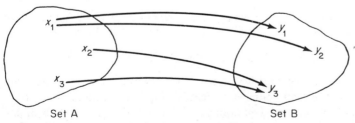

Set A Set B

Figure 2.11.

with two elements, y_1 and y_2, this requirement is not met. Consequently, the relationship in Fig. 2.11 is not a function.

A question that could easily arise from studying Fig. 2.11 is "Does the fact that x_2 and x_3 are both associated with y_3 violate the requirements of a function?" The answer is that it does not. The element x_2 in Fig. 2.11 is uniquely associated with y_3. The element x_3 is also uniquely associated with y_3. The fact that more than one element in the domain is associated with the same element in the range does not violate the requirement for a function.

A function is frequently represented by a formula or an equation. An example that immediately comes to mind is the linear function. Expressed in set notation, a linear function is the set

$$\{(x, f(x)) \mid f(x) = a + bx\} \qquad (2.4)$$

The expression is read "that set of ordered pairs $(x, f(x))$ such that $f(x) = a + bx$." The functional relationship between x and $f(x)$ is described by a formula which states that $f(x)$ is equal to a plus b times x.

Example. Use set terminology to describe the linear function

$$f(x) = 6 - 3x \qquad \text{for } -10 \le x \le 10$$

The function consists of the set of ordered pairs

$$\{(x, f(x)) \mid f(x) = 6 - 3x \qquad \text{for } -10 \le x \le 10\}$$

For each value of x in the domain $-10 \le x \le 10$, the value of $f(x)$ is determined by the rule of the function, $f(x) = 6 - 3x$. The domain of the function is the continuum $-10 \le x \le 10$, and the range of the function is the continuum $-24 \le f(x) \le 36$.

The rule used to establish a function need not, of course, be in the form of a formula or an equation. For instance, the set of ordered pairs

$$\{(1, 3), (2, 4), (4, 10), (8, -300)\}$$

is a function. One can verify that the set is a function by noting that each element in the domain is associated with a unique, or single, element in the range.

Example. Verify that the following set is a function. Give the domain and range of the function.

$$\{(0, 3), (1, 2), (3, 3), (4, 2), (5, -1)\}$$

The roster method is used to specify the ordered pairs $(x, f(x))$ that comprise the set. For each value of x there is only one value of $f(x)$; consequently, the set is a function. The domain of the function is $x = 0, 1, 3, 4, 5$ and the range of the function is $f(x) = -1, 2, 3$. This example illustrates that for each value of x there is only one value of $f(x)$. This value of $f(x)$ can, however, be associated with more than one x. In this function, $f(x) = 3$ when $x = 0$ and $f(x) = 3$ when $x = 3$.

Example. State why the following set is not a function:

$$\{(0, 0,), (0, 1), (2, 1), (3, 2)\}$$

This set does not qualify as a function, since the first two ordered pairs $(0, 0)$ and $(0, 1)$ indicate that $f(x)$ is 0 or 1 when $x = 0$. This violates the requirement that each value of x be associated with a unique, or single, value of $f(x)$.

Our discussion of functions has thus far been limited to rules that relate numbers. The general concept of a function is not, however, restricted to rules or procedures that relate numerical values. Instead, a function may relate any two sets. For instance, one set might consist of political candidates and another of political parties. If each candidate is associated with some definite political party, the pairing of candidates and political parties could be classified as a function.

Example. Let $A = \{\text{Smith, Jones, White}\}$ represent the set of candidates for governor and $B = \{\text{Republican, Democrat}\}$ represent the set of political parties. A function from A to B that describes political affiliation is

$$\{(\text{Smith, Republican}), (\text{Jones, Democrat}), (\text{White, Republican})\}$$

Example. Does the following set satisfy the requirements for a function?

$$\{(\text{Mantle, baseball}), (\text{West, basketball}), (\text{Simpson, football})\}$$

Each element in the domain is associated with a unique, or single, element in the range. Consequently, the requirement for a function is met.

The Cartesian product of two or more sets can be used in defining and illustrating functions.[3] The Cartesian product of two sets A and B is the set of ordered pairs (a, b), where $a \in A$ and $b \in B$. Utilizing set notation, we find that the Cartesian product $A \times B$ is the set

$$A \times B = \{(a, b) \mid a \in A, b \in B\}$$

The elements of the set A are termed the *domain of the Cartesian product* and the elements of the set B are termed the *range of the Cartesian product*.

For the set $A \times B$, a function is a subset in which each element of A occurs once and is associated with a unique element in B. The domain of the function consists of elements in A and the range of the function of elements in B. Although the elements in A must occur once in the function, there is no

---

[3] The Cartesian product was discussed in Chapter 1.

requirement that all elements in B be included in the function. Elements in B can occur zero, once, or repeatedly.

Example. Given the set $A = \{a, b, c\}$ and the set $B = \{1, 2\}$, the Cartesian product of A and B is the set

$$A \times B = \{(a, 1), (a, 2), (b, 1), (b, 2), (c, 1), (c, 2)\}$$

Examples of subsets that are functions include $\{(a, 1), (b, 1), (c, 1)\}$, $\{(a, 1),$ $(b, 2), (c, 1)\}$, and $\{(a, 2), (b, 1), (c, 1)\}$. The subset $\{(a, 1), (a, 2), (b, 1)\}$ is, however, not a function.

Example. Set $A = \{$boy, girl$\}$ and $B = \{$Yale, Princeton, Vassar, Smith$\}$. The Cartesian product of A and B is

$$A \times B = \{(\text{boy, Yale}), (\text{boy, Princeton}), (\text{boy, Vassar}), (\text{boy, Smith}),$$
$$(\text{girl, Yale}), (\text{girl, Princeton}), (\text{girl, Vassar}), (\text{girl, Smith})\}$$

Examples of subsets of the Cartesian product that satisfy the requirements of a function are $\{(\text{boy, Yale}), (\text{girl, Vassar})\}$ and $\{(\text{boy, Princeton}), (\text{girl,}$ Smith$)\}$.

2.3   Relations

   The preceding section introduced the concept of a function as a particular type of subset of the Cartesian product of two sets A and B. Specifically, for the set $A \times B$, a function was defined as a subset in which each element in A occurs once and is associated with a unique element in B. A function is but one of two important types of relationships between sets discussed in this chapter. The second is termed a *relation*.
   To introduce the concept of a relation, consider again two nonempty sets A and B. A relation is simply *any* rule whatever that serves to select a subset of the Cartesian product of A and B. That is,

   Let A and B be two nonempty sets. A relation defined on the set $A \times B$
is any rule that serves to define a subset of $A \times B$.

   To illustrate the definition of a relation, suppose that we have the sets $A = \{x_1, x_2\}$ and $B = \{y_1, y_2, y_3\}$. The Cartesian product of these two sets is

$$A \times B = \{(x_1, y_1), (x_1, y_2), (x_1, y_3), (x_2, y_1), (x_2, y_2), (x_2, y_3)\}$$

From the definition of a relation, we know that any rule that serves to define a subset of the Cartesian product is a relation. Thus, the subset

$$\{(x_1, y_1), (x_1, y_2), (x_2, y_2)\}$$

is a relation. Similarly, the subset

$$\{(x_1, y_1), (x_1, y_2)\}$$

satisfies the requirements for a relation. The elements in set A are termed the domain of the relation and those in set B are termed the range of the relation.

Two possible relations between sets A and B are shown in Fig. 2.12. Notice in Fig. 2.12(a) that x_1 is associated with both y_1 and y_2. Fig. 2.12(b) shows a relation in which one element in the domain, x_1, is associated with three elements in the range, y_1, y_2, y_3. The second element in the domain, x_2, is not associated with an element in the range. This nevertheless is a relation since the set $\{(x_1, y_1), (x_1, y_2), (x_1, y_3)\}$ is a subset of the Cartesian product of A and B.

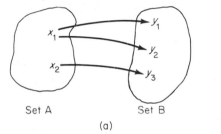 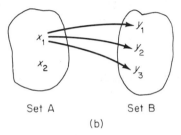

Set A   Set B   Set A   Set B
(a)   (b)

Figure 2.12.

Example. Let A be the set {boy, girl} and B be the set {Yale, Princeton, Vassar, Smith}. From the Cartesian product of these two sets, develop a relation that illustrates possible college attendance.

The Cartesian product is

$$A \times B = \{(\text{boy, Yale}), (\text{boy, Princeton}), (\text{boy, Vassar}), (\text{boy, Smith}),$$
$$(\text{girl, Yale}), (\text{girl, Princeton}), (\text{girl, Vassar}), (\text{girl, Smith})\}$$

A subset that shows possible college attendance is the relation

$$\{(\text{boy, Yale}), (\text{boy, Princeton}), (\text{girl, Vassar}), (\text{girl, Smith})\}$$

Example. Let $S = $ {all stocks traded on the New York Stock Exchange on a Friday} and let $A = \{+, 0, -\}$ represent the stock closing up, closing unchanged, or closing down. The Cartesian product $S \times A$ is the set of all possible movements in the stock prices. From the Cartesian product, the relation that describes the possible stock movements in a portfolio that contains Ford and IBM is

$$\{(\text{Ford}, +), (\text{Ford}, 0), (\text{Ford}, -), (\text{IBM}, +), (\text{IBM}, 0), (\text{IBM}, -)\}$$

A subset of the relation is the function which describes the actual price movement for a given Friday. This function is

$$\{(\text{Ford}, +), (\text{IBM}, -)\}$$

Notice that in the relation elements in the domain may appear more than once and may be associated with more than one element in the range. In the function, elements in the domain are associated with only one element in the range.

Example. Develop the relation between x and y such that x is greater than y and both x and y are positive integers.

The set I consists of all positive integers, i.e.,

$$I = \{1, 2, 3, 4, \ldots\}$$

The Cartesian product, $I \times I$, represents all integer pairs (x, y) in the first quadrant. The relation

$$R = \{(x, y) \mid x > y \text{ and } x, y \in I\}$$

is a subset of the Cartesian product $I \times I$. This relation is shown in Fig. 2.13. Only a few of the infinite number of ordered pairs that comprise the relation are shown in this figure.

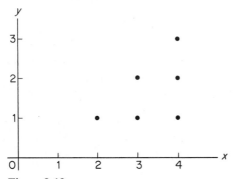

Figure 2.13.

A relation that will be of major importance in the chapters on linear programming is the linear inequality. A linear inequality is a relation of the form

$$y \geq kx + c \tag{2.5}$$

or alternatively of the form

$$y \leq kx + c \tag{2.6}$$

An inequality is a relation rather than a function, since for each value of x in the domain there is more than one value of y in the range. In fact, for any value of x in the linear inequality $y \geq kx + c$, there are an infinite number of values of y that satisfy the requirement that $y \geq kx + c$.

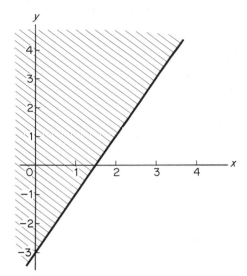

Figure 2.14.

As an example of a linear inequality, consider $y \geq 2x - 3$. This relation is shown by the shaded area in Fig. 2.14. Using set notation, we can express the linear inequality as $\{(x, y) \mid y \geq 2x - 3\}$.

Example. A university limits class enrollment to 64 students per teacher and cancels any class in which enrollment is six or less. Develop relations that describe the limitations.

If we represent the faculty size as x and the student enrollment as y, then the relation which describes the limitation is

$$\{y \leq 64x \text{ and } y \geq 7x \qquad \text{for } x, y \in I\}$$

The relation is a subset of the Cartesian product of $I \times I$, where I is the set of positive integers.

Example. Show the inequality $y \leq x - 1$ on a graph. The inequality is shown by the shaded area in Fig. 2.15.

Both functions and relations have been shown to be subsets of Cartesian products. In examples such as on p. 49 concerning the set $A = \{\text{boy, girl}\}$ and $B = \{\text{Yale, Princeton, Vassar, Smith}\}$, the Cartesian product was obvious. We have, however, not indicated the Cartesian product of which the function $f(x) = 2x + 3$ or the relation $y \leq 6x - 4$ are subsets. If x is unrestricted, the domain for both the function and the relation is $-\infty \leq x \leq \infty$.

The continuum $-\infty \leq x \leq \infty$ is described by the set of real numbers $R = \{-\infty \leq x \leq \infty\}$. The range of both the function and the relation is

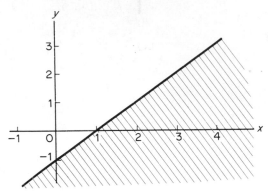

Figure 2.15.

similarly $R = \{-\infty \leq f(x) \leq \infty\}$. The Cartesian product $R \times R$ is the set of all ordered pairs of real numbers, i.e.,

$$R \times R = \{(x, f(x)) \mid x \in R \text{ and } f(x) \in R\}$$

or alternatively

$$R \times R = \{(x, y) \mid x \in R \text{ and } y \in R\}$$

Both the relation and the function are subsets of the Cartesian product of real numbers.

Example. Verify that the mathematical relationship

$$y^2 = x$$

is a relation rather than a function.

If $y^2 = x$, then $y = \pm\sqrt{x}$ and for each value of x in the domain $0 \leq x \leq \infty$, there are two values of y. This violates the requirement that there exist only one value of the dependent variable for each value of the independent variable. Consequently, $y = \pm\sqrt{x}$ is a relation rather than a function. This relation is graphed in Fig. 2.16.

2.4 Systems of Linear Equations

A function was defined earlier in this chapter as a mathematical relationship in which the value of a single dependent variable is determined from the values of one or more independent variables. In certain instances, it is not appropriate to consider one variable as "dependent" on other variables. Although the variables are explicitly related and the requirements for a function are satisfied, the relationship is referred to as an *equation* rather

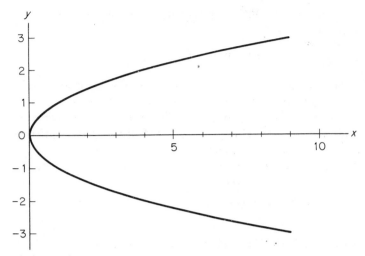

Figure 2.16.

than a function. Expressed in set notation, a linear equation with two variables is the set of ordered pairs (x, y) given by

$$\{(x, y) \mid ax + by = c\} \tag{2.7}$$

In comparing the definition of a linear equation given by Eq. (2.7) with
that of a linear function given by Eq. (2.4), the primary difference is that
neither variable is distinguished in Eq. (2.7) as the dependent or the independent variable.

A *system of equations* consists of one or more equations with one or more
variables. As an example, a system of two equations with two variables is

$$2x + 4y = 28$$
$$3x - 2y = 10$$

Similarly, a system of two equations with three variables is

$$20x + 15y - 10z = 20$$
$$30x - 25y + 35z = 85$$

The concept of a system of equations is illustrated by the following examples.

Example. A firm produces two products, A and B. Each unit of product
A requires 2 man-hours of labor and each unit of product B requires 3
man-hours of labor. Develop the equation describing the relationship between the two products, assuming that 110 man-hours of labor will be used.

The number of units of product A is represented by x and the number of units of product B is represented by y. The equation describing the relationship between the two variables is

$$2x + 3y = 110$$

Example. Products A and B discussed in the preceding example require machine time as well as labor time. Each unit of product A requires 1 hour of machine time, and each unit of product B requires 0.5 hour of machine time. Develop the equation describing the relationship between the two products, assuming that 35 hours of machine time will be used.

If we again represent the number of units of product A by x and the number of units of product B by y, the equation is

$$x + 0.5y = 35$$

2.4.1 SOLUTION SETS FOR SYSTEMS OF EQUATIONS

The preceding examples illustrate a system of two equations with two variables. The equation that relates units of products A and B with man-hours can be described by the set

$$L = \{(x, y) \mid 2x + 3y = 110\}$$

Similarly, the equation that relates units of products A and B with machine time can be described by the set

$$M = \{(x, y) \mid x + 0.5y = 35\}$$

There are an infinite number of ordered pairs (x, y) that are members of set L. For instance, (0, 36.67), (1, 36), (2, 35.33), (3, 34.67) are four ordered pairs which are solutions to the equation $2x + 3y = 110$. There are also an infinite number of ordered pairs (x, y) that are members of set M. These include (0, 70), (1, 68), (2, 66), (3, 64), ... The ordered pairs (x, y) that are members of the set L are termed *solutions* to the equation $2x + 3y = 110$. Similarly, the ordered pairs that are members of the set M are solutions to the equation $x + 0.5y = 35$. The sets L and M are referred to as *solution sets*.

The sets L and M describe the labor and machine constraints on the production of products A and B. The intersection of sets L and M contains those elements (x, y) that are members of both L and M. That is,

$$L \cap M = \{(x, y) \mid 2x + 3y = 110 \quad \text{and} \quad x + 0.5y = 35\}$$

The elements (x, y) represent the quantities of products A and B that are solutions for both the labor and the machine constraints.

The method of determining the solution set for a system of two equations is to determine the values of the variables that are solutions to both equations. These values are determined by solving the equations simultaneously. One of the methods commonly used in algebra for solving equations simultaneously is the method of substitution. This method involves solving for one variable in terms of the remaining variables and substituting this expression for the variable in the remaining equations. The two equations are

$$2x + 3y = 110$$

$$x + 0.5y = 35$$

Using the method of substitution, we solve for x in the second equation and obtain $x = 35 - 0.5y$. The expression $35 - 0.5y$ is substituted for x in the first equation to give

$$2(35 - 0.5y) + 3y = 110$$

$$2y = 40$$

$$y = 20$$

Substituting $y = 20$ in the second equation gives $x = 25$. The values of x and y that are solutions to both equations are $L \cap M = \{(25, 20)\}$, or 25 units of product A and 20 units of product B.

Example. A firm uses three processes to produce three products. Product 1 requires 2 hours of process A time, 4 hours of process B time, and 6 hours of process C time. Product 2 requires 1 hour of process A time and 2 hours of process B and C time. Product 3 requires 3 hours of process A time and 4 hours of process B and C time. One hundred hours of process A time, 160 hours of process B time, and 190 hours of process C time have been allocated to the production of the products. Determine the quantities of products 1, 2, and 3 that can be manufactured.

We let x_1, x_2, and x_3 represent the number of units of products 1, 2, and 3. The process A equation is given by

$$A = \{(x_1, x_2, x_3) \mid 2x_1 + x_2 + 3x_3 = 100\}$$

The process B equation is

$$B = \{(x_1, x_2, x_3) \mid 4x_1 + 2x_2 + 4x_3 = 160\}$$

The process C equation is

$$C = \{(x_1, x_2, x_3) \mid 6x_1 + 2x_2 + 4x_3 = 190\}$$

There are an infinite number of ordered triplets (x_1, x_2, x_3) that are solutions to each equation. We require, however, the set of ordered triplets (x_1, x_2, x_3) that are solutions to all three equations. This is the solution set for the system of simultaneous equations. The solution set is determined by solving the three equations simultaneously for x_1, x_2, and x_3. The solution set is

$A \cap B \cap C = \{(15, 10, 20)\}$, or $x_1 = 15$ units, $x_2 = 10$ units, and $x_3 = 20$ units.

The solution set for $A \cap B \cap C$ can be determined by using any one of the methods introduced in algebra. Using the method of substitution, we solve the first equation for x_1 to obtain $x_1 = -0.5x_2 - 1.5x_3 + 50$. Substituting this quantity for x_1 in the second and third equations gives

$$4(-0.5x_2 - 1.5x_3 + 50) + 2x_2 + 4x_3 = 160$$

and

$$6(-0.5x_2 - 1.5x_3 + 50) + 2x_2 + 4x_3 = 190$$

These two equations reduce to

$$2x_3 = 40$$

and

$$x_2 + 5x_3 = 110$$

These two equations are solved for x_2 and x_3 to give $x_2 = 10$ and $x_3 = 20$. Substituting these values in process equation A gives $x_1 = 15$.

2.4.2 CONSISTENT AND INCONSISTENT SYSTEMS OF LINEAR EQUATIONS

The solution set for the system of linear equations in the preceding two examples had only a single element. In the first example the element was $L \cap M = \{(25, 20)\}$ and in the second example the element was $A \cap B \cap C = \{(15, 10, 20)\}$. These examples illustrate the case in which a *unique* solution exists for the system of equations. A unique solution exists when there is only one element in the intersection of the sets describing the individual equations.

A unique solution is one of three possiblities for the solution to a system of linear equations. A system of linear equations may have (1) no solution; (2) exactly one solution (unique); or (3) an infinite number of solutions. A system that has no solution is termed *inconsistent*. A system that has one or more solutions is said to be *consistent*. These three possibilities are illustrated by Fig. 2.17.

The concepts presented in Fig. 2.17 can be explained by using set notation. Assume that $f(x_1, x_2, x_3)$, $g(x_1, x_2, x_3)$, and $h(x_1, x_2, x_3)$ represent three different equations. The set of ordered triplets (x_1, x_2, x_3) that are solutions to each equation is described by the sets

$$F = \{(x_1, x_2, x_3) \mid f(x_1, x_2, x_3)\}$$

$$G = \{(x_1, x_2, x_3) \mid g(x_1, x_2, x_3)\}$$

$$H = \{(x_1, x_2, x_3) \mid h(x_1, x_2, x_3)\}$$

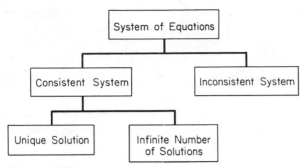

Figure 2.17.

The solution to the system of equations is defined as all ordered triplets $(x_1,$ $x_2, x_3)$ that are common to the three equations. This solution set is the set of ordered triplets contained in the intersection of F, G, and H, i.e., $F \cap G \cap H$. If this intersection is null (empty), the system of equations is inconsistent. If, however, the intersection contains one or more elements, the system is consistent. For those cases in which the intersection contains exactly one element, this element is the unique solution to the system of equations. If the intersection contains more than one element, the system of equations is consistent, but the solution is not unique.

Graphs of equations are useful as an aid to understanding the solution sets for systems of equations. We shall show a system of equations that is consistent and has a unique solution, a system of equations that is consistent with an infinite number of solutions, and a system of equations that is inconsistent and consequently has no solution.

2.4.3 CONSISTENT SYSTEM WITH A UNIQUE SOLUTION

As an illustration of a consistent system of equations with a unique solution, consider the equations $2x + y = 8$ and $3x - 2y = -2$. These equations are shown in Fig. 2.18. The equations can be described by the sets

$$A = \{(x, y) \mid 2x + y = 8\}$$
$$B = \{(x, y) \mid 3x - 2y = -2\}$$

The set A contains all ordered pairs (x, y) that are solutions to the equation $2x + y = 8$. These ordered pairs are shown in Fig. 2.18 by the linear equation $2x + y = 8$. The set B contains all ordered pairs (x, y) that are solutions to the equation $3x - 2y = -2$. These ordered pairs are shown by the linear equation $3x - 2y = -2$. The solution set for the system of equations contains all ordered pairs that are solutions to both A and B. This set

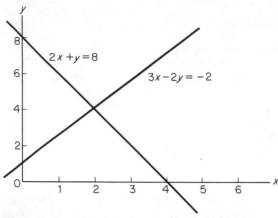

Figure 2.18.

is shown in Fig. 2.18 by the intersection of the two equations. The ordered pairs that are common to both equations are

$$A \cap B = \{(x, y) \mid 2x + y = 8 \quad \text{and} \quad 3x - 2y = -2\}$$

The values of the variables x and y are determined by solving the two equations simultaneously for x and y. Substituting $y = 8 - 2x$ from the first equation for y in the second equation gives

$$3x - 2(8 - 2x) = -2$$

$$7x = 14$$

$$x = 2$$

Substituting $x = 2$ in the first equation gives $y = 4$. The ordered pair $(2, 4)$ is the unique solution to the system of equations.

Example. Determine the unique solution to the following system of three linear equations and three variables:

$$A = \{(x, y, z) \mid 2x + y - 2z = -1\}$$

$$B = \{(x, y, z) \mid 4x - 2y + 3z = 14\}$$

$$C = \{(x, y, z) \mid x - y + 2z = 7\}$$

A unique solution exists for the system of equations if there is an ordered triplet (x, y, z) that is a common solution to all three equations. Solving the three equations simultaneously gives $x = 2$, $y = 3$, and $z = 4$. The solution set for the system of equations is $A \cap B \cap C = \{(2, 3, 4)\}$.

Example. Verify that the following system of three equations with two variables is consistent and has a unique solution:

$$A = \{(x, y) \mid 2x + y = 9\}$$
$$B = \{(x, y) \mid x - y = 3\}$$
$$C = \{(x, y) \mid x + 2y = 6\}$$

The system of three equations with two variables is consistent and has a unique solution if there is a single ordered pair (x, y) that is a common solution to all three equations. Solving the first two equations simultaneously gives $A \cap B = \{(4, 1)\}$ as the unique solution to the first two equations. To determine if $(4, 1)$ is also a solution to the third equation, the ordered pair is substituted into the third equation. Since $4 + 2(1) = 6$, the ordered pair $(4, 1)$ is a solution for the third equation. The solution for the system of equations is $A \cap B \cap C = \{4, 1)\}$. The system of equations is consistent with a unique solution.

The equations described by the sets A, B, and C are shown in Fig. 2.19. The unique solution $x = 4$, $y = 1$ occurs at the intersection of the three equations.

A system of three equations with two variables has a unique solution when one of the equations can be expressed as a linear combination of the remaining two equations. If we use the sets A, B, and C to represent the equations, a linear combination exists if there are numbers a and b such that $aA + bB = C$.

To determine if a linear combination exists in the preceding example, we first write the equations as $2x + y - 9 = 0$, $x - y - 3 = 0$, and $x + 2y - 6 = 0$. A linear combination exists if there are numbers a and b such that

$$a(2x + y - 9) + b(x - y - 3) = x + 2y - 6$$

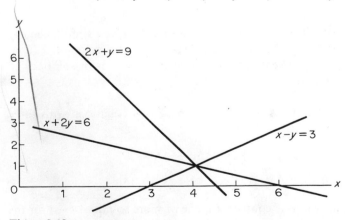

Figure 2.19.

In this example, equation C can be formed by $1A - 1B$. This is shown by

$$1(2x + y - 9) - 1(x - y - 3) = x + 2y - 6$$

The appropriate numbers are thus $a = 1$ and $b = -1$, and equation C is a linear combination of equations A and B.

2.4.4   CONSISTENT SYSTEM WITH AN INFINITE NUMBER OF SOLUTIONS

A system of linear equations has been defined as consistent if one or more solutions exist for the system of equations. As an illustration of a system of equations for which there is more than one solution, consider the equations

$$A = \{(x, y, z) \mid 2x - 4y + 4z = 20\}$$

$$B = \{(x, y, z) \mid 3x + 4y - 2z = 30\}$$

The solution to the system of two equations and three variables consists of all ordered triplets (x, y, z) that are solutions to both equations. From introductory algebra, we remember that it is impossible to determine a unique solution for three variables with two equations. We can, however, determine ordered triplets that are solutions to both equations. The procedure is to arbitrarily specify the value for one of the variables and to solve the two equations simultaneously for the remaining two variables. For instance, if the value of x is specified as 0, the solution set is $x = 0$, $y = 20$, and $z = 25$. These values are determined by solving the two equations

$$2(0) - 4y + 4z = 20$$

$$3(0) + 4y - 2z = 30$$

for y and z. Similarly, if $x = 5$, the equations can be solved simultaneously to obtain $y = 10$ and $z = 12.5$. Since an infinite number of values of x, y, or z could arbitrarily be specified, there are an infinite number of solutions to the system of two equations.

Example. Verify that the following system of two equations with three variables is consistent:

$$A = \{(x, y, z) \mid 2x + 3y + 4z = 20\}$$

$$B = \{(x, y, z) \mid 5x - 4y + 3z = 15\}$$

The system of equations is consistent if one or more solutions exist for the system. Equating $x = 0$ gives $y = 0$ and $z = 5$. Equating $x = 5$ gives $y = 2.8$ and $z = 0.4$. Since an infinite number of values of x, y, or z could be specified, the solution set $A \cap B$ contains an infinite number of triplets (x, y, z), and the system of equations is consistent.

Example. Verify that the following system of equations is consistent and has an infinite number of solutions:

$$A = \{(x, y, z) \mid 2x + 3y - 2z = 40\}$$
$$B = \{(x, y, z) \mid 3x - 2y + z = 50\}$$
$$C = \{(x, y, z) \mid x - 5y + 3z = 10\}$$

The system of three equations with three variables would appear at initial inspection to be consistent with a unique solution. When attempting to determine the solution, however, we find that this is not the case. Substituting the expression for x from the third equation in the first equation gives

$$2(10 - 3z + 5y) + 3y - 2z = 40$$
$$13y - 8z = 20$$

Substituting the same expression for x in the second equation gives

$$3(10 - 3z + 5y) - 2y + z - 50$$
$$13y - 8z = 20$$

Using the method of substitution, we would normally solve the two equations simultaneously for y and z. The equations that resulted from the original substitution are, however, the same. Consequently, there are an infinite number of ordered triplets (x, y, z) that are members of the solution set $A \cap B \cap C$. To obtain a solution to this system of equations, the value of one of the variables must be arbitrarily specified. If z is specified, the complete solution to the system of equations is

$$z = \text{specified}$$
$$y = \frac{20 + 8z}{13}$$
$$x = 10 - 3z + 5y$$

For instance, if $z = 0$, then $y = \frac{20}{13}$ and $x = \frac{230}{13}$. One solution is thus the ordered triplet $(\frac{230}{13}, \frac{20}{13}, 0)$. If $z = 1$, then $y = \frac{28}{13}$ and $x = \frac{231}{13}$. A second solution is $(\frac{231}{13}, \frac{28}{13}, 1)$. It is obvious that an infinite number of values of z could be specified and therefore an infinite number of solutions are possible.

Example. Verify that the equation $x - y = 6$ represents a consistent system of equations with an infinite number of solutions.

A system of equations is termed consistent if there are one or more solutions for the system of equations. The equation $x - y = 6$ is a system of one equation with two variables. The solution set consists of an infinite number of ordered pairs (x, y) that are solutions to the equation. One of the variables must be specified to determine a solution. For instance, if $x = 0$, then $y = -6$ and $(0, -6)$ is a solution. Similarly, if $x = 1$, then $y = -5$ and $(1, -5)$ is another solution. We thus conclude that the equation $x - y = 6$ is a consistent system of equations with an infinite number of solutions.

2.4.5 INCONSISTENT SYSTEM OF EQUATIONS

A system of equations is termed inconsistent when the solution set for the system of equations is null. As an example, consider the two equations

$$A = \{(x, y) \mid x + y = 6\}$$
$$B = \{(x, y) \mid 2x + 2y = 8\}$$

These equations are shown in Fig. 2.20. The solution to the system of two equations consists of the ordered pairs (x, y) that are common to both set A and set B, i.e., $A \cap B$. The equations described by the sets A and B are, however, parallel. Consequently, there is no ordered pair (x, y) that is a common solution to both equations. The solution set for the system of two equations with two variables is thus null, i.e., $A \cap B = \phi$.

Another example of an inconsistent system of equations is

$$A = \{(x, y) \mid x + y = 8\}$$
$$B = \{(x, y) \mid 2x - y = 2\}$$
$$C = \{(x, y) \mid x - 2y = -2\}$$

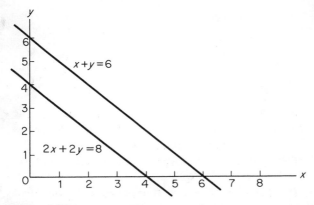

Figure 2.20.

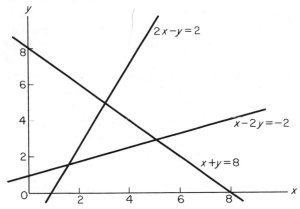

Figure 2.21.

These equations are shown in Fig. 2.21. The solution to the system of equations consists of the ordered pairs (x, y) that are common solutions to all three equations. Figure 2.21 shows that there are no ordered pairs (x, y) that are members of the solution set $A \cap B \cap C$. This can be verified algebraically by determining the solution to any two of the three equations and determining if this ordered pair is a solution to the third equation. For instance, the solution set for equation A and B is $A \cap B = \{(3\frac{1}{3}, 4\frac{2}{3})\}$. The ordered pair $(3\frac{1}{3}, 4\frac{2}{3})$ is, however, not a solution for equation C. Since there is no ordered pair that is a common solution for all three equations, we conclude that the equations are inconsistent.

2.5 Breakeven Analysis

The ability of a firm to survive in a competitive environment depends largely on the profitability of the firm. Since the manager must make decisions that affect the profitability of the firm, it is important that he understand the relationships between costs, sales revenue, and profit. This section provides a brief introduction to the cost–revenue–profit relationship. As the reader will discover, functions and systems of equations are useful for specifying the relationships between costs, sales revenue, and profit.

In order for a firm to make a profit, total revenue must exceed total cost. If total revenue is less than total cost during a specific period, the firm has incurred a loss during that period. If total revenue just equals total cost, the firm is said to *breakeven*. Since the survival of the firm depends on the firm's operating at a profit, it is important for the manager to be able to calculate the breakeven point.

The *breakeven point* is the volume or level of sales at which total revenue and total costs are exactly equal. Sales volume can be expressed in three ways. One way of expressing sales volume is by the number of units of the product sold. Another is the dollar volume of sales. A third is as a percentage of plant capacity. *Total revenue* is, of course, the number of dollars generated from sales. *Total costs* include all costs incurred in the production and marketing of the product along with the general administrative costs of operating the business.

In order to determine the breakeven point or the profit or loss associated with a given volume of sales, we must specify the revenue and cost functions. The total revenue depends on the price of the product and the number of units of the product sold. In calculating breakeven, price is normally established at a fixed level. Consequently, total revenue is a function of the sales volume. That is,

$$TR = f(SV) \tag{2.8}$$

where TR represents total revenue and SV represents sales volume.

Total costs are customarily separated into fixed costs and variable costs. Fixed costs are assumed to remain constant in total dollar amount regardless of the level of production. They include such costs as rent, interest, property taxes, certain administrative salaries, etc. Variable costs are incurred as a direct result of the production process. These costs are normally assumed to be constant per unit of production. These costs include such items as direct labor and materials. Thus, although variable costs are constant per unit, they vary in direct proportion with the number of units produced. Total costs can be described by the function

$$TC = FC + VC \tag{2.9}$$

where TC represents total cost, FC represents fixed costs, and VC represents variable costs.

To illustrate the relationship between revenue, cost, and profit, consider the case of a firm that manufactures a quadraphonic tape deck. This tape deck is sold through retailers at a list price of $150. The markup on the tape deck is 40 percent. This means that the manufacturer receives 60 percent of the retail price, i.e., $90 per unit.

The costs of manufacturing the tape deck include a fixed cost and a variable cost. The fixed cost consists of certain administrative charges, property taxes, insurance, depreciation, etc. These costs amount to $20,000 per year. The variable costs consist of the labor, materials, and components that are incurred as a direct result of manufacturing the quadraphonic tape deck. These costs amount to $40 per unit.

The difference between revenue per unit and variable cost per unit is termed the *contribution margin*. In our example, the revenue per unit is $90 and the variable cost per unit is $40. Subtracting variable cost from revenue gives a contribution margin of $50 per unit. This $50 represents the contribution per unit toward fixed costs and profit. In other words, of the $90 received for each tape deck, $50 remains after variable costs have been paid as a contribution toward fixed costs and profit.

The relationship between revenue, cost, and profit for the quadraphonic tape deck is shown in Fig. 2.22. The fixed costs of $20,000 are plotted as a horizontal line and remain constant regardless of the number of units of output. Total costs consist of fixed costs and variable costs. As shown by Fig. 2.22, total costs increase with the number of units of output. Total revenue is given by the product of the revenue per unit and the number of units. Breakeven occurs when total revenue and total costs are equal. In our example, 400 units must be sold to breakeven. This number may be obtained directly from the figure or, perhaps more conveniently, by recognizing that breakeven occurs when the contribution margin multiplied by the breakeven number of units is equal to the fixed cost. That is,

$$(\text{contribution margin})(\text{breakeven point}) = \text{fixed cost}$$

Alternatively, breakeven can be calculated by the formula

$$\text{breakeven point} = \frac{\text{fixed cost}}{\text{contribution margin}} \tag{2.10}$$

The profit (or loss) is, of course, equal to total revenue less total cost.

It is useful to be able to express the revenue–cost–profit relationship in a functional form. Total revenue is given by the product of the revenue per

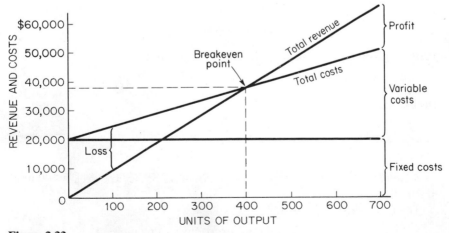

Figure 2.22.

unit and the number of units. Representing revenue per unit as r and the number of units as x, total revenue is

$$TR = rx \qquad (2.11)$$

Total cost consists of fixed costs and variable costs. Representing fixed cost as FC, variable cost per unit as v, and the number of units as x, total costs are

$$TC = FC + vx \qquad (2.12)$$

Profit is the difference between total revenue and total costs. Subtracting total revenue from total cost gives

$$P = TR - TC$$

$$P = rx - FC - vx$$

or

$$P = (r - v)x - FC \qquad (2.13)$$

Breakeven occurs when total revenue and total costs are equal, i.e., when profits are zero. Equating P with zero in Eq. (2.13) gives

$$x_{\text{breakeven}} = \frac{FC}{r - v} \qquad (2.14)$$

Since $r - v$ is the contribution margin, we see that Eq. (2.14) and Eq. (2.10) are the same.

The revenue, cost, and profit functions for our example problem can easily be determined. The revenue per unit from the quadraphonic tape deck was $90. Total revenue is thus

$$TR = 90x$$

where x is the number of units. The fixed costs allocated to the tape deck were $20,000 and the variable costs were $40 per unit. Total costs are

$$TC = 20,000 + 40x$$

Profit is given by total revenue less total cost and is

$$P = 50x - 20,000$$

Breakeven occurs when profits are zero and is

$$x_{\text{breakeven}} = \frac{20,000}{50} = 400 \text{ units}$$

Example. Determine the profit on the quadraphonic tape deck if 500 units are sold.

The profit from selling 500 units is

$$P = 50(500) - 20,000 = \$5000$$

Example. Assume that price competition has forced the manufacturer of the quadraphonic tape deck to reduce the wholesale price from $90 to $72. If variable and fixed costs remain unchanged, calculate the new breakeven point and determine the profit on sales of 700 units.

From Eq. (2.14), breakeven is

$$x_{breakeven} = \frac{FC}{(r - v)}$$

$$x_{breakeven} = \frac{20,000}{(72 - 40)} = 625 \text{ units}$$

Profit on the sale of 700 units is

$$P = (r - v)x - FC$$
$$P = (72 - 40)(700) - 20,000$$
$$P = \$2400$$

Example. Joe Miney is considering opening a small men's clothing store. Joe expects fixed costs such as rent, insurance, property taxes, depreciation of fixtures, utilities, and salaries to average $1000 per month. The typical markup on the kind of clothing Joe will carry is 40 percent. Joe will pay a commission of 5 percent of the retail price to his employees. Determine the volume of sales that results in breakeven. What volume of sales is necessary if Joe is to realize a profit of $2000 per month?

The contribution margin is 35 percent of the sales volume. In order to breakeven, monthly sales must be

$$x_{breakeven} = \frac{1000}{0.35} = \$2857$$

To realize a profit of $2000 per month requires sales of

$$x_{(breakeven + 2000)} = \frac{3000}{0.35} = \$8571$$

Example. Dennis Murphy is the owner of a small golf shop. Dennis sells golf clubs, golf balls, and other golf equipment. The monthly expenses associated with the golf shop are given in the following table:

Monthly Expenses	
Rent	$500.00
Salaries	1200.00
Utilities	200.00
Other	300.00
Total	$2200.00

Dennis realizes an average contribution margin of 25 percent on sales. Determine the sales volume necessary for breakeven.

$$x_{\text{breakeven}} = \frac{2200}{0.25} = \$8800$$

Sales of $8800 per month are needed for breakeven.

PROBLEMS

1. Plot the following functions and specify the slope of each function:
   (a) $f(x) = 4$ (b) $f(x) = x + 3$
   (c) $f(x) = 2x + 3$ (d) $f(x) = 4 - 3x$
   (e) $f(x) = -0.5x + 2$ (f) $f(x) = -x$
   (g) $f(x) = -3x - 4$ (h) $f(x) = \dfrac{x}{3} - 2$

2. Establish linear functions through the data points $(x, f(x))$.
   (a) $(3, 4)$ and $(5, 2)$ (b) $(9, 4)$ and $(4, -1)$
   (c) $(8, 2)$ and $(5, 5)$ (d) $(4, -3)$ and $(-5, 6)$
   (e) $(8, 4)$ and $(4, -3)$ (f) $(-4, -5)$ and $(4, 5)$
   (g) $(3, 3)$ with $b = 0.5$ (h) $(-5, 7)$ with $b = 0.75$
   (i) $(\frac{3}{4}, \frac{6}{5})$ with $b = -\frac{3}{2}$ (j) $(\frac{4}{3}, 70)$ with $b = \frac{3}{4}$
   (k) $(0.8, 0.5)$ with $b = 0.5$ (l) $(3, -4)$ with $b = 5$

3. Establish linear functions for the following problems:
   (a) Costs are $3000 when output is 0 and $4000 when output is 1500 units.
   (b) Fixed costs are $75,000 and variable costs are $3.50 per unit.
   (c) Costs are $5000 for 150 units and $6000 for 250 units.
   (d) If 1000 units can be sold at a price of $2.50 per unit and 2000 units can be sold at a price of $1.75 per unit, determine price as a function of the number of units sold.
   (e) If 1500 units can be sold at a price of $10 per unit and 2500 units can be sold at a price of $9 per unit, determine price as a function of the number of units sold.
   (f) A firm has 250 employees and expects to add to its work force by 10 employees per year.
   (g) An individual now earning $10,000 per year expects raises of $500 per year.
   (h) A dealer advertises that a car worth $3000 will be reduced in price by $100 per month until sold.

4. In a certain company, cost per unit is constant at $100 from 0 to 1000 units produced per month. From 1001 to 3000 units, costs decrease linearly to $50 per unit at 3000 units. Costs then increase by $0.01 per

unit produced up to 4000 units. Determine the functions describing the cost per unit for the product.

5. Determine the distance between the points (x_1, y_1) and (x_2, y_2).
   (a) (3, 1) and (5, 2)   (b) (6, 4) and (3, 7)
   (c) (8, 3) and (6, 1)   (d) (3, −3) and (4, 2)
   (e) (9, 4) and (−2, 6)   (f) (50, 10) and (30, 20)
   (g) (18, 9) and (13, 5)   (h) (−3, 4) and (2, −4)

6. A piece of machinery is to be shipped from Town A to Town C (see Fig. 2.23). The machinery can be shipped directly by air at a cost of $6 per mile or through Town B by truck at a cost of $5 per mile. Which way is the least expensive?

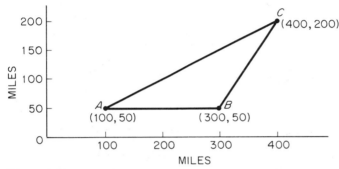

Figure 2.23.

7. An individual living in Town A plans on visiting Towns B and C (see Fig. 2.24). Determine the total round-trip distance.

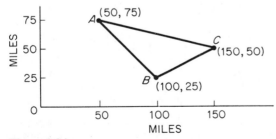

Figure 2.24.

8. Verify that the following set is a function. Give the domain and range of the function.

$$\{(3, 7), (4, 10), (5, 8), (0, 6)\}$$

9. Does the following set satisfy the requirements for a function?

$$\{(x, f(x)) \mid f(x) = 6 \quad \text{for } 0 \leq x \leq 10\}$$

10. Use set notation to specify the functions described in the following statements:
    (a) Smith is a Republican and Jones is a Democrat.
    (b) Bill and Jim were graduated from Harvard, but Frank was graduated from Yale.
    (c) Helena and Sandy received an A in the course, but Janet got a B.
    (d) Ben Hogan, Babe Ruth, and Bill Russell are well-known in their respective sports of golf, baseball, and basketball.

11. Specify which of the relationships in Fig. 2.25 are functions and which are relations.

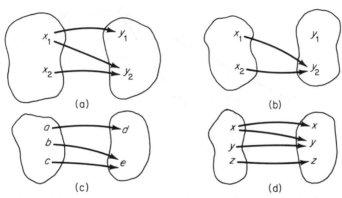

Figure 2.25.

12. Is the following set a function or a relation?

$$\{(A, 1), (A, 2), (B, 1), (B, 2)\}$$

13. Graph the following inequalities:
    (a) $y \geq 0.5x - 2$ (b) $y \leq 4 - 2x$
    (c) $y \leq 2x - 3$ (d) $y \geq x$

14. Graph the following inequalities:
    (a) $2x - y \geq 3$ (b) $x - 2y \leq 4$
    (c) $x + y \geq 2$ (d) $x \leq 2y - 4$

15. Develop equations for the following problems and give the domain of the variables:
    (a) Two products can be manufactured by using the turret lathe. One product requires 2 hours per unit on the lathe and the other product requires 3 hours per unit. The capacity of the lathe is 80 hours per week.

(b) An airplane has a useful load of 1800 pounds. Gasoline weighs 6 pounds per gallon, and the fuel cells have a maximum capcity of 140 gallons. The standard weight for passengers and crew is 170 pounds, and a maximum of six people can be carried on the airplane.

(c) A construction equipment leasing company is considering the purchase of dump trucks, road graders, and bulldozers at a cost of $20,000 per truck, $60,000 per grader, and $80,000 per bulldozer. A total of $1.2 million has been authorized for the equipment.

(d) An individual must have 300 units of a special dietary supplement each day. This supplement is available in two commercial products, A and B. Each ounce of A contains 25 units of the supplement and each ounce of B contains 40 units.

16. The total cost of four milk shakes and two coffees is $1.40. The cost of one milk shake and three coffees is $0.60. Find the price of a milk shake and a coffee.

17. A firm uses two machines in the manufacture of two products. Product A requires 3 hours on machine 1 and 5 hours on machine 2. Product B requires 6 hours on machine 1 and 4 hours on machine 2. If 36 hours of machine 1 time and 42 hours of machine 2 time are to be scheduled, determine the product mix that fully utilizes the machine time.

18. A company packages fruit in fancy boxes for sale as Christmas presents. One box contains six apples, four pears, and four oranges and sells for $1.68. A second box contains five apples, four pears, and six oranges and sells for $1.88. A third box contains eight apples, six pears, and ten oranges and sells for $3.02. Assuming that the cost of fruit is constant, determine the per unit cost of each type of fruit.

19. Determine whether the following consistent system of equations has a unique or an infinite number of solutions:

$$3x + 2y + 7z - 5 = 0$$
$$4x - y + 5z + 4 = 0$$
$$7x + y + 12z - 1 = 0$$

20. Verify by graphing that the following system of linear inequalities has an infinite number of solutions:

$$4x_1 + 5x_2 \le 20$$
$$6x_1 + 3x_2 \ge 18$$

21. Specify whether the following systems of equations are consistent or inconsistent. For those that are consistent, state whether there is a unique or an infinite number of solutions.

(a) $5x + 7y = 70$
    $x + y = 12$

(b) $6x + 15y = 90$
    $4x + 5y = 40$
    $11x + 8y = 87$

(c) $x + y = 10$
    $8x + 11y = 88$
    $7x + 14y = 98$

(d) $2x_1 + 3x_2 + 8x_3 = 56$
    $4x_1 - 3x_2 + 5x_3 = 48$
    $-2x_1 + 6x_2 + 3x_3 = 8$

22. Determine if the following system of equations has a consistent, unique solution:

$$x + y = 4$$

$$2x + y \geq 6$$

23. A firm is interested in determining the breakeven production for a new product. The cost of introducing the product is estimated to be $40,000. This cost includes initial advertising and promotion as well as the fixed cost necessary for 1 year of production. The variable cost per unit is $35, and the proposed selling price is $60.
    (a) Determine breakeven production.
    (b) Determine the profit on sales of 2000 units.

24. The sole proprietor of a rug-cleaning business has determined that he must work at least as many hours as it takes to cover his fixed cost of $45 per day. He averages $9 per hour and can work no more than 12 hours on any one day. Determine the profit function and the domain and range of the function. Also, determine his breakeven point in hours.

25. The Handley Company plans to produce and sell an item. The cost of production includes fixed cost of $50,000 and variable cost of $10 per item. They plan to sell the item for $25. How many items must be sold to obtain a profit of $100,000?

26. The Annual Report of Barnett Manufacturing Company, Inc. showed sales of $80,000, variable costs of $50,000, and fixed cost of $20,000 for the year 1975.
    (a) Determine the breakeven level of sales for Barnett.
    (b) Determine the cost function.
    (c) What would profits be on sales of $100,000?
    (d) Construct a breakeven chart similar to that shown in Fig. 2.22 for the Barnett Manufacturing Company.

27. Determine the effect of the following on breakeven for the Barnett Manufacturing Company in Problem 26:
    (a) A $5000 increase in fixed cost.
    (b) A 20 percent increase in variable cost.
    (c) A $5000 increase in fixed cost coupled with a 20 percent increase in variable cost.

28. The Kent Manufacturing Company sells a certain product at a price of $40 per unit. The variable costs of manufacturing this product are $25 per unit. Fixed costs during the past year were $540,000. Through a cost reduction program, management expects to reduce variable costs to $20 per unit and fixed costs to $500,000.
    (a) How many units had to be sold during the past year to breakeven?
    (b) How many units must be sold during the coming year to breakeven?
    (c) How many units must be sold during the coming year for gross profit to equal 20 percent of sales?

SUGGESTED REFERENCES

BURNS, CARL M., et al., *Algebra: An Introduction for College Students* (Menlo Park, Ca.: Cummings Publishing Company, Inc., 1972).

CHILDRESS, ROBERT L., *Mathematics for Managerial Decisions* (Englewood Cliffs, N.J.: Prentice-Hall, Inc., 1974).

FREUND, JOHN E., *College Mathematics with Business Applications* (Englewood Cliffs, N.J.: Prentice-Hall, Inc., 1975), 3, 4.

MOORE, GERALD E., *Algebra*, The Barnes and Noble College Outline Series (New York, N.Y.: Barnes and Noble, Inc., 1970).

NIELSEN, KAJ L., *Algebra: A Modern Approach*, The Barnes and Noble College Outline Series (New York, N.Y.: Barnes and Noble, Inc., 1969).

NIELSEN, KAJ L., *College Mathematics*, The Barnes and Noble College Outline Series (New York, N.Y.: Barnes and Noble, Inc., 1958), 1–6.

RICH, BARNETT, *Elementary Algebra*, Schaum's Outline Series (New York, N.Y.: McGraw-Hill Book Company, Inc., 1960).

THEODORE, CHRIS A., *Applied Mathematics: An Introduction* (Homewood, Ill.: Richard D. Irwin, Inc., 1975), 5–7.

Chapter Three

Nonlinear And Multivariate Functions

Our discussion of functions has, to this point, centered primarily on the univariate, linear function. As the reader can well imagine, there are many situations in business, economics, and the sciences in which a linear function cannot be used to describe the functional relationship between two variables. For this reason, it is important to be familiar with certain nonlinear functions. This chapter introduces two of the most commonly applied nonlinear functions—the quadratic function and the exponential function.

The quadratic and exponential functions are univariate. Like the linear function, they have only one independent variable. They differ, however, from the linear function in that they are nonlinear. This simply means that the graph of the function is not a straight line.

This chapter also introduces the multivariate function. A multivariate function is a functional relationship between a dependent variable and more than one independent variable. One of the most important applications of multivariate functions is in linear programming. We shall introduce the concept of the multivariate function in this chapter and postpone our discussion of applications of multivariate functions until the chapter on linear programming (Chap 11).

3.1   The Quadratic Function

The *quadratic function* is a univariate, nonlinear function that has important applications in business, economics, and the sciences. The function is termed univariate because it has only one independent variable, and it is termed nonlinear because the graph of the function does not plot as a straight line. The general form of the quadratic function is

$$f(x) = a + bx + cx^2, \qquad c \neq 0 \tag{3.1}$$

where $f(x)$ is the dependent variable, x is the independent variable, and a, b, and c are the parameters of the function. The shape of the quadratic function is determined by the magnitude and signs of the parameters a, b, and c. Common forms of the quadratic function are shown in Fig. 3.1.

From the diagrams in Fig. 3.1, it can be seen that the value of a positions the function vertically. In Fig. 3.1(a), for instance, if a were negative instead of positive, the function would intercept the vertical axis below the horizontal axis. This would have the effect of moving the entire function downward. The values of b and c determine the general shape of the function. For b negative and c positive, the function will have the concave shape illustrated by Fig. 3.1(a). For b positive and c negative, the function will be convex as shown in Fig. 3.1(b). If b and c are both positive, the function has the shape illustrated by Fig. 3.1(c). Conversely, if b and c are both negative, the function appears as illustrated by Fig. 3.1(d).

3.1.1 ESTABLISHING THE FUNCTION

It is important to be able to graph a quadratic function or, given a set of data points, to be able to establish the quadratic function that passes through

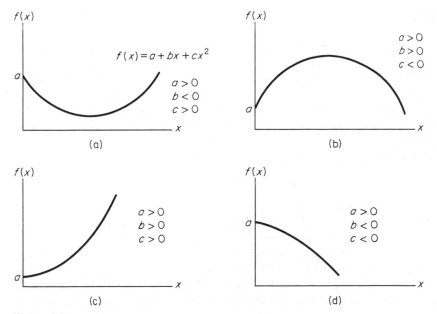

Figure 3.1.

the data points. To illustrate graphing a quadratic function, consider the function

$$f(x) = 12 - 8x + x^2$$

This function can be graphed by constructing a table of values of x and $f(x)$. The data points from the table are then plotted. The graph is completed by connecting the data points with a smooth line. This procedure is illustrated in Fig. 3.2.

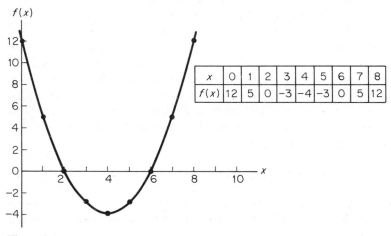

x	0	1	2	3	4	5	6	7	8
$f(x)$	12	5	0	-3	-4	-3	0	5	12

Figure 3.2.

Only selected values of x and $f(x)$ are plotted in Fig. 3.2. Since the domain of the function is not specified, one might assume that additional data points should be plotted. This, however, is not the case. Customarily, only enough data points are plotted to show the general shape of the function. Should it be necessary to determine the value of the dependent variable for values of x other than those included in the table, the function can easily be evaluated for those values of x.

Example. Construct a graph of the function

$$f(x) = -4 + 5x - x^2$$

The graph of the function is shown in Fig. 3.3.

In addition to being able to graph a quadratic function, it is important to be able to establish a quadratic function from a set of data points. The quadratic function has parameters a, b, and c. In order to establish the quadratic function, we must determine the values of the parameters. These parameters can be determined by substituting data points into the general form of the quadratic function and solving the resulting equations simulta-

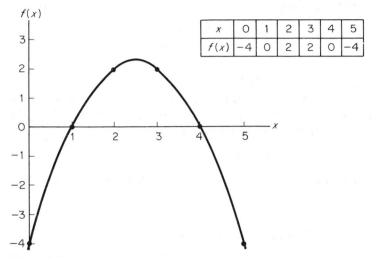

x	0	1	2	3	4	5
f(x)	-4	0	2	2	0	-4

Figure 3.3.

neously for the parameters. A general rule in establishing functions from data points is that the number of data points required to establish the function is equal to the number of parameters. Since there are three parameters in the quadratic function, three data points are required. The three data points are substituted into the general form of the quadratic function and the resulting three equations are solved simultaneously for the three parameters.

To illustrate, suppose that we establish a quadratic function through the data points (1, 4), (2, 2), and (4, 2). These data points are shown in Fig. 3.4. The three data points are substituted into the general form of the quadratic function,

$$f(x) = a + bx + cx^2$$

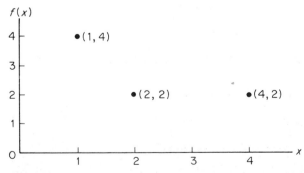

Figure 3.4.

The resulting three equations are

$$4 = a + b(1) + c(1)^2$$
$$2 = a + b(2) + c(2)^2$$
$$2 = a + b(4) + c(4)^2$$

These equations can be solved simultaneously by using the method of substitution or the method of subtraction or by using matrix algebra. Using the method of substitution, we obtain from the first equation

$$a = 4 - b - c$$

This expression for a is substituted into the second and third equations giving

$$2 = (4 - b - c) + 2b + 4c$$

and

$$2 = (4 - b - c) + 4b + 16c$$

Collecting the terms in these two equations results in

$$-2 = b + 3c$$

and

$$-2 = 3b + 15c$$

These two equations are now solved simultaneously for c. Substituting $b = -2 - 3c$ from the first equation into the second gives

$$-2 = 3(-2 - 3c) + 15c$$

or

$$4 = 6c$$

and

$$c = \tfrac{2}{3}$$

Substituting $c = \tfrac{2}{3}$ into the expression $b = -2 - 3c$ gives

$$b = -2 - 3(\tfrac{2}{3})$$
$$b = -4$$

Substituting $b = -4$ and $c = \tfrac{2}{3}$ for b and c in the expression $a = 4 - b - c$ gives

$$a = 4 - (-4) - \tfrac{2}{3} = 7\tfrac{1}{3}$$

The quadratic function is thus

$$f(x) = 7\tfrac{1}{3} - 4x + \tfrac{2}{3}x^2$$

The algebra can be checked by evaluating the function for $x = 1$, $x = 2$, and $x = 4$. This results in values of $f(x)$ of $f(1) = 4$, $f(2) = 2$, and $f(4) = 2$.

Example. Establish a quadratic function through the data points (1, 5), (3, -3), and (4, -4).

These data points come from the function plotted in Fig. 3.2. Substituting the data points into Eq. (3.1) gives

$$5 = a + b(1) + c(1)^2$$

$$-3 = a + b(3) + c(3)^2$$

$$-4 = a + b(4) + c(4)^2$$

Substituting $a = 5 - b - c$ from the first equation into the second and third equation gives

$$-8 = 2b + 8c$$

and

$$-9 = 3b + 15c$$

Solving for b in the new first equation, we obtain $b = -4 - 4c$. Substituting this expression for b in the new second equation gives

$$-9 = 3(-4 - 4c) + 15c$$

$$3 = 3c$$

$$c = 1$$

Since $b = -4 - 4c$ and $c = 1$, the value of b is

$$b = -4 - 4(1) = -8$$

Finally, from the original first equation

$$a = 5 - b - c$$

$$a = 5 - (-8) - 1$$

$$a = 12$$

The function is $f(x) = 12 - 8x + x^2$.

Example. Bill Cook owns a small machine shop. Because of the shop's limited capacity, average costs are related to the volume of output. Cost studies undertaken by Bill show that the average cost of producing 9 units of output is \$8.50 per unit, the average cost of producing 10 units is \$8.00 per unit, and the average cost per unit of producing 12 units is \$8.25 per unit. Determine the functional relationship between average cost and output.

The three data points, plotted in Fig. 3.5, show that a function that decreases and then increases describes the cost–output data. This functional relationship is described by the quadratic function. The procedure for determining the function is to substitute the data points into the general form of the

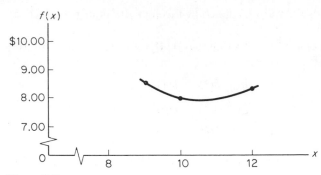

Figure 3.5.

quadratic function and solve the resulting three equations simultaneously for the parameters a, b, and c. The three equations are

$$8.50 = a + b(9) + c(9)^2$$
$$8.00 = a + b(10) + c(10)^2$$
$$8.25 = a + b(12) + c(12)^2$$

The solution of the three simultaneous equations is $a = 31.75$, $b = -4.458$, and $c = 0.2083$. The average cost function is

$$f(x) = 31.75 - 4.458x + 0.2083x^2$$

Example. James Mason is concerned with the effect of fatigue on the productivity of skilled machinists. He has observed that productivity, which he defines as standard units of output per man-hour, increases as the number of hours that an individual works increases up to approximately 40 hours per week and then begins to decline as fatigue begins to become a factor. Mason believes that three representative observations of the effect of man-hours on productivity are 9.3 units of output during the 35th hour of work, 9.5 units during the 40th hour, and 8.5 units during the 45th hour.

The observations are plotted in Fig. 3.6. Productivity as a function of the hour of work increases and then decreases. The relationship shown in the illustration is described by a quadratic function. We substitute the data points in the general form of the quadratic function and obtain

$$9.3 = a + b(35) + c(35)^2$$
$$9.5 = a + b(40) + c(40)^2$$
$$8.5 = a + b(45) + c(45)^2$$

Solving the three equations simultaneously for a, b, and c gives $a = -25.7$, $b = 1.84$, and $c = -0.024$. The productivity function is

$$f(x) = -25.7 + 1.84x - 0.024x^2$$

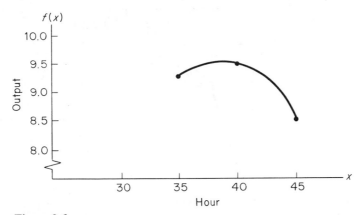

Figure 3.6.

3.1.2 THE QUADRATIC FORMULA

In certain applications it is necessary to determine the *roots* of a quadratic function. The roots of a quadratic function are defined as the values of the independent variable, x, for which the dependent variable, $f(x)$, is equal to zero. To illustrate, consider the quadratic function $f(x) = 4 - 5x + x^2$. This function is plotted in Fig. 3.7. The roots of the function are those values of x such that $f(x) = 0$. From Fig. 3.7 we see that $f(x)$ is equal to zero for $x = 1$ and $x = 4$. The roots of the function are thus $x = 1$ and $x = 4$.

The roots of a quadratic function can be determined by equating the quadratic function with zero and factoring the resulting quadratic equation. Equating the quadratic function plotted in Fig. 3.7 with zero gives the quadratic equation

$$x^2 - 5x + 4 = 0$$

This equation can be factored

$$(x - 1)(x - 4) = 0$$

and the roots are $x = 1$ and $x = 4$.

Unfortunately, not all quadratic equations can be factored as easily as the one above. If the quadratic equation cannot be readily factored or the reader is rusty on the algebra of factoring, the roots can be found by using the *quadratic formula*. The quadratic formula is used to determine the values of x such that

$$a + bx + cx^2 = 0 \tag{3.2}$$

The quadratic formula is

$$x = \frac{-b + \sqrt{b^2 - 4ac}}{2c} \quad \text{and} \quad x = \frac{-b - \sqrt{b^2 - 4ac}}{2c} \tag{3.3}$$

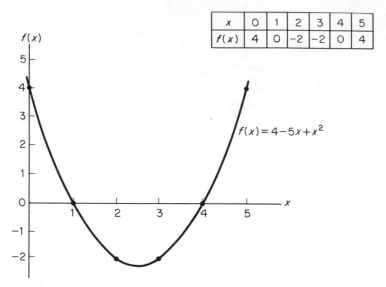

Figure 3.7.

The quadratic formula gives two roots. The roots are found by substituting the values of the parameters a, b, and c into the formula and solving for the values of x.

As an example, we shall determine the roots of the quadratic function plotted in Fig. 3.7. The parameters of the function are $a = 4$, $b = -5$, and $c = 1$. Substituting these values into the quadratic formula, Eq. (3.3), gives

$$x = \frac{-(-5) + \sqrt{(-5)^2 - 4(4)(1)}}{2(1)}$$

$$x = \frac{5 + \sqrt{25 - 16}}{2} = 4$$

and

$$x = \frac{-(-5) - \sqrt{(-5)^2 - 4(4)(1)}}{2(1)}$$

$$x = \frac{5 - \sqrt{25 - 16}}{2} = 1$$

The roots are $x = 4$ and $x = 1$.

Example. Determine the roots of the quadratic function

$$f(x) = 6 - 8x + 0.5x^2$$

Substituting $a = 6$, $b = -8$, and $c = 0.5$ into Eq. (3.3), we obtain

$$x = \frac{-(-8) + \sqrt{(-8)^2 - 4(6)(0.5)}}{2(0.5)}$$

$$x = \frac{8 + \sqrt{52}}{1} = 8 + 7.211 = 15.211$$

and

$$x = \frac{8 - \sqrt{52}}{1} = 8 - 7.211 = 0.789$$

The square root was obtained from Table A.8, Squares and Square Roots.

Our examples have thus far included only functions with ordinary numerical roots. As an illustration of a quadratic function that does not have these roots, consider the function

$$f(x) = 4 - 3x + x^2$$

This function is plotted in Fig. 3.8. From this plot we see that there are no values of x for which $f(x)$ equals zero. Consequently, this function does not have ordinary roots.[1]

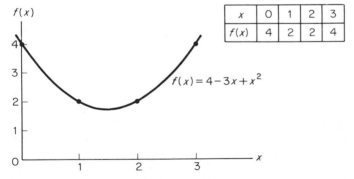

x	0	1	2	3
$f(x)$	4	2	2	4

$f(x) = 4 - 3x + x^2$

Figure 3.8.

We reach the same conclusion from the quadratic formula. Substituting $a = 4$, $b = -3$, and $c = 1$ into the quadratic formula, we obtain

$$x = \frac{-(-3) + \sqrt{(-3)^2 - 4(4)(1)}}{2(1)} = \frac{3 + \sqrt{-7}}{2}$$

---

[1] The roots of functions of the type plotted in Fig. 3.8 exist only in a mathematical sense. They involve the square root of a negative number and are classified as *imaginary* roots.

and

$$x = \frac{-(-3) - \sqrt{(-3)^2 - 4(4)(1)}}{2} = \frac{3 - \sqrt{-7}}{2}$$

The roots of this function include a term involving the square root of a negative number. When this occurs, the function does not cross the horizontal axis and ordinary numerical roots do not exist.

Example. Verify that $f(x) = 3 - 2x + 0.5x^2$ does not have ordinary numerical roots.

From the quadratic formula, we obtain

$$x = \frac{-(-2) + \sqrt{(-2)^2 - 4(3)(0.5)}}{2(0.5)} = 2 + \sqrt{-2}$$

and

$$x = \frac{-(-2) - \sqrt{(-2)^2 - 4(3)(0.5)}}{2(0.5)} = 2 - \sqrt{-2}$$

Since the square root of -2 is imaginary, the function does not have ordinary numerical roots. When roots of this type occur, the function does not cross the horizontal axis.

3.1.3   MAXIMUM OR MINIMUM

The quadratic function is used to describe a functional relationship in which the dependent variable reaches a maximum or a minimum value. Two examples of this type were included in Sec. 3.1.1. The first example concerned the relationship between average cost and the volume of output in a machine shop. In this example (p. 79), Bill Cook found that the relationship between average cost and output could be described by the quadratic function

$$f(x) = 31.75 - 4.458x + 0.2083x^2$$

where $f(x)$ represents average cost and x represents units of output. This function is plotted in Fig. 3.9.

The second example concerned the effect of fatigue on the productivity of skilled machinists. In this example (p. 80), James Mason found that productivity increases as the number of hours an individual works increases up to approximately 40 hours per week and then productivity begins to decline as fatigue becomes a factor. The relationship between productivity and hours was described by the quadratic function

$$f(x) = -25.7 + 1.84x - 0.024x^2$$

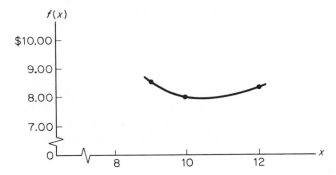

Figure 3.9.

where $f(x)$ represents output per man-hour and x represents the number of hours worked. This function is plotted in Fig. 3.10.

It is important to be able to determine the value of the independent variable that results in the maximum or the minimum value of a quadratic function. Bill Cook, for instance, would certainly be interested in knowing the level of output that results in minimum average cost. Similarly, James Mason would undoubtedly be interested in knowing the number of hours of work that leads to maximum productivity. In order to determine the level of output that results in minimum cost or the number of hours that leads to maximum productivity, we must introduce a method for determining that value of x for which $f(x)$ is a maximum or a minimum.

The general form of the quadratic function was given by Eq. (3.1) as

$$f(x) = a + bx + cx^2$$

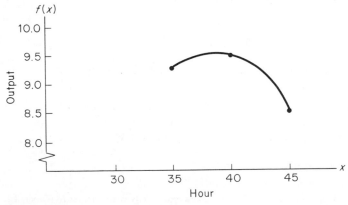

Figure 3.10.

Using differential calculus, it can be shown that the value of x for which $f(x)$ is an *optimum* (i.e., a maximum or a minimum) is

$$x = -\frac{b}{2c} \qquad (3.4)$$

and the optimum is a maximum if c is negative and a minimum if c is positive.[2] The optimum value of $f(x)$ is found by evaluating the function for $x = -b/2c$.

To illustrate Eq. (3.4), consider again the example of average cost in Bill Cook's machine shop. The quadratic function relating average cost, $f(x)$, to the volume of output, x, was given on p. 79 as

$$f(x) = 31.75 - 4.458x + 0.2083x^2$$

The value of x that results in the optimum value of $f(x)$ is given by Eq. (3.4) and is

$$x = -\frac{-4.458}{2(0.2083)} = 10.7$$

Since c is positive, average cost is a minimum. The average cost is found by evaluating the function for $x = 10.7$.

$$f(10.7) = 31.75 - 4.458(10.7) + 0.2083(10.7)^2$$

$$f(10.7) = 31.75 - 47.70 + 23.85 = 7.90$$

The minimum average cost is $7.90.

As an example of a quadratic function that reaches a maximum, consider again the relationship that James Mason found to exist between productivity $f(x)$, and hours worked, x. This relationship was given on p. 80 as

$$f(x) = -25.7 + 1.84x - 0.024x^2$$

The optimum value of the function occurs when

$$x = -\frac{1.84}{2(-0.024)} = 38.33$$

Since c is negative, the function reaches a maximum. The value of the dependent variable for $x = 38.33$ is

$$f(38.33) = -25.7 + 1.84(38.33) - 0.024(38.33)^2$$

$$f(38.33) = -25.7 + 70.53 - 35.26 = 9.57$$

The maximum productivity is 9.57 units.

---

[2] Calculus is not covered in this text. The interested reader may wish to refer to Robert L. Childress, *Mathematics for Managerial Decisions*.

Example. Determine the optimum value of the function

$$f(x) = 5 - 6x + x^2$$

The function reaches an optimum when

$$x = -\frac{(-6)}{2(1)} = 3$$

Since c is positive, the function is a minimum at $x = 3$. The value of the function at $x = 3$ is

$$f(3) = 5 - 6(3) + (3)^2 = -4$$

Example. Helena Turkel, the proprietor of a small Swedish massage parlor, has found that the total profit from her business can be described by the quadratic function

$$f(x) = -100 + 20x - 0.10x^2$$

where $f(x)$ is profit and x is the number of customers. Determine the number of customers that results in maximum profit.

The number of customers that results in maximum profit is

$$x = -\frac{20}{2(-0.10)} = 100$$

The profit is

$$f(100) = -100 + 20(100) - 0.10(100)^2 = \$900$$

Since c is negative, we know that profit is a maximum rather than a minimum.

3.2 The Exponential Function

The exponential function, like the quadratic function, is nonlinear and univariate. It differs from the quadratic function, however, in that it describes a relationship in which the dependent variable increases or decreases by a *constant percentage* as the independent variable increases by a *constant amount*. Thus, the value on an investment that increases by a constant percent each period, the sales of a company that increase by a constant percent each period, and the value of an asset that declines by a constant percent each period are examples of functional relationships that are described by the exponential function. Several of these applications are discussed in this section. Applications of exponential function in financial formulas are presented in Chap. 14. Others will be encountered by those who take courses in such subjects as production, economics, and statistics.

The exponential function has the form

$$f(x) = k(a)^{cx} \qquad (3.5)$$

where $f(x)$ is the dependent variable, x is the independent variable, a is a constant termed the *base* that is greater than 0 and not equal to 1, and k and c are parameters that position the function on the coordinate axes. The possible forms of the exponential function are graphed in Fig. 3.11.

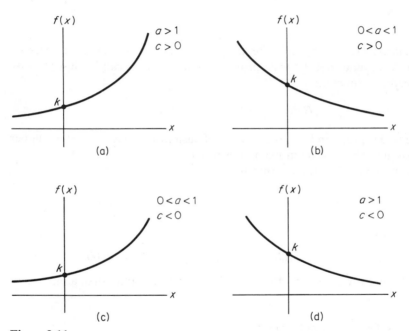

Figure 3.11.

Figures 3.11(a) and 3.11(c) both show an exponential function in which the dependent variable increases as the independent variable increases. It can be shown that by proper selection of the constants a and c in Fig. 3.11(a), the functional relationship in Fig. 3.11(c) is completely described. Similarly, the functional relationships shown in Figs. 3.11(b) and 3.11(d) are interchangeable. The most commonly used forms of the exponential function are those in which a is greater than 1, the functions shown in Figs. 3.11(a) and 3.11(d). Those shown in Figs. 3.11(b) and 3.11(c) are, however, also applied in business, economics, and the sciences.

3.2.1   ESTABLISHING THE FUNCTION

It is more difficult to plot an exponential function than a linear or a quadratic function because the base, a, must be raised to a variable power. As an example, consider the exponential function

$$f(x) = 2(1.25)^{1.5x}$$

The base in this function is 1.25. To plot this function, the base must be raised to the $1.5x$ power for alternative values of x and the result multiplied by the parameter $k = 2$. This can be done through the use of logarithms.[3]

The exponential function, Eq. (3.5), can be expressed in terms of logarithms. The logarithmic form is

$$\log f(x) = \log k + cx \log a \qquad (3.6)$$

Applying Eq. (3.6) to our example problem, we obtain

$$\log f(x) = \log 2 + 1.5x \log (1.25)$$

Values of $\log f(x)$ can now be calculated. Values of $f(x)$ are found by determining the antilogarithm of $\log f(x)$.

From Table A.5, Common Logarithms, we see that $\log (1.25)$ is 0.0969 and $\log 2$ is 0.3010. Substituting these values for $\log (1.25)$ and $\log 2$ gives

$$\log f(x) = 0.3010 + 0.0969(1.5x)$$

The logarithm of the dependent variable, $f(x)$, can now be determined for alternative values of the independent variable, x. For example,

$$\log f(0) = 0.3010$$

$$\log f(1) = 0.3010 + 0.1044 = 0.4054$$

$$\log f(2) = 0.3010 + 0.2088 = 0.5098$$

$$\log f(3) = 0.3010 + 0.3132 = 0.6142$$

etc.

Values of $f(x)$ are found by determining the antilogarithm of $\log f(x)$, i.e., the number whose logarithm is $\log f(x)$. The number whose logarithm is 0.3010 is 2. Thus, the antilogarithm of $\log f(0)$ is 2. Similarly, from Table A.5,

$$f(1) = \text{antilog} (0.4054) = 2.543$$

$$f(2) = \text{antilog} (0.5098) = 3.235$$

$$f(3) = \text{antilog} (0.6142) = 4.114$$

---

[3] Readers unfamiliar with logarithms should refer to the Appendix, Logarithms: Laws of Exponents.

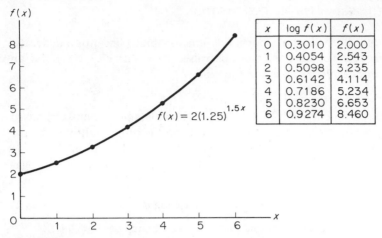

x	$\log f(x)$	$f(x)$
0	0.3010	2.000
1	0.4054	2.543
2	0.5098	3.235
3	0.6142	4.114
4	0.7186	5.234
5	0.8230	6.653
6	0.9274	8.460

$f(x) = 2(1.25)^{1.5x}$

Figure 3.12.

The third decimal for each of the values of $f(x)$ was found by linear interpolation. The function is plotted in Fig. 3.12.

Example. Plot the exponential function

$$f(x) = 5(1.18)^x$$

The values of $f(x)$ for alternative values of x can be found by using Table A.5, Common Logarithms. In terms of logarithms, the function is

$$\log f(x) = \log 5 + \log (1.18)x$$

or

$$\log f(x) = 0.6990 + 0.0719x$$

Values of x, $\log f(x)$, and $f(x)$ are shown in the table in Fig. 3.13. The function is plotted from the values in this table.

The exponential function, like the linear function, can be established from two data points. The procedure for establishing the function is to substitute the two data points into the logarithmic form of the exponential function. The resulting two equations are solved simultaneously for the parameters of the function.

The general form of the exponential function was given by Eq. (3.5) as

$$f(x) = k(a)^{cx}$$

In order to establish the function, we must specify the value of either the base, a, or the parameter, c. In the following examples the base of the function is specified. In Sec. 3.2.2 we specify the value of the parameter, c. The choice of specifying the base or the parameter is arbitrary. It is shown in Sec. 3.2.4

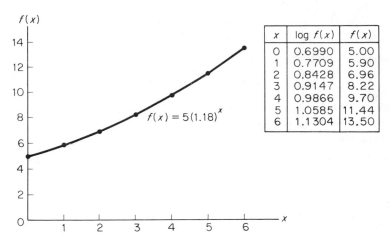

x	log $f(x)$	$f(x)$
0	0.6990	5.00
1	0.7709	5.90
2	0.8428	6.96
3	0.9147	8.22
4	0.9866	9.70
5	1.0585	11.44
6	1.1304	13.50

$f(x) = 5(1.18)^x$

Figure 3.13.

that both methods result in the same functional relationship between x and $f(x)$.

One of the most commonly used bases for the exponential function is the number e. The numerical value of e with accuracy of nine places is 2.718281828.[4] Replacing a by e in the exponential function, Eq. (3.5), we obtain

$$f(x) = ke^{cx} \qquad (3.7)$$

k and c are again parameters of the exponential function and e is the base.

As an example of establishing an exponential function, assume that we want to establish the function through the data points $(x = 2, f(x) = 3)$ and $(x = 4, f(x) = 4)$. Instead of substituting these data points into Eq. (3.7), we write Eq. (3.7) in terms of *natural logarithms* and substitute the data points into this form of the function.[5] Expressing both sides of Eq. (3.7) in terms of natural logarithms, we obtain

$$\ln f(x) = \ln k + cx \qquad (3.8)$$

To solve for $\ln k$ and c, values of $\ln f(x)$ and x are substituted into Eq. (3.8). From Table A.6, Natural or Naperian Logarithms, we find that $\ln 3 = 1.09861$ and $\ln 4 = 1.38629$. Substituting the data points (2, 1.09861) and (4, 1.38629) into Eq. (3.8) gives

$$1.09861 = \ln k + 2c$$

---

[4] The number e is derived by using the limiting process associated with the calculus. Readers interested in the derivation of e should refer to Robert L. Childress, *Mathematics for Managerial Decisions*.

[5] Readers unfamiliar with natural or *Naperian* logarithms should refer to the Appendix, Logarithms: Laws of Exponents.

and
$$1.38629 = \ln k + 4c$$

From the first equation, we obtain
$$\ln k = 1.09861 - 2c$$

Substituting this expression for $\ln k$ into the second equation gives
$$1.38629 = 1.09861 - 2c + 4c$$

or
$$2c = 0.28768$$

and
$$c = 0.14384$$

Substituting $2c = 0.28768$ into the expression $\ln k = 1.09861 - 2c$ gives
$$\ln k = 0.81093$$

The value of k is found by determining the antilogarithm of $\ln k$. From Table A.6,
$$\text{antilog}_e (0.81093) = 2.25$$

The exponential function is thus
$$f(x) = 2.25e^{0.14384x}$$

This function is plotted in Fig. 3.14. The values of $\ln f(x)$ given in the table in Fig. 3.14 were calculated by using
$$\ln f(x) = 0.81093 + 0.14384x$$

The values of $f(x)$ are found by determining the antilogarithms of $\ln f(x)$.

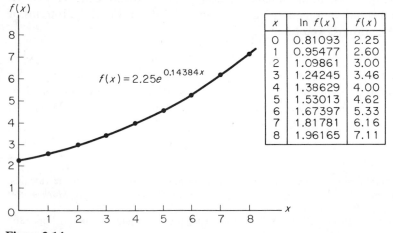

Figure 3.14.

x	$\ln f(x)$	$f(x)$
0	0.81093	2.25
1	0.95477	2.60
2	1.09861	3.00
3	1.24245	3.46
4	1.38629	4.00
5	1.53013	4.62
6	1.67397	5.33
7	1.81781	6.16
8	1.96165	7.11

Example. Establish an exponential function through the data points (3, 3) and (8, 6).

The parameters of the exponential function are determined by substituting values of x and $\ln f(x)$ into Eq. (3.8). From Table A.6, we see that $\ln 3 = 1.09861$ and $\ln 6 = 1.79179$. Substituting the data points (3, 1.09861) and (8, 1.79179) into Eq. (3.8) gives

$$1.09861 = \ln k + 3c$$

and

$$1.79179 = \ln k + 8c$$

These equations are solved simultaneously to give $c = 0.13864$ and $\ln k = 0.68270$. The function, in terms of logarithms, is

$$\ln f(x) = 0.68270 + 0.13864x$$

By taking the antilogarithm of both sides, we obtain

$$f(x) = 1.98e^{0.13864x}$$

This function is plotted in Fig. 3.15.

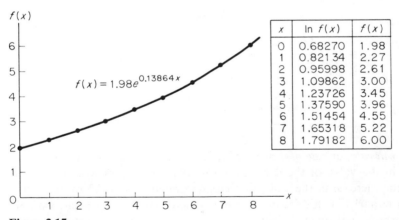

x	$\ln f(x)$	$f(x)$
0	0.68270	1.98
1	0.82134	2.27
2	0.95998	2.61
3	1.09862	3.00
4	1.23726	3.45
5	1.37590	3.96
6	1.51454	4.55
7	1.65318	5.22
8	1.79182	6.00

Figure 3.15.

Example. Determine the exponential function that passes through the data points (2, 6) and (4, 4).

Substituting the values of $\ln f(x)$ and x for the two data points into Eq. (3.8) gives

$$1.79176 = \ln k + 2c$$

and

$$1.38629 = \ln k + 4c$$

The two equations are solved simultaneously to give $c = -0.20274$ and $\ln k = 2.19723$. The function, expressed in terms of natural logarithms, is

$$\ln f(x) = 2.19723 - 0.20274x$$

The exponential function is

$$f(x) = 9.00e^{-0.20274x}$$

The function is plotted in Fig. 3.16.

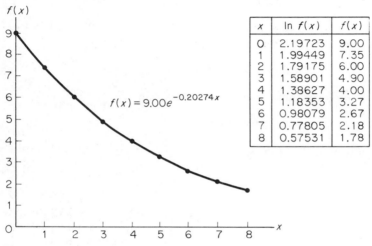

x	$\ln f(x)$	$f(x)$
0	2.19723	9.00
1	1.99449	7.35
2	1.79175	6.00
3	1.58901	4.90
4	1.38627	4.00
5	1.18353	3.27
6	0.98079	2.67
7	0.77805	2.18
8	0.57531	1.78

Figure 3.16.

3.2.2 GROWTH RATE

The *growth rate*, or *rate of growth*, of a function is defined as the percentage change in the value of the dependent variable that occurs with a one-unit arithmetic increase in the value of the independent variable.[6] In our introductory remarks in this section, we mentioned that the dependent variable in the exponential function increases or decreases by a constant percentage as the independent variable increases by a constant arithmetic amount. The exponential function is the only function that has a constant rate of growth. Because there are many variables in business and the sciences that increase or decrease in value by a constant percentage each period, the exponential function is widely used.

---

[6] Our discussion in this section is confined to periodic growth rates. For an explanation of continuous growth rates, see Robert L. Childress, *Mathematics for Managerial Decisions*, pp. 485–88.

To illustrate what is meant by a constant rate of growth, consider the exponential function

$$f(x) = 2(1.25)^{1.5x}$$

This function was plotted in Fig. 3.12. The rate of growth of this function is determined by calculating the percentage change in the dependent variable that occurs with a one-unit change in the value of the independent variable. For instance, the percentage change in $f(x)$ as x increases from $x = 1$ to $x = 2$ is

$$\frac{f(2) - f(1)}{f(1)} = \frac{3.235 - 2.543}{2.543} = 0.272 = 27.2\%$$

Similarly, the percentage change in $f(x)$ as x increases from $x = 2$ to $x = 3$ is

$$\frac{f(3) - f(2)}{f(2)} = \frac{4.114 - 3.235}{3.235} = 0.272 = 27.2\%$$

The percentage change in $f(x)$ as x increases from $x = 3$ to $x = 4$ is also 27.2 percent. In fact, the percentage increase or rate of growth remains constant for any one-unit increase in x. Consequently, we conclude that this exponential function has a constant rate of growth of 27.2 percent. The reader can easily verify this statement for alternative values of x.

Example. Verify that the dependent variable increases by a constant percentage as the independent variable increases by a constant amount for the exponential function.

$$f(x) = 1.98e^{0.13864x}$$

This function was plotted on p. 93. Values of x and $f(x)$ are given in the following table:

x	0	1	2	3	4	5	6	7	8
$f(x)$	1.98	2.27	2.61	3.00	3.45	3.96	4.55	5.22	6.00

The percentage increase in the dependent variable $f(x)$ as x increases from $x = 1$ to $x = 2$ is

$$\frac{f(4) - f(3)}{f(3)} = \frac{3.45 - 3.00}{3.00} = 0.15 = 15\%$$

The dependent variable increases by this same percentage, 15 percent, for any one-unit increase in x.

3.2.3 PROJECTING BUSINESS VARIABLES

The fact that the dependent variable in the exponential function increases by a constant percentage as the independent variable increases by a constant arithmetic amount makes the exponential function very useful in business and economics. For instance, consider the following statements:

> *Productivity in the United States is expected to increase by 4 percent per year.*

> *Sales of Acme Wire and Products are expected to continue to increase at an annual rate of 10 percent.*

> *The dollar continues to decline in value at a rate of 5 percent per year.*

Each of these statements describes a relationship in which the dependent variable changes by a constant percentage rate as the independent variable changes by a constant arithmetic amount. They can therefore be modeled by the exponential function.

The general form of the exponential function was given by Eq. (3.5) as

$$f(x) = k(a)^{cx}$$

Suppose that we use this form of the exponential function to model the following statement:

> *Sales of Acme Wire and Products are currently $100 and are expected to increase by 10 percent per year for the next 5 years.*

Assuming a 10 percent annual increase in sales and a current level of sales of $100, sales for the next 5 years can easily be projected. This projection is shown in the following table:

Year	0	1	2	3	4	5
Sales	$100.00	110.00	121.00	133.10	146.41	161.05

Year 0 is the current year, year 1 the period 1 year from now, year 2 the period 2 years from now, etc. The sales figures for each year were found by increasing the sales of the preceding year by 10 percent.

The exponential function is established from any two data points in Table 3.1. Suppose that we select year 1 and year 2. Values of x are 1 and 2, values of $f(x)$ are $110.00 and $121.00, and values of $\log f(x)$ are 2.0414 and 2.0828. Values of x and $\log f(x)$ are substituted into Eq. (3.6) to give

$$2.0414 = \log k + c \log a$$

and

$$2.0828 = \log k + 2c \log a$$

Since there are three variables (i.e., $\log k$, c, and $\log a$) and only two equations, we must specify the value of one of the variables. Assume, for the moment, that the value of c is $c = 1$. Based on this assumption, the equations reduce to

$$2.0414 = \log k + \log a$$

and

$$2.0828 = \log k + 2 \log a$$

The expression $\log a = 2.0414 - \log k$ from the first equation can be substituted into the second equation to give

$$2.0828 = \log k + 2(2.0414 - \log k)$$

$$2.0828 = \log k + 4.0828 - 2 \log k$$

or

$$\log k = 2.0000$$

From the first equation we obtain

$$\log a = 2.0414 - \log k = 2.0414 - 2.0000$$

$$\log a = 0.0414$$

The antilogarithm of $\log k$ is 100 and the antilogarithm of $\log a$ is 1.10. The exponential function is thus

$$f(x) = 100(1.10)^x$$

Several conclusions can be drawn from this example. First, note that the value of k in the exponential function is equal to the sales of Acme Wire and Products in year zero, i.e., $k = 100$. Next, note that the base, a, is equal to one plus the annual percentage change in sales, i.e., $a = 1 + 0.10$ or 1.10. Finally, note that these results were obtained by specifying $c = 1$.

Based on the above example, we can offer a function that can be used to model any similar problem. The function is

$$f(x) = k(1 + r)^x \tag{3.9}$$

where $f(x)$ is the dependent variable, x is the time period, k is the value of the dependent variable in time period zero, and r is the percentage change in $f(x)$ during one time period.

To illustrate Eq. (3.9), suppose that we are asked to develop a function that can be used to forecast profits of the Hall Manufacturing Company. We are told that the profits of this company were \$120,000 in 1975 and are expected to increase by 6 percent per year.

Since profits are expected to increase by a constant percentage, the profits for any year can be forecast by using Eq. (3.9). The parameters of the function are $k = 120{,}000$ and $r = 0.06$. The function is

$$f(x) = 120{,}000(1.06)^x$$

where $f(x)$ represents profit and x represents the year.

Example. The president of the Hall Manufacturing Company has requested a forecast of profits for the year 1985. Using the exponential function established in the preceding example, determine the forecast.

Based upon a current profit level of $120,000 and a growth rate of 6 percent per year, profits 10 years hence should be

$$f(10) = 120{,}000(1.06)^{10}$$

$$\log f(10) = \log (120{,}000) + 10 \log (1.06)$$

$$\log f(10) = 5.0792 + 10(0.0253)$$

$$\log f(10) = 5.3322$$

$$f(10) = \$215{,}000$$

Example. The cost of sending a student to a certain school has increased at an average rate of 6 percent per year. If the current cost is $1500, determine the cost in 10 years.

The functional relationship is

$$f(x) = 1500(1.06)^x$$

The cost in 10 years is

$$f(10) = 1500(1.06)^{10}$$

$$\log f(10) = \log (1500) + 10 \log (1.06)$$

$$\log f(10) = 3.1761 + 10(0.0253)$$

$$\log f(10) = 3.4291$$

Thus,

$$f(10) = \$2686$$

Example. A company that manufactures copying equipment has determined that the life of a copier is limited by obsolescence. If a copier whose initial value is $10,000 decreases in value by 20 percent of its value at the beginning of the preceding year, determine the value of the machine at the end of each of the 5 years.

The functional relationship is

$$f(x) = \$10,000(0.80)^x$$

The copier's value at the end of the first year is

$$f(1) = \$10,000(\tfrac{8}{10})^1 = \$8000$$

The copier's value at the end of the second year is

$$f(2) = \$10,000(\tfrac{8}{10})^2$$

$$\log f(2) = \log(10,000) + 2(\log 8 - \log 10)$$

$$\log f(2) = 4.000 + 2(0.9031 - 1.0000)$$

$$\log f(2) = 3.8062$$

$$f(2) = \$6400$$

The value at the end of each year is given in the following table:

x	1	2	3	4	5
$f(x)$	$8000	6400	5120	4096	3277

Example. The purchasing power of the dollar declined at an average rate of 8 percent during 1973, 1974, and 1975. If a continuation of this rate of decline is assumed, what will a 1975 dollar be worth in 1995?

The functional relationship is

$$f(x) = \$1(1 - 0.08)^x$$

or

$$f(x) = 1(0.92)^x$$

The value of a dollar in 1995, in terms of 1975 dollars, would be

$$f(20) = 1(0.92)^{20}$$

$$\log f(20) = \log 1 + 20[\log(92) - \log(100)]$$

$$\log f(20) = 0.0000 + 20[1.9638 - 2.0000]$$

$$\log f(20) = -0.7240$$

Since -0.7240 is equal to $0.2760 - 1.0000$,

$$\log f(20) = 0.2760 - 1.0000$$

$$f(20) = \frac{1.888}{10} = \$0.1888 \quad \text{or} \quad 18.88\cancel{c}$$

3.2.4   CHANGE OF BASE[7]

The example problems in the previous section have $(1 + r)$ as the base in the exponential function. The reason for this choice is that it permits one to establish an exponential function given only the initial value of the dependent variable, k, and the periodic growth rate of the dependent variable, r. As we explained on p. 90, the choice of a value for the base is arbitrary. Although the base $(1 + r)$ has obvious advantages for certain kinds of problems, we could also have used base e in the problems in this section.

To illustrate the use of base e for the problems discussed in this section, consider the exponential function, Eq. (3.9),

$$f(x) = k(1 + r)^x$$

This function can be changed to base e as follows. First, we equate Eq. (3.9) and Eq. (3.7).

$$ke^{cx} = k(1 + r)^x$$

Both sides of the equation are divided by k. This gives

$$e^{cx} = (1 + r)^x$$

The equation is now written in terms of natural logarithms. This gives

$$cx = x \ln (1 + r)$$

or

$$c = \ln (1 + r) \tag{3.10}$$

The exponential function is thus

$$f(x) = ke^{\ln (1 + r)x} \tag{3.11}$$

Example. Verify that $f(x) = 100(1.10)^x$ and $f(x) = 100e^{\ln (1.10)x}$ are alternative forms of the same functional relationship.

The fact that the two functions are alternative expressions of the same functional relationship can be demonstrated by writing both functions in terms of natural logarithms. Writing

$$f(x) = 100(1.10)^x$$

in terms of natural logarithms, we obtain

$$\ln f(x) = \ln (100) + \ln (1.10)x$$

Writing

$$f(x) = 100e^{\ln (1.10)x}$$

---

[7] This section is optional at the discretion of the instructor.

in terms of natural logarithms also gives

$$\ln f(x) = \ln (100) + \ln (1.10)x$$

The functional relationships are the same.

Example. Determine the rate of growth of the dependent variable, $f(x)$, in the function

$$f(x) = 3e^{0.157x}$$

The rate of growth of this function can be determined by using Eq. (3.10). Since the parameter c has the value 0.157 and the relationship between c and r is defined by Eq. (3.10), we merely solve Eq. (3.10) for r. Thus,

$$0.157 = \ln (1 + r)$$

or

$$\text{antilog}_e (0.157) = \text{antilog}_e \left[\ln (1 + r)\right]$$

and from Table A.6,

$$1.17 = 1 + r$$

and

$$r = 0.17 \quad \text{or} \quad 17\%$$

Example. Express the function

$$f(x) = 100(1.12)^x$$

in terms of base e.

From Eq. (3.11), the function is

$$f(x) = 100e^{\ln (1.12)x}$$

Since $\ln (1.12)$ is 0.11333, the function is

$$f(x) = 100e^{0.11333x}$$

3.3 Multivariate Functions

We have stated that a function has only one dependent variable but may have one or more independent variables. Functions in which the single dependent variable is related to more than one independent variable are termed *multivariate functions*. An example of a multivariate function is

$$f(x_1, x_2) = 2x_1 + 3x_1x_2 + 6x_2$$

where $f(x_1, x_2)$ is the dependent variable, x_1 is an independent variable, and x_2 is a second independent variable. Multivariate functions are difficult to plot, since they require one axis (or dimension) for each variable. A function with two independent variables and one dependent variable, for instance, requires three dimensions for plotting. Multivariate functions with three independent variables cannot be plotted, since four dimensions are required. In spite of the fact that multivariate functions with more than two independent variables cannot be plotted, they are necessary to describe many business and economic models. Multivariate functions with two independent variables are illustrated by the following examples. Additional examples are given in Chap. 11, Linear Programming.

Example. Sales of the Southern Distributing Company consist of sales made through retail and wholesale outlets. Profit per dollar of sales at retail is $0.15 and profit per dollar of sales at wholesale is $0.05. Determine the profit function.

If retail sales in dollars are represented by x_1, wholesale sales in dollars by x_2, and profit by $P(x_1 x_2)$, the profit function is

$$P(x_1, x_2) = 0.15x_1 + 0.05x_2$$

Example. Assume that the Neshay Candy Company makes net profit of $0.05 per almond bar and $0.07 per chocolate bar. If net profit P is represented as a function of monthly sales in units of the almond bar x_1 and the chocolate bar x_2, determine the appropriate functional relationship and calculate net profit for a month in which $x_1 = 10,000$ and $x_2 = 20,000$,

$$P = 0.05x_1 + 0.07x_2$$
$$P = 0.05(10,000) + 0.07(20,000)$$
$$P = \$1,900$$

Multivariate functions can be established through data points by using the same technique described for univariate functions. This technique involves substituting the appropriate data points into the specified general form of the multivariate function and solving the resulting equations simultaneously for the parameters of the function. The technique is illustrated by the following example.

Example. Research has demonstrated that the output Y of a firm is related to labor L and capital K. The general form of the function is

$$Y = aL + bK + c(LK)$$

It has been determined that $Y = 100,000$ when $L = 9.0$ and $K = 4.0$, $Y = 120,000$ when $L = 10.0$ and $K = 5.0$, and $Y = 150,000$ when $L = 11.5$ and $K = 7.0$. The domains of L and K are $9.0 \leq L \leq 11.5$ and $4.0 \leq K \leq 7.0$. Determine the values of a, b, and c.

The three data points are substituted into the general form of the function to give

$$100,000 = a(9.0) + b(4.0 + c(36.0)$$

$$120,000 = a(10.0) + b(5.0) + c(50.0)$$

$$150,000 = a(11.5) + b(7.0) + c(80.5)$$

These equations are solved simultaneously to give $a = -60,000$, $b = 304,000$, and $c = -16,000$. The function is

$$Y = -60,000L + 304,000K - 16,000(LK) \qquad \text{for } 9.0 \le L \le 11.5$$
$$4.0 \le K \le 7.0$$

PROBLEMS

1. Plot the following functions:
   (a) $f(x) = 3 - 4x + x^2$
   (b) $f(x) = -6 + 5x - x^2$
   (c) $f(x) = 12 - 8x + x^2$
   (d) $f(x) = -8 - 2x + x^2$
   (e) $f(x) = 9 - 6x + x^2$
   (f) $f(x) = 2 - 4.5x + x^2$
   (g) $f(x) = -4 + x^2$
   (h) $f(x) = 16 - 10x + x^2$

2. Plot the following functions:
   (a) $f(x) = 12 - 5x + 0.5x^2$
   (b) $f(x) = -3 + 2x + x^2$
   (c) $f(x) = 4 - 6x + 2x^2$
   (d) $f(x) = 16 - 4x - 2x^2$
   (e) $f(x) = -9 + x^2$
   (f) $f(x) = -12 + 11x - 2x^2$
   (g) $f(x) = x^2$
   (h) $f(x) = 4 - 4x + x^2$

3. Establish quadratic functions through the data points $(x, f(x))$.
   (a) $(1, 0)$, $(2, -1)$, and $(3, 0)$
   (b) $(2, 0)$, $(0, 16)$, and $(1, 7)$
   (c) $(3, -5)$, $(2, -10)$, and $(-2, 0)$
   (d) $(1, 4)$, $(0, 9)$, and $(2, 1)$
   (e) $(0, -4)$, $(2, 0)$, and $(1, -3)$
   (f) $(2, 0)$, $(3, -5)$, and $(4, -8)$
   (g) $(1, 10)$, $(2, 0)$, and $(-2, 16)$
   (h) $(-1, 12)$, $(2, 0)$, and $(1, 0)$

4. Determine the roots of the following quadratic functions:
   (a) $f(x) = 8 - 6x + x^2$
   (b) $f(x) = 6 - 7x + x^2$
   (c) $f(x) = -3 - 2x + x^2$
   (d) $f(x) = -100 + 10x + 2x^2$
   (e) $f(x) = 18 - 15x + 2x^2$
   (f) $f(x) = -20 - 3x + 2x^2$
   (g) $f(x) = -20 + 9x + 0.5x^2$
   (h) $f(x) = 36 - 15x + 1.5x^2$

5. Determine the roots of the functions in Problem 1.

6. Verify that the following quadratic functions do not have ordinary numerical roots, i.e., the roots are imaginary:
   (a) $f(x) = 16 - 6x + x^2$
   (b) $f(x) = 12 - 4x + x^2$
   (c) $f(x) = 15 - 5x + 0.5x^2$
   (d) $f(x) = -2 + 2x - x^2$
   (e) $f(x) = -8 + x - 0.05x^2$
   (f) $f(x) = 1 + x^2$

7. Determine the local optimum of the following quadratic functions. State whether the function reaches a maximum or minimum.
   (a) $f(x) = 16 - 4x - 2x^2$ (b) $f(x) = 12 - 8x + 2x^2$
   (c) $f(x) = -20 + 3x - 0.5x^2$ (d) $f(x) = 10 + 2x - 0.05x^2$
   (e) $f(x) = 8 - 12x + x^2$ (f) $f(x) = -14 + 2x - 0.01x^2$

8. The profit function of Graham Manufacturing Company is described by a quadratic function. The following data points were estimated by the chief accountant: output of zero units results in a loss of $10 million, output of 6000 units results in a profit of $8 million, and output of 8000 units results in a profit of $6 million. Determine the quadratic function that describes profit as a function of the number of units produced and the number of units that results in maximum profit.

9. The average cost of a certain assembly manufactured by Milwaukee Electronics Corporation is $4 for 300 units, $2.50 for 500 units, and $3 for 600 units. Determine the quadratic function that describes the relationship between average cost per unit and output and the output that results in minimum average cost.

10. In a controlled experiment, the productivity of a work group was found to be related to the number of workers assigned to the group. Specifically, the analyst found that with a group of 6 workers the productivity was 8 units per man. When the group size was increased to 8 workers, the productivity rose to 12 units per man. An additional increase in group size to 10 workers also resulted in productivity of 12 units per man. Determine the quadratic function that relates productivity to the group size and the group size that results in maximum productivity.

11. It is known that the earning ability of an individual increases as he grows older and then decreases as he approaches retirement age. In a certain profession, average earnings at age 40 are $25,000. The figure increases to $32,000 by age 55 and declines to $28,000 at age 60. Determine the quadratic function that describes the relationship between age and earnings and the age that leads to maximum earnings.

12. Sales of a certain product are known to be related to advertising expenditures. In a controlled experiment in which factors such as price, quality, etc., were held constant and advertising expenditures were varied, the following results were observed: Advertising expenditures of $100 resulted in sales of $10,000; advertising expenditures of $300 resulted in sales of $11,500; advertising expenditures of $500 resulted in sales of $12,500. Determine the function that describes the relationship between advertising and sales and the advertising expenditures that result in maximum sales.

13. The profit of Merril Equipment Company is determined by the number of units of equipment rented by Merril. Profits of $1000 are possible when 75 units are rented; profits of $1200 are possible when 85 units are

rented: profits of $1300 are possible with the rental of 100 units of equipment. Determine the functional relationship between profit and the number of units of equipment rented and the number of rental units that make possible maximum profits.

14. Plot the following functions (Hint: Use $x = 0, 1, 2, 3$, and 4 as values of the independent variable).

(a) $f(x) = 10^{0.2x}$ (b) $f(x) = 2^{0.5x}$

(c) $f(x) = 2(1.5)^x$ (d) $f(x) = 10^{-0.2x}$

(e) $f(x) = 3e^{0.5x}$ (f) $f(x) = 10e^{-0.5x}$

(g) $f(x) = 100(1.10)^x$ (h) $f(x) = 10(0.8)^x$

15. Establish exponential functions of the form $f(x) = ke^{cx}$ through the data points $(x, f(x))$.

(a) $(2, 2)$ and $(4, 3)$ (b) $(1, 2)$ and $(4, 4)$

(c) $(2, 10)$ and $(5, 6)$ (d) $(3, 2)$ and $(6, 4)$

(e) $(1, 4)$ and $(4, 2)$ (f) $(5, 75)$ and $(10, 56)$

(g) $(0, 100)$ and $(10, 148)$ (h) $(5, 147)$ and $(10, 216)$

16. Use Eq. (3.10) to determine the rate of growth of the exponential functions in Problem 14.

17. A company's sales grew from $10 million to $15 million in the 8-year period from 1968 to 1976. Establish the exponential function that relates sales and time. Assuming a continuation in the rate of growth, determine sales in 1980. What is the annual growth of sales for this company?

18. From 1940 to 1970 the price of eggs rose from $0.23 per dozen to $0.70. Determine the annual percentage increase in the price of eggs. If this rate of price increase continues, what would be the price of one dozen eggs in 1990?

19. The gross national product in the United States increased from $284.8 billion in 1950 to $503.7 billion in 1960. Determine the annual growth rate during this period.

20. Harvey Smith received a spendable income of $184.50 per week in 1970. His spendable income has increased to $225.00 per week in 1975. Determine the annual rate of increase in spendable income for Mr. Smith. If this rate of increase continues, what would Mr. Smith's income be in 1985?

21. A company sells three products, A, B, and C. The prices of the products are $5, $8, and $6, respectively. Establish the multivariate function that relates total sales to the number of units of the products sold. Determine total sales if 1000 units of A, 2000 units of B, and 1500 units of C are sold.

22. The sales of the Ace Novelty Company are a function of price p, advertising a, and the number of salesmen n. The functional relationship is

$$S = (10,000 - 700p)n^{2/3}a^{1/2}$$

If the current price is $5, advertising is $10,000 and 64 salesmen are employed, determine sales.

23. Profit derived from the sale of gasoline at a service station depends on the quantities sold of three types of gas: regular (R), unleaded (U), and premium (P).

   (a) State the profit function for the total profit on gallons sold of the three types of gas.

   It was determined from past experience that the profit derived from certain sales combinations is as follows:

   profit = {($440, 5000 gallons R, 1000 gallons U, 4000 gallons P),
   ($245, 3000 gallons R, 500 gallons U, 2000 gallons P),
   ($93, 1000 gallons R, 100 gallons U, 1000 gallons P)}

   (b) Determine the profit per gallon for each type of gas by substituting these data points into the profit function and solving for the parameters of the profit function.

SUGGESTED REFERENCES

The references for this chapter are listed in Chap. 2.

Chapter Four

Fundamental Rules For Counting

This chapter represents something of a departure from the concepts discussed in Chaps. 2 and 3. Those chapters were concerned with alternative forms of relationships that exist between variables. We now turn to a different problem, that of counting. Rather than merely counting the number of variables in a problem or the number of elements in a set, we shall develop rules or formulas that permit one to calculate the number of possible outcomes of some process or experiment. These formulas will prove useful in answering questions that often occur in business and the sciences.

As an example of the problems that will be discussed in this chapter, consider the following two questions:

1. Automobile license plates of the State of California consist of three digits followed by three letters. How many different license plates are possible?
2. The product line of a company includes exactly 100 different products. From these products, five are to be selected for a special advertising campaign. How many different combinations of the five products could be selected?

The answers to these questions could be determined by counting the possible number of license plates and the number of different combinations of five products. This would, however, be a very tedious task. There are over 17 million different possible license plates and over 75 million different combinations of five products. Clearly, it would not be practical to answer questions of this kind by listing the different possible outcomes and counting these possibilities.

Our objective in this chapter is to develop formulas for counting the number of possible outcomes in problems such as those posed above. These formulas will prove useful in solving various business and scientific problems. In addition, they will provide the basis for several of the probability distributions discussed in Chap. 6.

4.1   A Fundamental Rule for Counting

The number of possible outcomes of a process or experiment can often be determined by applying the following fundamental rule of counting:

> *If a process or experiment consists of* k *separate steps and the first step can occur in* n_1 *different ways, the second step in* n_2 *different ways, ..., and the kth step in* n_k *different ways, then the number of different possible outcomes is*

$$n_1 \cdot n_2 \cdot \cdots \cdot n_k \qquad (4.1)$$

To illustrate this fundamental rule, consider the selection of a committee comprised of one labor leader, one governmental official, and one community leader from a group of two labor leaders, three governmental officials, and two community leaders. There are two ways in which the selection of the labor leader can be made, three ways in which the selection of the governmental official can be made, and two ways in which the selection of the community leader can be made. From the fundamental rule of counting, the total possible number of ways of selecting the committee is $2 \cdot 3 \cdot 2 = 12$.

The fundamental rule of counting is merely a restatement of the formula for determining the number of members in the Cartesian product of two or more sets. The reader will remember from Chap. 1 that the Cartesian product of the sets A and B, designated by $A \times B$, is the set containing all possible ordered pairs (a, b) such that $a \in A$ and $b \in B$. The number of ordered pairs in $A \times B$ was given by the product of the number of elements in the two sets. Similarly, the number of members in the Cartesian product of the sets A_1, $A_2, \ldots, A_k$, containing $n_1, n_2, \ldots, n_k$ elements, respectively, is

$$n_1 \cdot n_2 \cdot \cdots \cdot n_k \qquad (4.2)$$

Returning to the problem of selecting a committee comprised of one labor leader, one governmental official, and one community leader from a group of two labor leaders, three governmental officials, and two community leaders, the 12 different possible committees can be shown by the Cartesian product of three sets. If $L = \{l_1, l_2\}$ represents the set of two labor leaders, $G = \{g_1, g_2, g_3\}$ represents the set of three governmental officials, and

$C = \{c_1, c_2\}$ represents the set of two community leaders, the Cartesian product of L, G, and C is

$$\{(l_1, g_1, c_1), (l_1, g_1, c_2), (l_1, g_2, c_1), (l_1, g_2, c_2), (l_1, g_3, c_1), (l_1, g_3, c_2),$$
$$(l_2, g_1, c_1), (l_2, g_1, c_2), (l_2, g_2, c_1), (l_2, g_2, c_2), (l_2, g_3, c_1), (l_2, g_3, c_2)\}$$

The Cartesian product contains $2 \cdot 3 \cdot 2 = 12$ elements, each element representing a possible committee selection. The number of possible outcomes is, of course, the same as was found by using the fundamental rule of counting.

The possible outcomes in the selection of the committee can also be shown by a *tree diagram*. A tree diagram is a representation of a sequence of events. In our example, the first event can be thought of as the selection of a labor leader. The two possibilities for the labor leader, l_1 and l_2, are shown by *branches* on the tree. Since we are considering the selection of the labor leader as the first event, branches l_1 and l_2 start at the *root* of the tree.

The second event is the selection of a governmental official. The three possibilities in this event, g_1, g_2, and g_3, are represented by branches that connect at the terminal point of the branches for the first event. Similarly, the two possibilities for the selection of the community leader, c_1 and c_2, are shown as branches that connect at the terminal point of the branches for the second event. The complete tree diagram is shown in Fig. 4.1.

The tree diagram in Fig. 4.1 terminates on the right with 12 different branches, each branch representing a possible committee. To determine the members of the committee, we select a path through the tree. Each path begins at the root of the tree and terminates at the tip of a branch. There are

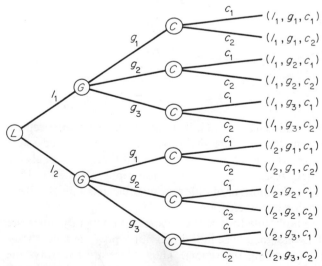

Figure 4.1.

12 different paths through the tree, each path representing a possible committee.

Example. Use the fundamental rule of counting to determine the number of possible outcomes from tossing a coin three times and recording the outcome of each toss as a head or a tail. Show these outcomes as the Cartesian product of three sets and by a tree diagram.

The experiment of tossing a coin three times consists of three steps. There are two possible outcomes on the first toss (i.e., heads and tails), two possible outcomes on the second toss, and two possible outcomes on the third toss. The total number of possible outcomes is thus $2 \cdot 2 \cdot 2 = 8$. If $A = \{H, T\}$ represents the set of possible outcomes on the first toss, $B = \{H, T\}$ the possible outcomes on the second toss, and $C = \{H, T\}$ the possible outcomes on the third toss, the Cartesian product of A, B, and C is

$$\{(HHH), (HHT), (HTH), (THH), (HTT), (THT), (TTH), (TTT)\}$$

The set contains $2 \cdot 2 \cdot 2 = 8$ elements. A tree diagram showing the eight possible outcomes is given in Fig. 4.2.

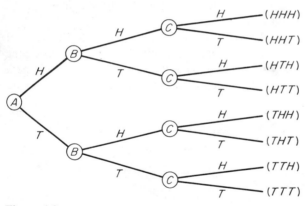

Figure 4.2.

Example. Jim, Fred, and Barbara are jointly preparing a research report. Three jobs must be completed before the report can be published—field research, drafting the report, and typing the report. All three individuals are equally qualified for field research and drafting the report. However, Fred does not type. Determine the number of different possible assignments. Show these assignments by using a tree diagram.

Jim and Barbara are both able to type the report. Consequently, there are two ways this job can be assigned. The individual not typing the report joins Fred as a candidate for drafting the report. Thus, there are two ways the assignment of drafting the report can be made. After this job is assigned,

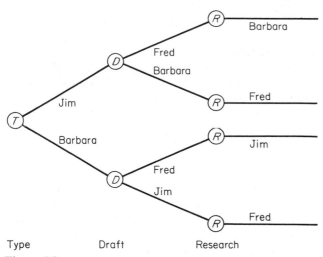

Figure 4.3.

the remaining individual is assigned the job of field research. From the fundamental rule of counting, the assignments can be made in $2 \cdot 2 \cdot 1$ or four different ways. These are shown by the tree diagram in Fig. 4.3.

It is often unnecessary to list all possible outcomes of the process or experiment. For instance, in the introductory remarks in this chapter a question was raised concerning the number of different automobile license plates possible in the State of California. Answering this question does not require a list of the different license plates. Instead, we merely need to know the total number of license plates possible. This number can be determined from the fundamental rule of counting.

The license plates in California consist of three digits followed by three letters. There are ten ways of selecting each digit and 26 ways of selecting each letter. From the fundamental rule of counting, the total number of license plates is

$$10 \cdot 10 \cdot 10 \cdot 26 \cdot 26 \cdot 26 = 17,576,000$$

Example. An experiment consists of drawing five cards from a well-shuffled deck of 52 playing cards. If it is assumed that a card once drawn from the deck is not replaced, in how many possible ways may the selection of five cards be made?

There are 52 possible outcomes from the first draw, 51 possible outcomes from the second draw, etc. The total number of possible outcomes is $52 \cdot 51 \cdot 50 \cdot 49 \cdot 48$, or 311,875,200.

Example. In a marketing survey, an individual is asked to taste four different wines and rank them in order of preference. How many different rankings are possible?

There are four possible choices for the first preference. After this selection is made, there are three possible choices for the second preference. Similarly, there are two choices for the third preference and one choice for the least preferred wine. The number of possible rankings is thus $4 \cdot 3 \cdot 2 \cdot 1$, or 24.

4.2 Permutations

The fundamental rule of counting was used in the previous example to determine the number of ways of ranking four kinds of wine. In determining the number of ways of ranking the wines, we assumed that the order in which the wines were ranked was important. This means that ranking a burgundy first followed by a rosé differs from ranking a rosé first followed by a burgundy. The ranking of the four wines in terms of taste preference is an example of a permutation.

The term *permutation* refers to an ordered arrangement of objects. In referring to a group of r objects selected from a group of n objects, each different possible arrangement of the r objects is a permutation.

In the example of ranking wines, the four wines were arranged or ranked in order of preference. Since the order in which the wines were ranked was important, each of the 12 different possible rankings represented a permutation. The selection of four objects, i.e., wines, from the group of four objects resulted in 12 permutations.

The concept of a permutation of objects can also be illustrated by considering alternative arrangements of the first three letters of the alphabet. Given the set of letters $\{a, b, c\}$, there are three possible choices for the first letter, two possible choices for the second letter, and one choice for the third letter. The number of possible arrangements or permutations of the letters, again if it is assumed that the order of the arrangement of the letters is important, is $3 \cdot 2 \cdot 1$ or 6 possible permutations. These are

$$abc \qquad bac \qquad cab$$

$$acb \qquad bca \qquad cba$$

Each of these arrangements is referred to as a permutation.

As another example of permutations, consider the problem of selecting five cards from a deck of 52 playing cards. In counting the number of different five-card hands, we assumed that the order of selection of the individual cards from the deck of cards was important. This means that the selection of the ace of hearts followed by the king of diamonds differs from the selection of the king of diamonds followed by the ace of hearts. This difference occurs because the order in which the cards were selected in the two hands differs. Each of the different five-card hands selected from the deck of 52 cards is a permutation. From the fundamental rule of counting, we found that there were $52 \cdot 51 \cdot 50 \cdot 49 \cdot 48$ or 311,875,200 different five-card permutations.

The formula for the number of possible permutations of all n objects in a set is derived from the fundamental rule of counting. Since there are n objects in the set, there are n ways in which the first item can be selected. Similarly, there are $(n - 1)$ ways of selecting the second item, $(n - 2)$ ways of selecting the third item, and finally, $(n - n + 1)$ or one way of selecting the final item. The formula for counting the number of possible permutations of n items in a set of n items is thus

$$n(n - 1)(n - 2) \cdot \cdots \cdot (3)(2)(1) \tag{4.3}$$

An alternative and more commonly used expression of this formula is

$$_nP_n = n! \tag{4.4}$$

where $n!$, read *n factorial*, is equal to $n(n - 1)(n - 2) \cdot \cdots \cdot (3)(2)(1)$.

Example. A management consulting firm has four contracts that must be assigned to four separate consultants. In how many ways can the assignments be made?

Each of the different assignments represents a permutation. From Eq. (4.4), the number of permutations is

$$_4P_4 = 4! = 24$$

Example. Determine the number of possible permutations in a set of six objects.

$$_6P_6 = 6! = 720$$

Equation (4.4) applies when one is counting the number of ways all n objects in a set can be selected or arranged. Very often we are interested in counting the total number of possible ways of selecting a subset of r of the n objects in a set. To illustrate, consider the problem of selecting a president, vice-president, secretary, and treasurer from a group of ten candidates. Although there are ten objects in the set (i.e., candidates), only four of the ten are to be selected. Again, using the fundamental rule of counting, the number of possible permutations is computed by recognizing that there are ten ways of selecting the president, nine ways of selecting the vice-president, eight ways of selecting the secretary, and seven ways of selecting the treasurer. The total number of possible permutations is thus $10 \cdot 9 \cdot 8 \cdot 7$, or 5040. There are 5040 possible ways of arranging four objects selected from a set of ten objects when the order of arrangement of the objects is important.

The formula for determining the number of possible permutations of r objects selected from a set of n objects is

$$n(n - 1)(n - 2) \cdots (n - r + 1) \tag{4.5}$$

An alternative and more commonly used expression of this equation is

$$_nP_r = \frac{n!}{(n - r)!} \tag{4.6}$$

Example. Using the preceding example of selecting four officers from a group of ten candidates, show that Eqs. (4.5) and (4.6) are equivalent.

From Eq. (4.5) with $n = 10$ and $r = 4$, the total number of ways of selecting a president, vice-president, secretary, and treasurer from a group of ten candidates is

$$10 \cdot 9 \cdot 8 \cdot 7 = 5040$$

By Eq. (4.6), the total number of possible ways is

$$_{10}P_4 = \frac{10!}{6!} = 10 \cdot 9 \cdot 8 \cdot 7 = 5040$$

It can be seen that the two formulas are equivalent. Eq. (4.6) is merely an alternative expression of Eq. (4.5).

Example. A retailing chain has 20 stores. Three of the stores are to be selected for testing three different products. If it is assumed that only one product will be tested at a specific store, how many different arrangements are possible?

$$_{20}P_3 = \frac{20!}{17!} = 20 \cdot 19 \cdot 18 = 6840$$

Example. An experiment consists of selecting two numbers from the set of digits, $D = \{0, 1, 2, 3, 4, 5, 6, 7, 8, 9\}$. If it is assumed that the first number selected is not replaced, how many different arrangements of the two numbers are possible?

$$_{10}P_2 = \frac{10!}{8!} = 90$$

Example. The judges of a beauty contest have the enviable task of selecting the first, second, third, and fourth place winners. If 16 girls are entered in the contest, in how many ways can the winners be selected?

Each possible selection of the first, second, third, and fourth place winners represents a permutation. From Eq. (4.6), the number of permutations is

$$_{16}P_4 = \frac{16!}{12!} = 16 \cdot 15 \cdot 14 \cdot 13 = 43,680$$

Example. Show that the formula for determining the number of permutations of r objects selected from n objects, Eq. (4.6), is the same as the formula for determining the number of permutations of all n objects in a set, Eq. (4.4), in the special case when $r = n$.

The formula for determining the number of permutations of r objects selected from n objects, Eq. (4.6), is

$$_nP_r = \frac{n!}{(n-r)!} \tag{4.6}$$

In the special case when $n = r$, the equation reduces to

$$_nP_n = \frac{n!}{(n-n)!} = \frac{n!}{0!}$$

or alternatively,

$$_nP_n = n! \tag{4.4}$$

Since $0! = 1$, the two formulas are equivalent in the special case when $r = n$.

4.2.1 PERMUTATIONS INVOLVING INDISTINGUISHABLE OBJECTS

In our discussion of the number of possible permutations of a set of n objects, it has been assumed that all n objects in the set are distinguishable. Based on this assumption, the formula for the number of permutations of the n objects in a set was given by Eq. (4.4) as

$$_nP_n = n! \tag{4.4}$$

We shall now consider a somewhat different problem, namely, the problem of determining the number of permutations of n objects in a group of n objects when some of the objects in the group are indistinguishable from other objects in the group. To illustrate, suppose that we have a group of three x's and two y's. If the x's and y's were distinguishable, there would be a total of 5! or 120 different permutations of the five objects. The x's and y's are not, however, distinguishable. Therefore, not all of the 5! or 120 permutations are distinguishably different. For any permutation of the five objects, there are 3! ways that the x's can be arranged or permuted among themselves and 2! ways that the y's can be arranged or permuted among themselves. Thus, for each permutation of the given objects, there are 3! 2! or 12 permutations of the five objects that are identical. The total number of distinguishable permutations of the five objects is therefore

$$\frac{5!}{3!\,2!} = \frac{120}{12} = 10$$

The ten distinguishable permutations are

xxxyy	*yyxxx*	*xxyyx*
xxyxy	*yxyxx*	*xyxyx*
xyxxy	*yxxyx*	
yxxxy	*xyyxx*	

Example. Assume in the preceding example that the *x*'s and *y*'s are distinguishable. Show that each permutation of three *x*'s and two *y*'s results in 3! 2! or 12 permutations of three distinguishable *x*'s and two distinguishable *y*'s.

Suppose that we rewrite the five letters as *x*, *X*, **x**, *y*, *Y*. The three *x*'s and two *y*'s are now distinguishable. The first permutation listed in the preceding example was *xxxyy*. Substituting the distinguishable letters in this permutation results in 12 permutations. These are

xXxyY	*XxxyY*	**xx**XYy
xxXyY	*XxxyY*	*xXxYy*
xxXyY	*xXxYy*	*Xxx*Yy
xXxyY	*xx*XYy	*Xxx*Yy

The same result would be true for any of the original ten permutations. We therefore conclude that the number of permutations of the five distinguishable objects, *x*, *X*, **x**, *y*, *Y*,

$$_5P_5 = 5! = 120$$

is 3! 2! or 12 times larger than the number of distinguishable permutations of three indistinguishable *x*'s and two indistinguishable *y*'s.

The same procedure applies when more than two objects in a group are identical. To illustrate, suppose that a group of nine objects consists of three *x*'s, two *y*'s, and four *z*'s. If all nine objects were distinguishable, there would be a total of 9! or 362,880 permutations of the nine objects. Since the *x*'s, *y*'s, and *z*'s are not, however, distinguishable, the 9! permutations must be divided by the number of ways the *x*'s, *y*'s, and *z*'s can be permuted among themselves. There are 3! ways that the *x*'s can be permuted among themselves, 2! ways that the *y*'s can be permuted among themselves, and 4! ways that the *z*'s can be permuted among themselves. Consequently, for each permutation of the nine objects, there are 3! 2! 4! or 288 permutations that are identical. The total number of distinguishable permutations is therefore

$$\frac{9!}{3!\,2!\,4!} = \frac{362,380}{288} = 1260$$

There are 1260 distinguishable permutations of three *x*'s, two *y*'s and four *z*'s.

Example. Determine the number of distinguishable permutations of the letters in the word "Mississippi."

The word Mississippi contains eleven letters. These include one m, four i's, four s's, and two p's. The number of distinguishable permutations of the eleven letters is

$$\frac{11!}{1!\,4!\,4!\,2!} = \frac{39{,}916{,}800}{(1)(24)(24)(2)} = 34{,}650$$

The energetic reader may want to list these 34,650 permutations to see if any permutation other than "Mississippi" results in a word!

Using the preceding examples, we can develop a formula for counting the number of distinguishable permutations. Given a group of n objects composed of n_1 identical objects of one kind, n_2 identical objects of a second kind, n_3 identical objects of a third kind, and so on for k kinds of objects, the number of distinguishably different permutations of all n objects is

$$_nP_n(n_1, n_2, \ldots, n_k) = \frac{n!}{n_1!\,n_2!\cdots n_k!} \qquad (4.7)$$

where $n_1 + n_2 + \cdots + n_k = n$.

Example. An individual has five pennies, three nickels, two quarters, and two half-dollars. In how many ways can these 12 coins be arranged so that the order of arrangement is distinguishably different?

From Eq. (4.7), the number of different arrangements is

$$_{12}P_{12}(5, 3, 2, 2) = \frac{12!}{5!\,3!\,2!\,2!} = 166{,}320$$

Example. A professor has three identical copies of an economics text, three identical copies of a history text, and two identical copies of a mathematics text on his desk. In how many ways can these eight textbooks be arranged so that the order of arrangement is distinguishably different?

$$_8P_8(3, 3, 2) = \frac{8!}{3!\,3!\,2!} = 560$$

Example. A binary number contains only the digits 0 and 1. How many 10-digit binary numbers can be formed from six 1's and four 0's? Assume that the first digit in the binary number is a 1.

The first digit must be a 1. Of the remaining digits, five are 1's and four are 0's. The number of distinguishably different permutations is thus

$$_9P_9(5, 4) = \frac{9!}{5!\,4!} = 126$$

A total of 126 different binary numbers can be formed.

4.2.2 PERMUTATIONS ALLOWING REPETITIONS

In deriving the formulas in this section, we have assumed that each object in a group of n objects appears only once in a permutation. For instance, given a set of elements $\{a, b, c\}$, the possible permutations of these three elements are

$$abc \qquad cab \qquad bca$$
$$acb \qquad bac \qquad cba$$

Notice that the elements in the set $\{a, b, c\}$ are included in each permutation only once, i.e., repetition of the elements is not allowed.

We shall now introduce a formula for counting the number of permutations in a group of n distinguishable objects assuming that repetitions are allowed. For instance, if repetitions of the elements in the set $\{a, b, c\}$ are allowed, the list of permutations given above must be expanded to include permutations such as aaa, aab, aac, aca, etc. These permutations differ from those discussed earlier in that any one of the n elements in the set may appear at any point in the permutation, regardless of whether or not it has previously occurred. In this example, repetition of the elements a, b, and c is allowed.

The formula for determining the number of permutations of r elements selected from a set of n elements assuming repetition of the r elements is derived from the fundamental rule of counting. Since there are n ways that the first element can be selected, n ways that the second element can be selected, n ways that the third element can be selected, and so forth, there are $n \cdot n \cdot n \cdots \cdot n$ ways that the r elements can be selected. The formula $n \cdot n \cdot n \cdots \cdot n$ is merely n multiplied by itself r times, i.e., n^r. The total number of permutations of r elements from a set of n distinguishable elements if repetitions of the r elements are allowed is

$$_nP^r = n^r \tag{4.8}$$

To illustrate, consider again the set of letters $\{a, b, c\}$. If repetitions are allowed, the number of permutations obtained by selecting two of the three elements in the set is given by Eq. (4.8) as

$$_3P^2 = 3^2 = 9$$

The nine permutations are

$$aa \qquad bb \qquad cc$$
$$ab \qquad bc \qquad cb$$
$$ac \qquad ba \qquad ca$$

Example. We define the set of digits as $\{0, 1, 2, 3, 4, 5, 6, 7, 8, 9\}$. Assuming that repetition of the digits is allowed, determine the number of permutations of three elements selected from this set of ten elements.

From Eq. (4.8), the number of permutations is

$$_{10}P^3 = 10^3 = 1000$$

The 1000 permutation can be described by the set

$$\{000, 001, 002, 003, \ldots, 998, 999\}$$

Example. A stock analyst ranks stocks for growth potential using the letters A, B, C, and D where A represents the highest potential and D the lowest. Six stocks are to be ranked. How many different permutations of rankings are possible?

$$_4P^6 = 4^6 = 4096$$

Example. A certain game of chance involves rolling four dice and observing the outcome of the roll. How many different outcomes are possible?

A die has six sides, designated as one through six. Since four dice are rolled, the number of possible outcomes is

$$_6P^4 = 6^4 = 1296$$

Example. A single card is withdrawn from a deck of 52 playing cards. The number and suit of the card are observed. The card is then returned to the deck and the deck is shuffled. The same procedure is repeated two additional times. How many different selections of three cards are possible? Assume that the order in which the cards are selected is important.

$$_{52}P^3 = 52^3 = 140,608$$

4.3   Combinations

In our discussion of permutations, we stated that a permutation was an ordered arrangement of objects. When ranking four wines, each different possible ranking is a permutation. When awarding first, second, and third place prizes in a contest in which there are ten entrants, each different possible selection of first, second, and third place winners is a permutation. In problems such as these, the order of selection or arrangement of the objects is important. When this is the case, we are dealing with permutations.

We now turn to a slightly different problem, namely, that of determining the number of ways r objects can be selected from a set of n objects when the *order* in which the r objects are selected *is not important*. For instance, consider an experiment that involves selecting three committee members from a list of ten candidates. In this case we are interested in the number of different groups of three objects selected from a set of ten objects. Since a committee consisting of, say, Jones, Smith, and Rogers is the same as a committee consisting of Smith, Jones, and Rogers, the order of selecting the individual members of the committee is of no importance.

A group of r objects selected from a set of n objects in which the order of selection is of no importance is referred to as a *combination* of the objects. To develop the formula for the number of combinations, consider the preceding illustration of selecting three committee members from a list of ten candidates. From Eq. (4.6), the number of different permutations of three objects selected with regard to order from a set of ten objects is

$$_{10}P_3 = \frac{10!}{7!} = 720$$

Assume that Smith, Jones, and Rogers are included in the list of ten candidates. The permutations that include Smith, Jones, and Rogers are $\{(S, J, R), (S, R, J), (R, S, J), (R, J, S), (J, R, S), (J, S, R)\}$.

The six permutations that include Smith, Jones, and Rogers represent, however, only one combination (i.e., committee). Since the same result would be true for any combination of three of the ten objects, the number of combinations is only one-sixth as large as the number of permutations. The number of different three-man committees is, therefore, $\frac{720}{6}$, or 120.

This result can be generalized by recognizing that the number of permutations per combination is $r!$. Thus, in the preceding problem there are $3! = 3 \cdot 2 \cdot 1$, or six permutations per combination. The number of possible combinations is therefore given by dividing the number of permutations by the number of permutations per combination. The formula is

$$_nC_r = \frac{n!}{(n - r)! \, r!} \tag{4.9}$$

Example. Use Eq. (4.9) to determine the number of ways of selecting three committee members from a list of ten candidates.

$$_{10}C_3 = \frac{10!}{7! \, 3!} = \frac{10 \cdot 9 \cdot 8}{3 \cdot 2 \cdot 1} = 120$$

In certain formulas used in probability, the notation $_nC_r$ is cumbersome. For this reason, an alternative notation is often used. The notation is $\binom{n}{r}$. The number of combinations of r objects selected from a set of n objects can thus also be expressed as

$$\binom{n}{r} = \frac{n!}{(n - r)! \, r!} \tag{4.10}$$

Equations (4.9) and (4.10) are, of course, the same.[1]

---

[1] The numbers given by Eq. (4.10) are termed *binomial coefficients*. Binomial coefficients are discussed in Sec. 4.4.

Example. Seven students wish to select four of their group to do research on a team project. How many different combinations are possible?

$$_7C_4 = \frac{7!}{3!\,4!} = 35$$

Example. A grocery chain has 14 stores. Management decides to try out a new product in four of the stores. How many different possible four-store samples are there?

$$_{14}C_4 = \frac{14!}{10!\,4!} = 1001$$

Example. From a list of 12 applicants for a job, five are to be selected. How many different combinations of five selected from 12 are possible?

$$_{12}C_5 = \frac{12!}{7!\,5!} = 792$$

Example. The product line of a certain company includes exactly 100 different products. From these products, five are to be selected for a special advertising campaign. How many different combinations of the five products could be selected?

This problem was posed in the introductory remarks in this chapter. The solution, from Eq. (4.9), is

$$_{100}C_5 = \frac{100!}{95!\,5!} = 75,287,520$$

Example. A poker hand of five cards is dealt from a deck of 52 playing cards. How many different five-card hands are possible?

Assuming that the order in which the cards are dealt is not important, the number of different five-card hands is

$$_{52}C_5 = \frac{52!}{47!\,5!} = 2,591,460$$

Example. A binary number contains only the digits 0 and 1. How many ten-digit binary numbers can be formed from six 1's and four 0's? Assume that the first digit is a 1.

This problem was solved on p. 117 by using the equation for permutations involving indistinguishable objects. The solution from Eq. (4.7) was

$$_9P_9(5, 4) = \frac{9!}{5! \, 4!} = 126$$

The problem can also be solved by using the formula for combinations, Eq. (4.9). To illustrate, consider the ten-digit binary number as ten separate slots into which are placed 0's and 1's. Since the first digit is a 1, five 1's and four 0's must be placed in the remaining nine slots. Suppose that we select five of the nine slots and place a 1 in each of these slots and 0's in the remaining four slots. The number of different ways of selecting five items (i.e., slots) from a group of nine items, assuming that the order of selection is not important, is given by Eq. (4.9) as

$$_9C_5 = \frac{9!}{4! \, 5!} = 126$$

There are 126 ways of distributing the 1's among five of the nine slots with 0's in the remaining four slots. Since each way is a separate binary number, there are 126 possible binary numbers. This solution is the same as was found by using the formula for permutations involving indistinguishable objects.

The preceding example illustrates that permutations involving two categories of indistinguishable objects can be considered a problem of combinations. In the special case of r objects of one type and $(n - r)$ objects of a second type, the number of distinguishable permutations is given by Eq. (4.7) as

$$_nP_n((n - r), r) = \frac{n!}{(n - r)! \, r!} \qquad (4.11)$$

Comparing Eq. (4.11) with Eq. (4.9),

$$_nC_r = \frac{n!}{(n - r)! \, r!} \qquad (4.9)$$

we see that the two formulas are the same. The choice of which formula to use thus depends on whether the reader interprets the problem as that of permutations involving two categories of indistinguishable objects or of combinations of r objects selected from n objects.

4.3.1 ORDERED PARTITIONS

As a variation in the problem of determining the number of combinations of r objects selected from a set of n objects, consider the following.

Example. A group of 11 professors is to be divided into two committees. The first committee, consisting of five professors, will discuss course content.

The second committee, consisting of the remaining six professors, will discuss grading procedures. In how many ways can the two committees be formed? The 11 professors are divided or partitioned into two groups, five in the first group and six in the second. The selection of five professors for the first committee automatically fixes the membership of the second committee. Consequently, we only need to determine the number of ways of selecting the first five. The number of ways of selecting five professors without regard to order from a group of 11 is given by Eq. (4.9).

$$_{11}C_5 = \frac{11!}{6!\,5!} = 462$$

The two committees can be formed in a total of 462 different ways.

In our example, 11 professors were divided into two committees. When a set of n objects is divided into two or more subsets, the set is said to be *partitioned*. Dividing the set of n objects into two subsets results in a twofold partition. Similarly, dividing a set of n objects into r subsets results in an r-fold partition. To illustrate a threefold partition, consider the following example.

Example. A group of 11 professors is to be divided into three committees. The first committee, consisting of four professors, will discuss course content. The second committee, consisting of three professors, will discuss grading procedures. The third committee, consisting of the remaining four professors, will discuss admission requirements. In how many ways can the committees be formed?

The number of ways of selecting four professors for the first committee from the set of 11 professors is $_{11}C_4$. Once these four have been selected, seven professors remain in the set. The number of ways of selecting three professors from the seven remaining in the set is $_7C_3$. Selection of the three professors for the second committee automatically leaves four professors for the third committee. From the fundamental rule of counting, the total number of ways of partitioning the set of 11 professors into three subsets is

$$_{11}C_4 \cdot {_7C_3} = \frac{11!}{7!\,4!} \cdot \frac{7!}{4!\,3!} = \frac{11!}{4!\,3!\,4!} = 11{,}550$$

The set of 11 professors can be partitioned into three subsets with four in the first subset, three in the second subset, and four in the third subset in 11,550 different ways.

The partitions of the set of professors into two subsets in the first example and three subsets in the second example are called *ordered partitions*. The term ordered partition applies when the association of elements with particular subsets is important. To illustrate, assume that we have the set

$$U = \{a, b, c, d\}$$

that is to be partitioned into subsets A and B with $n_1 = 2$ elements in A and $n_2 = 2$ elements in B. One possible partition is

$$A = \{a, b\} \qquad B = \{c, d\}$$

and another possible partition is

$$A = \{c, d\} \qquad B = \{a, b\}$$

Since we are concerned with the association of particular elements with particular subsets, the two partitions given above differ. The association of elements a and b with subset A in the first partition differs from the association of elements a and b with B in the second partition. When, as in the examples of dividing professors into subcommittees for discussing particular topics, the association of a particular professor with a specific topic is important, the partition is ordered. If this is not the case, i.e., if the association of particular elements together in any subset were all that mattered, the partition would be *unordered*. The two ordered partitions given above thus represent only one unordered partition.

For a set of n elements partitioned into k subsets with n_1 elements in the first subset, n_2 elements in the second subset, etc., and n_k elements in the kth subset, the possible number of ordered partitions is

$$_nC_{n_1, n_2, \ldots, n_k} = \frac{n!}{n_1! \, n_2! \cdots n_k!} \tag{4.12}$$

where $n_1 + n_2 + \cdots + n_k = n$. An alternative expression of this formula is

$$\binom{n}{n_1, n_2, \ldots n_k} = \frac{n!}{n_1! \, n_2! \cdots n_k!} \tag{4.13}$$

The two formulas are, of course, equivalent.[2]

As an illustration of Eq. (4.12), consider again the problem of dividing 11 professors into three committees with four professors on the first committee, three professors on the second committee, and four professors on the third committee. From the formula for ordered partitions, the possible number of different committees is

$$_{11}C_{4,3,4} = \frac{11!}{4! \, 3! \, 4!} = 11{,}550$$

Example. Nine persons are traveling in three cars from Los Angeles to San Francisco. If four persons travel in the first car, three in the second car,

---

[2] The numbers given by Eqs. (4.12) and (4.13) are called *multinomial coefficients*.

and two in the third car, how many different traveling assignments are possible?

$$_9C_{4,3,2} = \frac{9!}{4!\,3!\,2!} = 1260$$

Example. A student learned that his professor assigned three A's, four B's, four C's, and one D in a class of 12 students. In how many different ways could these grade assignments be made?

$$_{12}C_{3,4,4,1} = \frac{12!}{3!\,4!\,4!\,1!} = 138,600$$

Example. The football team at Thomas Jefferson High School has ten games scheduled next season. In how many different ways can the season end with six wins, three losses, and one tie?

$$_{10}C_{6,3,1} = \frac{10!}{6!\,3!\,1!} = 840$$

Example. A firm has eight customers and three salesmen, Smith, Jones, and Brown. The sales manager wants to assign three customers to Smith, three customers to Jones, and two customers to Brown. In how many different ways can the assignments be made?

$$_8C_{3,3,2} = \frac{8!}{3!\,3!\,2!} = 560$$

Example. A group of nine executives plans to divide into three subgroups of three each for the purpose of discussing methods of increasing profits. How many different subgroups are possible?

This problem seems similar to those discussed above, but there is one important difference. In the preceding examples, the association of the elements with a particular subset was important. This is not the case in our current problem. For any given subset of three executives, it makes no difference whether they are collectively in the first subgroup, the second subgroup, or the third subgroup. They are the same subset of executives and will carry on the same discussion regarding methods of increasing profits. Our problem, therefore, is that of determining the number of unordered partitions rather than the number of ordered partitions.

The number of unordered partitions can be determined by recognizing that there are 3! or six ways that the three subsets can be permuted among themselves. The total number of unordered partitions is thus one-sixth as large as the number of ordered partitions. The number of possible subgroups is

$$\tfrac{1}{6}(_9C_{3,3,3}) = \frac{1}{6}\left(\frac{9!}{3!\,3!\,3!}\right) = 280$$

Example. Modify the preceding example by assuming that ten executives break up into groups of four, three, and three.

We again have unordered partitions. This example differs from the preceding one, however, in that only the two subgroups of three executives can be permuted among themselves. The total number of ordered partitions is therefore 2! or two times larger than the number of unordered partitions. The number of possible subgroups is

$$\tfrac{1}{2}(_{10}C_{4,3,3}) = \frac{1}{2}\left(\frac{10!}{4!\,3!\,3!}\right) = 2100$$

4.3.2 SOME INTERESTING COMBINATORIAL PROBLEMS

Many students find a number of the combinatorial problems that involve games of chance very interesting. For instance, the aspiring poker player may be interested in the number of ways that a full house can be dealt. The bridge addict could certainly amaze his opponents by showing them how to calculate the number of different possible bridge hands. Even the coin tosser might benefit by finding the number of ways that three heads can occur in six tosses of a coin. These types of problems are discussed in this section. The reader may find that an understanding of these problems could allow him to recover the price of this textbook from his less knowledgeable opponents.

Suppose that we begin by determining the number of ways a full house can be dealt in the game of poker. For the nonpoker player, a full house occurs when the five cards held by the poker player include one pair (i.e., two cards of one denomination) and three of a kind (i.e., three cards of another denomination). As an example, one particular full house consisting of the jack of hearts, the jack of clubs, the five of spades, the five of diamonds, and the five of clubs is shown by shading in Fig. 4.4.

To determine the number of different ways a full house can be dealt, we observe that for any particular row in Fig. 4.4, three of a kind can be dealt in $_4C_3$ or four ways. Since there are 13 rows, the number of ways of dealing three of a kind is 13(4) or 52. The hand is completed by a pair from one of the remaining 12 rows. The number of ways of dealing this pair is $12(_4C_2)$ or 72 ways. From the fundamental rule of counting, the number of different ways a full house can be dealt is $52 \cdot 72$ or 3744.

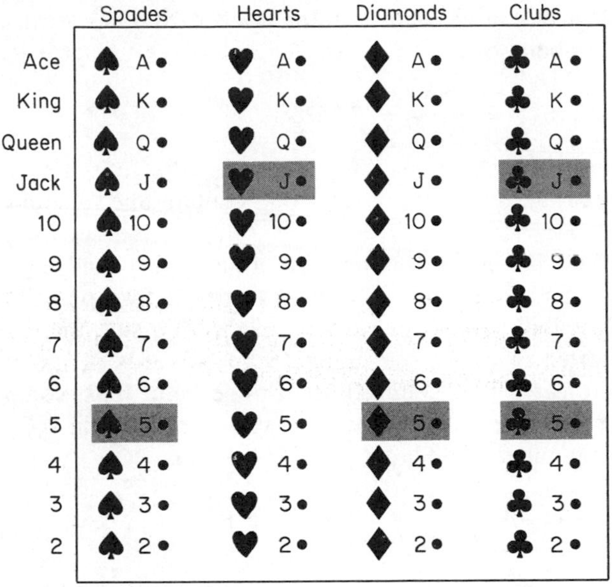

	Spades	Hearts	Diamonds	Clubs
Ace	♠ A •	♥ A •	♦ A •	♣ A •
King	♠ K •	♥ K •	♦ K •	♣ K •
Queen	♠ Q •	♥ Q •	♦ Q •	♣ Q •
Jack	♠ J •	♥ J •	♦ J •	♣ J •
10	♠ 10 •	♥ 10 •	♦ 10 •	♣ 10 •
9	♠ 9 •	♥ 9 •	♦ 9 •	♣ 9 •
8	♠ 8 •	♥ 8 •	♦ 8 •	♣ 8 •
7	♠ 7 •	♥ 7 •	♦ 7 •	♣ 7 •
6	♠ 6 •	♥ 6 •	♦ 6 •	♣ 6 •
5	♠ 5 •	♥ 5 •	♦ 5 •	♣ 5 •
4	♠ 4 •	♥ 4 •	♦ 4 •	♣ 4 •
3	♠ 3 •	♥ 3 •	♦ 3 •	♣ 3 •
2	♠ 2 •	♥ 2 •	♦ 2 •	♣ 2 •

Figure 4.4.

Example. Determine the number of different ways that four of a kind can be dealt in the game of poker.

Four of a kind occurs when four of the five cards in a poker player's hand have the same denomination, e.g., four jacks, four queens, etc. To determine the number of different poker hands of four of a kind, we observe that the four cards must come from one of the 13 rows in Fig. 4.4. Since the fifth card in the hand may be any one of the remaining 48 cards in the deck, there are $13 \cdot 48$ or 624 different poker hands of four of a kind.

Turning from poker to the game of bridge, suppose that we first calculate the number of different possible bridge hands. For the nonbridge player, a bridge hand consists of 13 cards dealt from a standard deck of 52 cards. Since the bridge player is not concerned with the order in which the cards are dealt, the number of different bridge hands is simply

$$_{52}C_{13} = \frac{52!}{39! \ 13!} = 635,013,559,600$$

or approximately 635 billion.

It might also be interesting to determine the possible number of different bridge hands that contain only honor cards. An honor card is a ten, jack, queen, king, or ace. Since there are 20 honor cards, we must determine the number of ways that 13 cards can be dealt from the total of 20 honor cards. From the formula for combinations,

$$_{20}C_{13} = \frac{20!}{7! \; 13!} = 77{,}520$$

Example. How many bridge hands are possible that contain one void and ten honor cards?

A void occurs when there are no cards of one suit in the bridge hand, e.g., no clubs, no spades, etc. Since there are four suits, there are four possible voids. Eliminating the possibility of honor cards from the void suit, the ten honor cards can be selected from the remaining 16 honor cards in $_{16}C_{10}$ different ways. The three cards that complete the bridge hand must come from the 23 remaining nonhonor cards. These cards can be selected in $_{23}C_3$ different ways. The total number of different possible hands is

$$4(_{16}C_{10})(_{23}C_3) = 4 \left(\frac{16!}{6! \; 10!} \right)\left(\frac{23!}{20! \; 3!} \right) = 8{,}103{,}096$$

We next consider a simple problem in tossing coins. Suppose that we toss a coin six times and record the outcome of each toss as either a head, H, or a tail, T. The total number of possible sequences of H and T is given from the equation for permutations allowing repetitions, Eq. (4.8), and is

$$_2P^6 = 2^6 = 64$$

It might be interesting to determine how many of these 64 different sequences contain exactly three heads. This problem is equivalent to determining the number of distinguishably different ways of arranging three H's and three T's. From the formula for permutations involving indistinguishable objects, Eq. (4.7), the number of distinguishably different permutations is

$$_6P_6(3, 3) = \frac{6!}{3! \; 3!} = 20$$

We shall show in Chap. 5 that the probability of getting three heads in six tosses is 20/64 or 0.3125.

Example. Suppose that you are offered the following game. You and a friend each put five quarters in a hat. The ten quarters are then tossed from the hat onto a flat surface. If exactly four, five, or six of the ten quarters come up heads, your friend pockets all ten quarters. If this does not happen, you keep all ten quarters. Would you like to play this game?

The total number of possible outcomes of tossing ten quarters is given by the formula for permutations allowing repetitions and is

$$_2P^{10} = 2^{10} = 1024$$

The number of ways that four heads can occur is given by the formula for permutations involving indistinguishable objects and is

$$_{10}P_{10}(4, 6) = \frac{10!}{4! \; 6!} = 210$$

Similarly, the number of ways of obtaining five heads is $_{10}P_{10}(5, 5)$ or 252 and the number of ways of obtaining six heads is $_{10}P_{10}(6, 4)$ or 210. The total number of ways of obtaining four, five, or six heads is thus (210 + 252 + 210) or 672. Since there are only (1024 − 672) or 352 ways of obtaining any other result, it would be best to pass up this game.

Since not everyone enjoys games of chance, suppose that we conclude this section with a problem involving astrological signs. The setting is a romantic spot, the music is turned low, and the conversation has drifted to astrology. You and your date had planned on spending the evening alone but, unfortunately, three friends have dropped by. One of these friends seems knowledgeable of the characteristics of people born under different astrological signs. As part of the conversation, he mentions that he has a feeling that at least two of the people in the group have the same sign. Since there are 12 astrological signs and only five people in the group, your first reaction is to doubt the likelihood of your friend's feeling being correct. But, is your first reaction right?

To answer this question, we first determine how many different sequences of astrological signs are possible. Since there are 12 signs, there are 12 possibilities for the sign of the first person, 12 possibilities for the sign of the second person, etc. From the formula for permutations allowing repetitions, there are

$$_{12}P^5 = 12^5 = 248,842$$

different sequences. Of these 248,842 different sequences, we next determine how many do not contain any repetitions. Assuming no repetitions, there are 12 possibilities for the sign of the first person, 11 possibilities for the sign of the second person, 10 possibilities for the sign of the third person, etc. From the formula for permutations, there are

$$_{12}P_5 = \frac{12!}{7!} = 95,040$$

sequences that exclude repetitions. This means that there are (248,842 − 95,040) or 153,792 sequences that include repetitions. Since there are 153,792 sequences that include at least two signs out of a total of 248,842 possible sequences, the chances are slightly better than 60 percent that your friend's feeling is right.

4.4 THE BINOMIAL THEOREM[3]

The *binomial theorem* provides a method for expanding the binomial expression. The binomial expression is

$$(x + y)^n \qquad (4.14)$$

where x and y are variables and n is any number. Because the binomial expression appears very often in probability as well as in other branches of mathematics, it is worthwhile to introduce the theorem that is used to evaluate the binomial expression. In this discussion, we shall consider only the most common case in which the exponent in the expression, n, is 0 or a positive integer.

To develop the binomial theorem, consider the following expansions of the binomial expressions:

$$
\begin{aligned}
(x + y)^0 &= 1 \\
(x + y)^1 &= x + y \\
(x + y)^2 &= x^2 + 2xy + y^2 \\
(x + y)^3 &= x^3 + 3x^2y + 3xy^2 + y^3 \\
(x + y)^4 &= x^4 + 4x^3y + 6x^2y^2 + 4xy^3 + y^4 \\
(x + y)^5 &= x^5 + 5x^4y + 10x^3y^2 + 10x^2y^3 + 5xy^4 + y^5 \\
(x + y)^6 &= x^6 + 6x^5y + 15x^4y^2 + 20x^3y^3 + 15x^2y^4 + 6xy^5 + y^6
\end{aligned}
$$

Since our objective is to develop a method of expanding the binomial expression for any positive integer value of n, we must determine if there are any discernible patterns in the expansions. One pattern is apparent, namely, in going from left to right the powers of x decrease from n to 0 and the powers of y increase from 0 to n. Thus, for any value of n, the expansion has the form

$$x^n + (\quad)x^{n-1}y + (\quad)x^{n-2}y^2 + \cdots + (\quad)x^{n-r}y^r + \cdots + y^n$$

where ( ) represents the missing coefficient.

The values of the coefficients of the individual terms in the expansion can be calculated by using the formula for the number of combinations of r objects selected from a set of n objects. This formula was developed in Sec. 4.3 and is

$$\binom{n}{r} = \frac{n!}{(n-r)!\,r!} \qquad (4.10)$$

The coefficients given by $\binom{n}{r}$ are called the *binomial coefficients*.

---

[3] This section may be omitted without loss of continuity.

The binomial expression $(x + y)^n$ can now be expanded. The expansion is

$$(x + y)^n = \binom{n}{0} x^n + \binom{n}{1} x^{n-1}y^1 + \binom{n}{2} x^{n-2}y^2 + \cdots$$

$$+ \binom{n}{r} x^{n-r}y^r + \cdots + \binom{n}{n-1} x^1 y^{n-1} + \binom{n}{n} y^n$$

$$(4.15)$$

An alternative and more compact way of writing Eq. (4.15) is

$$(x + y)^n = \sum_{r=0}^{n} \binom{n}{r} x^{n-r}y^r \qquad (4.16)$$

Equations (4.15) and (4.16) represent an algebraic statement of the binomial theorem.

Example. Use the binomial theorem to expand the binomial expression $(x + y)^6$.

The expansion is

$$(x + y)^6 = \binom{6}{0} x^6 + \binom{6}{1} x^5 y + \binom{6}{2} x^4 y^2$$

$$+ \binom{6}{3} x^3 y^3 + \binom{6}{4} x^2 y^4 + \binom{6}{5} xy^5 + \binom{6}{6} y^6$$

$$= x^6 + 6x^5 y + 15x^4 y^2 + 20x^3 y^3 + 15x^2 y^4 + 6xy^5 + y^6$$

Example. Use the binomial theorem to expand the binomial expression $(x + y)^7$.

$$(x + y)^7 = \binom{7}{0} x^7 + \binom{7}{1} x^6 y + \binom{7}{2} x^5 y^2 + \binom{7}{3} x^4 y^3$$

$$+ \binom{7}{4} x^3 y^4 + \binom{7}{5} x^2 y^5 + \binom{7}{6} xy^6 + \binom{7}{7} y^7$$

$$= x^7 + 7x^6 y + 21x^5 y^2 + 35x^4 y^3$$
$$+ 35x^3 y^4 + 21x^2 y^5 + 7xy^6 + y^7$$

The computations required to determine the binomial coefficients can be reduced considerably by recognizing that

$$\binom{n}{r+1} = \binom{n}{r}\binom{n-r}{r+1} \qquad (4.17)$$

Equation (4.17) states that the binomial coefficient $\binom{n}{r+1}$ is equal to the product of the preceding coefficient, $\binom{n}{r}$ and $\left(\frac{n-r}{r+1}\right)$. To illustrate, consider again the expansion of $(x+y)^7$. The binomial coefficient for the first term in the expansion is $\binom{7}{0}$ or 1. From Eq. (4.17), the binomial coefficient of the second term is $\binom{7}{1} = \binom{7}{0}\binom{7}{1}$ or 7. Similarly, the binomial coefficients of the remaining terms are

$$\binom{7}{2} = \binom{7}{1}\binom{6}{2} = 7(3) = 21$$

$$\binom{7}{3} = \binom{7}{2}\binom{5}{3} = 21\left(\frac{5}{3}\right) = 35$$

$$\binom{7}{4} = \binom{7}{3}\binom{4}{4} = 35(1) = 35$$

$$\binom{7}{5} = \binom{7}{4}\binom{3}{5} = 35\left(\frac{3}{5}\right) = 21$$

$$\binom{7}{6} = \binom{7}{5}\binom{2}{6} = 21\left(\frac{2}{6}\right) = 7$$

$$\binom{7}{7} = \binom{7}{6}\binom{1}{7} = 7\left(\frac{1}{7}\right) = 1$$

This relationship is especially easy to apply if one recognizes that the coefficient of the kth binomial term is equal to the coefficient of the $(k-1)$ term multiplied by the power of x in the $(k-1)$ term and divided by the power of y in the kth term. Thus, if the $(k-1)$ term is $21x^5y^2$, the coefficient of the kth term is 21(5)/3 or 35.

Example. Use Eq. (4.7) to determine the binomial coefficients for the binomial expression $(x+y)^6$.

$$(x+y)^6 = x^6 + 6x^5y + \frac{6(5)}{2}x^4y^2 + \frac{15(4)}{3}x^3y^3$$

$$+ \frac{20(3)}{4}x^2y^4 + \frac{15(2)}{5}xy^5 + \frac{6(1)}{6}y^6$$

or by simplifying,

$$(x+y)^6 = x^6 + 6x^5y + 15x^4y^2 + 20x^3y^3 + 15x^2y^4 + 6xy^5 + y^6$$

Example. Show that Eq. (4.17) is true.

Writing the left side of Eq. (4.17) in factorial notation, we obtain

$$\binom{n}{r+1} = \frac{n!}{(n-r-1)!\,(r+1)!}$$

The right side of Eq. (4.17), also expressed in factorial notation, is

$$\binom{n}{r}\binom{n-r}{r+1} = \frac{n!}{(n-r)!\,r!} \cdot \frac{(n-r)}{(r+1)} = \frac{(n-r)}{(n-r)!} \cdot \frac{n!}{(r+1)r!}$$

Since $(n-r)/(n-r)!$ is equal to $1/(n-r-1)!$ and $(r+1)r!$ is equal to $(r+1)!$, the right side of Eq. (4.17) reduces to

$$\frac{n!}{(n-r-1)!\,(r+1)!}$$

The left and right sides of Eq. (4.17) are the same; therefore, the equation is true.

An alternative method for calculating the binomial coefficients was developed by Blaise Pascal.[4] Pascal arranged the binomial coefficients in the manner shown in Fig. 4.5. This arrangement, termed *Pascal's triangle*, shows the values of the binomial coefficients for different values of n.

It is interesting to note that one need not expand the binomial expression to obtain the triangle. Instead, we observe that the border of the triangle is made up of 1's and that each interior number is equal to the sum of the two numbers which appear on the line above just to the left and to the right of the number. Examples are shown by the shaded triangles in Fig. 4.5.

Pascal's triangle provides a quick method of determining the binomial coefficients for relatively small values of n. For large values, the triangle

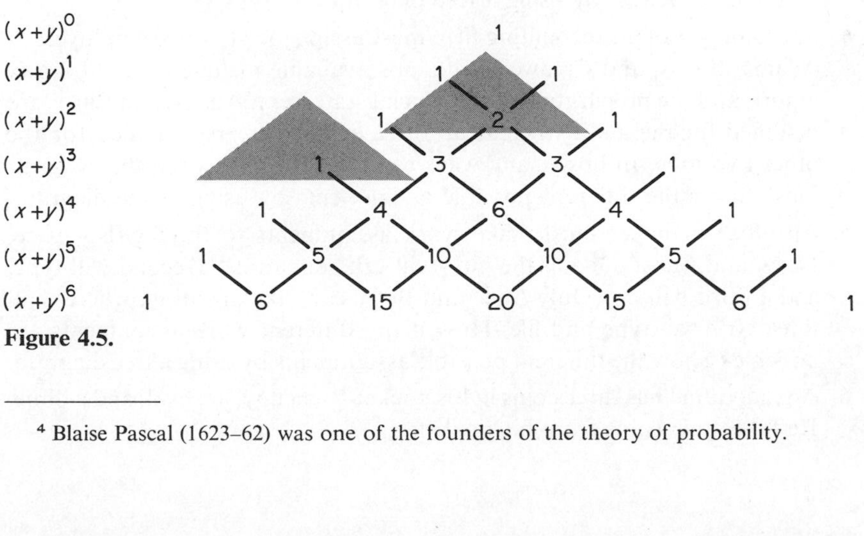

Figure 4.5.

---

[4] Blaise Pascal (1623–62) was one of the founders of the theory of probability.

becomes unwieldly and it is better to use Eq. (4.10). Of course, if one needs to calculate these values very often, there is nothing as practical as a table of binomial coefficients.

Example. Use Pascal's triangle to calculate the binomial coefficients for $(x + y)^7$.

From Fig. (4.5), the new entries are

$$1, 1 + 6 = 7, 6 + 15 = 21, 15 + 20 = 35,$$
$$20 + 15 = 35, 15 + 6 = 21, 6 + 1 = 7, 1$$

PROBLEMS

1. In a certain election, Smith and Jones are candidates for president, Brown, Rogers, and Pierce are candidates for vice-president, and Reed and Thomas are candidates for secretary. How many different slates of officers are possible? Show these different slates by using a tree diagram.

2. An employer wants to hire three college graduates—one in accounting, one in finance, and one in marketing. Allen and Andrews have applied for the job in accounting, Freeman and Fuller for the job in finance, and Mathews, Mayer, and Mendez for the job in marketing. In how many ways can the jobs be filled? Show the different possibilities as elements of the Cartesian product of three sets.

3. Bob and John agree to play a tennis match consisting of three sets. The player who wins two of the three sets is the winner of the match. In how many different ways can a player win the match? Show these different sequences by using a tree diagram.

4. The manager of an accounting firm must assign jobs to three employees— Adams, Burns, and Caldwell. The jobs available include an audit, a tax report, and the preparation of a financial statement. Adams and Burns are qualified for the audit whereas all three employees are qualified for the other two jobs. In how many ways can the three employees be assigned jobs? Show the different possible assignments by using a tree diagram.

5. An office manager must make work assignments to three girls—Alice, Betty, and Carol. One of the girls will take dictation, a second will type, and a third will file. Only Alice and Betty can take dictation, whereas all three girls can type and file. How many different work assignments are possible? Show the different possible assignments by using a tree diagram.

6. An individual has three coins in his pocket—a penny, a nickel, and a dime. He takes two coins from his pocket.

(a) Construct a tree diagram that shows the possible order of selection, assuming that the first coin is replaced before the second coin is selected.

(b) Construct a tree diagram showing the possible order of selection, assuming that the first coin is not replaced before the second coin is selected.

7. An individual is planning to purchase a high-fidelity music system. He is considering the following options:

   Make: Marantz, Harmon–Kardon
   Channels: 2-channel (stereo), 4-channel (multiplex)
   Auxiliary: tape deck, record changer, both tape deck and changer

   In how many ways may he specify the options?

8. A particularly unfortunate student has six assignments that must be completed over the weekend. The most pressing assignment is to read a chapter in his finite math text. After reading this chapter he can either work the problems that accompany the chapter or write a short term paper for his history class. After completing these assignments, he must study for exams in economics and political science and read a chapter in psychology. In how many different ways can the assignments be completed?

9. Before leaving on vacation, the manager of a men's shoe store told his inexperienced assistant to order ten pairs each of 30 different styles of shoes. In placing the order, the assistant assumed that the manager wanted ten pairs of each width and each length in each style. If the shoes were ordered in widths A, B, C, D, E, EE, and EEE and lengths 7, $7\frac{1}{2}$, 8, $\cdots$, 13, how many pairs of shoes were ordered?

10. A machine shop has two milling machines, three lathes, six drill presses, and three grinders. In how many ways can a part be routed that must first be milled, then turned on a lathe, then drilled, and then ground?

11. Determine the number of different ways a part can be routed in Problem 10 if the four operations can be performed in any order.

12. Four assignments are to be made to four equally qualified employees. In how many different possible ways can the four assignments be made?

13. Six company presidents have been asked to speak at an upcoming conference. How many different speaking orders are possible?

14. A personnel manager plans on visiting ten college campuses during the coming year to interview for prospective employees. In how many ways can the ten trips be ordered?

15. In how many ways can five people line up for a group photograph?

16. Four traveling salesmen are working from door to door in the same rural area. Assuming that no two of them arrive simultaneously, in how many different orders may they call on a particular farmer's daughter?

17. In a marketing survey, a respondent is asked to compare five different brands of coffee and to state her first three choices in order of preference. How many different rankings are possible?

18. A Scrabble player has eight different letters on his rack. He decides to test all six letter permutations of the eight letters before making the next move. If it takes one second to form each permutation, how long must his opponent wait before the next move?

19. How many fraternity names can be formed from the 24 letters of the Greek alphabet if each name consists of three letters and no letter is repeated in a name?

20. A secret code is to be designed so that a particular sequence of two different symbols conveys a certain message. For instance, the sequence %# might mean "the British are coming." If 156 messages are required, how many different symbols are needed to form the code?

21. A construction foreman has five jobs that must be completed and five men available to assign to the jobs. In how many different ways can the foreman assign the men to the jobs?

22. A company has divided the nation into ten different sales territories. The sales manager plans on visiting five of the different sales territories during the coming year. In how many different ways can the five visits be scheduled?

23. Which can be arranged in more distinguishable ways, the letters in the word "professor" or the letters in the word "student"?

24. In how many distinguishable ways can the following group of words be arranged?

    the, the, the, to, to, is, of, time, men, for, party, all, aid, come, now, good

25. How many nine-digit binary numbers that contain six 1's and three 0's are possible? Assume that the first digit is a 1.

26. In how many distinguishable ways can three x's, three y's, and three z's be arranged?

27. A football game is sponsored on television by a tire company, a razor blade company, and a clothing manufacture. The contracts between the television network and the companies call for 15 one-minute commercial announcements, five commercial announcements for the tire company, four for the razor blade company, and six for the clothing manufacture.
    (a) If each company provides only one advertisement for its product, in how many different sequences can the 15 commercials be aired?

(b) If each company provides different advertisements (i.e., five different tire ads, four different razor blade ads, and six different clothing ads), in how many different sequences can the 15 commercials be aired?

28. Four couples get together once a month for dinner followed by a game of bridge. If each couple acts as host three times each year, in how many different ways can the dinner and bridge games be scheduled?

29. A true–false examination has eight questions. Assuming that a student can mark each question true or false, in how many ways can a student guess at the answers?

30. If a student is totally unprepared for an examination, should he prefer a test with 12 true–false questions or a test with eight multiple choice questions? Assume that each multiple choice question has three possible answers.

31. A telephone number has seven digits. How many telephone numbers are possible? How many numbers are possible if the first digit cannot be a zero or a one?

32. Suppose that an individual prepares a chart that shows his direct ancestors for the past ten generations. Including himself, how many names would be on the chart?

33. From a group of 15 products, five are to be selected for a special promotional sale. How many different combinations of five products are possible?

34. Three products are to be selected from a group of ten products for market testing. How many different three-product combinations are possible?

35. A group of 12 men attend a conference. If each man shakes hands with every other man, how many handshakes are exchanged?

36. An airline serves 20 major cities. If the airline provides direct service between every possible pair of cities, how many different routes does the airline fly?

37. A group of eight bridge players must be arranged into two tables of bridge. In how many different ways can this be done? Assume that two arrangements are the same if a player has the same partner and the same pair of opponents.

38. A class of 12 students is to be divided into three groups of four students, each to work on different team projects. In how many different ways can the groups be arranged?

39. Suppose in Problem 38 that each group is assigned the same project. In how many different ways can the groups be arranged?

40. A police academy is graduating a class of 14 policemen. In how many ways can the policemen be assigned to four precincts if five policemen are

assigned to one precinct and three policemen are assigned to each of the others?

41. Yormark, McBride and Associates, a management consulting firm, has contracts to provide management consultants to seven different firms. Four consultants are available for assignments—Ralph, Ricardo, Wilhelm, and George. If management assigns two firms to Ralph, three firms to Ricardo, and one firm each to Wilhelm and George, in how many ways can the consultants be assigned?

42. Determine the number of ways three of a kind can be dealt in the game of poker.

43. Show that

$$\binom{n}{r} = \binom{n}{n-r}$$

44. Use Pascal's triangle to find the coefficients in the expansion of $(x + y)^9$.

45. Use the binomial theorem to expand $(A - 3B)^4$. (Hint: Let $x = A$ and $y = -3B$. Expand $(x + y)^4$ and substitute for x and y in the expansion.)

SUGGESTED REFERENCES

BUSH, GRACE A., and JOHN E. YOUNG, *Foundations of Mathematics* (New York, N.Y.: McGraw-Hill Book Company, Inc., 1973).

EWART, P., J. FORD, and C. LIN, *Probability for Statistical Decision Making* (Englewood Cliffs, N.J.: Prentice-Hall, Inc., 1974).

GOODMAN, A. W., and J. S. RATTI, *Finite Mathematics with Applications* (New York, N.Y.: The Macmillan Company, 1971).

KEMENY, JOHN G., et al., *Finite Mathematics with Business Applications* (Englewood Cliffs, N.J.: Prentice-Hall, Inc., 1972).

LIPSCHUTZ, SEYMOUR, *Finite Mathematics*, Schaum's Outline Series (New York, N.Y.: McGraw-Hill Book Company, Inc., 1966).

LIU, C. L., *Introduction to Combinatorial Mathematics* (New York, N.Y.: McGraw-Hill Book Company, Inc., 1968).

Chapter Five

Probability

The next two chapters of this text deal with probability and probability functions. Most students have an intuitive notion of the meaning of probability. For instance, the majority of students have placed a bet on the outcome of a football or basketball game. In placing this bet, the student either consciously or subconsciously calculated the probability or the "chances" of his team's winning. If the teams were evenly matched, one might say that the chance of winning the bet was fifty–fifty. An alternative statement is that the probability of winning was 0.5.

Most readers would agree that an understanding of probability would be very useful when placing bets in games of chance. In fact, the theory of probability was first developed to model the outcomes of games of chance (e.g., dice, roulette, cards, etc.). If, however, our purpose is not to study the intricacies of a gambling casino or the operation of a football betting pool, the question then arises as to the usefulness of probability in making decisions. The answer to this question is really quite simple. Many decisions must be based on predictions of future events. These predictions inevitably carry with them uncertainty and probable error. An understanding of the theory of probability better enables the administrator or decision maker to recognize the importance of the uncertainty in the prediction. It also enables him to take into account the effect of the probable error in making the decision.

Probability, like other mathematical techniques, is based on certain definitions and assumptions. Our task is to introduce the important definitions and assumptions and to provide realistic and important examples. We shall begin with the concept of a sample space.

5.1 Sample Space

The term *sample space* refers to the set of all possible outcomes of a statistical process or experiment. To illustrate, assume that the experiment involves tossing two coins and recording the outcome of the tosses. The sample space, or *set of possible outcomes,* is

$$S = \{(HH), (HT), (TH), (TT)\}$$

where (HH) means a head on the toss of the first coin and a head on the toss of the second coin, (HT) means a head on the toss of the first coin and a tail on the toss of the second, etc. The sample space is the set of possible outcomes of the tosses of the two coins.

The elements within the sample space are often referred to as *sample points.* If we designate the sample points by o_i, the sample space can be written as

$$S = \{o_1, o_2, o_3, \ldots, o_n\} \tag{5.1}$$

where the sample points $o_1, o_2, o_3, \ldots, o_n$ describe all possible outcomes of the statistical process or experiment. In the example of tossing two coins, the sample space was

$$S = \{(HH), (HT), (TH), (TT)\}$$

and the sample points were (HH), (HT), (TH), and (TT).

In defining the term sample space, we have used the terms *statistical process* and *experiment.* The terms statistical process and experiment are used interchangeably. Both terms refer to a process that leads to some well-defined outcome. In the preceding example, the process of tossing the two coins was the experiment or statistical process. The outcome of the experiment consisted of recording a head or a tail on the toss of the first coin and a head or a tail on the toss of the second coin. Other examples of experiments include observing the outcome from the roll of a pair of dice, recording the number of automobiles sold by a car dealer during a given week, observing whether the closing price of a particular stock is up, down, or unchanged from the previous day, and asking an individual his brand preference in color television. In each of these examples, we only require that the experiment leads to some well-defined outcome and that the set of all possible outcomes can be specified.

Example. Specify the sample space that results from the experiment of tossing three coins and recording the outcome of the toss of each coin as a head, H, or tail, T.

The sample space is

$$S = \{(HHH), (HHT), (HTH), (THH), (HTT), (THT), (TTH), (TTT)\}$$

Example. In designing a questionnaire, the income of the respondent is often important. Could the following income categories be considered as a sample space?

> Your Income (Check the appropriate box)?
>
> ☐ $0 - $4999   ☐ $5000 - $9999   ☐ $10,000 - $14,999   ☐ Over $15,000

The income categories would be a sample space. The sample space is

$$S = \{(0\text{–}4999), (5000\text{–}9999), (10,000\text{–}14,999), (over\ 15,000)\}$$

Example. Assume that the income categories in the questionnaire in the preceding example were as follows:

> Your Income (Check one box)?
>
> ☐ $0 - $4999   ☐ $4000 - $9999   ☐ $7500 - $14,999   ☐ Above $12,500

Do these categories represent a sample space?

These categories would not be a sample space. A sample space is defined as the *set* of all possible outcomes of the experiment. Since the income categories shown overlap, one of the requirements for a set is not satisfied. This requirement, it will be remembered, is that the objects in the set (i.e., the income categories) must be distinct. The categories in this example do not meet this requirement.

5.1.1   SAMPLE SPACE AND CARTESIAN PRODUCT

The concept of the Cartesian product of two or more sets was introduced in Chap. 1. The reader will remember that the Cartesian product of the sets A and B, designated as $A \times B$, was the set containing all ordered pairs (a, b) such that $a \in A$ and $b \in B$. The concept of the Cartesian product can sometimes be very useful in specifying the sample space of an experiment.

To illustrate, assume that our experiment again involves the toss of two coins. Suppose we let $A = \{H, T\}$ represent the sample space for the first toss and $B = \{H, T\}$ represent the sample space for the second toss. The sample space for the experiment of tossing the two coins is given by the Cartesian product of A and B. Thus,

$$S = A \times B = \{(HH), (HT), (TH), (TT)\}$$

It can be seen that the sample space formed by the Cartesian product of A and B is the same as that specified earlier for the toss of the two coins.

Example. Assume that an experiment involves surveying individuals with regard to their political affiliation. As a part of the survey, it is important that the following questions are answered:

1. Age
   ☐ Under 39   ☐ Over 39
2. Sex
   ☐ Male   ☐ Female
3. Political affiliation
   ☐ Democrat   ☐ Republican   ☐ Other

Use the Cartesian product of the three sets to form the sample space of the experiment.

Let A represent the set of all possible ages, B the set of possible sex, and C the set of political affiliation. Then

$$A = \{\text{under 39, over 39}\}$$

$$B = \{M, F\}$$

and

$$C = \{D, R, O\}$$

The sample space for the experiment is $S = A \times B \times C$, or

$S = \{$(under 39, M, D), (under 39, M, R), (under 39, M, O), (under 39, F, D), (under 39, F, R), (under 39, F, O), (over 39, M, D), (over 39, M, R), (over 39, M, O), (over 39, F, D), (over 39, F, R), (over 39, F, O)$\}$

Example. Assume that an experiment involves rolling a pair of dice. Use the Cartesian product of the outcome on each die to specify the sample space.

Table 5.1.

A \ B	1	2	Second Die 3	4	5	6
1	(1, 1)	(1, 2)	(1, 3)	(1, 4)	(1, 5)	(1, 6)
2	(2, 1)	(2, 2)	(2, 3)	(2, 4)	(2, 5)	(2, 6)
First 3	(3, 1)	(3, 2)	(3, 3)	(3, 4)	(3, 5)	(3, 6)
Die 4	(4, 1)	(4, 2)	(4, 3)	(4, 4)	(4, 5)	(4, 6)
5	(5, 1)	(5, 2)	(5, 3)	(5, 4)	(5, 5)	(5, 6)
6	(6, 1)	(6, 2)	(6, 3)	(6, 4)	(6, 5)	(6, 6)

Let $A = \{1, 2, 3, 4, 5, 6\}$ represent the six sides on the first die and $B = \{1, 2, 3, 4, 5, 6\}$ represent the six sides on the second die. The possible outcomes of rolling the pair of dice are given by the Cartesian product of A and B. The Cartesian product is shown by the box diagram in Table 5.1.

5.1.2   TREE DIAGRAMS

Rather than listing the elements in the sample space, it is sometimes helpful to show the possible outcomes of an experiment by the use of a tree diagram. To review the construction of a tree diagram, consider again the experiment of surveying individuals with regard to their age, sex, and political affiliation. The possible outcomes of the questions concerning age, sex, and political affiliation were described by the sets

$$A = \{\text{under 39, over 39}\}$$

$$B = \{M, F\}$$

and

$$C = \{D, R, O\}$$

The tree diagram for the sample space of this experiment is constructed by showing sets A, B, and C as branches on a tree. Starting from a common point, we draw two lines to represent the elements in set A. From each line, two additional lines are drawn to represent the two elements in set B. To complete the tree diagram, three lines are drawn to represent the elements in set C. The completed diagram is shown in Fig. 5.1. The branches on the tree represent the sample space of the experiment. The sample space is, of course, the same as that given in the earlier example.

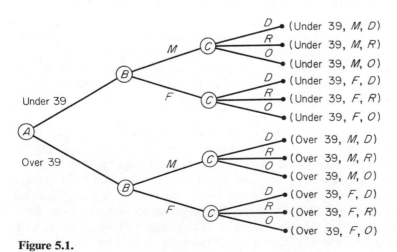

Figure 5.1.

Example. An experiment consists of observing the purchases of stock by a brokerage house. If stock is purchased on the New York Stock Exchange, the American Stock Exchange, or "over-the-counter" in "round lots" (i.e., 100 share blocks of stock) or "odd lots" (i.e., less than 100 shares), determine the sample space for the experiment.

The sample space is shown by the tree diagram in Fig. 5.2.

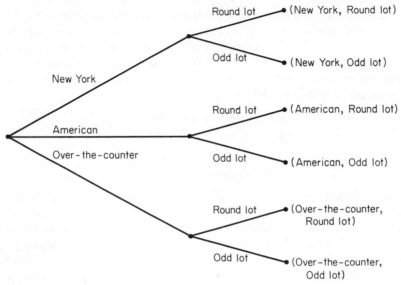

Figure 5.2.

5.1.3 FINITE AND INFINITE SAMPLE SPACE

The term *sample space* has been defined as the set of all possible outcomes of an experiment. In illustrating this definition, we have used experiments in which the sample space contained a *finite* number of possible outcomes. For instance, the sample space shown in Fig. 5.1 contains 12 sample points or possible outcomes, and the sample space in Fig. 5.2 contains six possible outcomes.

It is important to point out that in many experiments the sample space contains an *infinite*, rather than a finite, number of sample points. As a simple illustration, consider the experiment of determining the time required to complete a long-distance telephone call. The sample space for this experiment consists of all points along a line that begins at zero and continues to infinity. Since there are an infinite number of points on any continuous line, the sample space for this experiment would contain an infinite number of sample points. This would also be true if the upper bound on the time limit were 3 minutes rather than infinity. Again, we recognize that there is an infinite number of points along a continuous line that begins at 0 and continues to 3. We shall consider experiments in which the sample space is described by points along a continuum in Chap. 6.

It is also possible for the sample space to contain a countable, although infinite, number of sample points. To illustrate what is meant by a *countably infinite* number of sample points, consider an experiment that involves determining the number of defects in a bolt of cloth. The sample space for this experiment is

$$S = \{0, 1, 2, 3, \ldots\}$$

The first sample point in the sample space is 0, implying that there are no defects in the bolt of cloth. The second sample point is 1, meaning that one defect is a possible outcome from inspecting the cloth. Similarly, two defects are possible. In fact, there is no maximum number of defects in the bolt of cloth. Consequently, the sample space contains a countable, although infinite, number of sample points.

Example. Specify whether the sample space for the following experiments would contain a finite, countably infinite, or an infinite number of sample points:

1. Drawing the winning ticket for a sweepstakes.
   Answer: Finite. The sample space would be a list of all entrants.
2. Tossing a single coin until the occurrence of a "head."
   Answer: Countably infinite. The tosses of the coin are countable, and there is no upper limit on how many times the coin must be tossed until the occurrence of a head.
3. Determining the time before the arrival of the next car at a toll booth.
   Answer: Infinite. The sample space consists of the continuum between zero and infinity.

5.1.4 SAMPLE SPACE AND RULES FOR COUNTING

Constructing a complete list of the possible outcomes (i.e., sample points) of an experiment can be a time-consuming and burdensome task. Fortunately, there are many problems in probability in which a listing of the possible outcomes or sample points in an experiment is not required. In these problems we need know only the total number of sample points in the sample space of the experiment and the number of sample points in a specific event. The rules and formulas that are used in counting the number of possible outcomes of some statistical process or experiment were discussed in Chap. 4. The relationship between counting the number of possible outcomes and the probability of a specific event is discussed in Sec. 5.3.2.

5.2 Subsets of a Sample Space: Events

We have defined the term *sample space* as the set of all possible outcomes of a statistical process or experiment. Assume, for the moment, that we are specifically interested in one or more of the possible outcomes of the experiment. These possible outcomes would be described by a subset of the sample space. In the theory of probability, a subset of a sample space is defined as an *event*.

To illustrate an event as a subset of a sample space, consider the experiment of tossing two coins. The sample space for this experiment is

$$S = \{(HH), (HT), (TH), (TT)\}$$

Assume that we are interested in the event "two heads." The subset of the sample space that defines this event is

$$A = \{(HH)\}$$

Similarly, if our interest were in the event "at least one head," the appropriate subset would be

$$B = \{(HH), (HT), (TH)\}$$

The subsets A and B of the sample space S are examples of events.

Example. Consider the experiment of rolling a pair of dice. The sample space for this experiment was given by Table 5.1 and is repeated in Table 5.2.

Table 5.2.

A \ B		Second Die				
		1	2	3	4	5	6
1		(1, 1)	(1, 2)	(1, 3)	(1, 4)	(1, 5)	(1, 6)
2		(2, 1)	(2, 2)	(2, 3)	(2, 4)	(2, 5)	(2, 6)
First 3		(3, 1)	(3, 2)	(3, 3)	(3, 4)	(3, 5)	(3, 6)
Die 4		(4, 1)	(4, 2)	(4, 3)	(4, 4)	(4, 5)	(4, 6)
5		(5, 1)	(5, 2)	(5, 3)	(5, 4)	(5, 5)	(5, 6)
6		(6, 1)	(6, 2)	(6, 3)	(6, 4)	(6, 5)	(6, 6)

Using this table, determine the sample points in the following events.

1. The sum of the spots on the pair of dice is 7.

$$\{(6, 1), (5, 2), (4, 3), (3, 4), (2, 5), (1, 6)\}$$

2. The sum of the spots on the pair of dice is 12.

$$\{(6, 6)\}$$

3. The number on the second die is larger than that on the first die.

$$\{(1, 2), (1, 3), (1, 4), (1, 5), (1, 6), (2, 3), (2, 4), (2, 5),$$
$$(2, 6), (3, 4), (3, 5), (3, 6), (4, 5), (4, 6), (5, 6)\}$$

4. The sum of the spots on the pair of dice is 2.

$$\{(1, 1)\}$$

The subset or events in the preceding example can be used to introduce the terms *simple* and *compound events*. When a subset or event contains only one sample point, the event is termed a *simple event*. Thus, the events describing statements 2 and 4 in the preceding example are simple events. If the subset contains more than one sample point, the event is referred to as a *compound* (or *joint*) *event*. The events described by statements 1 and 3 are compound events.

Example. Using the sample space in Table 5.2, specify the following events and state whether the events are simple or compound:

1. The sum of the spots on the pair of dice is 3.

$$\{(1, 2), (2, 1)\} \qquad \text{compound}$$

2. Five plus the number on the first die equals the number on the second die.

$$\{(1, 6)\} \qquad \text{simple}$$

3. The number on the first die is 1.

$$\{(1, 1), (1, 2), (1, 3), (1, 4), (1, 5), (1, 6)\} \qquad \text{compound}$$

5.2.1   SET OPERATIONS AND EVENTS

The set operations of intersection, union, and complementation are useful in specifying events.[1] To illustrate, consider again the experiment of rolling

---

[1] The set operations of intersection, union, and complementation were introduced in Chap. 1.

Table 5.3.

A \ B		1	2	Second Die 3	4	5	6
1		(1, 1)	(1, 2)	(1, 3)	(1, 4)	(1, 5)	(1, 6)
2		(2, 1)	(2, 2)	(2, 3)	(2, 4)	(2, 5)	(2, 6)
First 3		(3, 1)	(3, 2)	(3, 3)	(3, 4)	(3, 5)	(3, 6)
Die 4		(4, 1)	(4, 2)	(4, 3)	(4, 4)	(4, 5)	(4, 6)
5		(5, 1)	(5, 2)	(5, 3)	(5, 4)	(5, 5)	(5, 6)
6		(6, 1)	(6, 2)	(6, 3)	(6, 4)	(6, 5)	(6, 6)

a pair of dice. The sample space for this experiment was given by Tables 5.1 and 5.2 and is repeated in Table 5.3. Assume that the following events are defined on this sample space:

1. The event "the sum of the spots on the two dice is 7."

$$A = \{(6, 1), (5, 2), (4, 3), (3, 4), (2, 5), (1, 6)\}$$

2. The event "the number on the second die is 3."

$$B = \{(1, 3), (2, 3), (3, 3), (4, 3), (5, 3), (6, 3)\}$$

3. The event "the sum of the spots on the two dice is 11."

$$C = \{(6, 5), (5, 6)\}$$

Given the events A, B, and C, we can specify additional events by using the operations of intersection, union, and complementation. For instance, consider the event "seven or eleven." In combining events the conjunction *or* represents the union of the events. The subset of sample points included in the event "seven or eleven" is thus

$$A \cup C = \{(6, 1), (5, 2), (4, 3), (3, 4), (2, 5), (1, 6), (6, 5), (5, 6)\}$$

Similarly, the event "the sum of the spots on the two dice is 7 and the number on the second die is 3" can be described by using set operations. The conjunction *and* refers to the intersection of the events. The subset of sample points included in the event is

$$A \cap B = \{(4, 3)\}$$

Finally, the event "the sum of the spots on the two dice is not 7 or 11" is described by the complement of the union of events A and C. The event is

$$(A \cup C)' = \{(1, 1), (1, 2), (1, 3), (1, 4), (1, 5), (2, 1), (2, 2), (2, 3), (2, 4), (2, 6),$$
$$(3, 1), (3, 2), (3, 3), (3, 5), (3, 6), (4, 1), (4, 2), (4, 4), (4, 5), (4, 6),$$
$$(5, 1), (5, 3), (5, 4), (5, 5), (6, 2), (6, 3), (6, 4), (6, 6)\}.$$

Example. Consider the experiment of tossing three coins and recording the outcome of each toss as a "head" or a "tail." The sample space for this experiment is

$$S = \{(HHH), (HHT), (HTH), (THH), (HTT), (THT), (TTH), (TTT)\}$$

Assume that the following events are defined in this sample space:

1. The event "three heads."

$$A = \{(HHH)\}$$

2. The event "two heads and one tail."

$$B = \{(HHT), (HTH), (THH)\}$$

3. The event "three tails."

$$C = \{(TTT)\}$$

Given the events *A*, *B*, and *C*, specify the following events by using the operations of intersection, union, and complementation:

1. The event "at least two heads." This event is

$$A \cup B = \{(HHH), (HHT), (HTH), (THH)\}$$

2. The event "at least two tails." The event is

$$(A \cup B)' = \{(HTT), (THT), (TTH), (TTT)\}$$

3. The event "three heads" and "two heads and one tail." The event "three heads" cannot occur simultaneously with the event "two heads and one tail." The subset is thus

$$A \cap B = \{\ \}$$

or, alternatively,

$$A \cap B = \phi$$

4. The event "three heads or not three heads." This event is given by the union of *A* and *A'* and is

$$A \cup A' = \{(HHH), (HHT), (HTH), (THH),$$
$$(HTT), (THT), (TTH), (TTT)\}$$

or, alternatively,

$$A \cup A' = S$$

The events described by statements 3 and 4 illustrate events that are, respectively, *mutually exclusive* and *exhaustive*. To illustrate, consider two events H and G that are defined on the sample space S. The events H and G are termed mutually exclusive if the subset defined by the intersection of H and G is null, i.e., if H and G have no common sample points. Symbolically, the two events H and G are mutually exclusive if

$$H \cap G = \phi \qquad (5.2)$$

The events "three heads" and "two heads and one tail" are clearly mutually exclusive, i.e., $A \cap B = \phi$. One cannot toss three coins and get both "three heads" and "two heads and one tail" as the outcome of the tosses.

To define events that are exhaustive, consider the events I and J that are defined on a sample space S. The events I and J are termed exhaustive if the subset formed by the union of I and J is equal to the sample space S. Thus, the events I and J are exhaustive if

$$I \cup J = S \qquad (5.3)$$

The events A and A' in the preceding example are, of course, exhaustive, i.e., $A \cup A' = S$. This merely says that the result of the toss of three coins will be "three heads" or "not three heads." In fact, we know from Chap. 1 that the subset formed by the union of any event A and its complement A' is equal to the universal set U. It therefore follows that, for any event A defined on the sample space S, the events A and A' are exhaustive. Similarly, from the definition of complementary events, we know that the intersection of two complementary events is null and the events must, therefore, also be mutually exclusive.

Example. Consider an experiment of drawing a card from a deck of playing cards. Indicate if the following events are mutually exclusive, exhaustive, or both:

1. $A = \{$black card$\}$, $B = \{$red card$\}$.
   Answer: $A \cap B = \phi$, $A \cup B = S$. The events are both mutually exclusive and exhaustive.
2. $A = \{$heart$\}$, $B = \{$spade$\}$, $C = \{$club$\}$.
   Answer: $A \cap B = \phi$, $A \cap C = \phi$, $B \cap C = \phi$, $A \cup B \cup C \neq S$. The events are mutually exclusive but not exhaustive.
3. $A = \{$black card$\}$, $B = \{$club$\}$.
   Answer: $A \cap B \neq \phi$, $A \cup B \neq S$. The events are neither mutually exclusive nor exhaustive.
4. $A = \{$red card$\}$, $B = \{$spade$\}$.
   Answer: $A \cap B = \phi$, $A \cup B \neq S$. The events are mutually exclusive but not exhaustive.
5. $A = \{$face card$\}$, $B = \{$not a face card$\}$.
   Answer: $A \cap B = \phi$, $A \cup B = S$. The events are both mutually exclusive and exhaustive.

5.2.2 PARTITION OF THE SAMPLE SPACE

Events defined on a sample space that are both mutually exclusive and exhaustive form a *partition* of the sample space. Thus, in the simplest case, given an event A defined on the sample space S, the events A and A' form a *twofold* partition of the sample space. Similarly, if the n events $A_1, A_2, \ldots, A_n$ are both mutually exclusive and exhaustive, the n events form an *n-fold* partition of the sample space.

As an example of partitioning a sample space, consider an experiment of selecting a panel of jurors from a list of registered voters. If we let A represent the event "male" and B the event "female," then $A \cap B = \phi$ and $A \cup B = S$, and the events A and B are mutually exclusive and exhaustive. The events A and B form a twofold partition of the sample space.

Continuing with the example, assume that C represents the event "under 25 years of age," D the event "25 to 40 years of age," and E the event "over 40 years of age." The events C, D, and E are mutually exclusive, since no two of the events can occur at once. Since one of the events must occur, they are also exhaustive. Thus, the events C, D, and E form a threefold partition of the sample space.

Example. Explain why the events C, and D in the preceding illustration concerning voter selection do not form a partition of the sample space.

The requirements for partitioning a sample space are that the events be both mutually exclusive and exhaustive. Given the events C and D in the preceding example, we note that $C \cap D = \phi$ but that $C \cup D \neq S$. Therefore, the events C and D do not form a partition of the sample space.

5.3   Assigning Probabilities

A *probability* is a numerical measure of the likelihood of the occurrence of an event. To illustrate, consider the experiment of tossing two coins and recording the outcome of the toss of each coin as a head, H, or tail, T. The sample space for the experiment is

$$S = \{(HH), (HT), (TH), (TT)\}$$

Assume that we want to determine the likelihood of obtaining two heads as the outcome of this experiment. Since each of the four sample points in S is equally likely, the probability of the event "two heads" is 0.25. Similarly, the probability of each of the remaining sample points in S is also 0.25.

Probabilities are assigned on the basis of either *objective* or *subjective* estimates of the likelihood of occurrence of an event. *Objective probabilities* are determined from theoretical analysis of the experiment or from actual observation of numerous repetitions of the experiment. In the experiment involving the toss of two coins, for instance, we reasoned that each of the outcomes is equally likely. Since there are four possible outcomes and each outcome is equally likely, the probability of two heads is 0.25. This probability was determined by theoretical analysis rather than from actually observing repeated tosses of two coins. The probability of the event was given by

$$P(E) = \frac{n(E)}{n(S)} \qquad (5.4)$$

where $n(E)$ represents the number of sample points in event E and $n(S)$ represents the total number of equally likely sample points in the sample space S. In the example, $n(E)$ is 1, $n(S)$ is 4, and $P(HH) = \frac{1}{4}$ or 0.25.

Objective probabilities determined from actually observing numerous repetitions of the experiment are referred to as *empirical probabilities*. An empirical probability is calculated by the formula

$$P(E) = \frac{f}{n} \qquad (5.5)$$

where f is the frequency or number of occurrences of event E and n is the total number of times the experiment has been observed. For example, consider the manufacture of a certain component. If, during a long production run, 100,000 components have been manufactured and 5000 of these have been defective, an empirical estimate of the probability of a defective unit is $P(E) = 5000/100,000 = 0.05$.

Example. In the game of roulette, a ball is dropped on a revolving disk. The disk contains 38 slots—18 red, 18 black, and two green. Assuming that the ball is equally likely to drop in any slot, determine the probability of the event "black."

There is a total of 38 equally likely slots. Since 18 of these are black, we reason that the probability of the event black is

$$P(\text{black}) = \frac{n(\text{black})}{n(S)} = \frac{18}{38}$$

Example. A manufacturer of transistors must determine the probability of a defective transistor. One thousand transistors are randomly selected from the production line and tested. Forty are found to be defective. Determine the probability of a defective transistor.

A total of 1000 transistors were tested and 40 were defective. The probability of a defective is

$$P(\text{defective}) = \frac{f}{n} = \frac{40}{1000} = 0.04$$

Subjective probabilities differ from objective probabilities in that a subjective probability is a measure of an individual's belief in the likelihood of the occurrence of some event. Subjective probabilities are normally based on the opinion of a "reasonable" or "knowledgeable" person. Most often, subjective probabilities are used when it is impossible or impractical to determine the objective probability of the event. To illustrate, consider the case of introducing a new product. A major food company plans to market a new breakfast product. In order for the product to earn a reasonable return on investment, sales must be at least $10 million. The probability of sales reaching this figure is subjectively estimated by the vice-president of marketing as $P(\text{sales} \geq \$10 \text{ million}) = 0.80$.

In determining the subjective probability, the vice-president called on the combined expertise of people in sales, finance, and marketing. It is, of course, impossible to theoretically determine this probability or to introduce this particular product a large number of times and thereby determine the frequency of occurrence of the event "sales $\geq$ $10 million." The probability, instead, represents a subjective estimate of a knowledgeable individual of the likelihood of the sales of the product exceeding $10 million.

Example. In a recent panel discussion of the prospects for developing a cure for cancer, Dr. Jefferies stated that he believed that the probability of finding a cure for cancer before the year 2000 was 0.80. Dr Baxter, however, estimated that the probability was only 0.50. Explain how two probabilities for the same event can differ.

The probabilities differ because they are subjective estimates based on the personal belief of the two doctors. One could not expect two or more knowledgeable individuals to have the same subjective belief about a highly uncertain event.

5.3.1 AXIOMS OF PROBABILITY

In assigning either objective or subjective probabilities to a set of events, three requirements must be satisfied. First, the probability of any event must be equal to or greater than zero and less than or equal to one. A probability of zero means that the event cannot occur, and a probability of one means that the event must occur. Second, the sum of the probabilities of the events must equal one. This means that the total of the probabilities of the events in the sample space is one, or, alternatively, that one of the events must occur as a result of the experiment. Finally, the probability of the occurrence of two or more mutually exclusive events is equal to the sum of the probabilities assigned to the individual events. Thus, in the example above, the probability

of two heads or two tails is equal to the sum of the probabilities assigned
to the events two heads and two tails, that is,

$$P((HH) \cup (TT)) = P(HH) + P(TT) = 0.50$$

The requirements for assigning probabilities are termed the three axioms
of probability. For a sample space S with the exhaustive events $A_1, A_2, \ldots,$
A_n, the axioms can be summarized as

Axiom 1. $0 \le P(A_j) \le 1$ (5.6)

Axiom 2. $P(S) = 1$ (5.7)

Axiom 3. $P(A_i \cup A_j) = P(A_i) + P(A_j)$ for $A_i \cap A_j = \phi$ (5.8)

Axiom 1 states that the probability assigned to an event must be a number
between 0 and 1, inclusive. Axiom 2 requires that the sum of the probabilities
must equal 1. Axiom 3 states that the probability of the occurrence of the
mutually exclusive events A_i or A_j is given by the sum of the probabilities
assigned to A_i and A_j.

Axioms 2 and 3 can be used to determine the probabilities of complementary events. The events A and A' being given, it follows from Axiom 2 and
from the definition of complementary events that

$$P(A) + P(A') = 1$$ (5.9)

Alternatively, this relationship can be expressed as

$$P(A') = 1 - P(A)$$ (5.10)

Example. At a recent stockholders' meeting of a large corporation, the
president stated that the probability of an increase in profits during the coming
year was 0.60. He also stated that the probability of profits remaining
unchanged was 0.30 and that the probability of a decline in profit was 0.10.
Are these statements consistent with the axioms of probability?

Assume that I represents the event that profits increase, U the event that
profits remain unchanged, and D the event that profits decline. Then, $P(I) =$
0.60, $P(U) = 0.30$, and $P(D) = 0.10$. The probabilities of I, U, and D are all
represented by numbers between 0 and 1. Furthermore, the sum of the probabilities for the three mutually exclusive and exhaustive events are $P(I) +$
$P(U) + P(D) = 0.60 + 0.30 + 0.10 = 1.0$. Finally, the probability of the
union of the two events I and D is $P(I \cup D) = P(I) + P(D) = 0.60 +$
$0.10 = 0.70$.

Example. Determine the probability that profits will not increase in the
preceding example.

The event that profits will not increase is described as I'. The probability
of I' is $P(I') = 1 - P(I) = 1 - 0.60 = 0.40$.

The third axiom of probability was defined for events that are mutually exclusive. Quite often, however, probabilities must be assigned to events that are not mutually exclusive. In these cases, Axioms 1 and 2 still apply. Axiom 3 must be modified to take into account the fact that $A_i \cap A_j \neq \phi$. The modification is

$$\text{Axiom 3'.} \quad P(A_i \cup A_j) = P(A_i) + P(A_j) - P(A_i \cap A_j) \quad (5.11)$$

where $A_i \cap A_j \neq \phi$.

To illustrate Axiom 3', consider the events A and B defined on the sample space S in Fig. 5.3. The events A and B in Fig. 5.3(a) are mutually exclusive, i.e., $A \cap B = \phi$. The probability of the events A or B occurring is thus $P(A \cup B) = P(A) + P(B)$. The events A and B in Fig. 5.3(b), however, are not mutually exclusive, i.e., $A \cap B \neq \phi$. The event $A \cap B$ is a subset of both the event A and the event B. Since the $P(A \cap B)$ is included in both $P(A)$ and $P(B)$, it must be subtracted from $P(A) + P(B)$ in order to avoid double counting. The probability of the events A or B occurring is, therefore, $P(A \cup B) = P(A) + P(B) - P(A \cap B)$.

As an example of assigning probabilities to events, consider the experiment of drawing a card from a well-shuffled deck of 52 playing cards. The sample space for this experiment consists of the 52 cards in the deck. Assume that the events "$R = $ red card," "$C = $ club," and "$S = $ spade" are defined on the sample space. The events R, C, and S are mutually exclusive and exhaustive. Since there are 26 red cards, 13 spades, and 13 clubs in the deck and the selection of any one of the 52 cards is equally likely, the probabilities of the events R, S, and C are $P(R) = \frac{26}{52}$, $P(C) = \frac{13}{52}$, and $P(S) = \frac{13}{52}$. Notice that the probabilities assigned to the events are numbers between 0 and 1 and that $P(R) + P(C) + P(S) = \frac{26}{52} + \frac{13}{52} + \frac{13}{52} = 1$, indicating that one of the events must occur. Also note that the probability of a red card or a spade is given by $P(R \cup S) = P(R) + P(S) = \frac{26}{52} + \frac{13}{52} = \frac{39}{52}$; that is, the probability is given by the sum of the probabilities of the two mutually exclusive events.

To illustrate Axiom 3', assume that event "$F = $ face card" is defined on the sample space. A deck of cards has four face cards (jack, queen, king, ace)

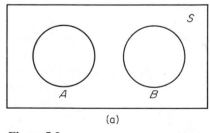
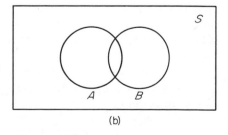

(a)   (b)

Figure 5.3.

in each suit and four suits (clubs, diamonds, hearts, spades). The probability of the event F is therefore $P(F) = \frac{16}{52}$.

Assume that we wish to determine the probability of the event "red card" or "face card." Since the events red card and face card are not mutually exclusive, the probability is $P(R \cup F) = P(R) + P(F) - P(R \cap F) = \frac{26}{52} + \frac{16}{52} - \frac{8}{52} = \frac{34}{52}$. The probability is found by subtracting the probability of a red face card from the sum of the probabilities of the events red card and face card.

Example. Consider two events A and B defined on a sample space S. If $P(A) = 0.40$, $P(B) = 0.30$, $P(A \cap B) = 0.10$, and $P(S) = 1.00$, determine $P(A \cup B)$, $P(A' \cap B')$, and $P((A' \cap B) \cup (A \cap B'))$. Is the assignment of probabilities consistent with the three axioms of probability?

The events are shown in the Venn diagram in Fig. 5.4. To verify that the probabilities are consistent with the axioms of probability, notice that the probabilities assigned are numbers between 0 and 1, that $P(S) = 1$, and that $P(A \cup B) = P(A) + P(B) - P(A \cap B)$.

The probabilities of $(A \cup B)$, $(A' \cap B')$, and $(A' \cap B) \cup (A \cap B')$ are $P(A \cup B) = 0.60$, $P(A' \cap B') = 0.40$, and $P((A' \cap B) \cup (A \cap B')) = 0.50$.

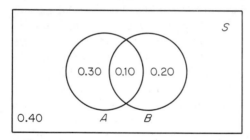

Figure 5.4.

5.3.2 ASSIGNING PROBABILITIES USING RULES FOR COUNTING

The sample space of certain experiments consists of a finite number of equally likely sample points or outcomes. In these cases, the rules for counting developed in Chap. 4 can be used in determining the probability of certain events.

As a simple example, consider the experiment of tossing two coins and recording the outcome of each toss as a head or a tail. The sample space for this experiment is

$$S = \{(HH), (HT), (TH), (TT)\}$$

Since each of the outcomes or sample points in S is equally likely, the probability of the event "two heads" can be calculated from Eq. (5.4) and is

$$P(E) = \frac{n(E)}{n(S)} = \frac{1}{4} = 0.25$$

where $P(E)$ is the probability of the event "two heads," $n(E)$ is the number of sample points comprising this event, and $n(S)$ is the number of equally likely sample points in S.

Instead of listing the sample points in S, suppose that we calculate the probability of the event "two heads" by using the rules for counting. The number of equally likely outcomes in the experiment, $n(S)$, is given by the formula for permutations allowing repetitions, Eq. (4.8). Since there are two possible outcomes from the first toss and two possible outcomes from the second toss, there are $_2P^2 = 2^2$ or four possible outcomes in the experiment. Each of these outcomes is equally likely. Only one of the outcomes represents the event "two heads." From Eq. (5.4), the probability of this event is again

$$P(E) = \frac{n(E)}{n(S)} = \frac{1}{4} = 0.25$$

As a somewhat more complicated and perhaps more realistic example, consider a manufacturer of air conditioning equipment who receives thermostats from a supplier in lots of 20. Suppose that the manufacturer wishes to test a sample of five thermostats selected "at random" from each lot. Under these conditions, if a particular lot of 20 thermostats contains two defective units, what is the probability that there will be exactly one defective among the five selected for testing?

To determine this probability, we let E represent the event that exactly one defective thermostat is included in the five selected for testing. The probability of E is given by Eq. (5.4),

$$P(E) = \frac{n(E)}{n(S)} \tag{5.4}$$

where $n(S)$ represents the number of different possible samples of five thermostats selected from a lot of 20 thermostats and $n(E)$ represents the number of different possible samples that contain exactly one defective thermostat and four nondefective thermostats.

The number of different possible samples of five thermostats selected from a lot of 20 can be calculated from the formula for combinations, Eq. (4.9),

$$_{20}C_5 = \frac{20!}{15!\,5!} = 15,504$$

The number of different possible samples that contain exactly one defective thermostat and four nondefective thermostats is found by recognizing that the one defective can be selected from the two defectives in the lot in

$$_2C_1 = \frac{2!}{1!\,1!} = 2$$

ways and the four nondefectives can be selected from the 18 nondefectives in the lot in

$$_{18}C_4 = \frac{18!}{14!\,4!} = 3060$$

ways. From the fundamental rule for counting, the number of different samples that contain one defective and four nondefectives is

$$_2C_1 \cdot _{18}C_4 = 2(3060) = 6120$$

The probability of obtaining a sample that contains exactly one defective is thus

$$P(E) = \frac{n(E)}{n(S)} = \frac{6120}{15,504} = 0.394$$

This problem will be discussed in more detail in Sec. 6.4.1, "The Hypergeometric Probability Distribution," in Chap. 6.

As another example of using the rules for counting for calculating the probability of events, consider the problem of determining the probability of obtaining exactly three heads in six tosses of a fair coin. It was shown in Chap. 4 that the number of possible outcomes of six tosses of a coin is calculated from the formula for permutations allowing repetitions and is $_2P^6 = 2^6$ or 64. The number of ways that three heads can occur in the six tosses is given by the formula for permutations involving indistinguishable objects and is

$$_6P_6(3,\ 3) = \frac{6!}{3!\,3!} = 20$$

From Eq. (5.4), the probability of getting exactly three heads in six tosses of a fair coin is

$$P(E) = \frac{n(E)}{n(S)} = \frac{20}{64} = 0.3125$$

As a final example of using rules of counting for determining the probability of events, suppose that we consider the "matching birthdays" problem. This problem involves determining the probability that two or more people in a group of r people have the same birthday. Most individuals believe that the number of people in the group must be relatively large, say $r = 100$, for the chances to be good that two people would have the same birthday. This, however, is not the case. As a matter of fact, the probability is greater than 0.50 that two people in a group of only 23 will have the same birthday. Because this is counter to most people's intuition, it is usually possible to make a small wager on matching birthdays. (Once again, you have the opportunity to recover the price of this text!)

An ideal setting for this problem is at a party with friends. Suppose that there are 50 people at this party and you propose to bet that at least two of the people have the same birthday. Since there are 365 possible birthdays, most people would think that the chances of two people's having the same birthday are fairly small. We shall show, however, that this is not at all the case.

To calculate the probability that two people in a group of 50 have the same birthday, we first find the probability that two people do not have the same birthday. Since there are 365 possible birthdays (neglecting February 29), there are 365 possible days for the first person's birthday, 365 possible days for the second person's birthday, etc. The number of different sequences for all 50 birthdays is given by the formula for permutations allowing repetitions, Eq. (4.8), and is $_{365}P^{50}$ or 365^{50}. If no two people have the same birthday, there are 365 possible days for the first person's birthday, 364 possible days for the second person's birthday, 363 possible days for the third person's birthday, etc., and 316 possible days for the 50th person's birthday. The number of sequences of different birthdays is given by the formula for permutations of r objects selected from a set of n objects, Eq. (4.6), and is $_{365}P_{50}$ or $365!/315!$. The probability that no two people have the same birthday can be calculated by using Eq. (5.4) and is

$$P(E') = \frac{n(E')}{n(S)} = \frac{_{365}P_{50}}{_{365}P^{50}} = 0.030$$

The probability that two people in the group of 50 have the same birthday is thus

$$P(E) = 1 - P(E') = 0.970$$

The probability that at least two people in a group of r people have the same birthday for several different values of r are given in Table 5.4.

5.4   Joint, Marginal, and Conditional Probability

One of the more useful methods of summarizing the probabilities of inter-related events is through the use of a *probability table*. To illustrate, consider an experiment recently conducted by a large department store. The purpose of this experiment was to determine the buying patterns of men and women shoppers in the department store. The experiment consisted of noting the

Table 5.4.

Number of People in the Group	Probability of at Least Two with the Same Birthday
5	0.027
10	0.117
15	0.253
20	0.411
25	0.569
30	0.706
40	0.891
50	0.970
60	0.994
70	0.9992
100	0.9999997

proportion of men and women shoppers who made at least one purchase before leaving the store. The results of this experiment are shown as probabilities in Table 5.5.

Four events are defined in the table. These are "Buyer (B)," "Nonbuyer (B')," "Men (M)," and "Women (W)." The probability of each of the events is read from the margins of the probability table. The probability of a buyer is thus $P(B) = 0.20$. Similarly, the probability of a nonbuyer is $P(B') = 0.80$, the probability of a man shopper is $P(M) = 0.30$, and the probability of a woman shopper is $P(W) = 0.70$. Probabilities such as $P(B)$, $P(B')$, $P(M)$, and $P(W)$ that appear in the margin of a probability table are referred to as *marginal probabilities*. The term marginal probability comes from the fact that the probabilities are shown in the margins of the table.

The probability of the joint occurrence of two events is shown in the body of the probability table. For example, the probability that a randomly selected individual entering the store is both a man and a buyer is $P(M \cap B) = 0.10$. The probability of the joint occurrence of the events man and buyer is termed a *joint probability*.

Table 5.5.   Buying Patterns of Men and Women Shoppers

	Men (M)	Women (W)	Total
Buyer (B)	0.10	0.10	0.20
Nonbuyer (B')	0.20	0.60	0.80
Total	0.30	0.70	1.00

Joint probabilities of two or more events are denoted as the probability of the intersection of the events. In the example, the probability of an individual's being a man and a buyer is $P(M \cap B) = 0.10$. Similarly, the probability of an individual's being a woman and a nonbuyer is $P(W \cap B') = 0.60$. The probability statement, $P(M \cap B)$, is read "the probability of the joint occurrence of the events M *and* B," or, alternatively, "the probability of M *and* B."

Example. For the example shown in Table 5.5, determine the probabilities of the events $(M \cap B')$, $(B \cap W)$, $(W \cup B)$, and $(M \cup B')$.

$P(M \cap B') = 0.20$

$P(B \cap W) = 0.10$

$P(W \cup B) = P(W) + P(B) - P(W \cap B) = 0.70 + 0.20 - 0.10 = 0.80$

$P(M \cup B') = P(M) + P(B') - P(M \cap B') = 0.30 + 0.80 - 0.20 = 0.90$

Example. The dean of a school of business recently authorized a study of the relative performance of individuals in the school's masters program. Specifically, the study classified individuals by undergraduate degree in business (B) and nonbusiness (B'), and by honor student (H) and nonhonor student (H'). The results of the study are shown in Table 5.6. Based on the entries in the probability table, determine the following:

1. Marginal probability of an honor student.

$$P(H) = 0.30$$

2. Joint probability of nonbusiness and nonhonor.

$$P(B' \cap H') = 0.40$$

Table 5.6.

	Honor	Nonhonor	Total
Undergraduate Degree	(H)	(H')	
Business (B)	0.10	0.30	0.40
Nonbusiness (B')	0.20	0.40	0.60
Total	0.30	0.70	1.00

3. The probability of nonbusiness or honor.

$$P(B' \cup H) = P(B') + P(H) - P(B' \cap H)$$
$$= 0.60 + 0.30 - 0.20 = 0.70$$

4. Are the events H and B mutually exclusive?

$$\text{No,} \qquad P(H \cap B) \neq 0$$

5. Do the events H and H' form a partition of the sample space? Yes, complementary events are both mutually exclusive and exhaustive. Thus, they partition the sample space.

$$P(H \cap H') = 0$$

and

$$P(H \cup H') = P(H) + P(H') = P(S)$$

5.4.1 CONDITIONAL PROBABILITY

In studying the buying patterns of men and women shoppers, we were able to determine the marginal probability of the event "man" and the joint probability of the event "buyer and man." Referring to Table 5.5 and our earlier calculations, we find that these probabilities were $P(M) = 0.30$ and $P(B \cap M) = 0.10$. Based on a relative frequency interpretation of probability, this means that 30 percent of the shoppers are men and 10 percent of the shoppers are both men and buyers.

Suppose that instead of determining the probability of a customer's being both a man and a buyer, we were asked to determine the probability of a man customer making a purchase. Referring to Table 5.5, we note that 10 percent of the customers are buyers and men and 30 percent of the customers are men. The proportion of the men customers who are buyers is $0.10/0.30 = 0.33$. Thus, 33 percent of the men customers are buyers or, alternatively, the probability of a buyer, given the fact that the customer is a man, is 0.33.

The probability of the event "buyer" given the event "man" is referred to as a *conditional probability*. The conditional probability of an event is the probability of the event's occurrence, given some specified condition. In our example, the probability of the event "buyer," given the condition that the event "man" has occurred, was 0.33. This is written as $P(B \mid M) = 0.33$, where the symbol "|" is read as "given."[2]

---

[2] Notice that the symbol "|" is read "such that" when used to define a set and "given" when used to express conditional probability.

The formula for the conditional probability of the event A given the occurrence of the event B is

$$P(A \mid B) = \frac{P(A \cap B)}{P(B)} \qquad (5.12)$$

where $P(A \cap B)$ represents the joint probability of the events A and B and $P(B)$ the marginal probability of B. The conditional probability, $P(A \mid B)$, is defined only if $P(B) > 0$.

Eq. (5.12) is illustrated by Fig. 5.5. The crosshatched area in the figure represents the event $A \cap B$. Since we are given the condition that event B has occurred, the sample space is reduced to the sample points in B. The probability of A given B, $P(A \mid B)$, is thus equal to the proportion of the sample points in A that are in the reduced sample space represented by B. Consequently, $P(A \mid B) = P(A \cap B)/P(B)$.

By multiplying both sides of Eq. (5.12) by $P(B)$, the joint probability of $A \cap B$ can be expressed as

$$P(A \cap B) = P(A \mid B)P(B) \qquad (5.13)$$

Since $P(A \cap B) = P(B \cap A)$, it also follows that

$$P(A \mid B)P(B) = P(B \mid A)P(A) \qquad (5.14)$$

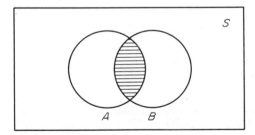

Figure 5.5.

Example. A large midwestern corporation has a savings plan available to its employees. Although there are no restrictions that limit participation, only 50 percent of the employees who are under 35 years of age have elected to participate in the plan. Of the total work force, 40 percent are under 35 years of age and 70 percent participate in the savings plan. Let U denote the event that an employee is "under 35 years of age" and P denote the event that an employee "participates" in the plan. Determine the following probabilities: $P(P' \mid U)$; $P(P \mid U')$; $P(P' \mid U')$; and $P(U \mid P)$.

In determining the probabilities, it will be helpful to construct a probability table. From the data we know that $P(U) = 0.40$, $P(U') = 1 - P(U) =$

Table 5.7.

	U	U′	Total
P	0.20		0.70
P′			0.30
Total	0.40	0.60	1.00

0.60, $P(P) = 0.70$, $P(P') = 1 - P(P) = 0.30$, and $P(P \mid U) = 0.50$. From the equation for conditional probability, we see that

$$P(P \cap U) = P(P \mid U)P(U) = 0.50(0.40) = 0.20$$

These probabilities are entered in Table 5.7.

The table can be completed by recognizing that

$$P(P' \cap U) = P(U) - P(P \cap U) = 0.40 - 0.20 = 0.20$$

$$P(P \cap U') = P(P) - P(P \cap U) = 0.70 - 0.20 = 0.50$$

$$P(P' \cap U') = P(P') - P(P' \cap U) = 0.30 - 0.20 = 0.10$$

These probabilities are shown in Table 5.8.

The conditional probabilities, $P(P' \mid U)$, $P(P \mid U')$, $P(P' \mid U')$, and $P(U \mid P)$, are

$$P(P' \mid U) = \frac{P(P' \cap U)}{P(U)} = \frac{0.20}{0.40} = 0.50$$

$$P(P \mid U') = \frac{P(P \cap U')}{P(U')} = \frac{0.50}{0.60} = 0.834$$

$$P(P' \mid U') = \frac{P(P' \cap U')}{P(U')} = \frac{0.10}{0.60} = 0.166$$

$$P(U \mid P) = \frac{P(U \cap P)}{P(P)} = \frac{0.20}{0.70} = 0.286$$

Table 5.8.

	U	U′	Total
P	0.20	0.50	0.70
P′	0.20	0.10	0.30
Total	0.40	0.60	1.00

The probability table in the preceding example can be used to illustrate an additional probability formula. Notice in Table 5.8 that

$$P(U) = P(P \cap U) + P(P' \cap U) = 0.40$$

A similar probability statement can be written for any row or column in the table. Although this relationship might seem obvious, it is worthwhile formally to state that the marginal probability of an event A can be determined from the fact that

$$P(A) = P(A \cap B) + P(A \cap B') \qquad (5.15)$$

or alternatively,

$$P(A) = P(A \mid B)P(B) + P(A \mid B')P(B') \qquad (5.16)$$

where A and B are events defined on the sample space S.

5.4.2   INDEPENDENCE

In illustrating the concept of the conditional probability of an event A given the occurrence of an event B, we used examples in which the conditional probability of A given B differed from the marginal probability of A, i.e.,

$$P(A \mid B) = \frac{P(A \cap B)}{P(B)} \neq P(A)$$

In problems of this type, the events A and B are termed *dependent events*.

When knowledge of the occurrence or nonoccurrence of an event B does not alter the probability of occurrence of the event A, the two events are said to be *independent*. In the case of independent events, the conditional probability of an event A given the occurrence of an event B is equal to the marginal probability of A. Thus, for independent events,

$$P(A \mid B) = P(A) \qquad (5.17)$$

For the special case in which A and B are independent, Eq. (5.13) for the joint probability of events,

$$P(A \cap B) = P(A \mid B)P(B) \qquad (5.13)$$

can be revised to read

$$P(A \cap B) = P(A)P(B) \qquad (5.18)$$

The joint probability of two independent events is thus equal to the product of the marginal probabilities of the events.

Example. An experiment consists of drawing a card from a deck of 52 playing cards. Let A represent the event the card is an ace and C the event the card is a club. Verify that the events A and C are independent.

Since there are four aces and thirteen clubs in the deck of cards, the marginal probability of an ace is

$$P(A) = \tfrac{4}{52} = \tfrac{1}{13}$$

and the marginal probability of a club is

$$P(C) = \tfrac{13}{52}$$

The joint probability of the "ace of clubs" is

$$P(A \cap C) = P(A)P(C) = \tfrac{1}{13} \cdot \tfrac{13}{52} = \tfrac{1}{52}$$

The conditional probability of an "ace" given a "club" is

$$P(A \mid C) = \frac{(A \cap C)}{P(C)} = \frac{\tfrac{1}{52}}{\tfrac{13}{52}} = \frac{1}{13}$$

The marginal probability of an "ace" equals the conditional probability of an "ace given a club," i.e.,

$$P(A) = P(A \mid C) = \tfrac{1}{13}$$

Consequently, the events are independent.

Example. Verify that the events given in Table 5.9 are independent.

From Eq. (5.18), we know that the events A and B are independent if $P(A \cap B) = P(A)P(B)$. We can verify that the events in Table 5.9 are independent by observing that

$$P(A \cap B) = P(A)P(B) = (0.60)(0.30) = 0.18$$

$$P(A' \cap B) = P(A')P(B) = (0.40)(0.30) = 0.12$$

$$P(A \cap B') = P(A)P(B') = (0.60)(0.70) = 0.42$$

and

$$P(A' \cap B') = P(A')P(B') = (0.40)(0.70) = 0.28$$

Table 5.9.

	A	A'	Total
B	0.18	0.12	0.30
B'	0.42	0.28	0.70
Total	0.60	0.40	1.00

The requirement for independence is given by Eq. (5.17). For any two events A and B to be independent, it is necessary that the events not be mutually exclusive, i.e., $A \cap B \neq \phi$. Although this is a necessary condition for independence, the fact that both events share some common sample points is not a sufficient condition. To illustrate, consider the events in Tables 5.10 and 5.11. The events are not mutually exclusive in either table. However, the events A and B in Table 5.10 are independent while the events F and U in Table 5.11 are dependent.

An example of events that are mutually exclusive is shown in Table 5.12. The events A and B in the table have no common sample points, i.e., $A \cap B = \phi$. Consequently, the occurrence of event B eliminates any possibility of the occurrence of event A, and $P(A \mid B) = 0$. Since the probability of A given B does not equal the marginal probability of event A, the events are dependent.

Table 5.10.

	A	A′	Total
B	0.12	0.28	0.40
B′	0.18	0.42	0.60
Total	0.30	0.70	1.00

A and B are not mutually exclusive but they are independent, i.e.,

$$P(A \mid B) = P(A) = 0.30$$

Table 5.11.

	F	F′	Total
U	0.20	0.20	0.40
U′	0.10	0.50	0.60
Total	0.30	0.70	1.00

F and U are not mutually exclusive but they are dependent, i.e.,

$$P(F \mid U) = \frac{P(F \cap U)}{P(U)} = 0.50$$
$$\neq P(F)$$

Table 5.12.

	A	A′	Total
B	0	0.40	0.40
B′	0.30	0.30	0.60
Total	0.30	0.70	1.00

A and B are mutually exclusive and are therefore dependent, i.e.,

$$P(A \mid B) = 0 \neq P(A)$$

In summary, if two events are mutually exclusive, they are dependent. If the events are not mutually exclusive, they can be independent or dependent. They are independent if

$$P(A \mid B) = P(A)$$

and dependent if

$$P(A \mid B) = \frac{P(A \cap B)}{P(B)} \neq P(A)$$

PROBLEMS

1. Specify the sample space for each of the following experiments:
   (a) In a survey of families with two children, the sex of each child in the family is recorded with the oldest child listed first.
   (b) In an experiment comparing purchasing habits, a respondent is asked whether he agrees (a) or disagrees (d) with each of three statements.
   (c) In a test of laundry soap, a housewife is asked to rank brands x, y, and z by order of preference.
   (d) Four parts turned out by a machine are inspected and denoted as good (g) or defective (d).

2. Three well-known economists are questioned regarding a proposed tax increase. Let F represent favoring the increase and N represent not favoring the increase. Specify the sample space of possible replies.

3. Ajax Corporation recently hired three new salesmen—Smith, Jones, and Clark. Each salesman must be assigned a sales territory. The territories are California, Oregon, and Arizona. If the salesmen are thought of as a set A and the territories a set B, the possible assignments can be viewed as a sample space with $S = A \times B$. List the elements in S.

4. Beginning next Monday, three commercials will be broadcast on a local radio station—one commercial on Monday, a second on Tuesday, and a third on Wednesday. If it is assumed that set A is the set of days and set B the set of different commercials, the possible assignments of commercials to days can be viewed as a sample space with $S = A \times B$. List the elements in S.

5. The proud father of a new baby boy anticipates his son's attending either Harvard or Yale and studying either law or medicine. Treating the possible choice of schools and course of study as an experiment, construct a tree diagram that represents the sample space of the experiment.

6. A shipment of parts must be sent from a manufacturer in London to a distributor in Kent, Ohio. The parts can be air freighted to New York, Philadelphia, or Chicago and then shipped by rail or truck to Kent.

Treating the possible routes as an experiment, construct a tree diagram that represents the sample space of the experiment.

7. A salesman must call on three customers, *A*, *B*, and *C*. Treating the possible order of calling on the customers as an experiment, construct a tree diagram that represents the sample space of the experiment.

8. A company is considering an expansion program that involves either adding to the current production facilities or building new production facilities. The decision on which action to take will be influenced by management's estimates of demand for the company's product. The possible levels of demand are high, moderate, and low. Treating the expansion decision and the possible levels of demand as an experiment, construct a tree diagram that represents the sample space for the experiment.

9. An experiment consists of counting the number of customers in a queue at a checkout counter. Give the sample space for this experiment. Does the sample space contain a finite or a countably infinite number of sample points?

10. For each of the following experiments, give the sample space and specify whether the sample space contains a finite or a countably infinite number of sample points:
    (a) The number of defects in a lot of five parts.
    (b) The number of flaws in a bolt of cloth.
    (c) The number of misspelled words in a list of ten words.

11. An experiment consists of inspecting three parts to determine if the parts are good (*g*) or defective (*d*). Give the sample points in each of the following events and state whether the event is a simple or a composite event:
    (a) None of the parts is defective.
    (b) No more than one part is defective.
    (c) At least one part is defective.
    (d) All three parts are defective.
    (e) The first part inspected is defective.

12. An avionics manufacturer produces an aircraft radio that gives both the distance to a station and the ground speed of the aircraft. Before delivering a radio, it must be inspected for accuracy with respect to both distance and ground speed. If the radio is within tolerance for both distance and ground speed, it is shipped to the customer. If not, it is retained for repairs. Defining the event *D* as within tolerance with respect to distance and *G* as within tolerance with respect to ground speed, interpret the following events:
    (a) $D' \cap G'$
    (b) $D \cap G'$
    (c) $D' \cap G$
    (d) $D \cap G$

13. The personnel manager of the Rogers Manufacturing Company has gathered the following statistics:

	Experience (Years)			
Department	Less than 3 (D)	3 to 5 (E)	More than 5 (F)	Total
Production (A)	8	9	33	50
Engineering (B)	2	6	12	20
Sales (C)	5	10	15	30
Total	15	25	60	100

Determine the number of employees in each of the following categories:

(a) $A \cap D$ (b) $A \cup D$

(c) $B \cup F$ (d) $B' \cap F'$

(e) $A \cup B \cup C$ (f) $A \cap B \cap C$

(g) $A \cap (D \cup E)$ (h) F'

14. Referring to the table in Problem 13, specify if the following events are mutually exclusive, exhaustive, both mutually exclusive and exhaustive, or neither mutually exclusive nor exhaustive.

(a) A, B

(b) A, A'

(c) C, F

15. Consider two events A and B defined on a sample space S. If $P(A) = 0.40$, $P(B) = 0.30$, $P(A \cap B) = 0.10$, and $P(S) = 1$, determine $P(A \cup B)$, $P(A' \cap B)$, and $P(A \cup B)'$. Is the assignment of probabilities consistent with the three axioms of probability?

16. A coin is tossed three times. Assuming that the coin is "fair," what is the probability of getting at least two heads? Is this a theoretical or an empirical (experimental) probability?

17. One chip is selected at random from 20 chips numbered from 1 through 20. What is the probability of selecting a chip that can be divided evenly by 4?

18. The "failure rate" for a certain electronic component is defined as the probability that the component will have a useful life of less than 2 hours. If a sample of 20 components is selected at random and three of the components fail during the first 2 hours of operation, what would be the empirically estimated failure rate of the component?

19. During 1970 deaths from motor vehicle accidents totaled 54,800 and deaths from aircraft accidents totaled 1400. The population of the

United States at that time was 203 million. Using these data, determine the empirical probability that a randomly selected individual will die in a motor vehicle accident. In an aircraft accident. Can we conclude that flying is approximately 40 times safer than driving? Why?

20. Letters are drawn one at a time from a box containing the letters *O, B, B*. What is the probability that the letters are drawn in an order such that they spell the word "*BOB*"?

21. What is the probability of being dealt a five-card poker hand consisting of all spades?

22. A four-volume encyclopedia is placed on a bookshelf in random order. What is the probability that one or more of the volumes is not in its correct order?

23. From a group of five management and four union representatives, three individuals are to be selected to form a grievance committee. If the selection of the three individuals for the committee is completely random, what is the probability that the committee will include exactly two union representatives?

24. A group of five people are gathered for lunch. What is the probability that at least two of the five people were born during the same month?

25. Mark Thomas, president of Thomas Ford, Inc., has gathered the following historical data on car sales:

		Method of Payment		
		Cash (C)	*Time* (T)	*Total*
Type of car purchased (percent)	New (*N*)	12	30	42%
	Used (*U*)	20	38	58%
	Total	32%	68%	100%

(a) Would the numbers in the table be classified as objective or subjective probabilities?

(b) What is the marginal probability of an individual's paying cash for a car?

(c) Determine the joint probability of a purchase's being a new car and on time.

(d) Determine the conditional probability that a used car purchaser will pay cash.

(e) Are the events "new car purchase" and "payment by cash" statistically independent?

26. On the basis of his knowledge of Ajax Corporation and the stock market, an investment counselor developed the following probability table:

**Expected Performance of Ajax Corporation Stock
and the Dow–Jones Industrial Index**

		Stock Price—Ajax Corp.		
		Increase (A)	Decrease (A')	Total
Dow–Jones Index	Increase (B)	0.50	0.10	0.60
	Decrease (B')	0.20	0.20	0.40
	Total	0.70	0.30	1.00

(a) Would the probabilities be classed as objective or subjective?
(b) What is the marginal probability of an increase in the price of Ajax Corporation stock?
(c) What is the probability of an increase in Ajax Corporation stock, given an increase in the Dow–Jones Index?
(d) What is the probability of a decrease in Ajax Corporation stock, given an increase in the Dow–Jones Index?
(e) Are the events "increase in Ajax Corporation stock" and "increase in the Dow–Jones Index" statistically independent?

27. Suppose that 40 percent of companies in a certain industry have at least one lawyer on the board of directors and 70 percent have at least one banker. Assuming that the events "lawyer" (L) and "banker" (B) are independent, determine the proportion of companies in the industry that have both a lawyer and a banker on their board of directors.

28. An analysis of the sales of Brooks Brothers Clothiers shows that 20 percent of the purchases were made by men and that 40 percent of the purchases were over $20 in value. The records also show that 40 percent of the purchases of over $20 were made by men.
(a) What is the probability of a purchase of over $20, given that the customer is a man?
(b) What is the probability of a purchase of less than $20, given that the customer is a woman?
(c) Are the events "man" and "over $20" dependent?

29. Management of the Caroline Corporation is currently negotiating for two separate contracts, A and B. Management believes that the probability of winning contract A is 0.60. Contract A will be decided before contract B, and the winner of A will have a definite advantage in the negotiations

for *B*. Management believes that the probability of winning *B* is 0.70 if they first win *A*. However, if they fail to win *A*, the probability of winning *B* drops to 0.10.

(a) What is the marginal probability of winning contract *B*?

(b) What is the joint probability of losing both contracts?

30. The Sutton Corporation must raise funds for expansion. Management believes that there is a 70 percent chance of raising the funds through a private offering of common stock. They also believe that there is a 90 percent chance of obtaining the necessary capital through a public offering, provided that a financial report on Sutton is favorable. If, however, the report is not favorable, the probability of raising the necessary funds from a public stock offering drops to 0.50. Management estimates that the probability of receiving a favorable report is 0.60. Should a public or a private stock offering be made?

SUGGESTED REFERENCES

CHOU, YA-LUN, *Statistical Analysis* (New York, N.Y.: Holt, Rinehart and Winston, Inc., 1969).

DeGROOT, MORRIS H., *Optimal Statistical Decisions* (New York, N.Y.: McGraw-Hill Book Company, Inc., 1970).

EWART, P., J. FORD, and C. LIN, *Probability for Statistical Decision Making* (Englewood Cliffs, N.J.: Prentice-Hall, Inc., 1974).

HAMBURG, MORRIS, *Statistical Analysis for Decision Making* (New York, N.Y.: Harcourt, Brace & World, Inc., 1970).

HAYS, WILLIAM L., and ROBERT L. WINKLER, *Statistics* (New York, N.Y.: Holt, Rinehart and Winston, Inc., 1971).

LAPIN, LAWRENCE L., *Statistics for Modern Business Decisions* (New York, N.Y.: Harcourt Brace Jovanovich, Inc., 1973).

PETERS, W. S., and G. W. SUMMERS, *Statistical Analysis for Business Decisions* (Englewood Cliffs, N.J.: Prentice-Hall, Inc., 1968).

RAIFFA, HOWARD, *Decision Analysis* (Reading, Mass.: Addison-Wesley Publishing Company, Inc., 1968).

SCHLAIFER, ROBERT, *Analysis of Decisions Under Uncertainty* (New York, N.Y.: McGraw-Hill Book Company, Inc., 1969).

SPURR, W. A., and C. P. BONINI, *Statistical Analysis for Business Decisions* (Homewood, Ill.: Richard D. Irwin, Inc., 1973).

Chapter Six

Probability Functions

In Chap. 5 we introduced the basic concepts of probability. In this chapter we extend these concepts to include selected *probability functions* or, alternatively, *probability distributions*.[1] The reader will remember from Chap. 2 that a function is a mathematical relationship in which the values of a single dependent variable are determined from the values of one or more independent variables. A probability function, like the functions discussed in Chap. 2, describes the relationship between the dependent and the independent variable. In a probability function, the dependent variable gives the probability of the occurrence of an event and the independent variable is a numerical value that is assigned to the event. A probability function is thus used to determine the probability of an event that is defined in terms of a numerical value.

As a simple example of a probability function, consider a problem involving testing flashbulbs for cameras. In this problem, a probability function could be used to determine the probability that out of a package of, say, ten flashbulbs, all ten flashbulbs will work. The dependent variable would be probability, and the independent variable would be the number of flashbulbs that work. The relationship between the number of flashbulbs that work and probability would be given by the probability function.

Certain probability functions are widely used as an aid in making decisions. Among the more important are the hypergeometric, the binomial, and the normal probability functions. These distributions, along with the basic concept of a probability function, are discussed in this chapter.

---

[1] The terms *probability function* and *probability distribution* can be used interchangeably.

6.1   Random Variables

We have stated that a probability function describes the relationship between the probability of an event and a numerical value that represents the event. In our introductory remarks the probability of the event was referred to as the dependent variable. Although the properties of functions that were discussed in Chap. 2 apply to probability functions, the terminology commonly used in probability functions differs from that introduced in Chap. 2. One of the major differences is in the use of the term "random variable" instead of the term "independent variable."

The term *random variable* refers to a rule for assigning numerical values to the outcome of a random process or an experiment. As an example, consider the experiment of tossing two coins and recording the outcome of each toss as a head or a tail. The sample space for this experiment is

$$S = \{(HH), (HT), (TH), (TT)\}$$

Suppose that we are interested in the number of heads resulting from the toss of the coins. If this were the case, we could define X as equaling the number of heads resulting from the random process. In this experiment, X would be the random variable. The rule for determining the value of the random variable would be to assign to X the number of heads resulting from the toss of the coins. The possible values of the random variable would thus be $x = 0$, $x = 1$, and $x = 2$. Notice that the random variable X is designated by an uppercase letter and values of the random variable are designated by lowercase letters.

It is important to note that a random variable refers to a rule that links the elements in the sample space with the set of real numbers. In the preceding example the sample space consisted of the set $S = \{(HH), (HT), (TH), (TT)\}$. The rule involved assigning as values of the random variable the number of heads resulting from the toss of the coins. The real numbers were the values of the random variable, namely, $x = 0$, $x = 1$, and $x = 2$.

Example. An experiment consists of inspecting three randomly selected parts. Each part is classified as good, g, or defective, d. The sample space for this experiment is

$$S = \{(ggg), (ggd), (gdg), (dgg), (gdd), (dgd), (ddg), (ddd)\}$$

Suppose that we are interested in the number of defective parts in our sample. The sample points can be converted into numerical values by defining the random variable X as equaling the number of defectives. The values of the random variable are thus $x = 0$, $x = 1$, $x = 2$, and $x = 3$.

Example. An experiment consists of rolling a pair of dice. The sample space for this experiment is

$S = \{(1, 1), (1, 2), (1, 3), (1, 4), (1, 5), (1, 6), (2, 1), (2, 2), (2, 3), (2, 4), (2, 5),$
$(2, 6), (3, 1), (3, 2), (3, 3), (3, 4), (3, 5), (3, 6), (4, 1), (4, 2), (4, 3), (4, 4),$
$(4, 5), (4, 6), (5, 1), (5, 2), (5, 3), (5, 4), (5, 5), (5, 6), (6, 1), (6, 2), (6, 3),$
$(6, 4), (6, 5), (6, 6)\}$

Suppose that we are interested in the sum of the numbers on the dice resulting from the roll. The random variable X would then be defined as the sum of the numbers on the pair of dice. The possible values of the random variable would be $x = 2, 3, \ldots, 12$.

6.1.1   TRADITIONAL DEFINITION OF A RANDOM VARIABLE

On the basis of the example in the preceding section, we can offer a formal definition of the term *random variable*. A random variable is traditionally defined as a function (or rule) whose domain is the sample space of an experiment and whose range is the set of real numbers. The rule of the function links (or maps) the elements in the domain (i.e., sample space) with the set of real numbers. In other words, the random variable is a rule that assigns numerical values to the elements of the sample space. The values that are assigned to the elements of the sample space are the values of the random variable.

The term *random* comes from the fact that the value of the variable is determined from the outcome of an experiment or a random process. Thus, the particular numerical value of the random variable is not known until the outcome of the experiment has been observed. The possible numerical values are, however, known before the experiment, and the probability of the occurrence of alternative values of the random variable can be calculated from the probability function.

6.1.2   DISCRETE AND CONTINUOUS RANDOM VARIABLES

Random variables are commonly classified as being either *discrete* or *continuous*. A random variable is termed discrete if the set of possible values of the random variable contains a finite or a countably infinite number of numerical values. For instance, in the experiment involving tossing two coins and recording the number of heads, the set of possible values of the random variable was

$$\text{(1)} \qquad \{x \mid x = 0, 1, 2\}$$

This set contains a finite number of numerical values. Consequently, the random variable is discrete.

As another example of a discrete random variable, consider an experiment that involves counting the number of defects in a bolt of cloth. If the random variable X represents the number of defects, the set of possible values of X is

$$\{x \mid x = 0, 1, 2, 3, \ldots\}$$

Since this set has a countably infinite number of numerical values, the random variable is discrete.

A random variable is termed continuous if the random variable can have any value along a continuum between specific limits. To illustrate, consider an experiment that involves measuring the time between arrivals of customers at a service facility. If the random variable X represents time between arrivals of customers, the possible values of the random variable are described by the set

$$\{x \mid 0 \leq x \leq \infty\}$$

In this example, the random variable can have any one of the infinite number of values along the continuum from $x = 0$ to $x = \infty$.

Suppose that we modify the above example by assuming that the maximum possible time between arrivals of customers is 60 minutes. Based on this assumption, the possible values of the random variable are described by the set

$$\{x \mid 0 \leq x \leq 60\}$$

Since the random variable can have any of the infinite number of possible values along the continuum between $x = 0$ and $x = 60$, the random variable X is again continuous.

In the preceding example, we assume that the random variable can have any value in the continuum between 0 and 60. Obviously, however, the values of the random variable will be limited by the accuracy with which we measure the outcome of the experiment. In the example, for instance, we may measure time between arrivals to the nearest one-tenth of a second. Although as a practical matter, a measurement to the nearest one-tenth second may be satisfactory, it would be theoretically possible to measure the time between arrivals to any desired precision. For this reason, random variables that can theoretically have any value in a continuum are classified as continuous rather than discrete.

To distinguish between the random variable and the value of the random variable, we customarily denote the random variable by a capital letter and possible values of the random variable by a lowercase letter. Thus, the random variable X represented time between arrivals in the preceding example and the possible values of the random variable were $0 \leq x \leq 60$.

Example. In testing automobile head lamps, an inspector selects a lot of five lamps and determines (1) if the lamps function properly and (2) the length of time the lamps will burn before failure. If we define the random variable D as the number of defectives in the sample, D is a discrete random variable. The possible values of D are $\{d \mid d = 0, 1, 2, 3, 4, 5\}$. If we define L as the length of time a randomly selected lamp will burn before failure, L is a continuous random variable. The possible values of L are $\{l \mid 0 \le l \le \infty\}$.

6.2   Probability Functions

To introduce the basic concept of a probability function, consider again the experiment of tossing two coins and recording the outcome of each toss as a head or a tail. The sample space for this experiment was

$$S = \{(HH), (HT), (TH), (TT)\}$$

Assume that we define the random variable X as representing the number of heads resulting from the toss of the coins. The possible values of X were $x = 0, 1, 2$. Since each outcome in the sample space is equally likely, the probability of no heads is $P(X = 0) = \frac{1}{4}$, the probability of one head is $P(X = 1) = \frac{1}{2}$, and the probability of two heads is $P(X = 2) = \frac{1}{4}$. The functional relationship between the value of the random variable and the probability of occurrence of that value can be described by the set

$$\{(0, \tfrac{1}{4}), (1, \tfrac{1}{2}), (2, \tfrac{1}{4})\}$$

This set is an example of a probability function.

The relation between the number of heads and the probability of occurrence can also be given in tabular or graphic form. The probability function is given in tabular form in Table 6.1 and in graphic form in Fig. 6.1. Both the table and the figure show each possible value of the random variable together with the probability of the occurrence of the value.

Table 6.1.   Probability Table for Number of Heads

x	P(X = x)
0	$\frac{1}{4}$
1	$\frac{1}{2}$
2	$\frac{1}{4}$

Example. Consider the experiment of tossing a pair of dice. Let the random variable X be defined as the sum of the spots on the pair of dice. Construct a probability table that describes the probability of each possible value of the random variable.

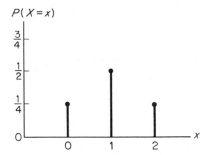

Figure 6.1.   Plot of probability function for number of heads.

The sample space for this experiment was given on p. 176. The possible values of the random variable, the number of sample points associated with each possible value of the random variable, and the probability of occurrence of each value of the random variable are shown in Table 6.2.

Table 6.2.   Probability Table for Roll of Dice

x	No. of Sample Points	P(X = x)
2	1	$\frac{1}{36}$
3	2	$\frac{2}{36}$
4	3	$\frac{3}{36}$
5	4	$\frac{4}{36}$
6	5	$\frac{5}{36}$
7	6	$\frac{6}{36}$
8	5	$\frac{5}{36}$
9	4	$\frac{4}{36}$
10	3	$\frac{3}{36}$
11	2	$\frac{2}{36}$
12	1	$\frac{1}{36}$
	$\overline{36}$	$\frac{36}{36} = 1.0$

6.2.1   DISCRETE PROBABILITY FUNCTIONS

The preceding examples illustrate that a probability function describes the relationship between the value of a random variable and the probability of the occurrence of the different possible values of the random variable. More formally, we can state that if X is a discrete random variable with values $x_1, x_2, \ldots, x_n$, then the probability function $p(x)$ is given by

$$p(x) = P(X = x_i) \qquad \text{for } i = 1, 2, \ldots, n \qquad (6.1)$$

A probability function for a discrete random variable is called a *discrete probability function* or, alternatively, a *probability mass function*. From the axioms of probability discussed in Chap. 5, we recognize that a discrete probability function has the following properties:

1. $p(x) \geq 0$ for all values of X, that is, $p(x)$ cannot be negative.
2. $\sum_{\text{all } x} p(x) = 1$, that is, the sum of the probabilities of each separate value of the random variable must be 1.

To illustrate a discrete probability function, consider an example involving testing a certain electronic part. From past experience it has been determined that the probability of a randomly selected part's being defective is 0.3. Five parts are selected for testing. The probability mass function that describes the probabilities of various numbers of defectives has been determined and is

$$p(x) = \binom{5}{x} (0.3)^x (0.7)^{5-x} \qquad \text{for } x = 0, 1, 2, \ldots, 5$$

The probability of no defectives in the lot of five parts is calculated by evaluating the probability mass function for $x = 0$. The probability is

$$P(X = 0) = p(0) = \frac{5!}{0! \, 5!} (0.3)^0 (0.7)^5 = 0.1681$$

Similarly, the probability of exactly one defective is

$$P(X = 1) = p(1) = \frac{5!}{1! \, 4!} (0.3)^1 (0.7)^4 = 0.3602$$

The probabilities of two, three, four, and five defectives are found in the same manner. The probabilities are shown in Table 6.3.

Table 6.3.

Number of defectives x	0	1	2	3	4	5
Probability $p(x)$	0.1681	0.3602	0.3087	0.1323	0.0283	0.0024

Notice in Table 6.3 that $p(x) \geq 0$ for all possible values of the random variable X. Furthermore, the sum of the probabilities of the random variables is equal to

$$\sum_{x=0}^{x=5} p(x) = 0.1681 + 0.3602 + 0.3087 + 0.1323 + 0.0283 + 0.0024 = 1.0000$$

These two properties illustrate the requirements that the probability of a random variable be greater than or equal to zero and that the sum of the probabilities of all possible values of the random variable equals one.

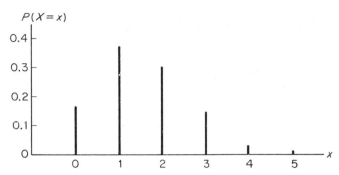

Figure 6.2.

The probabilities from Table 6.3 are plotted in Fig. 6.2. Since the random variable in the example is the number of defective units, probabilities are associated only with integer numbers. This is shown in the figure by representing the probabilities as straight lines rising from integer values along the horizontal axis. This method of presentation emphasizes the discrete nature of this distribution.

Example. Using the probability distribution in Table 6.3, determine $P(X \leq 1)$ and $P(X \geq 2)$.

$$P(X \leq 1) = P(X = 0) + P(X = 1) = 0.1681 + 0.3602 = 0.5283$$

$$P(X \geq 2) = P(X = 2) + P(X = 3) + P(X = 4) + P(X = 5)$$
$$= 0.3087 + 0.1323 + 0.0283 + 0.0024$$
$$= 0.4717$$

Example. Western Properties, Inc., has large holdings of land in northern California. This land has been subdivided and is being sold in small parcels to individuals for vacation and retirement homes. The sales manager of Western Properties has found that 40 percent of the prospective customers who actually visit the property purchase a parcel of land. A probability mass function that describes the probability of the number of purchases, X, in a group of n customers has been determined and is

$$p(x) = \binom{n}{x} (0.40)^x (0.60)^{n-x}$$

Suppose that four customers visit the property on a certain weekend. The probability of purchases by zero, one, two, three, or all four of the customers can be calculated from the probability mass function. The probabilities are

$$p(0) = \frac{4!}{0!\,4!}\,(0.40)^0(0.60)^4 = 0.1296$$

$$p(1) = \frac{4!}{1!\,3!}\,(0.40)^1(0.60)^3 = 0.3456$$

$$p(2) = \frac{4!}{2!\,2!}\,(0.40)^2(0.60)^2 = 0.3456$$

$$p(3) = \frac{4!}{3!\,1!}\,(0.40)^3(0.60)^1 = 0.1536$$

$$p(4) = \frac{4!}{4!\,0!}\,(0.40)^4(0.60)^0 = 0.0256$$

$$\sum p(x) = 1.0000$$

The reader will notice that the probability mass function used in this example is similar to the one used in the example on p. 180. This function, termed the *binomial distribution*, is widely applied and will be discussed later in this chapter. Based on the function, we see that the probability of land purchases by zero, one, two, three, or all four customers are, respectively, 0.1296, 0.3456, 0.3456, 0.1536, and 0.0256. Again, the probabilities are nonnegatives for all values of the random variable, and the sum of the probabilities is one.

Example. Using the probability distribution in the preceding example, determine the probability that at least two customers purchase parcels of land during a certain weekend.

$$P(X \geq 2) = P(X = 2) + P(X = 3) + P(X = 4)$$
$$= 0.3456 + 0.1536 + 0.0256 = 0.5248$$

6.2.2 CONTINUOUS PROBABILITY FUNCTIONS

Continuous random variables, the reader will remember, can have any value along a continuum between specific limits. If, for example, an experiment involved weighing individuals, then the weights of the individuals would be a continuous random variable. In this example, the probability of a randomly selected individual's weight being between a lower and an upper limit, say 140 to 160 pounds, is described by a *continuous probability function*, or alternatively, a *probability density function*.

Fig. 6.3 shows a hypothetical continuous probability function. The random variable in this figure is the weight of adult men. The continuous probability function, denoted in the figure as $f(x)$, describes the distribution of weights of adult men.

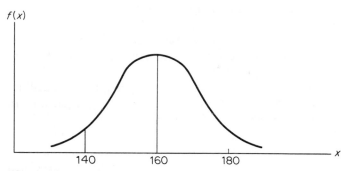

Figure 6.3.

A continuous random variable, such as the weight of adult men, is represented by a continuum of possible values. Because of this fact, *the probability associated with any specific value of the random variable is zero.* Instead, then, of determining the probability of a specific value of a continuous random variable, we determine the probability that the value of the random variable will be in some interval. In our example, this interval is from 140 to 160 pounds.

The probability that a randomly selected man would weigh between 140 and 160 pounds is represented in Fig. 6.4 by the shaded area. To determine this probability, we would either use tables that are available for certain continuous probability distributions or the methods of integral calculus. Regardless of which method is used, the shaded area represents the probability that the random variable falls in the specified interval. Our task, therefore, is merely to determine the area under the continuous probability function in the specified interval. The use of tables for determining probabilities described by the *normal probability function* is presented in Sec. 6.5. Probabilities that

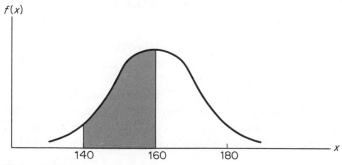

Figure 6.4.

must be calculated using integral calculus are, of course, beyond the scope of this text.[2]

The properties of a continuous probability function are, in many respects, similar to those of a discrete probability function. First, in both the continuous and discrete case, the probability function must be greater than or equal to zero for all values of the random variable. Second, the probabilities associated with all possible values of the random variable must equal one. In the continuous case, this means that the area under the continuous probability function is equal to one. In the discrete case, this means that the probabilities associated with each of the discrete values of the random variable must sum to one. Finally, probabilities for continuous random variables are associated with intervals, whereas probabilities for discrete random variables are associated with the finite or countably infinite number of discrete values of the random variable. These properties are summarized in Table 6.4.

Table 6.4.

Property	*Probability Function*	
	Discrete	*Continuous*
1. Probabilities are non-negative.	$p(x) \geq 0$	$f(x) \geq 0$
2. Probabilities sum to one.	$\sum_{\text{all } x} p(x) = 1$	Area under the density function for the continuum of possible values of the random variable is one.
3. $P(a \leq X \leq b)$.	Sum discrete probabilities in the interval $a \leq x \leq b$	Determine area in the interval $a \leq x \leq b$.

Example. On a certain assembly line, the time required in minutes to complete an assembly is described by the probability density function

$$f(t) = \tfrac{1}{8}t \qquad \text{for } 0 \leq t \leq 4$$

Verify that $f(t)$ has the properties of a probability density function and calculate the probability that the time required to complete an assembly will be less than 2 minutes.

The density function is plotted in Fig. 6.5, which shows that $f(t) \geq 0$ for $0 \leq t \leq 4$. Using the formula for the area of a triangle, we note that the area under the triangular shaped curve between $t = 0$ and $t = 4$ is 1. Thus, the requirements for a continuous probability function are satisfied.

---

[2] The interested reader should refer to Robert L. Childress, *Mathematics for Managerial Decisions* (Englewood Cliffs, N.J.: Prentice-Hall, Inc., 1974).

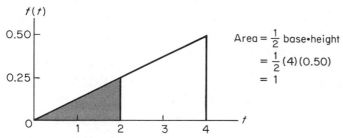

Figure 6.5.

The probability that an assembly will be completed in less than two minutes is given by the shaded area in the figure. Since this area is triangular shaped, the probability can be calculated from the formula for the area of a triangle, which is

$$\text{area} = \tfrac{1}{2}\,\text{base} \cdot \text{height}$$

The probability is

$$P(T \le 2) = \tfrac{1}{2}(2)(0.25) = 0.25$$

6.3   Measures of Central Tendency and Variability

There are several measures of central tendency and variability of a random variable that are widely used in decision making. Among the important measures of central tendency are the mean, or expected value, the median, and the mode. Two important measures of variability of a random variable are the variance and the standard deviation. Our objective in this section is to introduce these measures of central tendency and variability and to illustrate their importance as summary measures that describe the behavior of a random variable. Since this text deals with finite mathematics, our discussion will be limited to measures of central tendency and variability for discrete random variables.[3]

6.3.1   MEAN OR EXPECTED VALUE

Perhaps the most widely used measure of central tendency is the *mean*, or *expected value*, of a random variable. The expected value of a random variable can be interpreted as the average value of the random variable. Thus, if the

[3] Formulas for central tendency and variability for continuous random variables are similar to those for discrete random variables. The formulas for continuous random variables do, however, involve integral calculus and are thus beyond the scope of this text. Readers who have a background in calculus should refer to Robert L. Childress, *Mathematics for Managerial Decisions*, Chap. 16.

random variable X represents the number of defective parts in a lot of five parts, then the expected value or mean of the random variable is the average number of defectives one would find from repeated inspection of similar lots of five parts. Similarly, if the random variable represents the number of cars sold per week by a dealership, then the expected value of the random variable can be interpreted as the average number of cars sold per week by the dealership.

The mean, or expected value, of a random variable is denoted by the Greek letter μ (mu) or, alternatively, by $E(X)$. The expected value of a discrete random variable is defined as

$$\mu = E(X) = \sum_{\text{all } x} xp(x) \tag{6.2}$$

where $p(x)$ is the probability mass function, x is the random variable, and $\sum$ is the mathematical symbol for summation. The expected value of a random variable is, from Eq. (6.2), merely equal to summation of the products of each value of the random variable multiplied by the probability of that value.

To illustrate the procedure for determining the expected value of a discrete random variable, consider the example on p. 180 of testing electronic parts. From a group of parts, five were selected for testing. The probabilities of zero, one, two, three, four, or five defectives in the lot of five were given in Table 6.3 and are repeated in Table 6.5. The expected value of the random variable is calculated from Eq. (6.2) and is

$$E(X) = \sum_{\text{all } x} xp(x) = 0(0.1681) + 1(0.3602) + 2(0.3087) + 3(0.1323)$$
$$+ 4(0.0283) + 5(0.0024) = 1.500$$

Thus, if this experiment were repeated a large number of times, we would expect an average of 1.500 defectives in the lots of five items. This does not mean that we would expect 1.500 defectives in any one lot. Obviously, this is impossible. Instead, after numerous repetitions of this experiment the average number of defectives would be 1.500 per lot.

Table 6.5.

Number of defectives x	0	1	2	3	4	5
Probability $p(x)$	0.1681	0.3602	0.3087	0.1323	0.0283	0.0024

Example. Jerry Clark, the western representative of Mountain Rose Wine, calls on large markets and liquor stores in an attempt to obtain orders for Mountain Rose. He has found that the probability of obtaining orders for Mountain Rose can be described by the discrete probability function given in Table 6.6. Determine the expected value of the random variable.

Table 6.6.

Number of cases x	0	1	2	3	4	5
Probability $p(x)$	0.30	0.20	0.20	0.10	0.10	0.10

The random variable X represents the number of cases ordered. The expected value of X is

$$E(X) = \sum_{\text{all } x} xp(x) = 0(0.30) + 1(0.20) + 2(0.20) + 3(0.10)$$
$$+ 4(0.10) + 5(0.10) = 1.8$$

Thus, the long-run average number of cases ordered per store is 1.8.

Example. Bill Smith, owner of Smith Chevrolet, has gathered data for daily new car sales for the past 100 days. These data are shown in Table 6.7. Using these data, develop an empirical probability distribution for car sales per day and determine the mean sales per day.

Table 6.7.

Cars sold	0	1	2	3	4	5	6
No. of days	5	20	25	20	15	10	5

Letting X represent car sales per day, the discrete probability function is shown in Table 6.8. The expected value or mean of the random variable is

$$E(X) = \sum_{\text{all } x} xp(x) = 0(0.05) + 1(0.20) + 2(0.25) + 3(0.20)$$
$$+ 4(0.15) + 5(0.10) + 6(0.05) = 2.7$$

Thus, the average sales during the 100-day period were 2.7 cars per day.

Table 6.8.

Cars sold x	0	1	2	3	4	5	6
Probability $p(x)$	0.05	0.20	0.25	0.20	0.15	0.10	0.05

6.3.2 MEDIAN

A second important measure of central tendency in a probability distribution is the median of the distribution. Ideally, the median of a probability distribution is the value of the random variable that divides the distribution in half.

This value should be such that the probability is 0.50 that an experiment will result with the value of the random variable being either above or below the median. In a sense, then, the median is the 50th percentile of the distribution, i.e., numerous repetition of the experiment would result in 50 percent of the observations falling below the median and 50 percent falling above.

Although the above discussion presents the generally accepted concept of the median, it is not always possible in a discrete distribution to specify a value for the median such that the probability is exactly 0.50 that a random variable has a value either above or below the median. Therefore, a formal definition is in order. For a discrete random variable, the median is that value of the random variable x_m such that the probability is at least 0.50 that X is less than or equal to x_m and at least 0.50 that X is greater than or equal to x_m. This definition can be expressed as

$$P(X \leq x_m) \geq 0.50 \quad \text{and} \quad P(X \geq x_m) \geq 0.50 \tag{6.3}$$

where x_m is the median of the discrete distribution.

To illustrate determining the median of a discrete probability distribution, consider the distribution shown in Table 6.9. This distribution was first given in Table 6.3 and was repeated in Table 6.5. Notice from the table that

$$P(X \leq 1) = 0.5283 \quad \text{and} \quad P(X \geq 1) = 0.8319$$

Since both requirements of Eq. (6.3) are satisfied, the median of the distribution is $x_m = 1$.

Table 6.9.

Number of defectives x	0	1	2	3	4	5
Probability $p(x)$	0.1681	0.3602	0.3087	0.1323	0.0283	0.0024

As a second example, consider the probability distribution for new car sales at Smith Chevrolet. This distribution was given in Table 6.8 and is repeated in Table 6.10. From the table, we see that

$$P(X \leq 2) = 0.50 \quad \text{and} \quad P(X \geq 2) = 0.75$$

and from Eq. (6.3), $x = 2$ qualifies as the median.

Table 6.10.

Cars sold x	0	1	2	3	4	5	6
Probability $p(x)$	0.05	0.20	0.25	0.20	0.15	0.10	0.05

Closer inspection of Table 6.10 shows, however, that

$$P(X \leq 3) = 0.70 \quad \text{and} \quad P(X \geq 3) = 0.50$$

Thus, $x = 3$ also qualifies as the median. In fact, $P(X \leq x_m) \geq 0.50$ and $P(X \geq x_m) \geq 0.50$ are satisfied for any value of x_m such that

$$2 \leq x_m \leq 3$$

If we are asked to give one value for the median, a reasonable choice is to take the midpoint of the interval from 2 to 3. Using this method, the median is $x_m = 2.5$.

The preceding example illustrates that the median of a discrete probability distribution need not be a unique value. Rather, it can sometimes assume any value in an interval. When this is the case, we recommend the convention of selecting the midpoint of the interval as the median of the distribution.

Example. The probability distribution for sales of Mountain Rose Wine was given in Table 6.6 and is repeated in Table 6.11. Determine the median of this distribution.

Applying Eq. (6.3), we see that both $x = 1$ and $x = 2$ satisfy the requirement for the median. Based on the convention of selecting the midpoint between the two possible values, the median of this distribution is $x_m = 1.5$.

Table 6.11.

Number of cases x	0	1	2	3	4	5
Probability $p(x)$	0.30	0.20	0.20	0.10	0.10	0.10

Example. The probability distribution for land sales by Western Properties, Inc., was given on p. 181 and is repeated in Table 6.12. Determine the median of this distribution.

Table 6.12.

Number of parcels sold x	0	1	2	3	4
Probability $p(x)$	0.1296	0.3456	0.3456	0.1536	0.0256

Since $P(X \leq 2) = 0.8208$ and $P(X \geq 2) = 0.5248$, the median number of parcels sold is $x_m = 2$.

6.3.3 MODE

A third measure of central tendency in a probability distribution is the mode. The *mode* is roughly defined as the value of the random variable most likely to occur. For a discrete probability distribution, the mode is that value of the random variable with the highest probability of occurrence. To illustrate, consider the probability distribution given in Table 6.13. The value of the random variable that has the greatest probability of occurrence is $x = 1$. Consequently, the modal value of the random variable in Table 6.13 is $x = 1$.

Table 6.13.

Number of defectives x	0	1	2	3	4	5
Probability $p(x)$	0.1681	0.3602	0.3087	0.1323	0.0283	0.0024

In certain cases, it is possible to have more than one modal value of the random variable. To illustrate, consider the plot of the discrete probability distribution in Fig. 6.6. In this distribution the values $x = 2$ and $x = 5$ are both relatively likely to occur. In cases such as this, both values of the random variable are considered to be modes, and the distribution is said to be *bimodal*.

To take into account the possibility of more than one modal value in a probability distribution, an alternative and more precise definition of the mode must be given. More formally, a mode of a probability distribution is defined as a value of a random variable at which the probability function reaches a local maximum. For a discrete probability function, a local maximum occurs at any value of the random variable that has a greater probability than adjacent values of the random variable. Thus, in Fig. 6.7, the probability of $x = 2$ is greater than that of the adjacent values of the random variable. Therefore, $x = 2$ is a local maximum and a mode. Similarly, $x = 5$ has a greater probability of occurring than $x = 4$ or $x = 6$. Thus, $x = 5$ is also a mode.

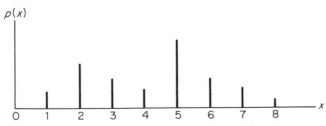

Figure 6.6.

Example. The probability distribution for sales of Mountain Rose Wine is given in Table 6.14. Give the mode for this distribution.

The mode for this distribution is $x = 0$.

Table 6.14.

Number of cases x	0	1	2	3	4	5
Probability $p(x)$	0.30	0.20	0.20	0.10	0.10	0.10

Example. The probability distribution for land sales of Western Properties, Inc., is given in Table 6.15. Determine the mode of this distribution.

Table 6.15.

Number of parcels sold x	0	1	2	3	4
Probability $p(x)$	0.1296	0.3456	0.3456	0.1536	0.0256

The random variables $x = 1$ and $x = 2$ are equally likely to occur. Since the probability of occurrence of these values is greater than any other values, the distribution is bimodal and the modes are $x = 1$ and $x = 2$.

6.3.4 VARIANCE AND STANDARD DEVIATION

The variance of a random variable provides a measure of the magnitude of variation of the values of the random variable about the expected value or mean of the random variable. Fig. 6.7 illustrates the concept of variance for a discrete random variable. The distributions in Figs. 6.7(a) and 6.7(b) both

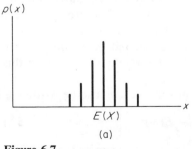

Figure 6.7.

have the same expected value. The values of the random variable in Fig. 6.7(a) are, however, more closely distributed about the expected value than those in Fig. 6.7(b). Since the variance provides a measure of the magnitude of the variation of the values of the random variable about the expected value, the variance of the distribution in Fig. 6.7(a) is less than that of the distribution in Fig. 6.7(b).

The formula for the variance of a discrete random variable is

$$\text{Var}\,(X) = \sum_{\text{all } x} (x - E(X))^2 p(x) \tag{6.4}$$

From the formula we note that the variance provides a measure of the squared deviation of the values of the random variable from the expected value of the random variable. Obviously, the variance of a random variable that is widely distributed about the expected value will be larger than that of a random variable that is closely distributed about the expected value.

The calculations for the variance of a discrete random variable using Eq. (6.4) are shown in Table 6.16. The expected value of the random variable

Table 6.16.

x	p(x)	xp(x)	x − E(x)	(x − E(x))²	(x − E(x))²p(x)
0	0.1	0	− 2.3	5.29	0.529
1	0.2	0.2	− 1.3	1.69	0.338
2	0.3	0.6	− 0.3	0.09	0.027
3	0.2	0.6	0.7	0.49	0.098
4	0.1	0.4	1.7	2.89	0.289
5	0.1	0.5	2.7	7.29	0.729
		$E(X) = \overline{2.3}$			$\text{Var}\,(X) = \overline{2.010}$

is calculated in the third column of the table and is $E(X) = 2.3$. The deviation of the random variable from the expected value is given in the fourth column, and the squared deviation is given in the fifth column. The product of the squared deviation and the probability mass function is given in the sixth column. The summation of the product of the squared deviation and the probability mass function gives the variance of the random variable. In this example the variance is $\text{Var}\,(X) = 2.01$.

The formula for the variance of a random variable can also be written as

$$\text{Var}\,(X) = E(X^2) - E(X)^2 \tag{6.5}$$

The formula shows that the variance can be calculated in two steps. The expected value of X^2, i.e., $E(X^2)$, is first determined. The square of the expected value, $E(X)^2$, is then subtracted from $E(X^2)$. For a discrete distribution, the expected value of X^2 is given by

$$E(X^2) = \sum_{\text{all } x} x^2 p(x) \tag{6.6}$$

In order to compare the two formulas for the variance, the variance of the discrete probability function given in Table 6.16 is calculated by using Eq. (6.5) in Table 6.17. The expected value of X is calculated in the third column of the table, and the expected value of X^2 is calculated in the fourth column. Using Eq. (6.5), we find that the variance is again Var $(X) = 2.01$.

Table 6.17.

x	p(x)	xp(x)	x²p(x)
0	0.1	0	0
1	0.2	0.2	0.2
2	0.3	0.6	1.2
3	0.2	0.6	1.8
4	0.1	0.4	1.6
5	0.1	0.5	2.5
		$E(X) = 2.3$	$E(X^2) = 7.3$

$$\text{Var }(X) = E(X^2) - E(X)^2 = 7.3 - (2.3)^2 = 2.01$$

Example. Use Eq. (6.4) to calculate the variance of new car sales at Smith Chevrolet.

The probability distribution for sales of new cars at Smith Chevrolet was given in Table 6.8 and is repeated in the first two columns in Table 6.18. The calculations necessary to determine the variance using Eq. (6.4) are given in the remaining columns of the table. The variance of new car sales is Var $(X) = 2.41$.

Table 6.18.

x	p(x)	xp(x)	x − E(X)	(x − E(X))²	(x − E(X))²p(x)
0	0.05	0	−2.70	7.29	0.3645
1	0.20	0.20	−1.70	2.89	0.5780
2	0.25	0.50	−0.70	0.49	0.1225
3	0.20	0.60	0.30	0.09	0.0180
4	0.15	0.60	1.30	1.69	0.2535
5	0.10	0.50	2.30	5.29	0.5290
6	0.05	0.30	3.30	10.89	0.5445
	$E(X) = 2.70$				Var $(X) = 2.4100$

Example. The probability distribution for sales of Mountain Rose Wine was first given in Table 6.6. Use Eq. (6.5) to determine the variance of the random variable in this distribution.

The calculations necessary to determine the variance using Eq. (6.5) are given in Table 6.19. The variance is Var $(X) = 2.76$.

Table 6.19.

x	p(x)	xp(x)	x^2p(x)
0	0.30	0	0
1	0.20	0.20	0.20
2	0.20	0.40	0.80
3	0.10	0.30	0.90
4	0.10	0.40	1.60
5	0.10	0.50	2.50
		$E(X) = 1.80$	$E(X^2) = 6.00$

$$Var\ (X) = E(X^2) - E(X)^2 = 6.00 - 3.24 = 2.76$$

The *standard deviation* is a second important measure of the variability of a random variable. The standard deviation, usually denoted by the Greek letter σ (sigma), is merely the square root of the variance. That is,

$$\text{standard deviation} = \sigma_x = \sqrt{Var\ (X)} \qquad (6.7)$$

Thus, if the variance of a random variable is Var $(X) = 4$, then the standard deviation is $\sigma_x = 2$. Similarly, if the variance of a random variable is Var $(X) = 0.49$, then the standard deviation is $\sigma_x = 0.7$.

Example. The variance of the random variable T is Var $(T) = 0.6$. Determine the standard deviation.

$$\sigma_t = \sqrt{Var\ (t)} = \sqrt{0.6}$$
$$= 0.77$$

Since the standard deviation is simply the square root of the variance, the reader may question the need for both the standard deviation and the variance as measures of the variability of a random variable. The variance has certain properties that are important in statistical analysis. These properties are beyond the scope of this text but are included in any of the referenced texts in statistics. Because the variance is calculated by squaring the deviations of the random variable from the expected value, it is difficult to interpret the meaning of the variance. For this reason, the standard deviation is important as a descriptive measure of the variability of a random variable.

6.4   Two Important Discrete Distributions

This section introduces two of the more commonly used discrete probability functions, the *hypergeometric distribution* and the *binomial distribution*.

6.4.1   THE HYPERGEOMETRIC PROBABILITY DISTRIBUTION

Many statistical experiments involve sampling from a finite population. If the finite population is *dichotomous*, i.e., consists of elements that may be classified into two *mutually exclusive categories*, and the sampling is performed *without replacement*, the *hypergeometric* probability distribution applies. To illustrate populations that are both finite and dichotomous, consider the following:

1. A lot of ten parts, of which eight are good and two are defective.
2. A group of 15 students, of which ten are men and five are women.
3. A company of 1000 employees, of which 200 are nonunion and 800 are union members.

When one is sampling without replacement from a dichotomous population, the actual number of elements in each of the two mutually exclusive classes is normally unknown. In fact, the reason for sampling from the finite population is to obtain an estimate of the number of elements in each of the two categories. This estimate can then be used in making decisions regarding the population.

In introducing the hypergeometric distribution, it is helpful to assume that the number of elements in each of the two mutually exclusive categories is known. This assumption permits one to determine the probability of obtaining a specified number of elements in the sample from each of the two mutually exclusive categories in the population. In other words, we shall limit our discussion to the hypergeometric probability distribution rather than the related problem of using the hypergeometric distribution in statistical inference. The use of a distribution, such as the hypergeometric distribution, in statistical inference is one of the major topics in a course in statistics.

To illustrate the hypergeometric distribution, consider a lot of ten electronic parts. Suppose that seven of the parts are good and three are defective. A sample of four parts is randomly selected from the lot of ten parts. Since the population is finite (i.e., the population contains a total of ten parts) and dichotomous (i.e., the elements in the population can be classified as good or defective), the hypergeometric distribution applies. Suppose we let X represent the number of defective parts in the sample. The possible values of X are $x = 0$, $x = 1$, $x = 2$, and $x = 3$. Our task is to determine the probability that $x = 0$, $x = 1$, $x = 2$, and $x = 3$.

The probabilities of the events zero defective, one defective, two defectives, and three defectives are found by using Eq. (5.4),

$$P(E) = \frac{n(E)}{n(S)} \qquad (5.4)$$

In this equation, $n(E)$ represents the number of ways in which X defectives can occur, and $n(S)$ represents the total number of ways in which the sample can be selected from the population. To determine the probability of zero defectives in the sample of size four, we first determine the number of ways in which four good parts can be selected from the seven good parts in the population. This is

$$_7C_4 = \binom{7}{4} = \frac{7!}{4!\,3!} = 35$$

The total number of ways in which the sample of size four can be selected is

$$_{10}C_4 = \binom{10}{4} = \frac{10!}{6!\,4!} = 210$$

Since the event zero defectives can occur in 35 different ways and the sample can be selected in 210 different ways, the probability of the event $X = 0$ defectives is

$$P(X = 0) = \frac{\binom{7}{4}}{\binom{10}{4}} = \frac{35}{210}$$

The probability of one defective in the sample of size four is calculated in a similar fashion. The number of ways of selecting one defective part from the three defective parts in the population is $_3C_1$, or, alternatively, $\binom{3}{1}$. The number of ways of selecting three good parts from the seven good parts in the population is $\binom{7}{3}$. Thus, the total number of ways of selecting one defective part and three good parts is

$$\binom{3}{1}\binom{7}{3} = \frac{3!}{1!\,2!} \cdot \frac{7!}{3!\,4!} = 105$$

The total number of ways of selecting a sample of size four from a population of size ten is again $\binom{10}{4}$, or 210. The probability of the event $X = 1$ is thus

$$P(X = 1) = \frac{\binom{3}{1}\binom{7}{3}}{\binom{10}{4}} = \frac{105}{210}$$

The probabilities of the events two defectives and three defectives are calculated in the same manner. The probability of two defectives is

$$P(X = 2) = \frac{\binom{3}{2}\binom{7}{2}}{\binom{10}{4}} = \frac{63}{210}$$

and the probability of three defectives is

$$P(X = 3) = \frac{\binom{3}{3}\binom{7}{1}}{\binom{10}{4}} = \frac{7}{210}$$

Since 0, 1, 2, and 3 account for all possible values of the random variable X, we would expect the probabilities associated with the values of the random variables to sum to 1.0. By summing the probabilities we see that this is indeed true, i.e.,

$$\sum_{x=0}^{x=3} p(x) = \tfrac{35}{210} + \tfrac{105}{210} + \tfrac{63}{210} + \tfrac{7}{210} = \tfrac{210}{210} = 1.0$$

The preceding example illustrates the hypergeometric distribution. To formalize the probability model, suppose we let

$N =$ the total number of elements in a finite, dichotomous population

$n =$ the size of the random sample selected from the population

$D =$ the number of elements in the population in one of the two mutually exclusive categories

$X =$ the number of elements in the sample with the same characteristic as the elements in D

The hypergeometric mass function is

$$h(x \mid N, n, D) = \frac{\binom{D}{x}\binom{N - D}{n - x}}{\binom{N}{n}} \qquad \text{for } x = 0, 1, 2, \ldots, k \qquad (6.8)$$

Notice in the formula that $\binom{D}{x}$ is the number of ways of selecting x elements from the D elements in the population, $\binom{N - D}{n - x}$ is the number of ways of selecting the remaining $n - x$ elements in the sample from the $N - D$ elements in the population, and $\binom{N}{n}$ is the total number of ways a sample of size n can be selected. The number k in the formula is equal to either n or D, whichever is smaller.

Example. In a group of twelve council members, five are known to be Republicans and seven are known to be Democrats. If a committee of four is selected by lottery, what is the probability of the committee's consisting of three Democrats and one Republican?

In this problem, $N = 12$, $n = 4$, $D = 7$, and $X = 3$. The probability is

$$P(X = 3 \mid 12, 4, 7) = \frac{\binom{7}{3}\binom{5}{1}}{\binom{12}{4}} = \frac{\dfrac{7!}{3!\,4!}\dfrac{5!}{1!\,4!}}{\dfrac{12!}{4!\,8!}} = \frac{175}{495}$$

Example. A particular manufacturer orders a certain electronic part in lots of ten. From the lot of ten parts, three are selected for testing. The lot is accepted if none of the three parts tested is found to be defective. If the lot of ten parts actually contains four defective parts, what is the probability of the lot's being accepted?

In this problem, $N = 10$, $n = 3$, $D = 4$, and $X = 0$. The probability is

$$P(X = 0 \mid 10, 3, 4) = \frac{\binom{4}{0}\binom{6}{3}}{\binom{10}{3}} = \frac{\dfrac{6!}{3!\,3!}}{\dfrac{10!}{7!\,3!}} = \frac{1}{6}$$

The expected value of a random variable that is distributed according to the hypergeometric distribution is

$$\mu = E(X) = n\left(\frac{D}{N}\right) \tag{6.9}$$

and the variance is

$$\text{Var}(X) = n\left(\frac{D}{N}\right)\left(\frac{N - D}{N}\right)\left(\frac{N - n}{N - 1}\right) \tag{6.10}$$

These formulas were derived by substituting the hypergeometric mass function for $f(x)$ in Eqs. (6.2) and (6.4). The use of these formulas makes it possible to determine the mean and the variance directly from the parameters of the distribution. This, of course, has the advantage of eliminating the calculations required by Eqs. (6.2) and (6.4).

Example. Determine the expected value and variance of a random variable that is distributed according to the hypergeometric distribution with parameters $N = 10$, $n = 3$, and $D = 4$.

$$E(X) = n\left(\frac{D}{N}\right) = 3\left(\frac{4}{10}\right) = 1.20$$

$$\text{Var}(X) = n\left(\frac{D}{N}\right)\left(\frac{N-D}{N}\right)\left(\frac{N-n}{N-1}\right) = 3\left(\frac{4}{10}\right)\left(\frac{6}{10}\right)\left(\frac{7}{9}\right)$$

$$= 0.55$$

6.4.2 THE BINOMIAL PROBABILITY DISTRIBUTION

In the preceding section we explained that the hypergeometric distribution applies when one is sampling from a finite, dichotomous population. If, as is true in many problems, the population is either infinite or so large that it can effectively be considered as infinite, the hypergeometric distribution no longer applies. Provided that certain requirements are met, the random variable is instead distributed according to the binomial probability distribution.

To illustrate, consider the population of registered voters in the State of California. Suppose that 40 percent of these voters favor candidate A. If a sample of size 20 is randomly selected from this population, the binomial probability distribution can be used to determine the probability of $x = 0$, $x = 1$, $x = 2, \ldots, x = 20$ voters in the sample favoring candidate A.

Notice in this example that the population, although finite, is large enough to essentially be considered infinite. Furthermore, the population is dichotomous; i.e., each person interviewed may be classified as favoring or not favoring candidate A. Although it is not stated in the example, we assume that the interviews of the voters are conducted so as to be independent, i.e., the results of the previous interviews have no effect on the outcome of subsequent interviews. Finally, the probability of a person's favoring candidate A is $p = 0.40$ and the probability of the person's not favoring candidate A is $(1 - p) = 0.60$. Since the results of each interview are independent, these probabilities remain constant throughout the experiment.

A sampling experiment that is conducted under conditions such as those described above is termed a Bernoulli process (named after Jacques Bernoulli, 1654–1705). The requirements for a Bernoulli process can be summarized as follows:

1. Each trial of the experiment results in one of two possible outcomes. These outcomes are often termed "success" or "failure," "defective" or "nondefective," "yes" or "no," etc.

2. Each trial is independent of all other trials. This means that the probability of a "success" on any given trial is the same, regardless of the outcomes of the preceding trials.
3. The probability of an outcome, p, remains constant from trial to trial.

Given that the requirements for a Bernoulli process are met, the random variable X that represents the number of "successes" in a sample of size n taken from the population is distributed according to the binomial probability distribution.

To develop the binomial mass function, consider a simple experiment that involves randomly selecting a sample of size five from an infinite population. Assume that the elements in the population can be classified as "success" or "failure" and that the probability of a success on any given trial is 0.20. Let the random variable X represent the number of successes in the sample of size 5. The possible values of X are 0, 1, 2, 3, 4, and 5.

The event zero successes can occur only if all five trials result in failures. Since the trials are independent and the probability of a failure on any single trial is 0.80, the probability of five successive failures is

$$P(X = 0) = (0.80)^5 = 0.3277$$

The event one success occurs when one trial results in a success and four trials result in failures. Since the success can occur on any of the five trials, there are $\binom{5}{1}$ or 5 different sequences in which the success can occur. These are

$$\{(s, f, f, f, f), (f, s, f, f, f), (f, f, s, f, f), (f, f, f, s, f), (f, f, f, f, s)\}$$

The probability of any one of these sequences is $(0.20)^1(0.80)^4$. Consequently, the probability of exactly one success is

$$P(X = 1) = \binom{5}{1} (0.20)^1(0.80)^4 = \frac{5!}{1! \, 4!} (0.20)^1(0.80)^4 = 0.4096$$

The event two successes occurs when two trials result in successes and three trials result in failures. The number of sequences in which two successes can occur is $\binom{5}{2}$, or ten different sequences. The probability of two successes and three failures on any sequence is $(0.20)^2(0.80)^3$. The probability of the event two successes is thus

$$P(X = 2) = \binom{5}{2} (0.20)^2(0.80)^3 = \frac{5!}{2! \, 3!} (0.20)^2(0.80)^3 = 0.2048$$

The probabilities of the events three successes, four successes, and five successes are found in the same manner. The probabilities are

$$P(X = 3) = \binom{5}{3}(0.20)^3(0.80)^2 = \frac{5!}{3!\,2!}(0.20)^3(0.80)^2 = 0.0512$$

$$P(X = 4) = \binom{5}{4}(0.20)^4(0.80)^1 = \frac{5!}{4!\,1!}(0.20)^4(0.80)^1 = 0.0064$$

$$P(X = 5) = \binom{5}{5}(0.20)^5(0.80)^0 = \frac{5!}{5!\,0!}(0.20)^5(0.80)^0 = 0.0003$$

On the basis of the preceding example, we can offer a probability mass function for a random variable that is distributed according to the binomial distribution. Suppose we let

$$p = \text{probability of ``success'' on a single trial}$$

$$n = \text{the number of trials (i.e., the sample size)}$$

$$X = \text{the number of ``successes'' in the } n \text{ trials}$$

The probability mass function for the random variable X is

$$b(x \mid n, p) = \binom{n}{x} p^x (1 - p)^{n-x} \qquad \text{for } x = 0, 1, \ldots, n \qquad (6.11)$$

This formula gives the probability of x successes in a series of n Bernoulli trials. The resulting probability distribution is called the *binomial probability distribution*.

Example. The probability that an item produced by a certain manufacturing process is defective is $p = 0.10$. If a sample of four items is selected for testing, what is the probability of no defectives? Of one defective? Of two defectives?

The probability of no defectives is

$$P(X = 0 \mid 4, 0.10) = \binom{4}{0}(0.10)^0(0.90)^4 = 0.6561$$

the probability of one defective is

$$P(X = 1 \mid 4, 0.10) = \binom{4}{1}(0.10)^1(0.90)^3 = 0.2916$$

and the probability of two defectives is

$$P(X = 2 \mid 4, 0.10) = \binom{4}{2}(0.10)^2(0.90)^2 = 0.0486$$

Eq. (6.11) provides the mathematical definition of the binomial mass function. Rather than actually computing the binomial probabilities, however, tables of the binomial distribution are used. Table A.9 in the Appendix contains the binomial probabilities for selected values of n and p. The use of this table is illustrated by the following examples.

Example. In a nationwide survey of business executives, 50 percent indicated that they believed that corporate profits would increase during the coming year. If a sample of size 12 is selected from this population, determine the probability of exactly four executives stating that profits will increase.

For $X = 4$, $n = 12$, and $p = 0.40$, the binomial probability is

$$P(X = 4 \mid 12, 0.40) = \binom{12}{4} (0.40)^4 (0.60)^8$$

Instead of calculating the probability, we can read the probability directly from Table A.9. The probability is 0.2128.

Example. In the preceding example, determine the probability of between three and five executives stating that they believe profits will increase.

The parameters of the distribution are again $n = 12$ and $p = 0.40$. The probability of between three and five executives favoring the increase is

$$\begin{aligned} P(3 \leq X \leq 5) &= P(X = 3) + P(X = 4) + P(X = 5) \\ &= 0.1419 + 0.2128 + 0.2270 \\ &= 0.5817 \end{aligned}$$

Example. A survey indicated that 70 percent of the television viewers watched the past year's Super Bowl game. If five television viewers are selected at random, what is the probability that exactly four of the five watched the game?

For $X = 4$, $n = 5$, and $p = 0.70$, the binomial probability is

$$P(X = 4 \mid 5, 0.70) = \binom{5}{4} (0.70)^4 (0.30)^1$$

Since Table A.9 includes values of p only up to $p = 0.50$, we must recognize that the probability of four "successes" in five trials with a probability of 0.70 of success is equivalent to the probability of one failure in five trials with a probability of 0.30 of failure. The binomial probability is thus

$$P(X = 1 \mid 5, 0.30) = \binom{5}{1} (0.30)^1 (0.70)^4$$

From Table A.9, the answer is 0.3602.

The mean or expected value of a random variable X that is distributed according to the binomial distribution is

$$\mu = E(X) = np \tag{6.12}$$

and the variance is

$$\text{Var}(X) = np(1 - p) \tag{6.13}$$

These formulas were derived by substituting $b(x \mid n, p)$ for $p(x)$ in Eqs. (6.2) and (6.4).

Example. Determine the expected value and variance of a random variable that is distributed according to the binomial distribution with $n = 12$ and $p = 0.40$.

From Eq. (6.12), the expected value is

$$E(X) = np = 12(0.40) = 4.8$$

and from Eq. (6.13), the variance is

$$\text{Var}(X) = np(1 - p) = 12(0.40)(0.60) = 2.88$$

6.5   The Normal Distribution

In this final section we introduce one of the most widely used continuous probability functions. This distribution, termed the *normal distribution*, describes many continuous random variables in business as well as in the social, physical, and biological sciences. In addition, as those readers will discover who study statistics, it forms the basis for many problems in statistical analysis.

The density function for the normal distribution is

$$f(x) = \frac{1}{\sigma\sqrt{2\pi}}\, e^{-\frac{1}{2}[(x-\mu)/\sigma]^2} \qquad \text{for } -\infty < x < \infty \tag{6.14}$$

where σ and μ are parameters of the distribution representing the standard deviation and the mean, or expected value, respectively. The distribution is symmetrical about the mean μ and is asymptotic to the horizontal axis. These features lead to a description of the normal distribution as bell-shaped, as shown by Fig. 6.8.

As we mentioned earlier, probabilities are determined for intervals for continuous random variables rather than for discrete values of the random variable. The probability of a random variable X having a value in the interval $a \le x \le b$ is thus given by the area under the density function in the

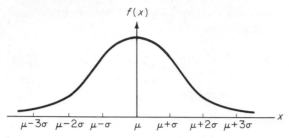

Figure 6.8.

interval from a to b. To determine this area, we use a *standard normal* probability table.

A standard normal probability table contains cumulative probabilities for a normally distributed random variable with $\mu = 0$ and $\sigma = 1$. A standard normal probability table is included as Table A.10. To illustrate the use of the table, assume that a random variable Z is normally distributed with $\mu = 0$ and $\sigma = 1$. The probability that $a \leq Z \leq b$ is found by determining the cumulative probabilities that $Z \leq a$ and $Z \leq b$. These probabilities are read directly from the table. Since the table gives cumulative probabilities, the $P(a \leq Z \leq b)$ is

$$P(a \leq Z \leq b) = P(Z \leq b) - P(Z \leq a) \qquad (6.15)$$

As an example, assume that $a = -2$ and $b = 1$. From Table A.10,

$$P(-2 \leq Z \leq 1) = P(Z \leq 1) - P(Z \leq -2)$$
$$= 0.8413 - 0.0228 = 0.8185$$

Example. A random variable Z is normally distributed with $\mu = 0$ and $\sigma = 1$. Determine the probability that $-0.52 \leq Z \leq 1.07$.

$$P(-0.52 \leq Z \leq 1.07) = P(Z \leq 1.07) - P(Z \leq -0.52)$$
$$= 0.8577 - 0.3015 = 0.5562$$

Example. A random variable Z is normally distributed with $\mu = 0$ and $\sigma = 1$. Determine the probability that $-1.65 \leq Z \leq -0.22$.

$$P(-1.65 \leq Z \leq -0.22) = P(Z \leq -0.22) - P(Z \leq -1.65)$$
$$= 0.4129 - 0.0495 = 0.3634$$

To use a standard normal probability table to determine the probability that a normally distributed random variable X assumes a value in the interval $a \leq X \leq b$, the values of a and b must be *normalized*. This is done by converting a and b into z values. The normalized values of a and b are

$$z_a = \frac{a - \mu}{\sigma} \quad \text{and} \quad z_b = \frac{b - \mu}{\sigma} \qquad (6.16)$$

The probability that $a \leq X \leq b$ is then equal to the probability that $z_a \leq Z \leq z_b$, i.e.,

$$P(a \leq X \leq b) = P\left(\frac{a - \mu}{\sigma} \leq Z \leq \frac{b - \mu}{\sigma}\right)$$

Example. A random variable X is normally distributed with $\mu = 100$ and $\sigma = 10$. Determine the probability that $80 \leq X \leq 110$.

$$P(80 \leq X \leq 110) = P\left(\frac{80 - 100}{10} \leq Z \leq \frac{110 - 100}{10}\right)$$
$$= P(-2 \leq Z \leq 1)$$
$$= P(Z \leq 1) - P(Z \leq -2)$$
$$= 0.8413 - 0.0228 = 0.8185$$

Example. A random variable X is normally distributed with mean 60 and standard deviation 12. What is the probability that $X \geq 75$?

The probability that $X \geq 75$ is

$$P(X \geq 75) = 1 - P(X \leq 75)$$
$$= 1 - P\left(Z \leq \frac{75 - 60}{12}\right)$$
$$= 1 - P(Z \leq 1.25)$$
$$= 1 - 0.8943 = 0.1057$$

The area representing this probability is shown in Fig. 6.9,

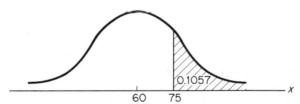

Figure 6.9.

Example. A manufacturer of a certain breakfast cereal states that a box of cereal contains 18.0 ounces. If the weight of the cereal is actually normally distributed with $\mu = 18.0$ and $\sigma = 0.2$, determine the probability that the contents of the box weigh more than 17.7 ounces.

$$P(X \geq 17.7) = P\left(Z \geq \frac{17.7 - 18.0}{0.2}\right)$$
$$= P(Z \geq -1.5)$$
$$= 1 - P(Z \leq -1.5)$$
$$= 1 - 0.0668 = 0.9332$$

Example. Suppose that a manufacturer of fishing line claims that there is a 0.1587 probability that the line will have a breaking strength of less than 9 pounds and a 0.9772 probability that the line will have a breaking strength of less than 14 pounds. If the breaking strength of the line is normally distributed, determine the mean and standard deviation of the distribution.

To solve for the two parameters of the distribution, μ and σ, we first determine the z values for the probabilities 0.1587 and 0.9772. From Table A.10, these are -1.00 and 2.00, respectively. We next recognize that

$$-1.00 = \frac{9 - \mu}{\sigma}$$

and

$$2.00 = \frac{14 - \mu}{\sigma}$$

Solving these two equations simultaneously gives $\mu = 10.67$ and $\sigma = 1.67$.

PROBLEMS

1. Specify the set of possible values of the random variable in the following experiments. State whether the random variable is discrete or continuous.
   (a) Four individuals are interviewed regarding their intent to vote for candidate A in an upcoming election.
   (b) In a time and motion study, an efficiency expert records the time required to complete a certain assembly.
   (c) An inspection process involves measuring the diameter of a certain steel shaft.
   (d) The number of defective parts in a lot of ten parts that are selected from a shipment of 100 parts is recorded.

2. A coin is tossed three times and the outcome of each toss is recorded as a head H or a tail T. Based on this experiment, give the set of possible values of the random variables described by the following statements.
   (a) The random variable X represents the number of heads.
   (b) The random variable X represents the square of the number of heads.
   (c) The random variable X represents the number of heads minus the number of tails.

3. Develop probability functions for the random variables in Problem 2.

4. A salesman schedules four calls per day. The probability mass function that gives the probability of 0, 1, 2, 3, and 4 sales is

$$p(x) = \frac{(5 - x)}{15} \qquad \text{for } x = 0, 1, 2, 3, 4$$

Verify that $p(x)$ satisfies the requirements for a probability mass function and determine the probabilities, $P(X = x)$.

5. A discrete random variable X has the probability mass function

$$p(x) = \frac{(x^2 + 4)}{50} \quad \text{for } x = 0, 1, 2, 3, 4$$

Determine the following probabilities:
(a) $P(X = 2)$
(b) $P(X \geq 3)$
(c) $P(X < 2)$
(d) $P(X > 2)$

6. A discrete random variable X has the probability mass function

$$p(x) = \frac{(2x + 1)}{25} \quad \text{for } x = 0, 1, 2, 3, 4$$

Determine the cumulative probabilities, $P(X \leq x)$, for $x = 0, 1, 2, 3, 4$.

7. The useful life of a certain highly sensitive component can be described by the probability density function

$$f(x) = \frac{x}{18} \quad \text{for } 0 \leq x \leq 6$$

where the random variable X represents useful life in hours.
(a) Verify that $f(x)$ satisfies the requirements for a density function.
(b) Determine the probability that the useful life of the component will be between 2 and 4 hours.
(c) Determine the probability that the useful life of the component will be less than 2 hours.
(d) Determine the probability that the useful life of the component will be more than 4 hours.

8. The time required to complete an assembly can be described by the probability density function

$$f(t) = \frac{t}{4} \quad \text{for } 1 \leq t \leq 3$$

where the random variable T represents time in minutes.
(a) Verify that $f(t)$ satisfies the requirements for a density function.
(b) Determine the probability that the time required to complete an assembly will be greater than 2 minutes.
(c) Determine the probability that the time required will be between 1 and 2 minutes.

9. Explain why probabilities are not associated with discrete values of a random variable in a continuous probability function.

10. The probability distribution for a discrete random variable X is given below.

x	0	1	2	3	4
$p(x)$	0.10	0.20	0.20	0.30	0.20

Determine the mean and variance of the random variable.

11. Determine the mean and variance for the discrete probability distribution in Problem 4.

12. Determine the mean and variance for the discrete probability distribution in Problem 5.

13. Determine the mean and variance for the discrete probability distribution in Problem 6.

14. A discrete random variable X has the probability distribution given below.

x	0	1	2	3	4	5	6
$p(x)$	0.02	0.07	0.12	0.17	0.22	0.25	0.15

(a) Determine the median value of X.
(b) Determine the modal value of X.

15. A discrete random variable X has the probability mass function

$$p(x) = \begin{cases} 0.15 & \text{for } x = 0 \\ 0.05x^2 & \text{for } x = 1, 2 \\ 0.10(6 - x) & \text{for } x = 3, 4, 5 \\ 0 & \text{otherwise} \end{cases}$$

(a) Determine the median value of X.
(b) Determine the modal value of X.

16. Determine the median and the mode for the discrete probability distribution in Problem 5.

17. Determine the median and the mode for the discrete probability distribution in Problem 6.

18. A manufacturer purchases a certain component in lots of 20. Before accepting the lot, a random sample of five components is tested. If the lot contains three defective components, what is the probability that the sample will contain one defective?

19. The Bryant Corporation has developed an acceptance sampling plan in which three components from each 12 received by the company are tested. If no more than one component of the three tested is found to be defective, the lot of 12 is accepted. If the lot of 12 components contains four defectives, what is the probability of accepting the lot?

20. The manager of a small firm has decided to create an employee council to handle employee grievances. The council will consist of the plant manager, the personnel manager, and four employee representatives. The four employees are to be selected at random from a list of 15 nominees. If this list contains nine men and six women, what is the probability of two men and two women's being selected as the employee representatives?

21. A recent public opinion poll showed that 30 percent of the population favored an isolationist policy in foreign relations. If a sample of size 10 is selected from this population, what is the expected number that favors an isolation policy? What is the probability that exactly three individuals will favor the policy?

22. Five identical transistors are used in the assembly of a sophisticated navigational radio. All five transistors must work for the radio to function normally. If the probability of this type of transistor's failing is 0.01, what is the probability that the radio will function normally?

23. The probability that a certain type of air-to-air missile will hit a target is 0.40. How many missiles must be fired if it is desired that the probability of at least one hit be greater than or equal to 0.95?

24. Five people are chosen at random and asked if they favor a certain proposal. If only 40 percent of the population favors the proposal, what is the probability that the majority of the five people chosen will favor the proposal?

25. American Foods, Inc., is interviewing applicants for a cake mix testing panel. The applicants are selected for the panel based on their ability to discriminate between different cake mixes. Each applicant is asked to taste four cakes, three of the cakes baked from Mix A and one from Mix B. The applicant is asked to identify the cake from Mix B. This procedure is repeated ten times and the applicant is selected for the panel if she is able to identify the cake made from Mix B in at least seven of the trials. If an applicant cannot discriminate but guesses each time, what is the probability that she will be selected for the panel?

26. A random variable Z is distributed according to the standard normal distribution with $\mu = 0$ and $\sigma = 1$. Determine the following probabilities:
    (a) $P(Z \geq 1)$ (b) $P(-1 \leq Z \leq 1)$
    (c) $P(-0.75 \leq Z \leq 0.25)$ (d) $P(0.25 \leq Z \leq 1.50)$

27. A random variable X is distributed according to the normal distribution with $\mu = 20$ and $\sigma = 4$. Determine the following probabilities:
    (a) $P(X \geq 22)$ (b) $P(16 \leq X \leq 22)$
    (c) $P(X \leq 15)$ (d) $P(21 \leq X \leq 23)$

28. Assume that the weight of adult men is normally distributed with $\mu = 160$ and $\sigma = 20$. What is the probability that a randomly selected man will weigh more than 190 pounds? Less than 120 pounds?

29. The life of a certain component is distributed according to the normal distribution with a mean life of 15 months and a standard deviation of 2 months. Determine the probability of the component's failing between the twelfth and seventeenth month.

30. A gasoline company has two million credit card holders. During the preceding month the average billing to the card holders was $\mu = \$22.00$ and the standard deviation was $\sigma = \$6.00$. If it is assumed that the billings are normally distributed, determine the number of customers who received bills exceeding $26.

31. Two different processes, A and B, are available for producing wire fasteners. The fasteners produced by process A have a mean breaking strength of 50 pounds with a standard deviation of 5 pounds. The fasteners produced by process B have a mean breaking strength of 60 pounds with a standard deviation of 12 pounds. Suppose that the distributions of breaking strength are normal for both processes.
    (a) Which process will yield the greater proportion of fasteners with breaking strengths above 46 pounds?
    (b) Which process will yield the greater proportion of fasteners with breaking strengths less than 40 pounds?

32. An investigation of the life of a particular electronic component disclosed that there was a 0.10 probability that the component would last 25 months or longer and a 0.67 probability that it would last 17 months or longer. Assume that the life of the component is a normally distributed random variable. What are the mean and the standard deviation of the life of this component?

SUGGESTED REFERENCES

The references for this chapter are listed in Chap. 5.

Chapter Seven

Selected Topics in Probability and Statistics

In introducing the concepts of probability and probability functions, it was necessary to assume that the probability distribution was known. For instance, in discussing the binomial probability distribution, we assumed that the probability of a success on any trial of an experiment was known. Our task, then, was to determine the probability of x successes in n trials of the experiment. Similarly, in introducing the normal probability distribution, it was assumed that the random variable was normally distributed and that the mean and standard deviation of the distribution were known. Based on this assumption, we were able to calculate the probability that the random variable x would have a value in a certain interval.

Although problems such as those discussed in the preceding two chapters are important, one is not always fortunate enough to know the parameters of the distribution. As a matter of fact, there are many real-world problems in which not even the form of the distribution is known. Fortunately, techniques are available that enable us to make probabilistic statements concerning certain of these types of problems. Our task in this chapter is to introduce several of the more important of these techniques. We begin with a technique that applies regardless of the form of the distribution. This technique, developed by the Russian mathematician P. L. Chebyshev (1821–94), is termed Chebyshev's inequality.

7.1 Chebyshev's Inequality

Several of the examples in Chap. 6 dealt with the problem of calculating the probability that a random variable would have a value that is within a certain number of units of the mean. For example, suppose that a random variable is normally distributed with mean $\mu = 100$ and standard deviation $\sigma = 10$. Given this distribution, we can determine the probability that a randomly selected observation is within, say, 20 units of the mean. Specifically,

$$P(|X - \mu| \le 20) = P(80 \le X \le 120)$$

$$P(|X - \mu| \le 20) = P\left(\frac{80 - 100}{10} \le Z \le \frac{120 - 100}{10}\right)$$

$$P(|X - \mu| \le 20) = P(-2 \le Z \le 2) = 0.9544$$

The probabilities were calculated by using Eq. (6.16) and Table A.10, Standard Normal Cumulative Distribution.

If, as in the preceding example, both the form (i.e., normal distribution) and the parameters of the distribution are known, we can easily calculate the probability that a random variable is within a certain number of units of the mean. Suppose, however, that only the parameters of the distribution are known, i.e., we are able to estimate μ and σ but are unable to determine the exact form of the probability distribution.[1] Given only the parameters μ and σ, bounds can still be set on the probability that a randomly observed value will be within a certain interval of the mean. These probabilities are calculated by using *Chebyshev's inequality*. The inequality may be stated for any $k \ge 1$ as

$$P(|X - \mu| < k\sigma) \ge 1 - \frac{1}{k^2} \tag{7.1}$$

or, alternatively,

$$P(|X - \mu| \ge k\sigma) \le \frac{1}{k^2} \tag{7.2}$$

The inequality applies for all probability distributions.

Although Eqs. (7.1) and (7.2) appear formidable, they are relatively easy to interpret. Eq. (7.1) says that the probability that the value of a random variable will be less than k standard deviations from the mean is at least equal to $1 - 1/k^2$. Conversely, Eq. (7.2) says that the probability that the value of a random variable will be k or more standard deviations from the mean in either direction is at most equal to $1/k^2$.

To illustrate Chebyshev's inequality, consider a random variable with mean $\mu = 100$ and standard deviation $\sigma = 10$. Assume that the form of the distribution is unknown. We can determine the minimum probability that a

---

[1] We assume that estimates of μ and σ are known from historical data or are obtained from samples.

randomly selected observation is within 20 units of the mean. Since 20 units represent two standard deviations, the probability from Eq. (7.1) is

$$P(|X - \mu| < 2\sigma) \geq 1 - \frac{1}{2^2}$$

or

$$P(|X - \mu| < 2\sigma) \geq 0.75$$

Regardless of the form of the distribution, the probability that a randomly selected observation is within two standard deviations of the mean is at least 0.75.

Chebyshev's inequality is important because it can be used even though the actual probability distribution of a random variable is unknown. If, as in the preceding example, we are able to estimate the mean and standard deviation of a random variable, the inequality can be used to set bounds on the probability that a randomly selected observation will be within k standard deviations of the mean. The ability to establish such bounds can, in certain cases, lead to a savings in both the time and effort required to determine the actual probability distribution.

Example. The mean and standard deviation of a random variable are $\mu = 60$ and $\sigma = 2$. Use Chebyshev's inequality to determine the maximum probability that a randomly selected observation will be less than 54 or greater than 66.

We note that 54 is 3σ below the mean and 66 is 3σ above the mean. From Eq. (7.2),

$$P(|X - \mu| \geq 3\sigma) \leq \frac{1}{3^2}$$

$$P(|X - \mu| \geq 3\sigma) \leq \tfrac{1}{9}$$

Regardless of the form of the distribution, the probability is at most 1/9 that the value of the random variable will be less than 54 or greater than 66.

Example. Assume that a random variable is normally distributed with mean $\mu = 60$ and standard deviation $\sigma = 2$. Determine the probability that a randomly selected observation will not be in the interval from 54 to 66. Compare this probability with the probability determined by using Chebyshev's inequality.

Assuming a normal distribution, the probability is

$$P(|X - \mu| \geq 3\sigma) = P(z \leq -3) + P(z \geq 3)$$

$$P(|X - \mu| \geq 3\sigma) = 0.0014 + 0.0014 = 0.0028$$

The probability calculated by using Chebyshev's inequality was determined in the previous example as $\frac{1}{9}$ or 0.1111. Comparing these two probabilities, we see that the probability calculated by using Chebyshev's inequality is greater than the exact probability found by using the normal distribution. Since Chebyshev's inequality applies regardless of the form of the distribution, the results are what we would expect.

Example. A census of households in a certain large city shows that the average annual household income is $\mu = \$12,000$ and the standard deviation is $\sigma = \$2000$. Information on the probability distribution of incomes is not available. Determine the maximum probability that the annual income of a randomly selected household would be either less than \$5000 or more than \$19,000.

If X represents household income with $\mu = \$12,000$ and $\sigma = \$2000$, then \$5000 and \$19,000 are 3.5 standard deviations on either side of the mean. From Chebyshev's inequality,

$$P(|X - \mu| \geq 3.5\sigma) \leq \frac{1}{(3.5)^2}$$

$$P(|X - \mu| \geq 3.5\sigma) \leq 0.0816$$

Example. A government economist is preparing a report on the increase in the cost of labor during the past year. He has been able to determine from various labor indices that the average wage increase was 8 percent and the standard deviation was 2 percent. He has not, however, been able to determine a probability distribution that describes the increase in the cost of labor. As a part of the report, the economist must give the percentage of workers who received wage increases of between 2 and 14 percent. Use Chebyshev's inequality to determine this percentage.

The probability distribution that describes the increase in the cost of labor is not known. Therefore, we cannot determine the exact percentage of workers who received increases of between 2 and 14 percent. However, since the interval between 2 and 14 percent represents three standard deviations on either side of the mean, Chebyshev's inequality can be used to determine the minimum probability that an individual's wage increase is within 3σ of the mean. From Chebyshev's inequality,

$$P(|X - \mu| < 3\sigma) \geq 1 - \frac{1}{3^2}$$

$$P(|X - \mu| < 3\sigma) \geq 0.8889$$

The minimum probability of a randomly selected individual's receiving a wage increase within three standard deviations of the mean is 0.8889. Therefore, the economist can state that at least 88.89 percent of the workers received wage increases of between 2 and 14 percent.

7.2   Sampling and Sample Statistics

In our introductory remarks in this chapter, we emphasized that one is not always fortunate enough to know the form and parameters of a probability distribution. In certain cases, such as those discussed in the preceding section, Chebyshev's inequality can be applied to make useful statements on probable values of a random variable. Chebyshev's inequality, however, is useful only if the mean and standard deviation of the distribution are known. In those cases in which the parameters of the distribution are not known, it is often possible to estimate these parameters by sampling. The objective in this section is to introduce the relationship between sample statistics and population parameters. The section begins with a discussion of the general concept of sampling. Two important statistics—the sample mean and the sample variance—are then introduced. The section concludes with a discussion of the probability distribution of sample means, a distribution of considerable importance in inferential statistics.

7.2.1   SAMPLING

A *sample* is a subset of objects selected from some *population*. For example, if the population consists of all college students, then any subset of these college students can be considered a sample. Similarly, if the population consists of the accounts receivable of a certain business firm, a subset of the accounts receivable can be considered a sample. A population is thus some universe of objects—people, products, workers, companies, etc.—and a sample is a subset of objects selected from this population.

Sampling is usually conducted because of the expense or time required to obtain a *census* of the entire population. To illustrate, suppose that an environmental protection agency is interested in determining the quantity and nature of pollutants emitted in the exhaust of automobiles. Testing the exhaust of every automobile in the United States would be extremely costly as well as time-consuming. An alternative would be to select a sample of automobiles and test the exhaust of the cars in this sample. This would, of course, save a great deal of time and expense.

As another example, suppose that a manufacturer of fluorescent bulbs wants to determine the life-span of this product. The life-span is measured by the time a bulb burns before failing. Obviously, the manufacturer cannot test every bulb made. Doing so might result in very complete data; however, it would also result in a warehouse full of burnt-out light bulbs. The practical alternative is to select a sample of bulbs from the population of fluorescent bulbs currently in inventory. As we will show, the average life-span of the bulbs in the population can be estimated accurately from the life-span of bulbs in the sample.

The objects included in the sample should be representative of those in the population. For instance, consider again the problem of determining the quantity and nature of pollutants emitted in the exhaust of automobiles. If our sample included only new cars, the results of the test would not apply to the entire population of automobiles. New cars have anti-pollution devices that are not found on older cars. A sample that includes only new cars would thus not be representative of the total population of automobiles.

There are many different methods of selecting a sample. These methods can be grouped under the general headings of *probabilistic* and *nonprobabilistic* sampling procedures. Probabilistic sampling has considerable theoretical appeal in that the laws of probability determine which objects from the population are included in the sample. In nonprobabilistic sampling, criteria other than the laws of probability are used in selecting the objects, for example, the opinion of experts, the accessibility of the elements, or merely convenience to the researcher. Since most of the statistical theory underlying sampling is based on the assumption of a probabilistic sample, the method used to select the objects that are included in the sample is important. If at all possible the sample data should be collected by using a probabilistic sampling technique.

One of the most straightforward probabilistic sampling procedures is the *simple random sample*. In the simple random sample, all objects in the population have an equal chance of being included in the sample. Thus, if the population contains N objects and the sample is to contain n objects, then each object in the population has probability n/N of being included in the sample. The objects included in the sample are determined by numbering all objects in the population and selecting a subset of these objects. The subset is selected by some chance process.[2] In the simplest of cases, the chance process might merely be drawing numbered slips of paper from a hat. The numbers randomly selected determine the choice of objects included in the sample.

The objects included in a simple random sample are determined by the laws of probability. Consequently, we are able to make probabilistic statements concerning population parameters, such as the mean and variance, based on the sample. Thus, decisions can be made on the population without

---

[2] Random number tables or computer generated random numbers are often used in selecting the objects to be included in a simple random sample. A complete discussion of sampling is given by W. Edwards Deming, *Sample Design in Business Research* (New York, N.Y.: John Wiley and Sons, Inc., 1960).

actually knowing the form or parameters of the population probability distribution. For this reason, we shall assume that all sample data are gathered by using a probabilistic sampling procedure.

7.2.2 SAMPLE MEAN AND VARIANCE

The sample mean and variance are similar in many respects to the mean and variance of a discrete random variable. The mean of a random variable was defined in Chap. 6 as the average value of the random variable. The sample mean is merely the average of those values of the random variable included in the sample.

To illustrate, suppose that we must estimate the average age of corporation presidents. The random variable in this example is age and the population is the set of corporation presidents. Instead of contacting all corporation presidents and determining the distribution of their ages, a sample is selected from the population. The sample mean is the average of the values of the random variable included in the sample, i.e., the average age of the presidents included in the sample.

The sample mean is customarily denoted by $\bar{x}$ (read x bar). Letting x_1 represent the value of the first observation, x_2 the value of the second observation, and so forth, with x_n the value of the nth observation, the sample mean is the arithmetic average of the n objects. The sample mean is

$$\bar{x} = \frac{x_1 + x_2 + \cdots + x_n}{n} = \frac{1}{n} \sum_{i=1}^{n} x_i \tag{7.3}$$

As an example, assume that a sample of six corporation presidents was randomly selected and that the ages of the six presidents are 35, 61, 52, 48, 57, and 65. From Eq. (7.3), the sample mean is

$$\bar{x} = \frac{35 + 61 + 52 + 48 + 57 + 65}{6} = \frac{1}{6}(318) = 53$$

The average age of the presidents in the sample is 53 years.

Example. Ten students were randomly selected from the class of graduating seniors at the University of Southern California. The students in this sample reported receiving job offers with the following annual starting salaries:

$12,500	$14,000	$13,500	$11,500	$13,000
$11,500	$11,000	$12,500	$13,000	$12,500

Determine the sample mean.

The mean starting salary is found by using Eq. (7.3) and is

$$\bar{x} = \frac{1}{n} \sum_{i=1}^{10} x_i = \frac{1}{10} \sum_{i=1}^{10} x_i = \$12,500$$

To show the similarity between the formula for the sample mean and the formula for the mean of a discrete random variable, suppose that we group those salaries in the previous example that are identical. Notice that two graduates received salary offers of \$11,500, three received offers of \$12,500, and two received offers of \$13,000. The remaining three offers were \$11,000, \$13,500, and \$14,000. An alternative way of calculating the sample mean is

$$\bar{x} = \frac{\$11,000 + \$11,500(2) + \$12,500(3) + \$13,000(2) + \$13,500 + \$14,000}{10}$$

$\bar{x} = \$12,500$

By using simple algebra, the above expression can be rewritten as

$$\bar{x} = \$11,000(\tfrac{1}{10}) + \$11,500(\tfrac{2}{10}) + \$12,500(\tfrac{3}{10}) + \$13,000(\tfrac{2}{10})$$
$$+ \$13,500(\tfrac{1}{10}) + \$14,000(\tfrac{1}{10})$$

$\bar{x} = \$12,500$

The fractions in the expression represent the proportion of graduates who received the same starting salary. One of the ten received an offer of \$11,000, two of the ten received offers of \$11,500, and so forth. For data grouped in this manner, the sample mean is calculated by using the formula

$$\bar{x} = \sum_{i=1}^{n} x_i f(x_i) \tag{7.4}$$

where x_i represents the value of the random variable and $f(x_i)$ represents the proportion of objects in the sample with value x_i. Comparing Eq. (7.4) with the formula for the mean of a discrete random variable, Eq. (6.2), we see that the only difference in the formulas is that $f(x_i)$ represents a proportion of objects in the sample with value x whereas $p(x)$ represents the probability of the occurrence of a random variable with value x.

Example. A sample of 100 families was taken to determine the average number of school-age children in a certain city. The results of the sample are shown in Table 7.1. Determine the average number of children per family.

Table 7.1.

Number of children	0	1	2	3	4	5	6
Families	30	20	15	15	10	8	2

$$\bar{x} = 0(0.30) + 1(0.20) + 2(0.15) + 3(0.15) + 4(0.10)$$
$$+ 5(0.08) + 6(0.02) = 1.87$$

The sample mean is 1.87 children per family.

The sample variance provides a measure of the variability of the values of the random variable included in the sample. The sample variance, like the variance of a discrete random variable, is measured by the average of the squared deviations of the values of the random variable about the mean. The sample variance is represented by the symbol s^2 and is

$$s^2 = \frac{1}{n} \sum_{i=1}^{n} (x_i - \bar{x})^2 \tag{7.5}$$

If the sample data are grouped by identical values, Eq. (7.5) can be rewritten as

$$s^2 = \sum_{i=1}^{n} (x_i - \bar{x})^2 f(x_i) \tag{7.6}$$

where $f(x_i)$ represents the proportion of objects in the sample with value x_i.

As an example of sample variance, consider again the sample of corporate presidents. The ages of the six presidents included in the sample were 35, 61, 52, 48, 57, and 65. The sample mean was $\bar{x} = 53$ years. From Eq. (7.5), the sample variance is

$$s^2 = \tfrac{1}{6}[(35 - 53)^2 + (61 - 53)^2 + (52 - 53)^2 + (48 - 53)^2$$
$$+ (57 - 53)^2 + (65 - 53)^2]$$

$$s^2 = \tfrac{1}{6}(571) = 95.17$$

The variance of ages of corporate presidents included in the sample is 95.17.

As an example of sample variance of data grouped according to identical values, consider the sample of ten graduating seniors. The mean salary offered the students in this sample was $\bar{x} = \$12,500$. The variance in salaries can be calculated by using Eq. (7.6). The sample variance is

$$s^2 = (11,000 - 12,500)^2(\tfrac{1}{10}) + (11,500 - 12,500)^2(\tfrac{2}{10})$$
$$+ (12,500 - 12,500)^2(\tfrac{3}{10}) + (13,000 - 12,500)^2(\tfrac{2}{10})$$
$$+ (13,500 - 12,500)^2(\tfrac{1}{10}) + (14,000 - 12,500)^2(\tfrac{1}{10})$$

$$s^2 = 800,000$$

The variance in starting salaries is $s^2 = 800,000$.

The sample standard deviation, rather than the sample variance, is often used to measure variability in sample data. The sample standard deviation is denoted by s and is the square root of the sample variance, i.e.,

$$s = \sqrt{s^2} \tag{7.7}$$

The fact that the variance is in squared units while the standard deviation is in the same units as the data makes the standard deviation a more easily understandable measure of the variation in the sample data.

Example. Determine the standard deviation of salary offers received by the sample of ten recent college graduates.

The variance was calculated on p. 219 and was $s^2 = 800,000$. The standard deviation is the square root of the variance and is

$$s = \sqrt{800,000} = 894$$

Example. The Acme Rubber Company, a manufacturer of sporting goods equipment, is considering replacing its current top line of golf balls with a new ball. The new ball differs from the current ball in two respects: It has a solid core instead of a liquid core and it has a more durable cover.

Acme management recognizes the competitive advantage of a golf ball with a durable cover. Although this feature is important, distance and variability in flight are even more important in the design of a golf ball. When the ball is struck with a certain force, it should consistently travel as far as possible. To test the distance and variability of flight of the two golf balls, management employed a professional golfer to hit 50 shots with the current ball and 50 shots with the proposed new ball. The mean distance traveled using the current ball was $\bar{x}_1 = 237$ yards and the standard deviation was $s_1 = 16$ yards. Comparable results for the new ball were $\bar{x}_2 = 233$ yards and $s_2 = 20$ yards. Should the new golf ball be introduced?

The test results show that the current ball is superior in both distance and variability. Therefore, unless management believes that the durable cover will provide a significant competitive advantage, the new ball should not be introduced in place of the current top line ball.

7.2.3 DISTRIBUTION OF SAMPLE MEANS

The purpose of sampling is to estimate population parameters from sample statistics. For example, the sample mean is an estimate of the population mean. Similarly, the sample variance is an estimate of the population variance. In order to make reasonable estimates of population parameters, it is important to have some idea of the reliability or accuracy with which the sample data describe the population. This reliability is customarily measured by the sampling distribution of the statistic. Our objective in this section is to introduce the sampling distribution of one of the more important statistics, the sample mean.

The *distribution of sample means* is a probability distribution that relates different possible values of the sample mean and the probability of the occurrence of these values. Since a sample represents only a subset of the population, many different samples are possible. Each of the different samples has a sample mean. To illustrate, assume that a sample of size ten is to be selected from a population containing 100 objects. If the objects are not replaced after selection, the number of different possible samples can be calculated from the formula for combinations, Eq. (4.9). The number of ways of selecting ten objects from a set of 100 objects is

$$_{100}C_{10} = \frac{100!}{90! \; 10!} = 3,934,161,240,100$$

or approximately 3.9 trillion different samples. Since a sample mean can be calculated for each of the samples, there are 3.9 trillion possible sample means. The distribution that relates values of the sample mean and the probability of the occurrence of these values is termed the sampling distribution of $\bar{x}$ or, alternatively, the distribution of sample means.

To continue with the preceding example, assume that we are actually going to select a sample of size ten from a finite population containing 100 objects. In keeping with our example, the objects in the sample are to be selected without replacement. Moreover, the purpose of the sample is to estimate the population mean. Recognizing that a total of 3.9 trillion different samples are possible, it would be very natural to speculate on the reliability of a single sample mean as an estimate of the population mean. Fortunately, the reliability or accuracy of the sample mean as an estimate of the population mean can be determined from the sampling distribution of $\bar{x}$.

The sampling distribution of $\bar{x}$ is, as we have previously stated, the probability distribution of all values of $\bar{x}$ that can occur from random samples of size n selected from a population. Since the sampling distribution of $\bar{x}$ is in itself a probability distribution, it has a mean and a variance. These parameters are customarily denoted by $\mu_{\bar{x}}$ and $\sigma_{\bar{x}}^2$. The mean of the sample means, $\mu_{\bar{x}}$ is equal to the population mean μ, i.e.,

$$\mu_{\bar{x}} = \mu \tag{7.8}$$

If the samples come from an infinite population or are selected with replacement, the variance of the sampling distribution of $\bar{x}$ is equal to the variance of the population σ^2 divided by the sample size n, i.e.,

$$\sigma_{\bar{x}}^2 = \frac{\sigma^2}{n} \tag{7.9}$$

A simple example will serve to illustrate Eqs. (7.8) and (7.9). Suppose that a population consists of only five objects and that the objects have numerical values 1, 2, 3, 4, and 5. The objects occur with equal frequency. Therefore, the probability distribution of the population is described by the set

$$\{(x, p(x)) \mid (1, \tfrac{1}{5}), (2, \tfrac{1}{5}), (3, \tfrac{1}{5}), (4, \tfrac{1}{5}), (5, \tfrac{1}{5})\}$$

The mean of the population can be calculated from Eq. (6.2) and is

$$\mu = \sum_{\text{all } x} xp(x) = 1(\tfrac{1}{5}) + 2(\tfrac{1}{5}) + 3(\tfrac{1}{5}) + 4(\tfrac{1}{5}) + 5(\tfrac{1}{5}) = 3$$

The variance of the population, from Eq. (6.4), is

$$\sigma^2 = \sum_{\text{all } x} (x - \mu)^2 p(x) = (1 - 3)^2(\tfrac{1}{5}) + (2 - 3)^2(\tfrac{1}{5}) + (3 - 3)^2(\tfrac{1}{5})$$
$$+ (4 - 3)^2(\tfrac{1}{5}) + (5 - 3)^2(\tfrac{1}{5}) = 2$$

Suppose that samples of size $n = 2$ are selected from the population. If the first object is replaced before the second object is selected, there are five possibilities for the first object and five possibilities for the second. The number of different possible samples is therefore 5·5 or 25 different possible samples. These samples are given in Table 7.2.

Table 7.2.

First Object

		1	2	3	4	5
	1	1, 1	1, 2	1, 3	1, 4	1, 5
	2	2, 1	2, 2	2, 3	2, 4	2, 5
Second Object	3	3, 1	3, 2	3, 3	3, 4	3, 5
	4	4, 1	4, 2	4, 3	4, 4	4, 5
	5	5, 1	5, 2	5, 3	5, 4	5, 5

The mean of each of the 25 possible samples can be calculated. For instance, the mean of the sample that consists of the numbers (3, 1) is $\bar{x} = 2$. Similarly, the mean of the sample that consists of the numbers (3, 5) is $\bar{x} = 4$. The sampling distribution of $\bar{x}$ is the probability distribution of the different possible sample means. This probability distribution is given in Table 7.3.

Table 7.3.

$\bar{x}$	1.0	1.5	2.0	2.5	3.0	3.5	4.0	4.5	5.0
$p(\bar{x})$	0.04	0.08	0.12	0.16	0.20	0.16	0.12	0.08	0.04

The probabilities in Table 7.3 are determined from the fact that the probability of each possible sample is 0.04. One sample has a mean of $\bar{x} = 1.0$; therefore, $P(\bar{x} = 1.0) = 0.04$. Similarly, two samples have means of $\bar{x} = 1.5$. Consequently, $P(\bar{x} = 1.5) = 0.08$. The probabilities that $\bar{x} = 2.0$, $\bar{x} = 2.5$, etc., are determined in the same manner.

According to Eq. (7.8), the mean of the sampling distribution of $\bar{x}$ should equal the mean of the population. The mean of the sampling distribution of $\bar{x}$ is

$$\mu_{\bar{x}} = \sum_{\text{all } \bar{x}} \bar{x}p(\bar{x}) = 1.0(0.04) + 1.5(0.08) + 2.0(0.12) + 2.5(0.16)$$
$$+ 3.0(0.20) + 3.5(0.16) + 4.0(0.12) + 4.5(0.08) + 5.0(0.04) = 3.0$$

The mean of the sampling distribution of $\bar{x}$ thus equals the mean of the population.

Equation (7.9) states that the variance of the sampling distribution of $\bar{x}$ is equal to the variance of the population divided by the sample size, n. To verify this relationship, the variance of the sampling distribution of $\bar{x}$ is

$$\sigma_{\bar{x}}^2 = \sum_{\text{all } \bar{x}} (\bar{x} - \mu_{\bar{x}})^2 p(\bar{x}) = (1.0 - 3.0)^2 0.04 + (1.5 - 3.0)^2 0.08$$
$$+ (2.0 - 3.0)^2 0.12 + (2.5 - 3.0)^2 0.16 + (3.0 - 3.0)^2 0.20$$
$$+ (3.5 - 3.0)^2 0.16 + (4.0 - 3.0)^2 0.12 + (4.5 - 3.0)^2 0.08$$
$$+ (5.0 - 3.0)^2 0.04 = 1.0$$

The variance of the sampling distribution of $\bar{x}$ is $\sigma_{\bar{x}}^2 = 1.0$ and the variance of the population is $\sigma^2 = 2$. For samples of size $n = 2$,

$$\sigma_{\bar{x}}^2 = \frac{\sigma^2}{n}$$

$$\sigma_{\bar{x}}^2 = \frac{2}{2} = 1$$

and, as shown by the example, Eq. (7.9) is true.

Example. Assume that a sample of size $n = 2$ is selected from the population described above. Determine the probability that the sample mean, $\bar{x}$, is within one unit of the population mean μ.

From the sampling distribution of $\bar{x}$ given in Table 7.3,

$$P(2.0 \leq \bar{x} \leq 4.0) = 0.12 + 0.16 + 0.20 + 0.16 + 0.12$$

$$P(2.0 \leq \bar{x} \leq 4.0) = 0.76$$

The probability that $\bar{x}$ will have a value between 2.0 and 4.0, inclusive, is 0.76.

It is interesting to compare the probability distribution of a population with the probability distribution of sample means. This kind of comparison shows, as we would expect from Eq. (7.9), that sample means are more closely distributed about the mean of the distribution than population values. To illustrate, the probability distributions of both x and $\bar{x}$ from the preceding example are plotted in Fig. 7.1. Figure 7.1(a) shows that each of the five values of x in the population is equally likely. In Fig. 7.1(b) we see that the

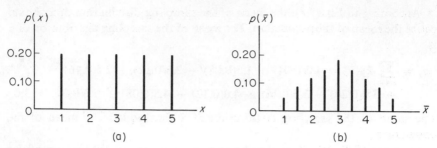

Figure 7.1.

more probable values of $\bar{x}$ are those that are close to the mean. The fact that sample means are more closely distributed about the population mean than individual values of the random variable is very important in statistical inference. This relationship will be explored in more detail in the following section.

Example. Based on extensive intelligence testing, scientists have concluded that the mean I.Q. of the individuals in a certain population is $\mu = 100$ and the variance is $\sigma^2 = 72$. A sample of eight individuals is selected from this population and the intelligence of these individuals is tested. Determine the probability that the mean intelligence of the individuals in the sample is between 91 and 109.

A minimum probability can be calculated by using Eq. (7.9) and Chebyshev's inequality. From Eq. (7.9),

$$\sigma_{\bar{x}}^2 = \frac{72}{8} = 9$$

and $\sigma_{\bar{x}} = \sqrt{9} = 3$. Since 91 and 109 are both three standard deviations from the population mean $\mu = 100$,

$$P(|\bar{x} - \mu| < 3\sigma_{\bar{x}}) \geq 1 - \frac{1}{(3)^2}$$

$$P(|\bar{x} - \mu| < 3\sigma_{\bar{x}}) \geq \tfrac{8}{9}$$

The probability that the sample mean is in the interval from 91 to 109 is at least 8/9.

Example. A manufacturer of automobile tires has made the claim that motorists can expect an average of 40,000 miles from the firm's radial tire. If the standard deviation of the life of the tire is $\sigma = 5000$ miles, determine the probability that the average mileage from a sample of 16 tires will be less than 35,000 miles or more than 45,000 miles.

The parameters of the sampling distribution of $\bar{x}$ are $\mu_{\bar{x}} = 40{,}000$ and $\sigma_{\bar{x}} = \sigma/\sqrt{n} = 5000/\sqrt{16}$ or 1250. Since 35,000 and 45,000 are both four standard deviations from the mean, the probability is

$$P(|\bar{x} - \mu| \geq 4\sigma_{\bar{x}}) \leq \frac{1}{(4)^2}$$

$$P(|\bar{x} - \mu| \geq 4\sigma_{\bar{x}}) \leq \tfrac{1}{16}$$

The probability that the sample mean is not in the interval from 35,000 to 45,000 is at most 1/16.

7.3 The Central Limit Theorem

We have shown that the mean of the sampling distribution of $\bar{x}$ is equal to the mean of the population and that the variance of this distribution is equal to the variance of the population divided by the sample size n. The form of the sampling distribution has, however, not yet been mentioned. There is a very remarkable theorem that specifies the form of the sampling distribution of $\bar{x}$. This theorem, termed the *central limit theorem*, states that as the sample size becomes large, the distribution of sample means will tend to become normally distributed. The theorem is remarkable in that it applies *regardless of the distribution of the parent population*. Specifically, the central limit theorem states the following:

> *The distribution of means of random samples taken from a population having mean μ and finite variance σ^2 approaches the normal distribution with mean μ and variance σ^2/n as the sample size n approaches infinity.*

The central limit theorem states that the distribution of sample means tends toward a normal distribution regardless of the form of the population from which the samples are selected. The theorem also states that for this phenomenon to occur, the size of the sample must increase toward infinity. Since it is both expensive and impractical to take extremely large samples, it is logical to question the usefulness of the theorem. Specifically, what size samples are necessary before the theorem begins to apply? The fact is that the normal distribution provides an excellent approximation of the distribution of sample means for relatively small sample sizes. Statisticians have shown that the approximation will be very good if $n \geq 30$. If the population is symmetrical, satisfactory results will be obtained for sample sizes of even less than 30.

To illustrate the central limit theorem, consider again the population discussed earlier that contained five equally probable objects. The objects

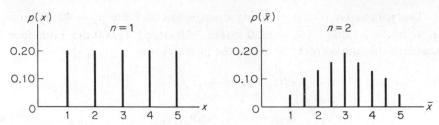

Figure 7.2. Sampling distributions of $\bar{x}$ for $n = 1$ and $n = 2$.

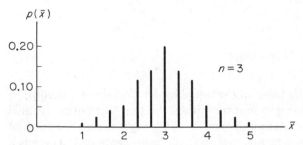

Figure 7.3. Sampling distribution of $\bar{x}$ for $n = 3$.

were numbered 1, 2, 3, 4, and 5 and the probability of selecting any of the objects was $p(x) = 0.20$. The probability distribution of the population and that of the sampling distribution of $\bar{x}$ for $n = 2$ was plotted in Fig. 7.1. These distributions are plotted again in Fig. 7.2. Suppose that instead of samples of size $n = 2$, samples of size $n = 3$ are selected from the population. Since each object from the population is replaced before the next object in the sample is selected, a total of $5 \cdot 5 \cdot 5$ or 125 different samples is possible. The sample mean of each of the 125 different samples can be calculated and the probability of each of these sample means determined. The probability distribution of $\bar{x}$ for $n = 3$ is plotted in Fig. 7.3 and the distribution is given in Table 7.4. Comparing the sampling distributions in Figs. 7.2 and 7.3, we see that increasing the sample size from $n = 2$ to $n = 3$ has had a pronounced effect on the sampling distribution of $\bar{x}$. The distribution of sample means shown in Fig. 7.3 has a marked similarity to the normal distribution.

To extend this example, suppose that samples of size $n = 5$ are selected from the parent population. Each object in the sample is replaced before the next object is selected; thus, the total number of different possible samples is

Table 7.4.

$\bar{x}$	1	$1\frac{1}{3}$	$1\frac{2}{3}$	2	$2\frac{1}{3}$	$2\frac{2}{3}$	3	$3\frac{1}{3}$	$3\frac{2}{3}$	4	$4\frac{1}{3}$	$4\frac{2}{3}$	5
$p(\bar{x})$	.008	.024	.048	.056	.120	.144	.200	.144	.120	.056	.048	.024	.008

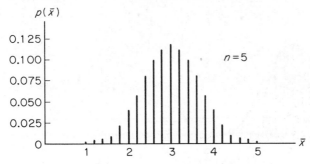

Figure 7.4.   Sampling distribution of $\bar{x}$ for $n = 5$.

5^5 or 3125. The mean of each of these samples can be calculated and the distribution of these means plotted. This distribution is plotted in Fig. 7.4. The numerical values of $\bar{x}$ and $p(\bar{x})$ are given in Table 7.5. Figure 7.4 shows that the sampling distribution of $\bar{x}$ closely resembles a normal distribution.

Table 7.5.

$\bar{x}$	$p(\bar{x})$	$\bar{x}$	$p(\bar{x})$	$\bar{x}$	$p(\bar{x})$
1.0	0.0003	2.4	0.0816	3.8	0.0592
1.2	0.0016	2.6	0.1024	4.0	0.0387
1.4	0.0048	2.8	0.1168	4.2	0.0224
1.6	0.0112	3.0	0.1220	4.4	0.0112
1.8	0.0224	3.2	0.1168	4.6	0.0048
2.0	0.0387	3.4	0.1024	4.8	0.0016
2.2	0.0592	3.6	0.0816	5.0	0.0003

The fact that the sampling distribution of $\bar{x}$ tends to become normally distributed as the sample size n increases allows the normal distribution to be used to determine the approximate probability that a sample mean will be within a certain interval. To illustrate, suppose that we determine the probability that the mean of a sample of size $n = 5$ selected from our population is in the interval from 2.4 to 3.6, inclusive. The variance of the sampling distribution of $\bar{x}$ is $\sigma_{\bar{x}}^2 = \sigma^2/n$ or 2/5 and the standard deviation is $\sigma_{\bar{x}} = \sqrt{2/5}$ or $\sigma_{\bar{x}} = 0.63$. A normal distribution with parameters $\mu = 3.0$ and $\sigma_{\bar{x}} = 0.63$ is superimposed on the actual sampling distribution of $\bar{x}$. This is shown in Fig. 7.5.

The probability that $\bar{x}$ has a value between 2.4 and 3.6 is shown by the shaded area in Fig. 7.5. Since we are approximating a discrete distribution, i.e., the sampling distribution of $\bar{x}$ given in Table 7.5, by a continuous

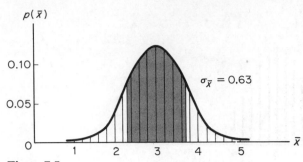

Figure 7.5.

distribution, each value of $\bar{x}$ is considered as the midpoint of an interval of width 0.2. (The width of the interval is equal to the distance between adjacent values of the discrete random variable.) From the central limit theorem, the approximate probability that $\bar{x}$ has a value between 2.4 and 3.6, inclusive, is

$$P(2.3 \leq \bar{x} \leq 3.7) = P\left(\frac{2.3 - 3.0}{0.63} \leq z \leq \frac{3.7 - 3.0}{0.63}\right)$$

$$P(2.3 \leq \bar{x} \leq 3.7) = P(-1.1 \leq z \leq 1.1)$$

$$P(2.3 \leq \bar{x} \leq 3.7) = 0.7286$$

The probability 0.7286 was calculated by using Eq. (6.15).

The actual probability that $\bar{x}$ has a value between 2.4 and 3.6 can be determined from Table 7.5. The probability is

$$P(2.4 \leq \bar{x} \leq 3.6) = \sum_{\bar{x}=2.4}^{\bar{x}=3.6} p(\bar{x}) = 0.7236$$

The probabilities calculated by using the normal approximation to the sampling distribution of $\bar{x}$ and the actual sampling distribution of $\bar{x}$ differ by only 0.0050. The approximation is therefore relatively accurate. The reader should be aware that this level of accuracy is due to the fact that the population is symmetrical. If this were not the case, samples of size 30 or more would be necessary to obtain similar results.

Example. A random sample of size $n = 36$ is selected from a continuous population with mean $\mu = 200$ and standard deviation $\sigma = 60$. Determine the probability that the sample mean, $\bar{x}$, has a value between 185 and 215.

The sampling distribution of $\bar{x}$ is normally distributed with mean $\mu_{\bar{x}} = \mu = 200$ and $\sigma_{\bar{x}} = \sigma/\sqrt{n} = 60/\sqrt{36} = 10$. The probability that a sample mean has a value in the interval from 185 to 215 is

$$P(185 \le \bar{x} \le 215) = P\left(\frac{185 - 200}{10} \le z \le \frac{215 - 200}{10}\right)$$

$$P(185 \le \bar{x} \le 215) = P(-1.5 \le z \le 1.5) = 0.8664$$

Example. A professor has given the same final examination for a number of semesters. The mean grade on the final has been $\mu = 75$ with standard deviation $\sigma = 10$. The professor, having just completed grading the final for his current class of 25 students, is somewhat puzzled by the fact that the average grade for this class is 85. Treating the 25 students as a sample, determine the probability that a sample mean of 85 or more comes from a population with mean $\mu = 75$ and standard deviation $\sigma = 10$.

The sampling distribution of $\bar{x}$ is normally distributed with mean $\mu_{\bar{x}} = 75$ and $\sigma_{\bar{x}} = 10/\sqrt{25} = 2$. The probability that a sample mean is 85 or more is

$$P(\bar{x} \ge 85) = P\left(z \ge \frac{85 - 75}{2}\right)$$

$$P(\bar{x} \ge 85) = P(z \ge 5) \simeq 0$$

The probability of a sample mean of 85 or greater is approximately zero. The professor might therefore be justified in concluding that copies of the examination are available to students in the class. In any case, it would be good teaching practice to give a new examination each semester.

Example. A certain machine produces parts that have a mean diameter of 0.250 inches and a standard deviation of 0.004 inches. A sample of 64 parts is selected for inspection. Determine the probability that the mean diameter of the parts in this sample is in the interval from 0.249 to 0.251.

The sampling distribution of $\bar{x}$ is normally distributed with mean $\mu_{\bar{x}} = 0.250$ and $\sigma_{\bar{x}} = 0.004/\sqrt{64} = 0.0005$. The probability that the sample mean has a value in the interval from 0.249 to 0.251 is

$$P(0.249 \le \bar{x} \le 0.251) = P\left(\frac{0.249 - 0.250}{0.0005} \le z \le \frac{0.251 - 0.250}{0.0005}\right)$$

$$P(0.249 \le \bar{x} \le 0.251) = P(-2 \le z \le 2) = 0.9544$$

7.4 Estimating Population Parameters

In the examples in the preceding section, it was assumed that the mean and standard deviation of the population are known. Obviously, this will not always be the case. The population mean and standard deviation, can, however, be estimated from the results of a sample. These estimates customarily take one of two forms—point estimates or interval estimates. Our objective in this section is to introduce these methods of estimating population parameters.

7.4.1 POINT ESTIMATES

Suppose that it is necessary to estimate the mean score of all applicants on the admissions test required for law school. The estimate could be made in one of two forms, either a single number, say 500, or an interval, say 470 to 530. Estimates that specify a single value, such as 500, are called *point estimates*. Those that specify a range of values, such as 470 to 530, are called *interval estimates*.

Point or interval estimates can be made for any population parameter. The most common estimates are those of the population mean μ and the variance σ^2. It can be shown that the sample mean $\bar{x}$ is, on the average, the best estimate of the population mean μ. The sample mean is both *unbiased*, meaning that $\bar{x}$ is consistently neither larger nor smaller than μ, and *efficient*, meaning that different possible values of $\bar{x}$ are closely concentrated about the population mean μ. For these reasons, the sample mean is the most commonly used estimate of the population mean.

Although the reader might intuitively expect that the sample variance provides an unbiased and efficient estimate of the population variance, this is not the case. It can be shown that, on the average, the sample variance s^2 is smaller than the population variance σ^2 by a factor of $(n - 1)/n$, where n is the sample size. This bias can be eliminated by multiplying the sample variance by $n/(n - 1)$. An unbiased estimate of the population variance is thus

$$S^2 = \frac{ns^2}{n - 1} \tag{7.10}$$

The sample statistic S^2 is both an unbiased and efficient estimator of σ^2.

As an example of point estimates, suppose that the director of the law school admissions examination must make an estimate of the mean and variance of scores received on the most recent admissions examination by law school applicants. Instead of tabulating the test scores of the thousands of applicants, the director randomly selects 50 applicants. The mean test score in this sample of 50 is $\bar{x} = 498$ and the variance is $s^2 = 9800$. The best estimate of the population mean is the sample mean, $\bar{x} = 498$. The population variance is estimated by using the statistic S^2 and is

$$S^2 = \frac{ns^2}{n - 1} = \frac{50(9800)}{49} = 10,000$$

7.4.2 CONFIDENCE INTERVAL

An interval estimate offers certain advantages as a method of estimating population parameters. To illustrate, suppose that the director of the law school admissions test is asked to provide some measure of the reliability of his estimate of the mean score of law school applicants. Although the sample mean ($\bar{x} = 498$) provides the best estimate of the population mean, the director would be unlikely to assert with much confidence that the population mean is exactly $\mu = 498$. He could, however, state that he is reasonably confident that the population mean is in some interval, say 468 to 528. An interval estimate of this kind is termed a *confidence interval*. Given such an interval, one can calculate the subjective probability or, alternatively, the *degree of confidence* that the population mean is contained within the interval.

A confidence interval is established by first specifying the desired degree of confidence. For instance, the director could specify a degree of confidence of 95 percent that the population mean μ is contained within the interval. The degree of confidence is arbitrary, 90 percent and 95 percent being commonly used values. From an objective point of view, a degree of confidence of 95 percent implies that if 100 intervals were established, 95 of these would contain the mean. Similarly, a confidence level of 90 percent implies that the mean should be contained in 90 of every 100 intervals. From a subjective point of view, a 95 percent confidence interval contains the mean with probability 0.95. Similarly, a 90 percent confidence interval would contain the mean with probability 0.90.

Confidence intervals for the population mean are derived by using the central limit theorem. For large samples, the central limit theorem states that the sampling distribution of $\bar{x}$ approaches a normal distribution with mean μ and variance σ^2/n. The distribution of sample means as described by the central limit theorem is shown in Fig. 7.6.

Figure 7.6 shows that the mean of a sample of size n will be in the interval from $\mu - k$ to $\mu + k$ with probability $1 - \alpha$. This can be written as

$$P[\mu - k \leq \bar{x} \leq \mu + k] = 1 - \alpha$$

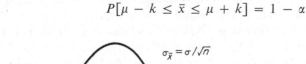

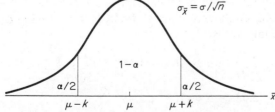

Figure 7.6.

Converting this expression to the standard normal distribution gives

$$P\left[z_{\alpha/2} \leq \frac{\bar{x} - \mu}{\sigma_{\bar{x}}} \leq z_{1-\alpha/2} \right] = 1 - \alpha$$

where $z_{\alpha/2}$ is the z value for a cumulative probability of $\alpha/2$ and $z_{1-\alpha/2}$ is the cumulative probability for $1 - \alpha/2$. Multiplying the terms inside the braces by $\sigma_{\bar{x}}$ gives

$$z_{\alpha/2}\sigma_{\bar{x}} \leq \bar{x} - \mu \leq z_{1-\alpha/2}\sigma_{\bar{x}}$$

Subtracting $\bar{x}$ from each term inside the braces gives

$$z_{\alpha/2}\sigma_{\bar{x}} - \bar{x} \leq -\mu \leq z_{1-\alpha/2}\sigma_{\bar{x}} - \bar{x}$$

Multiplying each term within the braces by -1 gives the $1 - \alpha$ confidence interval. The confidence interval is

$$\bar{x} - z_{\alpha/2}\sigma_{\bar{x}} \geq \mu \geq \bar{x} - z_{1-\alpha/2}\sigma_{\bar{x}} \tag{7.11}$$

To illustrate Eq. (7.11), suppose that we establish a 95 percent confidence interval for the mean admissions test score for law students. The results from a sample of 50 scores were $\bar{x} = 498$ and $s^2 = 9800$. Since the sample is large, it is acceptable to use the sample statistic S^2 to estimate the population variance σ^2. S^2 was calculated on p. 230 and is $S^2 = 10{,}000$. The standard deviation of the sampling distribution of $\bar{x}$ is $\sigma_{\bar{x}} = S/\sqrt{n} = 100/\sqrt{50}$ or 14.14. The z values for a 95 percent confidence interval are $z_{0.025} = -1.96$ and $z_{0.975} = 1.96$. The confidence interval is

$$498 - (-1.96)(14.14) \geq \mu \geq 498 - 1.96(14.14)$$

$$498 + 27.7 \geq \mu \geq 498 - 27.7$$

or
$$525.7 \geq \mu \geq 470.3$$

The director can thus assert that μ is contained in the interval from 470.3 to 525.7 with subjective probability of 0.95.

Example. The average weight of a randomly selected sample of 256 enlistees in the armed forces is 160 pounds. The unbiased estimate of the population standard deviation is $\sigma = 24$ pounds. Construct a 90 percent confidence interval for the average weight of all enlistees.

The z values for a 90 percent confidence interval are $z_{0.05} = -1.65$ and $z_{0.95} = 1.65$. The confidence interval is

$$165 - (-1.65)\left(\frac{24}{\sqrt{256}}\right) \geq \mu \geq 165 - 1.65\left(\frac{24}{\sqrt{256}}\right)$$

$$165 + 2.48 \geq \mu \geq 165 - 2.48$$

$$167.48 \geq \mu \geq 162.52$$

Confidence intervals can be established for parameters other than the mean of the population. For instance, it is often necessary to construct a confidence interval for the population variance. Although the procedure for constructing confidence intervals for population parameters such as the variance is similar to that of the mean, a detailed description of these procedures is beyond the scope of this text. The interested reader will find descriptions of the procedures in the statistics texts referenced at the end of this chapter.

7.5   Tests of Hypotheses

The discussion in the preceding sections concerned sample statistics as estimates of population parameters. In certain cases, the primary purpose of sampling is not to estimate population parameters but rather to test the validity of *assumed* values of these parameters. An assumption concerning the value of a population parameter is termed a *statistical hypothesis*. The statistical hypothesis, or assumed value of the population parameter, is either accepted or rejected based on a *test of the hypothesis*.

To illustrate, suppose that a psychologist is interested in the mean intelligence of a certain population of individuals. The psychologist is willing to assume that the mean I.Q. of this population is equal to the national average, i.e., $\mu = 100$. In order to test this assumption, the psychologist randomly selects 144 individuals from the population. An I.Q. test is given to each of the individuals in this sample and the average I.Q. is determined. The statistical hypothesis that $\mu = 100$ is thus being subjected to a test. The statistical hypothesis, i.e., the assumption that $\mu = 100$, is either accepted or rejected based upon the results of this test.

The hypothesized value of μ is termed the *null hypothesis* and is denoted by H_0. In our example

$$H_0: \mu = 100$$

is the null hypothesis. It is, of course, possible that the mean of the population is not $\mu = 100$. If this is the case, μ has some alternative value. The possible alternative values of μ are stated in the form of the *alternative hypothesis*. If there is only one alternative value of μ, say $\mu = 120$, then the alternative hypothesis would be stated as such. If a unique alternative value of μ is not known, then the alternative hypothesis should be stated as the complement of the null hypothesis. The alternative hypothesis in our example, denoted by H_1, is

$$H_1: \mu \neq 100$$

There are numerous forms that the null and alternative hypotheses can take. For example, had the psychologist hypothesized that the mean intelligence of the population is no greater than the national average, the null and alternative hypotheses would be

$$H_0: \mu \leq 100$$

$$H_1: \mu > 100$$

Conversely, had the null hypothesis been that the mean intelligence was equal to or greater than the national average, the two hypotheses would be

$$H_0: \mu \geq 100$$

$$H_1: \mu < 100$$

Regardless of the form of the null and alternative hypotheses, both hypotheses must be included in a statistical test. The purpose of the test is to determine which hypothesis to accept. Rejecting the null hypothesis automatically implies accepting the alternative hypothesis and vice versa. Consequently, the alternative hypothesis must be stated so that all possible values of μ not included in the null hypothesis are included in the alternative hypothesis. For instance, if $H_0: \mu = 100$ and $H_1: \mu = 120$, the only two possible values of μ are 100 and 120. If, however, $H_0: \mu = 100$ and $H_1: \mu \neq 100$, there is a continuum of possible values of μ. In either case, one hypothesis will be accepted on the basis of sample information and the other hypothesis will be rejected.

The test of a hypothesis is based upon the sampling distribution of the test statistic. The test statistic in our example is the sample mean $\bar{x}$. To test the null hypothesis

$$H_0: \mu = 100$$

against the alternative hypothesis

$$H_1: \mu \neq 100$$

the psychologist plans to determine the average intelligence of 144 different individuals. From the central limit theorem, we know that the different possible values of the sample mean $\bar{x}$ are normally distributed about the population mean μ and that the standard deviation of this distribution is $\sigma_{\bar{x}} = \sigma/\sqrt{n}$. If the null hypothesis is true, i.e., μ does in fact equal 100, the distribution of sample means would plot as shown in Fig. 7.7.

Assuming that the null hypothesis is true, we would not expect the sample mean $\bar{x}$ to differ significantly from the population mean μ. The probability that the sample mean $\bar{x}$ differs from the population mean μ by more than k units is shown in Fig. 7.7 as α. By specifying a value for α, we can determine values of $\bar{x}$ that we consider to differ significantly from μ. These values are shown in Fig. 7.7 as $\bar{x} < \mu - k$ and $\bar{x} > \mu + k$. Should the sample mean

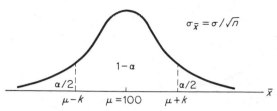

Figure 7.7.

have a value either less than $\mu - k$ or greater than $\mu + k$, the null hypothesis is rejected in favor of the alternative hypothesis.

The values of $\bar{x}$ that lead to rejecting the null hypothesis are called the *critical values* of $\bar{x}$ or, alternatively, the *critical region*. The critical values of $\bar{x}$ in Fig. 7.7 are $\bar{x} < \mu - k$ and $\bar{x} > \mu + k$. In order to calculate the critical values of $\bar{x}$, we must be willing to accept the risk of rejecting a true null hypothesis. The risk occurs because of the possibility of the sample mean's having a value in the critical region even though the null hypothesis is true. The probability of making this error is shown in Fig. 7.7 as α. Because the possibility of rejecting a true null hypothesis is present in all statistical tests, this error has been given a special designation. The error of rejecting a true null hypothesis is called a *Type I* error. The probability of making this error, α, must be specified before a decision can be made either to accept or reject the null hypothesis.

Before summarizing the steps involved in testing a hypothesis, we shall complete the test of the hypothesis concerning the mean intelligence of the population of individuals. The null and alternative hypotheses are

$$H_0 : \mu = 100$$

and

$$H_1 : \mu \neq 100$$

The probability of rejecting a true null hypothesis, α, must be specified. Commonly used values of α are 0.01, 0.05, and 0.10. We shall arbitrarily specify α as $\alpha = 0.10$. In order to calculate the critical values of $\bar{x}$, the standard deviation of the population must either be known or estimated. We shall assume that the standard deviation of the population is known and is $\sigma = 24$.[3] Given the standard deviation of the population ($\sigma = 24$), the sample size ($n = 144$), the assumed population mean ($\alpha = 100$), and the probability of a Type I error ($\alpha = 0.10$), the sampling distribution of $\bar{x}$ and the critical values of $\bar{x}$ can be determined. These are shown in Fig. 7.8.

The critical values of $\bar{x}$, $\bar{x} < 96.71$ and $\bar{x} > 103.29$, are determined by using the standard normal distribution, Table A.10, and the fact that $z =$

---

[3] The standard deviation of the population could be known from past studies or could be estimated from the sample of 144 individuals in this problem.

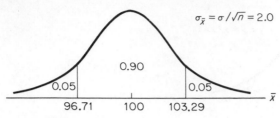

Figure 7.8.

$(\bar{x} - \mu)/\sigma_{\bar{x}}$. Since a Type I error occurs in this example if $\bar{x}$ is either significantly less or significantly greater than μ, the probability of a Type I error must be evenly divided between the lower and upper tails of the distribution. This is shown in Fig. 7.8 by placing $\alpha/2 = 0.05$ in each tail of the distribution. The critical z value for the lower tail of the distribution is $z = -1.645$ and that for the upper tail of the distribution is $z = 1.645$. Given these values of z, the critical value of $\bar{x}$ for the lower tail of the distribution is

$$-1.645 = \frac{\bar{x} - 100}{2}$$

$$\bar{x} = 96.71$$

Similarly, the critical value of $\bar{x}$ for the upper tail of the distribution is

$$1.645 = \frac{\bar{x} - 100}{2}$$

$$\bar{x} = 103.29$$

Given the critical values of $\bar{x}$, the psychologist can now either accept or reject the null hypothesis. The null hypothesis will be accepted if the mean I.Q. of the 144 individuals tested is between 96.71 and 103.29. If the mean of the sample is not in this interval, the null hypothesis that $\mu = 100$ will be rejected in favor of the alternative hypothesis that $\mu \neq 100$.

The steps required to test a hypothesis can be summarized as follows:

1. The null and alternative hypotheses are stated. Care must be taken to ensure that all possible values of the population parameter are included in these hypotheses.
2. The appropriate sample statistic and the sampling distribution of the statistic are specified.
3. The probability, α, of a Type I error is specified. Any value for α may be specified; however, commonly used values are 0.01, 0.05, and 0.10.
4. Given the sampling distribution of the statistic and the probability of a Type I error, the critical values of the sample statistic are

calculated. These critical values are used to establish the rejection region for the null hypothesis.

5. The sample is obtained. If the value of the test statistic falls in the rejection region, the null hypothesis is rejected. Otherwise, it is accepted. A sample result that falls in the rejection region is said to be significant or to differ significantly from the result expected if the null hypothesis is true.

Tests of hypotheses concerning population parameters other than the mean are beyond the scope of this text. In order to better understand tests of hypotheses concerning the mean, the reader will want to study the following examples carefully.

Example. A manufacturer of filter-tip cigarettes reports that his cigarettes contain 10.0 milligrams of tar. To test this claim, a random sample of 225 cigarettes is selected and the tar content of the cigarettes measured. The results of the test show an average of 10.20 milligrams in the cigarettes included in the sample. The standard deviation of tar content is available from previous tests and is $\sigma = 0.90$. Using $\alpha = 0.01$, test the null hypothesis that $\mu \leq 10.0$ against the alternative hypothesis that $\mu > 10.0$.

The null and alternative hypotheses in this example are stated in a manner such that the null hypothesis will be rejected in favor of the alternative only if the sample mean $\bar{x}$ is significantly greater than the population mean μ. Hypotheses are stated in this manner when deviations in one direction are deemed the most critical. In terms of our example, it is important that the tar content not be reported as less than it actually is. Tests of hypotheses in which deviations in only one direction are critical are called *one-tail tests*.

The sampling distribution of $\bar{x}$ is shown in Fig. 7.9. Although the null hypothesis is that μ is less than or equal to 10.0, the critical values of $\bar{x}$ are calculated for only one value of μ, namely, $\mu = 10.0$. Using $\mu = 10.0$ as the mean of the sampling distribution minimizes the probability of a Type I error. A Type I error, the reader will remember, occurs when a true null hypothesis is rejected. If the critical values of $\bar{x}$ were calculated by using a population mean of less than 10.0, say $\mu = 9.0$, the probability of incorrectly

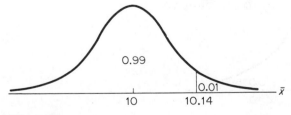

Figure 7.9.

rejecting the null hypothesis would be unduly large. For this reason, the critical values of $\bar{x}$ are always calculated under the assumption that μ is equal to the upper or lower bound specified by the null hypothesis. In our example, the upper bound on μ is 10.0. Consequently, the critical value of $\bar{x}$ is calculated assuming that μ is 10.0.

Using the standard normal distribution with $\alpha = 0.01$, the critical value of $\bar{x}$ is

$$2.326 = \frac{\bar{x} - 10.0}{0.90/\sqrt{225}}$$

$$\bar{x} = \frac{2.32(60.90)}{15} + 10.0$$

$$\bar{x} = 10.14$$

The critical values of $\bar{x}$ are $\bar{x} > 10.14$. Since the sample mean is $\bar{x} = 10.20$, the null hypothesis that $\mu \leq 10.0$ is rejected in favor of the alternative hypothesis that $\mu > 10.0$.

Example. A soft-drink manufacturer sells a cola in 12-ounce bottles. In order to determine if the cola bottles are filled to the proper level, the contents of a randomly selected sample of 36 bottles are periodically measured. The standard deviation of the bottling process is known from past tests and is $\sigma = 0.30$ ounces per bottle. Using an α value of 0.05, determine the critical values of $\bar{x}$ for the null hypothesis $H_0: \mu = 12.0$. Assume that the alternative hypothesis is $H_1: \mu \neq 12.0$.

The sampling distribution of $\bar{x}$ based on the assumption that $\mu = 12.0$ is shown in Fig. 7.10. The z values for a two-tail test with $\alpha = 0.05$ are -1.96 and 1.96. The critical values of $\bar{x}$ are

$$\bar{x} = 12.0 - 1.96(0.05) = 11.90$$

and

$$\bar{x} = 12.0 + 1.96(0.05) = 12.10$$

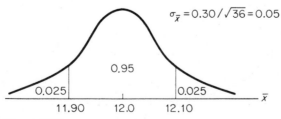

$$\sigma_{\bar{x}} = 0.30/\sqrt{36} = 0.05$$

0.95

0.025

0.025

11.90   12.0   12.10

Figure 7.10.

For a sample of 36 bottles, the null hypothesis should be rejected in favor of the alternative if the average content is less than 11.90 ounces or more than 12.10 ounces.

Example. The insurance commissioner in a certain state is interested in determining whether or not a significant reduction in the cost of automobile insurance occurs for policyholders in states that adopt "no fault" insurance laws. The mean cost of insurance under the traditional system is $300 per year and the standard deviation is $200. To determine if a reduction has occurred, the commissioner plans to take a random sample of 100 policyholders in "no fault" states. Assuming a one-tail test with $\alpha = 0.05$, determine the sample average that the commissioner should consider as significant.

The null and alternative hypotheses are

$$H_0 : \mu \geq 300$$

and
$$H_1 : \mu < 300$$

The z value for the one-tail test is $z_{0.05} = -1.645$ and the standard deviation of the sampling distribution of $\bar{x}$ is $\sigma_{\bar{x}} = 200/\sqrt{100} = 20$. The critical value of $\bar{x}$ is

$$\bar{x} = 300 - 1.645(20) = 267.1$$

If the average cost of insurance of the 100 individuals included in the sample is less than $267.10, the commissioner should conclude that no fault insurance has resulted in a significant reduction in the cost of automobile insurance.

PROBLEMS

1. The mean and standard deviation of a random variable are $\mu = 10$ and $\sigma = 2$. Use Chebyshev's inequality to determine the probability that a randomly selected observation is in the interval from 6 to 14.

2. Assume that the random variable in Problem 1 is normally distributed. Calculate the probability that a randomly selected observation is in the interval from 6 to 14. What is meant by the statement that Chebyshev's inequality provides a bound on the probability that a random variable is in a certain interval?

3. The mean and standard deviation of a random variable are $\mu = 200$ and $\sigma = 10$. Use Chebyshev's inequality to determine the following probabilities:
   (a) $P(180 < x < 220)$
   (b) $P(|x - \mu| \geq 30)$

4. The registrar at a certain large university stated that the mean grade point average of incoming freshman students is $\mu = 3.0$ and the standard deviation is $\sigma = 0.3$. Use Chebyshev's inequality to determine the maximum percentage of students with grade point averages below 2.4 and above 3.6.

5. A study of telephone company records revealed that the average telephone bill during a certain month was $16.00 and the standard deviation was $2.00. If the telephone company had 100,000 customers, determine the minimum number of customers whose bill was between $10.00 and $22.00.

6. An airline advertises that 75 percent of its flights arrive within 10 minutes of the scheduled arrival time. Assuming that this statement is true, use Chebyshev's inequality to estimate the standard deviation of arrival times. What percentage of flights arrive within 20 minutes of the scheduled arrival time?

7. In 1936 the now defunct *Literary Digest* predicted (based on a sample of over 2 million names selected from telephone directories and automobile registrations) that Landon would defeat Roosevelt in the presidential election. Instead, Roosevelt won by a substantial majority. Can you give a reason for this error?

8. Explain why a simple random sample has such intuitive appeal for use in statistics.

9. A random sample of ten students was selected from a class of 300. The final exam scores of these students were 80, 75, 90, 65, 70, 80, 70, 60, 95, 65. Determine the mean and variance of scores in this sample.

10. A sample of 50 boat dealers was taken to determine the current inventory of used boats. The results of this sample are shown in Table 7.6. Determine the sample mean and variance.

Table 7.6.

Inventory	0	1	2	3	4	5
Dealers	5	10	10	15	5	5

11. A contestant on a television quiz show can select from three plain white envelopes. One envelope contains nothing, the second envelope contains a $500 bill, and the third contains a $1000 bill. Determine the mean and standard deviation of possible winnings.

12. After selecting an envelope, the contestant in Problem 11 is asked to select from three additional plain white envelopes, again containing $0, $500, and $1000. The prize is the average of the amounts contained in

the first and the second envelopes. The contestant is not allowed to open the first envelope until both envelopes have been selected. Determine the mean and standard deviation of possible winnings.

13. Would a contestant who is "risk adverse" prefer the contest described in Problem 11 or that described in Problem 12?

14. Compare the variances in Problems 11 and 12. Are the results what would be expected from Eq. (7.9)?

15. State in your own words the difference between a probability distribution and a sampling distribution.

16. Why is the central limit theorem of special importance in inferential statistics?

17. A certain machine, when properly adjusted, produces parts with a mean diameter of 0.500 inches and a standard deviation of 0.008 inches. What is the probability that the mean diameter of a sample of 64 parts would be as large as 0.502?

18. (a) Given a normally distributed random variable with $\mu = 50$ and $\sigma = 6$, find

$$P(x \leq 41) \qquad P(x \geq 56) \qquad P(47 \leq x \leq 53)$$

(b) If a random sample of size $n = 36$ is selected from this population, find

$$P(\bar{x} \leq 48.5) \qquad P(\bar{x} \geq 51) \qquad P(49.5 \leq \bar{x} \leq 50.5)$$

(c) Sketch the distribution of x and the distribution of $\bar{x}$.

19. A random sample will be selected from the population of construction workers. The mean hourly wage of the population is $\mu = \$10.00$ and the standard deviation is $\sigma = \$2.00$. Assume that the wage rate is normally distributed.

(a) Calculate the probability that the wage rate of a randomly selected worker will be between $7.00 and $13.00.

(b) A sample of 100 workers is randomly selected from this population. Calculate the probability that the mean wage rate of the workers in the sample is between $9.75 and $10.25. Does your answer depend on the assumption of normally distributed wage rates?

20. A random sample of 100 is selected from the production records of a large firm. The objective of the sample is to estimate the mean production of a large group of employees. The results of the sample are $\bar{x} = 120$ units per day and $S = 15$ units per day. Calculate a 95 percent confidence interval for the mean production of all employees.

21. In order to determine the average dollar amount per sale, a random sample of 225 sales made during the year is selected from the records of a

large department store. The results of this sample are $\bar{x} = \$16.00$ and $S = \$7.50$. Calculate a 90 percent confidence interval for the mean sales of the store.

22. A random sample of 256 students is selected from the population of graduating high-school seniors in the state of California. The mean grade point average of the students in this sample is $\bar{x} = 2.98$ and the unbiased estimate of the population standard deviation is $S = 0.32$. Calculate a 98 percent confidence interval for the mean grade point average of the population of graduating seniors.

23. The mean strength of a large batch of steel bars is estimated by destructive tests of a sample of 81 randomly selected bars. The results of the tests were $\bar{x} = 59,000$ pounds and $S = 3000$ pounds. Calculate a 90 percent confidence interval for the mean strength of the population of steel bars.

24. A government agency reported that the average family income in a certain "poverty area" was $4560 and the standard deviation was $S = \$300.00$. Assuming that these statistics were obtained from a random sample of 250 families, calculate a 99 percent confidence interval for the family income of the population in this area.

25. A recent study gave a 95 percent confidence interval for the average per capita gasoline usage as $980.4 \leq \mu \leq 1019.6$ gallons. The confidence interval was calculated from a randomly selected sample of 900 individuals. Use this data to estimate the population mean μ and standard deviation σ.

26. For $\alpha = 0.05$, $\sigma = 16$, and $n = 64$, determine the critical values of $\bar{x}$ for the following null and alternative hypotheses:
    (a) $H_0: \mu = 50$, $H_1: \mu \neq 50$
    (b) $H_0: \mu \leq 50$, $H_1: \mu > 50$
    (c) $H_0: \mu \geq 50$, $H_1: \mu < 50$

27. A government contract calls for certain electronic parts that have an average useful life of at least 2000 hours. The standard deviation of the life of the part is known from past tests and is $\sigma = 100$ hours. Based upon a sample of size 100 and $\alpha = 0.01$, state the null and alternative hypotheses and values of $\bar{x}$ that would lead to acceptance of each hypothesis.

28. Morris Automobile Supply Company, one of the nation's largest automobile supply firms, regularly purchases automobile batteries from manufacturers for resale under its own name. The mean life of its top line battery is specified in the purchase agreement as no less than 38 months. The standard deviation of battery life has historically been $\sigma = 2$ months. Assuming samples of size $n = 36$ and $\alpha = 0.05$, state the null and alternative hypotheses and values of $\bar{x}$ that would lead to acceptance of each hypothesis.

29. Johnson Breweries, Inc., must regularly check the alcoholic content of its beer. The beer is supposed to have an alcoholic content of 5.00 percent. Some variation in content is, however, expected. This variation is measured by the standard deviation and has historically been 0.25 percent. In order to determine the content, the brewmaster randomly selects 25 containers of beer. If the average alcoholic content of this sample is less than 4.90 percent or more than 5.10 percent, the brewing process is halted and an adjustment is made.

    (a) State the null and alternative hypothesis for this procedure.

    (b) Calculate the probability of a Type I error.

30. Helen E. Hornquist, a certified public accountant, was engaged to verify the profit and loss statement of a certain firm. As a part of the verification procedure, Ms. Hornquist randomly selected the accounts of 100 customers of the firm in order to determine the average accounts receivable balance. The sample mean was $\bar{x} = \$92.00$ and the standard deviation was $S = \$24.00$. The firm records show that the average accounts receivable balance is $95.00. Should Ms. Hornquist accept the firm's statement that the average balance is $95.00 or should she request a complete audit of all accounts. Assume that accounting standards permit a probability of 10 percent of making a Type I error.

SUGGESTED REFERENCES

The references for this chapter are listed in Chap. 5.

Chapter Eight

Introduction
to
Decision Theory

Our objective in this chapter is to introduce an important kind of quantitative approach to decision making. As the chapter title indicates, this approach is termed decision theory. *Decision theory* is concerned with the problem of making decisions when less than complete information concerning the consequences of the decision is available. Our first task in this chapter is to examine the various situations under which decisions are made. These situations include certainty, risk, uncertainty, and conflict. This is followed by a discussion of the alternative criteria that are used in making decisions. The chapter concludes with a discussion of the value of information in decision making and a brief look at incremental analysis.

8.1   Conditions Under Which Decisions Are Made

A problem in decision making occurs when there are two or more alternative courses of action that can be followed in attempting to achieve some objective. The selection from among the several alternative courses of action is based upon some *criterion*. The *decision* involves selecting that course of action that is optimal in terms of the specified criterion.

Suppose, however, that the decision maker does not know exactly what consequences will result from each of the several courses of action. This can occur because the consequence of a specific action may be the result of events that are probabilistic in nature and are beyond the control of the decision maker. In decision theory, the different possible courses of action are referred to as the *action space* and the different possible probabilistic events are termed the *state space* or the *state of nature*.

To illustrate, consider a highly simplified example of a retailer who sells a perishable product. The product costs $3 and is sold for $5, leaving a profit of $2 per unit. The demand for the product is variable, ranging between one and three units per day. Unsold units have no salvage value.

The state space consists of the possible levels of demand, i.e., $q_1 = 1$, $q_2 = 2$, and $q_3 = 3$ units. The action space includes all possible courses of action available to the decision maker. These are to stock one, two, or three units of the product, i.e., $a_1 = 1$, $a_2 = 2$, and $a_3 = 3$ units. From the example, we see that the consequences of the decision depend not only on which of the three possible courses of action is selected but also on which of the three states of nature occurs. In other words, the profit associated with each course of action is conditional upon the state of nature, an event that is beyond the control of the decision maker. The profits that result from each combination of act and state of nature can be shown in a table. This table is referred to as a *conditional profit* or a *payoff table*. The conditional profit table for our retailer is shown in Table 8.1.

Table 8.1. Conditional Profit Table

State of Nature Demand Is:	Act		
	a_1 Stock 1	a_2 Stock 2	a_3 Stock 3
$q_1 = 1$	$2	−$1	−$4
$q_2 = 2$	2	4	1
$q_3 = 3$	2	4	6

The entries in the conditional profit table represent the different possible consequences of each act. Should the retailer elect to stock one unit, he is assured of a profit of $2. The profit associated with the decision to stock two units is conditional upon the state of nature and is −$1 if one unit is demanded and it is $4 if either two or three units are demanded. The profit from stocking three units also depends on the state of nature and varies from −$4 if one unit is demanded to $6 if three units are demanded.

Although the decision maker in our example knows the possible levels of profit associated with each act, he cannot uniquely associate each act with a particular profit. This inability comes from not knowing the state of nature. Depending on the information available concerning the state of nature, decision conditions can be classified into several categories. The decision conditions that we shall examine include: (a) certainty, (b) risk, (c) uncertainty, and (d) conflict.

8.1.1 DECISION MAKING UNDER CERTAINTY

Decision making involves evaluating the various alternatives and selecting the alternative that is optimal in terms of some criterion. If the decision maker can obtain complete information concerning the state of nature, the decision is said to be made under *conditions of certainty*.

To illustrate, suppose that the retailer stocks an item only on the basis of a previous order. The demand for the item is now known. In other words, the retailer has complete information and is therefore certain which state of nature will occur. He is now able to determine with certainty the consequences of his decision. The decision problem thus becomes that of computing the consequence of each course of action for the known state of nature and selecting the act that results in the optimal consequence. If, for instance, two units are ordered, the retailer elects to stock two units and makes a certain profit of $4. Similarly, if three units are demanded, the retailer stocks three units and realizes a profit of $6.

Decision problems under conditions of certainty are not always as trivial as our example might indicate. The problems of breakeven analysis discussed in Chap. 2 involved certainty. The presentation in Chaps. 11, 12, and 13 on linear programming and that in Chap. 14 on mathematics of finance also involve decisions made under conditions of certainty. In studying these chapters, the reader will find numerous examples of decision problems that occur under conditions of certainty.

8.1.2 DECISION MAKING UNDER RISK

Decision making under conditions of *risk* occurs when the different possible states of nature can be described by a probability distribution.[1] The consequences of the alternative courses of action available to the decision maker are again dependent on the state of nature. Although the decision maker does not know the true state of nature, he does have partial information concerning the various states. This partial information is in the form of an objective or a subjective probability distribution that describes the likelihood of occurrence of the different states. Given such a distribution, the decision problem becomes that of calculating the consequences of the various possible acts, evaluating these consequences in terms of the probabilities of occur-

---

[1] Many authors include decision making under risk and decision making under uncertainty as one category. Their reasoning is that there are very few decision problems in which a subjective probability distribution of the states of nature cannot be determined. Although this point of view is well established, we believe that decisions must often be made when information is not available to establish a probability distribution. For this reason, the decision conditions of risk and uncertainty are considered separately in this text.

rence of the several states of nature, and selecting the optimal act in accordance with some decision criterion.

Suppose that we return to our example of the retailer. The retailer is faced with the problem of ordering a stock of a perishable item prior to knowing how many units of the item will be demanded. In checking sales records for the past 100 days, the retailer finds that one unit was demanded on ten days, two units were demanded on 60 days, and three units were demanded on 30 days. The demand has never been less than one unit nor more than three units. Assuming that the same pattern of demand continues, the retailer can develop a probability distribution that describes the likelihood of occurrence of the several states of nature. This distribution is shown in Table 8.2.

Table 8.2. Probability Distribution

Demand	Number of Days	Probability
$q_1 = 1$	10	0.10
$q_2 = 2$	60	0.60
$q_3 = 3$	30	0.30
	100	1.00

The probability distribution from Table 8.2 and the profits associated with each act and state of nature from Table 8.1 can be included in a single table. This table is again termed a conditional profit or a payoff table. The conditional profit table for our retailer is shown as Table 8.3.

The conditional profit table contains the information needed to make a decision on the number of units to stock. The decision is made by using this table together with a decision criterion. As a preview, suppose that the decision criterion involves considering only the most likely state of nature. Using this criterion, the retailer assumes that two units will be demanded and therefore he stocks two units. If this state of nature occurs, the retailer will make a profit of $4. As another example, suppose that the decision criterion

Table 8.3. Conditional Profit Table

State of Nature		Act		
Demand q_i	Probability $p(q_i)$	a_1 Stock 1	a_2 Stock 2	a_3 Stock 3
$q_1 = 1$	0.10	$2	− $1	− $4
$q_2 = 2$	0.60	2	4	1
$q_3 = 3$	0.30	2	4	6
	1.00			

is to select that act which offers a chance for the maximum possible profit. Using this criterion, the retailer would select act a_3 and hope for a profit of $6. These criteria along with several others are discussed in Sec. 8.2.

8.1.3 DECISION MAKING UNDER UNCERTAINTY

Decision conditions involving *uncertainty* occur when the decision maker has *no information* concerning the occurrence of the various possible states of nature. In this case, the decision maker does not know the true state of nature and he cannot determine the probabilities of occurrence of the various states. He is, so to speak, completely "in the dark" regarding which state of nature will occur.

To illustrate, suppose that our retailer is adding a new item to his product line. For convenience, we shall assume that the product has the same costs and profit as before and that the item is again perishable. Since the item is sufficiently unique, there is no information available on potential demand. The retailer must order at least one unit of the product and can order no more than three units. To make this problem comparable to the earlier example, we assume that at least one unit can be sold. The conditional profits associated with the various possible acts and states of nature were given in Table 8.1 and are repeated in Table 8.4.

Assuming a complete lack of information on the likelihood of occurrence of different levels of demand, the decision must be made by using only the information in Table 8.4. The decision must therefore be made under conditions of uncertainty.

In the large majority of decision problems, the decision maker can obtain some information on the likelihood of occurrence of the different states. When such information is available, it should be used to develop a probability distribution for the states of nature. The problem becomes one of decision making under risk rather than under uncertainty.

Although problems involving complete uncertainty are relatively rare, they nevertheless do occur. For this reason, decision criteria that are par-

Table 8.4. Conditional Profit Table

State of Nature Demand Is:	Act		
	a_1 Stock 1	a_2 Stock 2	a_3 Stock 3
$q_1 = 1$	$2	− $1	− $4
$q_2 = 2$	2	4	1
$q_3 = 3$	2	4	6

ticularly applicable to decision making under conditions of uncertainty have been included in Sec. 8.2.

8.1.4 DECISION MAKING UNDER CONFLICT

Decision conditions that involve conflict occur when the consequences of a specific course of action no longer depend on the state of nature. Instead, they depend on the action taken by a competitive decision maker. The other decision maker, being an *adverse intellect* rather than nature, can be expected to purposely select a course of action that is to his benefit. If, as is often the case, the gain to one competitor is a loss to the other, a conflict occurs because each of the opposing decision makers is attempting to optimize his decision at the other's expense. Decisions that must be made under conditions of conflict can arise in business and other competitive environments.

As an example, consider the case of two competitive gasoline service stations, A and B. The owners of the two stations have been contacted by a trading stamp company. The manager of station A has had previous experience with trading stamps and figures that if his station begins to give stamps with gasoline purchases and station B does not, his monthly profits will increase from $1000 to $1100. On the other hand, if station B gives the stamps and station A doesn't, the profits of station A will drop to $950. If both stations give stamps, the profits of station A will fall to $980.

There are two alternatives available to the manager of station A, to give trading stamps or not give the stamps. The consequence of A's decision depends not only on the course of action chosen by A but also on the action taken by station B. The conditional profit table representing A's alternatives is given in Table 8.5. With additional information, a similar table could be constructed for B.

The techniques used for the analysis of decisions made under conditions of conflict are presented in a branch of applied mathematics known as *game theory*. Although game theory is not beyond the capabilities of the reader, the methods used to determine strategies for the more complicated decisions

Table 8.5. Conditional Profit Table for Station A

	A's Action	
B's Action	Not to Adopt the Stamps	To Adopt the Stamps
Not to adopt the stamps	$1000	$1100
To adopt the stamps	950	980

made under conditions of conflict are beyond the scope of this text.[2] We shall, however, discuss in the following section decision criteria that are used for the more elementary decision problems that occur under conditions of conflict.

8.2 Commonly Used Decision Criteria

When a decision is made under conditions of certainty, the actual consequences of the decision should be exactly as predicted from the analysis of the decision problem. If, however, the decision is made under conditions of risk, uncertainty, or conflict, the consequences of the decision are conditional upon the state of nature and are impossible to predict. This is true regardless of whether the state of nature can be described by a probability distribution, is completely unknown, or occurs because of the action of a competitor. Under such conditions, the decision maker must select a specific course of action from the available alternatives. He hopes that the decision will lead to maximum profits, but he knows that any decision has a certain element of risk. When making decisions under conditions of risk, uncertainty, or conflict, the decision maker is in effect gambling against nature or an opponent. He places his bet by making a certain decision, hoping that the decision will be optimal but knowing that it may not.

Since any decision made with less than complete information on the state of nature is something of a gamble, the decision maker is well-advised to understand alternative decision criteria. Many different decision criteria have been proposed. Some are based on monetary consequences alone and do not consider the probability of such consequences. Others are based mainly on the probabilities of the various states of nature and make only limited use of the monetary consequences of the acts. Still other criteria make use of both the monetary consequences and the probabilities of the states of nature. Our objective in this section is to explain and offer examples of the more commonly used decision criteria. The reader will see that criteria that are reasonable in making decisions under conditions of uncertainty or conflict leave much to be desired for decisions made under conditions of risk. Criteria that are appropriate under conditions of risk, however, cannot be used for decisions made under conditions of uncertainty or conflict.

8.2.1 CRITERIA BASED ON MONETARY CONSEQUENCES

In this section we shall consider two decision criteria that are based solely on the monetary consequences of the different courses of action. These

---

[2] The interested reader should refer to Melvin Dresher, *Games of Strategy: Theory and Applications* (Englewood Cliffs, N.J.: Prentice-Hall, Inc., 1961).

criteria are primarily used for selecting from among the alternative courses of action in problems in which probabilities of the states of nature are unknown, i.e., when decisions must be made under uncertainty or conflict. As a basis for decision making, the two criteria represent two extremes. At one extreme the decision maker hopes for the best of the possible consequences and selects that act which offers a chance for maximum possible profit. At the other extreme the decision maker expects the worst of the possible consequences and therefore selects an act which guarantees a certain minimum profit.

Maximax. An optimistic decision maker can be expected to select the alternative that offers some chance for the maximum possible profit. This alternative is found by determining the maximum possible profit for each act. These maximum profits are compared and the decision maker selects the act that provides an opportunity of achieving the *maxi*mum of the *maxi*mum possible profits. In so doing, the decision maker has used the *maximax* decision criterion.

As an example, consider the conditional profit table shown in Table 8.6. The reader will remember that these are the profits available to our retailer from stocking one, two, or three units of a certain product. The retailer might compare the three alternative acts and note that the maximum possible profit from a_1 is \$2, from a_2 is \$4, and from a_3 is \$6. If he is optimistic or for some reason must take a chance on making the greatest possible profit, he would use the maximax criterion. Comparing the maximum payoffs of \$2, \$4, and \$6, he would select the act that offers a chance of realizing the maximum possible profit. Using the maximax criterion, the retailer would select act a_3 and hope for a profit of \$6.

Table 8.6.   Conditional Profit Table

State of Nature Demand Is:	Act		
	a_1 Stock 1	a_2 Stock 2	a_3 Stock 3
$q_1 = 1$	\$2	-\$1	-\$4
$q_2 = 2$	2	4	1
$q_3 = 3$	2	4	6

Example. A decision maker has developed the conditional profit table shown in Table 8.7. Assuming the maximax criterion, determine the optimal decision.

The maximum profit possible under a_1 is \$8, under a_2 is \$9, under a_3 is \$10, and under a_4 is \$9. Using the maximax criterion, the decision maker selects

Table 8.7.

State of Nature	Act			
	a_1	a_2	a_3	a_4
q_1	− $2	$6	$5	$3
q_2	4	7	7	6
q_3	6	8	2	8
q_4	8	9	10	9

the act that offers a chance for the maximum possible profit. The decision maker selects act a_3 and hopes for a profit of $10.

Although the maximax criterion makes it possible to obtain the largest possible payoff, it is not a procedure that is generally followed. One reason is that the possibilities of low payoffs or losses are not considered. In order to remain in business, the decision maker must certainly include the possibility of a loss in his analysis of a decision problem. This possibility is considered by the following criterion.

Maximin. A pessimistic decision maker dares not hope for the best of all possible consequences. Instead, he tends to select the act that assures a certain minimum payoff regardless of which state of nature occurs. The selection is made by determining the minimum payoff possible for each course of action. These minimum payoffs are compared and the decision maker selects the act that gives the *maxi*mum of the *mini*mum possible payoffs. In choosing this act, the decision maker has employed the *maximin* decision criterion.[3]

To illustrate the maximin decision criterion, consider again the example of the retailer. The conditional profits for each act and state of nature are given in Table 8.6. If the retailer stocks one unit, he is assured of a minimum profit of $2. If he stocks two units, there is a possibility of a loss of $1. If he stocks three units, he could possibly lose $4. The retailer compares the minimum payoffs of $2, − $1, and − $4 and selects the maximum of the minimums. Using the maximin criterion, the retailer would select a_1 and thus be assured of a profit of $2.

The maximin criterion is one of the more conservative decision criteria. It does have the advantage of guaranteeing a certain minimal profit, but in so doing it precludes the opportunity of realizing higher profits. In terms of an ongoing business, repeated use of such a criterion could lead to stagnation.

---

[3] When applied to losses rather than to profits, the decision maker uses the *minimax* criterion, i.e., he selects the act that minimizes the maximum possible loss.

The business would almost certainly be overcome by its more innovative competitors.

In decision making under conditions of conflict, the decision maker is often forced to use the maximin criterion. The reader will remember that under conditions of conflict, the opponent is no longer nature but is an adverse intellect instead. If, as is often the case, one opponent's loss is the other's gain, the decision maker has no choice but to select the act that assures a certain minimum profit. To illustrate, consider the decision problem facing the gasoline station owner. The owner of station A must decide whether or not to adopt trading stamps. The conditional profits for station A were given in Table 8.5 and are repeated in Table 8.8. If A does not adopt the stamps, B most certainly will, thereby causing A's profit to fall to $950. If A adopts the stamps, B must also adopt the stamps and A's profit will fall to $980. Using the maximin criterion, A would elect to adopt the stamps, thus guaranteeing a profit of at least $980.

Table 8.8.   Conditional Profit Table for Station A

	A's Action	
B's Action	Not to Adopt the Stamps	To Adopt the Stamps
Not to adopt the stamps	$1000	$1100
To adopt the stamps	950	980

Example. In Table 8.9, A's gain is B's loss. Using the maximin criterion, determine the optimal strategy for both A and B.

The minimum gain from act a_1 is 3, from a_2 is 5, and from a_3 is 4. Using the maximin criterion, A selects the maximum of these minimum gains. A selects a_2 and is assured a gain of at least 5.

Decision maker B views the entries in Table 8.9 as losses. His strategy is to determine the maximum possible loss from each act and to select the

Table 8.9.

	A's Action		
B's Action	a_1	a_2	a_3
b_1	3	6	14
b_2	3	5	4
b_3	10	8	5

act that results in the minimum of the maximum losses. The maximum loss from act b_1 is 14, from b_2 is 5, and from b_3 is 10. Since B's strategy is to minimize the maximum loss, he uses the *minimax* criterion and selects act b_2. This guarantees B that his loss will be no more than 5.[4]

8.2.2 CRITERIA BASED PRIMARILY ON PROBABILITIES

In making decisions using either the maximax or the maximin criteria, it is assumed that the probabilities for the states of nature are not available. Because we were unable to determine even a subjective probability distribution for the states, the decision maker was forced to base his decision on monetary consequences alone. We will now consider two criteria that are based primarily on the probabilities of the various states of nature. One uses the concept of the mode and the other uses the concept of expected value. The two criteria are termed *maximum likelihood* and *expected state of nature*.

Maximum Likelihood. When employing the *maximum likelihood* decision criterion, the decision maker considers only that state of nature most likely to occur. From the probability distribution, the decision maker identifies the modal state, i.e., the state with the maximum likelihood of occurring. The conditional profits for the various courses of action are examined and the act that has the most desirable consequences is selected.

To illustrate, consider the decision problem facing our retailer. The conditional profits together with the probability distribution for the states of nature were given earlier and are repeated in Table 8.10. The most likely state

Table 8.10. Conditional Profit Table

State of Nature		Act		
Demand q_i	Probability $p(q_i)$	a_1 Stock 1	a_2 Stock 2	a_3 Stock 3
$q_1 = 1$	0.10	$2	− $1	− $4
$q_2 = 2$	0.60	2	4	1
$q_3 = 3$	0.30	2	4	6
	1.00			

---

[4] In game theory, a game in which the maximin and minimax solutions for the two players coincide is called a *strictly determined game*. The common solution is known as the *saddle point* and represents the value of the game. In the example, the saddle point is a_2, b_2. The value of the game is 5.

of nature is q_2, i.e., two units are demanded. Should this state of nature occur, the retailer would make $2 from a_1, $4 from a_2, and $1 from a_3. Using the maximum likelihood criterion, the retailer would consider only the state of nature most likely to occur. He would then select a_2, thereby hoping for a profit of $4.

The maximum likelihood criterion is reasonable for our example problem. This, however, is not always the case. For instance, assume that the probability of demand of one unit is 0.49 and of demand for two units is 0.51. Using the maximum likelihood criterion, the retailer would still elect to stock two units. In this case, the retailer would enjoy a profit of $4 and experience a loss of $1 with almost equal regularity. Most readers will agree that the decision maker would be better advised to stock only one unit. This decision would assure a long-term profit of $2. Should the retailer stock two units, the long-term profit would be only slightly greater than $1.50.

Expected State of Nature. Under the *expected state of nature* criterion, the decision maker calculates the expected or the mean state of nature. He then selects the act that has the maximum payoff under the assumption that the true state of nature will be that most nearly equal to the expected state of nature. The reader will remember that the expected value of a discrete random variable was discussed in Chap. 6. Representing the state of nature by the random variable Q, the formula for the expected state of nature is

$$E(Q) = \sum_{\text{all } q} q \cdot p(q) \tag{8.1}$$

where q is the value of the random variable and $p(q)$ is the probability of the occurrence of that value.

In the example of the retailer, the possible states of nature and the probability distribution for these states were given in Table 8.2. One unit was demanded with probability 0.10, two units with probability 0.60, and three units with probability 0.30. From Eq. (8.1), the expected state of nature is

$$E(Q) = 1(0.10) + 2(0.60) + 3(0.30) = 2.2$$

The retailer selects a course of action based on the assumption that that state of nature most nearly equal to the expected state of nature will occur. Based on this assumption, the retailer examines the monetary consequences associated with a demand of two units and selects act a_2. Should the state of nature most nearly equaling the expected state of nature actually occur, the retailer will realize a profit of $4.

In using either the maximum likelihood or the expected state of nature criteria, the monetary consequences of the decision are considered only after the state of nature is selected. The decision is thus primarily based on the probability of occurrence of the different states of nature. We now turn to a decision criterion in which both monetary consequence and probability are considered jointly. This criterion, termed expected monetary value, allows the decision maker to include both the payoff and the likelihood of obtaining this payoff in selecting the appropriate course of action.

8.2.3   EXPECTED MONETARY VALUE

When making decisions under conditions of risk, the decision maker should take into account both the probable occurrence of the different states and the associated monetary consequences. Our objective in this section is to introduce a decision criterion that explicitly includes both monetary consequences and probability. This criterion is termed *expected monetary value*.[5]

When using the expected monetary value criterion, abbreviated EMV, the decision maker computes the expected monetary value of each course of action and selects the act that offers the greatest expected monetary value. The expected monetary value or, alternatively, the expected payoff for a certain act is calculated by multiplying the conditional profit for each state of nature by the probability of occurrence of the state of nature and summing these products. This sum gives the expected monetary value for the act. These calculations are repeated for each of the alternative courses of action. The course of action that has the greatest expected monetary value is selected by the decision maker.[6]

Suppose that we let Q be a random variable representing the state of nature and let $v(a, Q)$ represent the conditional profit associated with act a and state of nature q. The formula for the expected monetary value of act a is

$$E[v(a, Q)] = \sum_{\text{all } q} v(a, q)p(q) \qquad (8.2)$$

where $p(q)$ is the probability of occurrence of state of nature q.

To illustrate the expected monetary value criterion, let us return to the decision problem facing the retailer. Consider the act of stocking two units. The conditional profits associated with this act along with the probability distribution for the states of nature were given in Table 8.3 and are repeated

---

[5] Some authors refer to the expected monetary value criterion as *Bayes' decision rule*.

[6] An alternative procedure is to base the decision on *opportunity loss*. Opportunity loss is defined as the profit foregone by not selecting the true state of nature. Under this procedure, the conditional profit table is converted to an opportunity loss table by subtracting each payoff for a given state of nature from the maximum payoff possible for that state of nature. The expected opportunity loss for each course of action is computed and the decision maker selects that act with the smallest expected opportunity loss. Since the expected opportunity loss and the expected monetary value criteria always result in exactly the same course of action, the expected opportunity loss criterion is not considered in detail in this text.

Table 8.11.   Expected Monetary Value of Stocking Two Units

| State of Nature | | Conditional Payoff | Calculation of Expected Payoff |
Demand q	Probability p(q)	v(2, q)	v(2, q)p(q)
1	0.10	− $1	− $0.10
2	0.60	4	2.40
3	0.30	4	1.20
	1.00	Expected monetary value =	$3.50

in Table 8.11. The conditional profits are multiplied by the probability of occurrence of the states of nature and these products are summed. The expected monetary value of stocking two units is computed in Table 8.11 and is $3.50.

The expected monetary value of stocking one unit and of stocking three units must also be computed. The reader should verify that the expected monetary value of stocking one unit is $2. By coincidence, the expected monetary value of stocking three units is also $2. Using the expected monetary value criterion, the decision maker would select the course of action that gives the greatest expected monetary value. In our example, the retailer would stock two units and would realize an average profit of $3.50 per day.

Example. A large newsstand in Los Angeles stocks the Sunday edition of various metropolitan newspapers. One of these papers is *The New York Times.* The dealer pays $0.60 for each copy of the *Times* and sells the paper for $1.00. Copies left unsold are nonreturnable and have no value. In order to determine how many copies of the paper to stock, the dealer has tabulated sales records for the past 100 weeks. These records are shown in Table 8.12.

Table 8.12.   Demand for the Sunday edition of *The New York Times*

Number of Copies Demanded per Week	Number of Weeks
18	10
19	16
20	20
21	24
22	16
23	8
24	6
	100

Table 8.13. Conditional Profit Table

| State of Nature | | Course of Action Number of Copies Stocked | | | | | | |
Demand q	Probability p(q)	18	19	20	21	22	23	24
18	0.10	$7.20	$6.60	$6.00	$5.40	$4.80	$4.20	$3.60
19	0.16	7.20	7.60	7.00	6.40	5.80	5.20	4.60
20	0.20	7.20	7.60	8.00	7.40	6.80	6.20	5.60
21	0.24	7.20	7.60	8.00	8.40	7.80	7.20	6.60
22	0.16	7.20	7.60	8.00	8.40	8.80	8.20	7.60
23	0.08	7.20	7.60	8.00	8.40	8.80	9.20	8.60
24	0.06	7.20	7.60	8.00	8.40	8.80	9.20	9.60
	1.00							

Assuming that no substantial changes occur in the demand for the paper, determine the number of copies that should be stocked in order to maximize the expected monetary value.

From Table 8.12, we see that demand has never been less than 18 nor more than 24. Since the dealer foresees no substantial changes in demand, his course of action is to stock some number of copies between 18 and 24, inclusive. The conditional profits for each combination of act and state of nature are shown in Table 8.13. The probability of each state of nature is also shown.

The expected monetary value of each course of action is calculated by using Eq. (8.2). The results of these calculations are shown in Table 8.14. The reader should verify that the optimal course of action is to stock 20 copies of the paper. The expected profit of this course of action is $7.64.

Table 8.14. Expected Monetary Value

| | Course of Action Number of Copies Stocked | | | | | | |
	18	19	20	21	22	23	24
Expected Monetary Value	$7.20	$7.50	$7.64	$7.58	$7.28	$6.82	$6.28

8.3 The Value of Information

In many decision situations, the decision maker has the opportunity to purchase information that can be used to predict the state of nature. Such information could come from market surveys, statistical studies, consultants, or various other sources. This information would be valuable to the decision maker because its use should lead to an increase in the expected profit. For instance, suppose that the decision maker is informed that one unit of a product will be demanded. He will respond by stocking one unit. Should he learn that demand will be two units, his response will be to stock two units. Similarly, should he learn that demand is k units, the optimal decision would be to stock k units. Because all uncertainty regarding the state of nature has been removed, the decision maker is able to increase the expected profit. He does this by merely selecting that course of action that is optimal for the known state of nature.

Our objective in this section is to introduce a method for determining the expected value of perfect information. We realize that the information available to the decision maker would seldom be perfect. Nevertheless, if the decision maker is able to calculate the expected value of perfect information, he is in a better position to judge the value of the less than perfect information that is available in the marketplace.

8.3.1 EXPECTED PROFIT UNDER CONDITIONS OF CERTAINTY

To determine the expected value of perfect information, we first calculate the *expected profit under conditions of certainty*. The expected profit under conditions of certainty, abbreviated EPCC, is the expected profit assuming prior knowledge of the state of nature.

Returning to our example of the retailer, suppose that each day's demand can be predicted with certainty. With advance knowledge of demand, the retailer simply stocks the number of units that will be demanded. Should one unit be demanded, the retailer stocks one unit and makes a $2 profit. Should the demand be two units, the retailer stocks two units and makes a $4 profit, etc. Given the availability of this perfect predictor and the probability of different levels of demand, we can calculate the expected profit under conditions of certainty. This is found by multiplying the conditional profit for a given state of nature by the frequency (i.e., probability) of occurrence of the states of nature and summing these products. The calculations for EPCC are shown in Table 8.15. With perfect information, the retailer would make an average daily profit of $4.40.

8.3.2 EXPECTED VALUE OF PERFECT INFORMATION

The expected value of perfect information, abbreviated EVPI, is calculated by subtracting the expected monetary value for the optimal decision from the expected profit under conditions of certainty, i.e.,

$$EVPI = EPCC - EMV \qquad (8.3)$$

Table 8.15. Expected Profit Under Conditions of Certainty

State of Nature		Conditional Profit	Expected Profit
Demand	Probability	Under Certainty	Under Certainty
$q_1 = 1$	0.10	$2	$0.20
$q_2 = 2$	0.60	4	2.40
$q_3 = 3$	0.30	6	1.80
	1.00		EPCC = $4.40

In terms of our example, the retailer can earn $3.50 per day by stocking two units (see Table 8.11). If perfect information is available, the expected profits increase to $4.40 (see Table 8.15). The difference between the EPCC and the EMV of the best decision is $0.90. This difference of $0.90 is the maximum amount the retailer could pay in order to obtain a perfect predictor. If he were to pay more than $0.90, he would actually decrease the expected profit. The expected value of perfect information is therefore $0.90.

Example. Determine the expected value of perfect information for the newsstand dealer referred to in the example on p. 257.

To determine the expected value of perfect information, we first calculate the expected profit under conditions of certainty. The conditional profit for each act and state of nature was given in Table 8.13. If the uncertainty regarding the state of nature is eliminated, the dealer will stock exactly the number of copies demanded. Should 18 copies be demanded, the dealer will stock 18 copies and make a profit of $7.20. Should 19 copies be demanded, the dealer will stock 19 copies and make a profit of $7.60, and so forth. These conditional profits along with the probability of the different levels of demand are shown in Table 8.16. The expected profit under conditions of certainty is

Table 8.16. EPCC for Newsstand Dealer

State of Nature		Conditional Profit	Expected Profit
Demand	Probability	Under Certainty	Under Certainty
18	0.10	$7.20	$0.720
19	0.16	7.60	1.216
20	0.20	8.00	1.600
21	0.24	8.40	2.016
22	0.16	8.80	1.408
23	0.08	9.20	0.736
24	0.06	9.60	0.576
	1.00		EPCC = $8.272

calculated by multiplying the conditional profits by the probability of occurrence of the various levels of demand. Referring to Table 8.16, we see that the expected profit under conditions of certainty is $8.272. This means that with prior knowledge of demand, the average weekly profit would be $8.272.

The expected value of perfect information is obtained by subtracting the expected monetary value of the optimal act from the expected profit under conditions of certainty. Without prior knowledge of the level of demand, the news dealer stocks 20 copies. The expected monetary value of this act is $7.64 (see Table 8.14). With perfect information, the expected profit is $8.272. The expected value of perfect information is thus $8.272 − $7.640 or $0.632. The newsstand operator can therefore pay up to $0.632 to obtain perfect information on the state of nature.

8.4   Incremental Analysis

In using the expected monetary value criterion, the decision maker must compute the EMV of each alternative course of action. Although the procedure for calculating the EMV is straightforward, the calculations are time-consuming, particularly if there are many alternative courses of action. For instance, consider an inventory stocking problem in which there are 20 possible levels of demand. In such a problem there would be 20 possible courses of action and a total of 400 entries in the conditional profit table. In problems such as this, it is not practical to compute the expected monetary value for all the alternative acts. Instead, an approach referred to as *incremental analysis* has been developed from the EMV criterion. This approach can be used in lieu of EMV to determine the optimal course of action. Incremental analysis gives exactly the same solution as EMV and offers a considerable reduction in computations.

To develop the formula for incremental analysis, we consider the profit and loss associated with selling a marginal or incremental unit of a product. Specifically, let

$$p = \text{profit from selling an incremental unit}$$

$$l = \text{loss incurred if the incremental unit is not sold}$$

In our example of the newsstand operator, p represents the profit from selling one additional copy of *The New York Times* and l the loss if this copy is stocked but not sold. Since the newsstand dealer pays $0.60 per copy and sells the paper for $1.00, the profit from selling one additional copy is $p = $0.40 and the loss if the paper is not sold is $l = $0.60.

A solution for the optimum stock level can now be obtained. We represent the state of nature by the random variable Q. From among the alternative courses of action, represented by $a_1, a_2, \ldots, a_n$, the optimal act is the smallest a_j that satisfies the following inequality:

$$P(Q \leq a_j) \geq \frac{p}{p + l} \tag{8.4}$$

where $P(Q \leq a_j)$ represents the probability that the demand will be less than or equal to the number of units stocked.

To illustrate Eq. (8.4), consider the example of *The New York Times* Sunday paper. To determine the optimal stock level we proceed as follows:

Step 1:   Determine p and l. In this example the newsstand dealer pays \$0.60 for a paper which he then sells for \$1.00. The profit is $p = \$0.40$ and the loss is $l = \$0.60$.

Step 2:   Calculate the ratio $r = p/(p + l)$.

$$r = \frac{0.40}{0.40 + 0.60} = 0.40$$

Step 3:   Determine the cumulative probability distribution for demand. The cumulative probability distribution is determined from the probability distribution originally given in Table 8.12. The cumulative distribution is given in Table 8.17. Notice that the cumulative probabilities merely represent the probability that demand is less than or equal to some value of q. For instance, the probability that demand is less than or equal to 22 copies is 0.86.

Table 8.17.   Cumulative Probabilities for *The New York Times*

State of Nature		*Cumulative*
Demand q	*Probability* $p(q)$	*Probability* $P(Q \leq q)$
18	0.10	0.10
19	0.16	0.26
20	0.20	0.46
21	0.24	0.70
22	0.16	0.86
23	0.08	0.94
24	0.06	1.00
	1.00	

Step 4:   From the cumulative probability distribution, find the smallest stock level a_j such that $P(Q \leq a_j) \geq r$. The prob-

ability that q is less than or equal to 20 copies is 0.46. From Eq. (8.4), we stock the smallest number of copies such that the probability that demand is less than or equal to a_j is greater than or equal to r. Since $r = 0.40$ and $P(Q \leq 20) = 0.46$, the optimal decision is to stock 20 copies of the Sunday *New York Times*. Notice that this is the same solution obtained by using EMV in Table 8.14.

Example. A retailer buys an item for $2.00 which he resells for $3.00. The item is perishable and those items left unsold at the end of the day have no value. The cumulative probability distribution for demand is shown in Table 8.18. Use incremental analysis to determine the optimum stock level.

The profit on each item is $p = \$1.00$ and the loss on an item not sold is $l = \$2.00$. The ratio of $p/(p + l)$ is $r = 0.33$. From Eq. (8.4), the optimum stock level is the smallest a_j such that $P(Q \leq a_j) \geq 0.33$. Since $P(Q \leq 29) = 0.43$, we see that the optimum stock level is 29 units.

Table 8.18.

State of Nature		Cumulative
Demand q	Probability $p(q)$	Probability $P(Q \leq q)$
26	0.05	0.05
27	0.10	0.15
28	0.12	0.27
29	0.16	0.43
30	0.16	0.59
31	0.20	0.79
32	0.10	0.89
33	0.06	0.95
34	0.05	1.00
	1.00	

PROBLEMS

1. An individual wishes to invest a sum of money for a period of one year. Three alternatives are being considered:

   a_1:   Invest in a local real estate venture.
   a_2:   Purchase common stock of Western Industries, Inc.
   a_3:   Deposit the money in an insured savings account.

The return from alternatives a_1 and a_2 largely depends on whether Western Industries receives a pending government contract. Western Industries is the largest local employer. Consequently, receipt of the contract would be an advantage to both the real estate venture and the purchase of common stock. The possible states of nature that affect the profits on the investments are thus:

q_1:   Western Industries receives the contract.
q_2:   Western Industries does not receive the contract.

The profits associated with each course of action and state of nature (in thousands of dollars) are shown in the following conditional profit table:

	Act		
State of Nature	a_1	a_2	a_3
q_1	$8	$9	$4
q_2	3	-2	4

(a) Determine the optimal course of action by using the maximax criterion.
(b) Determine the optimal course of action by using the maximin criterion.

2. A student must take a required English course. Two sections are offered. One section is taught by Professor Baker and the other will be taught by either Professor Anderson or Professor Clark. Rumor has it that Baker will give a B for a reasonable amount of classroom effort. It is said that the same effort is good for an A from Anderson but only a C from Clark. Professor Anderson or Clark will be named as instructor after the deadline for enrollment.
(a) Determine the optimal course of action by using the maximax criterion.
(b) Determine the optimal course of action by using the maximin criterion.

3. A commuter has two alternative routes between home and work. One route is by surface streets and the other by freeway. The time required to travel from home to work by surface streets is 40 minutes. The travel time is 30 minutes if the freeway is clear and 1 hour if the freeway is congested.
(a) Construct a conditional payoff table for the commuter's decision problem. (Hint: Remember that small payoffs in this problem are more desirable than large payoffs.)

(b) Determine the optimal route by using the maximax criterion.

(c) Determine the optimal route by using the maximin criterion.

4. An investment counselor has developed the three following alternative investment plans for a certain client:

a_1:   Invest in highly speculative growth stocks.

a_2:   Invest in moderately speculative income stocks.

a_3:   Invest in high-yield "blue chip" stocks.

The return from each plan depends on the general status of the stock market. According to the investment counselor, there are three possibilities. These are:

q_1:   The market will experience a sharp upward movement.

q_2:   The market will drift with no major upward or downward movement.

q_3:   The market will experience a sharp downward movement.

The profits associated with each investment alternative and state of nature are shown in the following conditional profit table:

	Act		
State of Nature	a_1	a_2	a_3
q_1	$20	$12	$ 8
q_2	5	16	10
q_3	−15	4	6

(a) Determine the optimal alternative by using the maximax criterion.

(b) Determine the optimal alternative by using the maximin criterion.

(c) Assume that the investment counselor learns that his client has a conservative investment philosophy. Which alternative is the client likely to select?

5. The Clark Manufacturing Company and the Roberts Machinery Company are direct competitors in a certain product line. Currently, each company is considering an expansion of its manufacturing facilities. Clark's management estimates that if it expands and Roberts does not, net income will increase by $1.5 million. If Roberts expands and Clark does not, Clark's net income will decrease by $1.2 million. If both companies expand, Clark's net income will decrease by $1.5 million. If the decision criterion is maximin, should the expansion be undertaken by Clark?

6. Company A plans to develop one of two new products. Unfortunately, A's competitor B also plans to develop one of the two products. The potential profits for A depend on the product developed by B. The profits are shown in the following payoff table:

	Company A	
Company B	Product 1	Product 2
Product 1	$500,000	$2,000,000
Product 2	$1,000,000	$300,000

If A uses the maximin decision criterion, which of the two products would they develop?

7. Firms A and B are the sole competitors in a certain market. Firm A has developed a new procedure that will lead to an increase in its market share. The percentage increase depends on the strategy A uses in introducing the procedure and the strategy employed by B to counteract A's competitive advantage. Each firm has two strategies. The payoffs associated with the strategies are shown in the following table:

	Firm A	
Firm B	Strategy 1	Strategy 2
Strategy 1	10%	14%
Strategy 2	7%	6%

(a) Determine the optimal strategy for Firm A by using the maximin criterion.

(b) Assume that Firm B selects its strategy by using the minimax criterion. Determine B's maximum percentage loss.

8. An employment agency provides temporary help on a daily basis. From past experience, the manager of the agency has determined the following probabilities for the demand for a certain kind of worker:

Daily Demand	Probability
1 worker	0.10
2 workers	0.40
3 workers	0.30
4 workers	0.20

The workers must be notified on the preceding day if their services are required. If the worker is told to report for work, he receives a salary of $40 regardless of whether or not he is sent out on a job. The agency charges $56 and pays the $40 salary of the worker.

(a) Construct the conditional profit table for the employment agency.

(b) If the manager uses the maximum likelihood criterion, how many workers should be available?

(c) If the manager uses the expected state of nature criterion, how many workers should be available?

9. A newsstand operator has determined that the probabilities of the demand for the Sunday *Los Angeles Times* are as follows:

Demand	*Probability*
10 copies	0.10
11 copies	0.15
12 copies	0.25
13 copies	0.30
14 copies	0.20

The paper sells for $0.50 and costs $0.30. Assume that unsold copies have no salvage value.

(a) Prepare a conditional profit table for the decision problem.

(b) Determine the optimal number of copies to stock by using the maximum likelihood criterion.

(c) Determine the optimal number of copies to stock by using the expected state of nature criterion.

(d) Determine the optimal number of copies to stock by using the expected monetary value criterion.

10. The manager of a small department store must place his order for women's bathing suits. Each suit costs $15 and sells for $25. Suits that are not sold by the end of the summer are reduced to $10. The demand for bathing suits has been estimated as shown below:

Demand	*Probability*
25	0.10
26	0.15
27	0.30
28	0.20
29	0.15
30	0.10

(a) Prepare a conditional profit table for the decision problem.

(b) Determine the optimal number of suits by using the expected monetary value criterion.

(c) Determine the expected value of perfect information.

11. A wholesaler has an opportunity to buy up to 5000 pairs of men's golf shoes that have been marked as "seconds" by a major shoe manufacturer. The wholesaler will pay $5 a pair and can sell the shoes for $10 a pair to retailers. Any shoes that cannot be sold to retailers can be sold through a foreign outlet at $2 a pair. The wholesaler assigns probabilities of demand by the retailers for the shoes as follows:

Demand	Probability
1000 pairs	0.30
2000 pairs	0.30
3000 pairs	0.20
4000 pairs	0.10
5000 pairs	0.10

(a) Prepare a conditional profit table for the decision problem.

(b) Determine the optimal number of pairs to purchase by using the expected monetary value criterion.

(c) Determine the expected value of perfect information.

12. Judy Atwood, the marketing manager of the Tiger Golf Club Company, must decide whether or not to introduce a new line of ladies' golf clubs called Lady Tigress. On the basis of studies of sales potential, Judy estimates the following probabilities of demand during the first year:

Demand Level	Probability
High	0.30
Moderate	0.50
Low	0.20

Judy estimates that profit will be $180,000 if demand is high and $60,000 if demand is moderate. If demand is low, however, the firm will incur a $210,000 loss.

(a) Prepare a conditional profit table for Judy's decision problem.

(b) On the basis of the expected monetary value criterion, should Judy recommend that the product be introduced?

(c) Determine the expected value of perfect information.

13. Bob Baker has an opportunity to invest $4000 in a small company. The success of the company depends on the ability of the company to obtain a patent on a certain device. If the company is successful in obtaining the patent, Bob will receive a net return of $16,000. If, however, the company does not obtain the patent, the entire investment will be lost. What is the minimum probability Bob should require for obtaining the patent in order for the investment to be desirable? Use the expected monetary value criterion.

14. An individual has been approached by a major oil company regarding the oil rights on a certain piece of land. The oil company has reason to believe that there is a possibility of oil under the land and has offered $200,000 for the outright purchase of the oil rights. Rather than immediately accepting the offer, the individual contacted an independent oil company. The independent company offered to drill for oil on the land and pay the individual a royalty. If oil is found, the royalties will be worth $1 million. What must the probability of discovering oil be for the landowner to be indifferent between the two offers? Use the expected monetary value criterion.

15. The proprietor of a newsstand believes that the number of copies demanded for a certain magazine can be described by the following probability distribution:

Demand	Probability
4	0.20
5	0.20
6	0.35
7	0.25

Assume that each copy costs $0.50 and sells for $1.00. Copies not sold have no value.

(a) Prepare a conditional profit table for this decision problem.

(b) Determine the optimal number of copies to stock by using the expected monetary value criterion.

(c) Determine the expected value of perfect information.

(d) Determine the optimal number of copies to stock by using incremental analysis.

16. Based on past records, a grocer has determined that the weekly demand for avocados can be described by the following probability distribution:

Cartons per Week	Probability
5	0.10
6	0.10
7	0.15
8	0.15
9	0.15
10	0.20
11	0.10
12	0.05

The grocer pays $30 per carton and sells the avocados for $50 per carton. Avocados not sold by the end of the week have no value. Use the method of incremental analysis to determine the optimal number of cases to stock.

SUGGESTED REFERENCES

BIERMAN, H., JR., C. BONINI, and W. HAUSMAN, *Quantitative Analysis for Business Decisions* (Homewood, Ill.: Richard D. Irwin, Inc., 1973).

EWART, P., J. FORD, and C. LIN, *Probability for Statistical Decision Making* (Englewood Cliffs, N.J.: Prentice-Hall, Inc., 1974).

KEMENY, JOHN G., et al., *Finite Mathematics with Business Applications* (Englewood Cliffs, N.J.: Prentice-Hall, Inc., 1972).

LEVIN, R. I., and C. A. KIRKPATRICK, *Quantitative Approaches to Management* (New York, N.Y.: McGraw-Hill Book Company, Inc., 1975).

TRUEMAN, RICHARD E., *An Introduction to Quantitative Methods for Decision Making* (New York, N.Y.: Holt, Rinehart and Winston, Inc., 1964).

Chapter Nine

Matrices and Matrix Algebra

Matrices and matrix algebra have begun to play an increasingly important role in modern techniques for mathematical analysis of business and social problems. The matrix, coupled with the electronic computer, provides a method for handling vast quantities of data. Matrix equations provide a compact method of representing the large systems of equations found in many of the newer mathematical models. The solutions to these systems of equations are obtained by using matrix algebra, very often with the aid of an electronic computer.

This chapter provides an introduction to the important topics of matrices and matrix algebra. Although certain problems are included to illustrate the topics, the discussion is largely theoretical. The reader who is anxious for practical applications of matrices and matrix algebra need not be overly concerned. The quantitative techniques presented in the remainder of this text use either wholly or partly the various aspects of matrices and matrix algebra for their representation and solution.

Before beginning a formal discussion of matrices and matrix algebra, let us define a *matrix* as an *array or group of numbers*. To illustrate, the quantity of inventory by product line at three local car dealers can be described by a matrix. This matrix might have the form shown below:

Ford	Chevrolet	Plymouth	
6	8	5	*Compact*
10	10	8	*Sedan*
3	4	2	*Convertible*
4	5	3	*Station wagon*

The matrix shows that the Ford dealer has an inventory of six compacts, ten sedans, three convertibles, and four station wagons. Similarly, the inventories of the Chevrolet and Plymouth dealers can be read directly from the matrix.

An important characteristic of a matrix is that the *position within the matrix of each number* as well as the *magnitude of the number* is important. For instance, the fact that the local Chevrolet dealer has four convertibles in inventory is described by the entry in the matrix at the intersection of the convertible row and the Chevrolet column. The position of the number is important in that a specific location within the matrix is reserved for the inventory of Chevrolet convertibles. The magnitude of the number is important, since it gives the number of Chevrolet convertibles in inventory.

Certain algebraic operations can be performed on matrices. One such operation might involve determining the value of the inventories of the car dealers. The value of the inventories could be found by multiplying the inventory matrix by a price matrix. This would be an example of matrix multiplication. Other examples of matrix algebra might involve the addition or subtraction of two or more matrices or the multiplication of each of the numbers in the matrix by a common number.

9.1 Vectors

A *vector* is defined as a row or column of numbers in which the position of each number within the row or column is of importance. A vector is a special case of a matrix in that it has only a single row or column rather than multiple rows and columns. Column vectors are written vertically and row vectors horizontally. Examples of column vectors are

$$\begin{pmatrix} 2 \\ 1 \end{pmatrix}, \quad \begin{pmatrix} -3 \\ 6 \end{pmatrix}, \quad \begin{pmatrix} 1 \\ 0 \\ 2 \\ 3 \end{pmatrix}, \quad \begin{pmatrix} 3 \\ -4 \\ 7 \end{pmatrix}.$$

Examples of row vectors are $(3, 2)$, $(6, 0, 4)$, and $(-6, 4, \frac{1}{2}, 7)$.

To illustrate that the position of the number within the vector is important, the row vector $(3, 2)$ and the row vector $(2, 3)$ have the same components but are not equal. Vectors are equal only if they are exactly the same. Thus, for two vectors to be equal, each component of the first vector must be exactly the same and occupy the same position as each component of the second vector. Row vectors can, therefore, equal other row vectors but not column vectors. Similarly, column vectors can equal other column vectors but not row vectors.

Example. Are the vectors $(6, 4)$ and $\begin{pmatrix} 6 \\ 4 \end{pmatrix}$ equal?

Since $(6, 4)$ is a row vector and $\begin{pmatrix} 6 \\ 4 \end{pmatrix}$ a column vector, the two vectors are not equal.

Example. Are the vectors $(6, 1, 0)$ and $(6, 1)$ equal?

Since the first vector has the element 0 while the second does not, the two row vectors are not equal.

Both row and column vectors are designated by capital letters. The numbers that comprise the vector are called the *elements* or *components* of the vector. We could thus write $A = (6, 10, 12)$, where A designates the row vector with components or elements 6, 10, and 12.

9.2   Vector Algebra

The basic rules for the addition, subtraction, and multiplication of vectors are analogous to the rules of ordinary algebra. The concept of division is not directly extendable to either vectors or matrices. The matrix counterpart of division will be treated in the discussion of matrix algebra.

9.2.1   VECTOR ADDITION AND SUBTRACTION

Row or column vectors may be added or subtracted to other like row or column vectors by adding or subtracting components. If, for instance, $A = (6, 10, 12)$ and $B = (4, 6, -3)$, then

$$A + B = (6 + 4, 10 + 6, 12 - 3) = (10, 16, 9)$$

and

$$A - B = (6 - 4, 10 - 6, 12 + 3) = (2, 4, 15)$$

Similarly, for the column vectors $G = \begin{pmatrix} 3 \\ 2 \end{pmatrix}$ and $H = \begin{pmatrix} -1 \\ 3 \end{pmatrix}$,

$$G + H = \begin{pmatrix} 3 \\ 2 \end{pmatrix} + \begin{pmatrix} -1 \\ 3 \end{pmatrix} = \begin{pmatrix} 2 \\ 5 \end{pmatrix}$$

It is important to recognize that the addition or subtraction of vectors can take place only if the vectors are of the same type, i.e., row vectors may be added to or subtracted from row vectors and column vectors may be added to or subtracted from column vectors. Furthermore, the vectors to be added or subtracted must have the same number of components. The vector $A = $

(6, 10, 12) cannot be added to the vector $E = (3, 2)$ nor the vector $F = (5, 3, 7, 10)$. Vectors that are the same type and have the same number of components are said to be *conformable* for addition or subtraction.

The commutative and associative laws of addition apply to vector addition. The reader may remember that the *commutative law* of addition states that $a + b = b + a$. If this applied to vector addition, the commutative law of addition states that $A + B = B + A$. This is easily verified, since for $A = (6, 10, 12)$ and $B = (4, 6, -3)$, $A + B = (10, 16, 9)$ and $B + A = (10, 16, 9)$.

The *associative law* of addition states that $a + (b + c) = (a + b) + c$. Applying this law to the vectors $A = (6, 10, 12)$, $B = (4, 6, -3)$, and $C = (-4, 8, 6)$, we have

$$A + (B + C) = (A + B) + C$$

$$(6, 10, 12) + (4 - 4, 6 + 8, -3 + 6) = (6 + 4, 10 + 6, 12 - 3) + (-4, 8, 6)$$

$$(6, 10, 12) + (0, 14, 3) = (10, 16, 9) + (-4, 8, 6)$$

$$(6 + 0, 10 + 14, 12 + 3) = (10 - 4, 16 + 8, 9 + 6)$$

$$(6, 24, 15) = (6, 24, 15)$$

This verifies that the associative law applies to vectors.

Example. A student has 96 semester credits of which 24 units are A, 36 units are B, 30 units are C, 6 units are D, and 0 units are F. If he receives 6 units of A, 6 units of B, 3 units of C, 0 units of D, and 3 units of F, determine the vector that describes his grades.

Let $G = (24, 36, 30, 6, 0)$ represent grades excluding the current semester and $S = (6, 6, 3, 0, 3)$ represent the current semester's grades. Then $G + S = (30, 42, 33, 6, 3)$ is the vector that describes his grades including the current semester.

9.2.2 VECTOR MULTIPLICATION

Vector multiplication includes the multiplication of a vector by a vector or a vector by a scalar. A *scalar* is any single number, such as 3, or -10, or 1.10, etc. A vector may be multiplied by a scalar simply by multiplying each component of the vector by the scalar and listing the resulting products as a new vector. To demonstrate, let

$$X = \begin{pmatrix} 3 \\ -6 \\ 2 \end{pmatrix}$$

and $a = -3$. The product of the vector X and the scalar a is

$$aX = -3 \cdot \begin{pmatrix} 3 \\ -6 \\ 2 \end{pmatrix} = \begin{pmatrix} -9 \\ 18 \\ -6 \end{pmatrix}$$

Similarly, if we multiply the row vector $A = (6, 10, 12)$ by the scalar $b = 1.5$, we obtain

$$bA = 1.5(6, 10 \ 12) = (9, 15, 18)$$

Example. The vector $P = (\$10, \$8, \$6)$ represents the prices of three items. If prices increase by 10 percent, determine the new price vector.

The new price vector is formed by the product of the scalar 1.10 and the vector P. Thus, $1.10P = 1.10(\$10, \$8, \$6) = (\$11, \$8.80, \$6.60)$.

Two vectors may be multiplied together, provided that (1) one of the vectors is a row vector and the other a column vector and (2) the row vector and column vector each contain the same number of components. To illustrate, consider the row vector $R = (r_1, r_2, r_3)$ and the column vector

$$C = \begin{pmatrix} c_1 \\ c_2 \\ c_3 \end{pmatrix}$$

Since R is a row vector and C a column vector, and since each vector has three components, it is possible to determine the product of R and C. The vector multiplication is made by placing the row vector to the left of the column vector. The first component of the row vector is then multiplied by the first component of the column vector, the second component of the row vector is multiplied by the second component of the column vector, etc. These products are summed and the resulting sum represents the product of the two vectors. The product of R and C is

$$RC = (r_1, r_2, r_3) \begin{pmatrix} c_1 \\ c_2 \\ c_3 \end{pmatrix} = r_1 c_1 + r_2 c_2 + r_3 c_3 \tag{9.1}$$

The result is termed the *inner product* of R and C and is a scalar.

Example. Determine the inner product of $A = (6, 10, 12)$ and

$$X = \begin{pmatrix} 3 \\ -6 \\ 2 \end{pmatrix}$$

$$AX = 6 \cdot 3 + 10(-6) + 12 \cdot 2 = -18$$

Example. After four semesters, a student has 30 units of A, 20 units of B, 10 units of C, 4 units of D, and 0 units of F. If an A is worth 4 grade points, a B is worth 3 grade points, a C is worth 2 grade points, a D is worth 1 grade point, and an F is worth 0 grade points, determine the student's total number of grade points.

Let

$$A = (30, 20, 10, 4, 0) \quad \text{and} \quad G = \begin{pmatrix} 4 \\ 3 \\ 2 \\ 1 \\ 0 \end{pmatrix}$$

The total number of grade points is given by the inner product of A and G. Thus,

$$AG = (30, 20, 10, 4, 0) \begin{pmatrix} 4 \\ 3 \\ 2 \\ 1 \\ 0 \end{pmatrix} = 120 + 60 + 20 + 4 + 0 = 204$$

The *commutative law* of multiplication states that $a \cdot b = b \cdot a$. Although this result is true in ordinary algebra, it is not true that $AB = BA$ in vector algebra. If A is a row vector and B a column vector, then AB is a scalar. As we demonstrate later in this chapter, the product of the column vector B and the row vector A, i.e., BA, is a matrix. It will become obvious that the scalar AB is not equal to the matrix BA; thus, the commutative law of multiplication does not apply to vector multiplication.

The *distributive law* of algebra applies to vector algebra. The distributive law states that $a(b + c) = ab + ac$. Applying this to vector algebra gives the result that $P(B + E) = PB + PE$, where P is a row vector and B and E are column vectors. The distributive law is demonstrated in the following example.

Example. A retailer stocks four brands of a product. The costs of these four brands are given by the row vector $P = (60, 70, 90, 100)$. The beginning inventory of the product is given by the column vector

$$B = \begin{pmatrix} 30 \\ 20 \\ 15 \\ 10 \end{pmatrix}$$

and the ending inventory by

$$E = \begin{pmatrix} 20 \\ 15 \\ 10 \\ 5 \end{pmatrix}$$

Assuming no purchases of inventory, determine the cost of the products sold during the period.

To illustrate the distributive law, determine $PB - PE$ and $P(B - E)$.

$$PB - PE = (60, 70, 90, 100) \begin{pmatrix} 30 \\ 20 \\ 15 \\ 10 \end{pmatrix} - (60, 70, 90, 100) \begin{pmatrix} 20 \\ 15 \\ 10 \\ 5 \end{pmatrix}$$

$$PB - P = (60 \cdot 30 + 70 \cdot 20 + 90 \cdot 15 + 100 \cdot 10)$$
$$- (60 \cdot 20 + 70 \cdot 15 + 90 \cdot 10 + 100 \cdot 5)$$

$$PB - PE = \$1900$$

and

$$P(B - E) = (60, 70, 90, 100) \begin{pmatrix} 30 - 20 \\ 20 - 15 \\ 15 - 10 \\ 10 - 5 \end{pmatrix}$$

$$P(B - E) = (60, 70, 90, 100) \begin{pmatrix} 10 \\ 5 \\ 5 \\ 5 \end{pmatrix}$$

$$= 60 \cdot 10 + 70 \cdot 5 + 90 \cdot 5 + 100 \cdot 5$$

$$P(B - E) = \$1900$$

The \$1900 obtained by both methods verifies the distributive law for vector algebra.

To summarize, the commutative and associative laws of addition apply to vector addition, i.e., $A + B = B + A$ and $A + (B + C) = (A + B) + C$. The distributive law of algebra applies to vector algebra, i.e., $A(B + C) = AB + AC$. The commutative law of multiplication, however, does not apply to vector algebra, i.e., $AB \neq BA$.

9.3 Matrices

A matrix is an array of numbers. The array is said to be *ordered*, meaning that the position of each number is important. The ordered array of numbers consists of m rows and n columns that are enclosed in parentheses and designated by a capital letter. The general form of a matrix is

$$A = \begin{pmatrix} a_{11} & a_{12} & a_{13} & \cdots & a_{1n} \\ a_{21} & a_{22} & a_{23} & \cdots & a_{2n} \\ a_{31} & a_{32} & a_{33} & \cdots & a_{3n} \\ \vdots & \vdots & \vdots & & \vdots \\ a_{m1} & a_{m2} & a_{m3} & \cdots & a_{mn} \end{pmatrix} \tag{9.2}$$

The matrix A consists of $m \cdot n$ components or elements a_{ij}. A component a_{ij} is designated by its position in the matrix, i referring to the row and j to the column. Component a_{24} thus refers to the component at the intersection of the second row and fourth column.

The matrix A is said to be of order m by n or, alternatively, is said to have dimensions of m by n, m and n again referring to the number of rows and number of columns.

Example.

$$A = \begin{pmatrix} 3 & 1 \\ 2 & 4 \\ 1 & 6 \end{pmatrix} \text{ is a 3 by 2 matrix}$$

$$B = \begin{pmatrix} 2 & 4 & 1 \\ 3 & 7 & 0 \end{pmatrix} \text{ is a 2 by 3 matrix}$$

$$C = \begin{pmatrix} 3 & 2 & 6 \\ 1 & 5 & -2 \\ -1 & 6 & 1 \end{pmatrix} \text{ is a 3 by 3 matrix}$$

$D = (3, 5, 2)$ is a 1 by 3 matrix or row vector

$$E = \begin{pmatrix} 2 \\ 1 \end{pmatrix} \text{ is a 2 by 1 matrix or column vector}$$

Matrices provide a convenient method of storing large quantities of data. As an example, consider a firm with multiple retail outlets. The inventory of the firm can be represented by the matrix G, where the columns represent the outlets and the rows represent products.

$$
\begin{array}{c}
\textit{Outlets} \\
\begin{array}{ccc} 1 & 2 & 3 \end{array}
\end{array}
$$

$$
G = \begin{pmatrix} 50 & 60 & 40 \\ 175 & 200 & 125 \\ 40 & 25 & 30 \\ 205 & 235 & 275 \end{pmatrix} \begin{array}{c} \textit{Products} \\ 1 \\ 2 \\ 3 \\ 4 \end{array}
$$

Component a_{21} of matrix G is 175 units of product 2 at outlet 1.
Component a_{42} is 235 units of product 4 at outlet 2.

9.4   Matrix Algebra

Matrix algebra, also termed *linear algebra*, provides a set of rules for the addition, subtraction, multiplication, and inversion of matrices. The rules for addition and subtraction of matrices are analogous to those for ordinary algebra. The multiplication of matrices requires the introduction of a straight-forward rule for matrix multiplication. Division of one matrix by another is not possible in matrix algebra. The process of matrix inversion, however, has many similarities with division in ordinary algebra. Matrix addition, subtraction, and multiplication are discussed in this section. Matrix inversion is presented in Sec. 9.5.

9.4.1 MATRIX ADDITION AND SUBTRACTION

Matrices can be added or subtracted, provided that they are conformable for addition or subtraction. Matrices are conformable for addition or subtraction if they have the same dimensions. A 2 by 3 matrix may be added to or subtracted from another 2 by 3 matrix, a 4 by 2 matrix added to or subtracted from another 4 by 2 matrix, etc. The addition or subtraction is performed by adding or subtracting corresponding components from the two matrices. The result is a new matrix of the same order or dimension as the original matrices.

Example. Let

$$A = \begin{pmatrix} 3 & 4 \\ 6 & 5 \\ 4 & 0 \end{pmatrix} \quad \text{and} \quad B = \begin{pmatrix} 2 & 5 \\ 3 & 6 \\ 4 & 2 \end{pmatrix}$$

Determine $A + B$ and $A - B$.

$$A + B = \begin{pmatrix} 3 + 2 & 4 + 5 \\ 6 + 3 & 5 + 6 \\ 4 + 4 & 0 + 2 \end{pmatrix} = \begin{pmatrix} 5 & 9 \\ 9 & 11 \\ 8 & 2 \end{pmatrix}$$

$$A - B = \begin{pmatrix} 3 - 2 & 4 - 5 \\ 6 - 3 & 5 - 6 \\ 4 - 4 & 0 - 2 \end{pmatrix} = \begin{pmatrix} 1 & -1 \\ 3 & -1 \\ 0 & -2 \end{pmatrix}$$

Example. Let

$$C = \begin{pmatrix} 1 & 3 & 4 \\ 2 & 6 & 5 \\ 1 & -1 & 6 \end{pmatrix} \quad D = \begin{pmatrix} 2 & 3 & 5 \\ 4 & 1 & 3 \\ 5 & 2 & 1 \end{pmatrix} \quad E = \begin{pmatrix} 1 & -2 & 5 \\ 3 & -1 & 4 \\ 6 & 5 & 3 \end{pmatrix}$$

Show that $C + (D + E) = (C + D) + E$.

$$C + (D + E) = \begin{pmatrix} 1 & 3 & 4 \\ 2 & 6 & 5 \\ 1 & -1 & 6 \end{pmatrix} + \begin{pmatrix} 3 & 1 & 10 \\ 7 & 0 & 7 \\ 11 & 7 & 4 \end{pmatrix} = \begin{pmatrix} 4 & 4 & 14 \\ 9 & 6 & 12 \\ 12 & 6 & 10 \end{pmatrix}$$

$$(C + D) + E = \begin{pmatrix} 3 & 6 & 9 \\ 6 & 7 & 8 \\ 6 & 1 & 7 \end{pmatrix} + \begin{pmatrix} 1 & -2 & 5 \\ 3 & -1 & 5 \\ 6 & 5 & 3 \end{pmatrix} = \begin{pmatrix} 4 & 4 & 14 \\ 9 & 6 & 12 \\ 12 & 6 & 10 \end{pmatrix}$$

These examples illustrate the commutative and associative laws of addition. The commutative law states that $A + B = B + A$. The associative law states that $C + (D + E) = (C + D) + E$. The examples show that these laws apply to the addition and subtraction of matrices.

9.4.2 MATRIX MULTIPLICATION

Matrix multiplication is subdivided into two cases: (1) the multiplication of a matrix by a scalar and (2) the multiplication of a matrix by a matrix. The simplest of these cases is the multiplication of a matrix by a scalar. A matrix may be multiplied by a scalar by multiplying each component in the matrix by the scalar. The resulting matrix will be of the same order as the original matrix. To illustrate, the 3 by 3 matrix

$$A = \begin{pmatrix} a_{11} & a_{12} & a_{13} \\ a_{21} & a_{22} & a_{23} \\ a_{31} & a_{32} & a_{33} \end{pmatrix}$$

can be multiplied by the scalar k to give the 3 by 3 matrix

$$kA = \begin{pmatrix} ka_{11} & ka_{12} & ka_{13} \\ ka_{21} & ka_{22} & ka_{23} \\ ka_{31} & ka_{32} & ka_{33} \end{pmatrix}$$

Similarly, if a 2 by 4 matrix were multiplied by a scalar, the resulting matrix would be a 2 by 4 matrix.

Example. Multiply the matrix

$$Y = \begin{pmatrix} 2 & 3 & 5 \\ 4 & 6 & -2 \end{pmatrix}$$

by the scalar $k = 3$.

$$kY = 3 \begin{pmatrix} 2 & 3 & 5 \\ 4 & 6 & -2 \end{pmatrix} = \begin{pmatrix} 6 & 9 & 15 \\ 12 & 18 & -6 \end{pmatrix}$$

Example. Multiply the matrix

$$B = \begin{pmatrix} 2 & 1 \\ 1 & -3 \end{pmatrix}$$

by the scalar $k = -2$.

$$kB = -2 \begin{pmatrix} 2 & 1 \\ 1 & -3 \end{pmatrix} = \begin{pmatrix} -4 & -2 \\ -2 & 6 \end{pmatrix}$$

Example. Multiply the matrix

$$A = \begin{pmatrix} 1 & 3 & -2 \\ 4 & 7 & 1 \end{pmatrix}$$

by the scalar $k = 0.1$.

$$kA = 0.1 \begin{pmatrix} 1 & 3 & -2 \\ 4 & 7 & 1 \end{pmatrix} = \begin{pmatrix} 0.1 & 0.3 & -0.2 \\ 0.4 & 0.7 & 0.1 \end{pmatrix}$$

The multiplication of two matrices is somewhat more complicated than the multiplication of a matrix by a scalar. Two matrices may be multiplied together only if the number of columns in the first matrix equals the number of rows in the second matrix. If matrix A has dimensions a by b and matrix B has dimensions c by d, the matrix product AB is defined only if b is equal to c. Thus, a 2 by 3 matrix can be multiplied by a 3 by 4 matrix but not, for example, by a 2 by 4 matrix. The matrix resulting from the multiplication has dimensions equivalent to the number of rows in the first matrix and the number of columns in the second matrix, i.e., a by d. To summarize, matrix A with dimensions a by b may be multiplied by B with dimensions c by d to form matrix AB with dimensions a by d, provided that b equals c. If the number of columns in the first matrix equals the number of rows in the second matrix, the matrices are said to be conformable for multiplication.

To illustrate the multiplication of two matrices, multiply

$$A = \begin{pmatrix} a_{11} & a_{12} & a_{13} \\ a_{21} & a_{22} & a_{23} \end{pmatrix} \text{ by } B = \begin{pmatrix} b_{11} & b_{12} & b_{13} \\ b_{21} & b_{22} & b_{23} \\ b_{31} & b_{32} & b_{33} \end{pmatrix}$$

Since A is a 2 by 3 matrix and B is 3 by 3 matrix, the matrices are conformable for multiplication. The resulting matrix AB has dimensions 2 by 3. The multiplication is performed as follows:

$$AB = \begin{pmatrix} a_{11} & a_{12} & a_{13} \\ a_{21} & a_{22} & a_{23} \end{pmatrix} \begin{pmatrix} b_{11} & b_{12} & b_{13} \\ b_{21} & b_{22} & b_{23} \\ b_{31} & b_{32} & b_{33} \end{pmatrix}$$

$$AB = \begin{pmatrix} a_{11}b_{11} + a_{12}b_{21} + a_{13}b_{31} & a_{11}b_{12} + a_{12}b_{22} + a_{13}b_{32} \\ a_{21}b_{11} + a_{22}b_{21} + a_{23}b_{31} & a_{21}b_{12} + a_{22}b_{22} + a_{23}b_{32} \end{pmatrix}$$

$$\begin{pmatrix} a_{11}b_{13} + a_{12}b_{23} + a_{13}b_{33} \\ a_{21}b_{13} + a_{22}b_{23} + a_{23}b_{33} \end{pmatrix}$$

Matrix AB in the preceding example is a 2 by 3 matrix. The components of the matrix are formed by the sum of the products of the rows of the first matrix and columns of the second matrix. The sum of the products of the components in the first row of A and the first column of B give the first row, first column of AB. Similarly, the sum of the products of the components in the first row of A and second column of B gives the first row, second column component of AB. This procedure is continued until the matrix AB is formed.

Example. Determine the product of

$$A = \begin{pmatrix} 6 & 2 & 4 \\ 1 & 2 & 2 \end{pmatrix} \quad \text{and} \quad B = \begin{pmatrix} 3 & 2 \\ 2 & 4 \\ 4 & 5 \end{pmatrix}$$

$$AB = \begin{pmatrix} 6 & 2 & 4 \\ 1 & 2 & 2 \end{pmatrix} \begin{pmatrix} 3 & 2 \\ 2 & 4 \\ 4 & 5 \end{pmatrix} = \begin{pmatrix} 6 \cdot 3 + 2 \cdot 2 + 4 \cdot 4 & 6 \cdot 2 + 2 \cdot 4 + 4 \cdot 5 \\ 1 \cdot 3 + 2 \cdot 2 + 2 \cdot 4 & 1 \cdot 2 + 2 \cdot 4 + 2 \cdot 5 \end{pmatrix}$$

$$AB = \begin{pmatrix} 18 + 4 + 16 & 12 + 8 + 20 \\ 3 + 4 + 8 & 2 + 8 + 10 \end{pmatrix} = \begin{pmatrix} 38 & 40 \\ 15 & 20 \end{pmatrix}$$

Example. Determine the product of

$$A = \begin{pmatrix} 6 & 1 \\ 3 & 7 \end{pmatrix} \quad \text{and} \quad B = \begin{pmatrix} 2 & 4 & 7 \\ 3 & 1 & 5 \end{pmatrix}$$

$$AB = \begin{pmatrix} 6 & 1 \\ 3 & 7 \end{pmatrix} \begin{pmatrix} 2 & 4 & 7 \\ 3 & 1 & 5 \end{pmatrix}$$

$$= \begin{pmatrix} 6 \cdot 2 + 1 \cdot 3 & 6 \cdot 4 + 1 \cdot 1 & 6 \cdot 7 + 1 \cdot 5 \\ 3 \cdot 2 + 7 \cdot 3 & 3 \cdot 4 + 7 \cdot 1 & 3 \cdot 7 + 7 \cdot 5 \end{pmatrix} = \begin{pmatrix} 15 & 25 & 47 \\ 27 & 19 & 56 \end{pmatrix}$$

In the first example, a 2 by 3 matrix is multiplied by a 3 by 2 matrix. The matrices are conformable for multiplication, since the number of columns in the first matrix equals the number of rows in the second. The dimensions of the resulting matrix are equal to the number of rows of the first matrix and the number of columns of the second matrix. The second example shows that a 2 by 2 matrix and a 2 by 3 matrix are conformable for multiplication. The product of the matrices is a 2 by 3 matrix.

Example. The inventory of a firm with multiple retail outlets is given by the matrix G

$$\begin{array}{ccc} \textit{Outlets} & & \textit{Products} \\ 1 \quad 2 \quad 3 & & \end{array}$$

$$G = \begin{pmatrix} 50 & 60 & 40 \\ 175 & 200 & 125 \\ 40 & 25 & 30 \\ 205 & 235 & 275 \end{pmatrix} \quad \begin{array}{c} 1 \\ 2 \\ 3 \\ 4 \end{array}$$

If the cost of inventory is given by the vector $C = (20, 10, 30, 40)$, determine the value of inventory at each outlet.

The value of inventory at each outlet is given by the matrix CG. Since C is a 1 by 4 matrix and G is a 4 by 3 matrix, the matrices are conformable for multiplication. The product is

$$CG = (20, 10, 30, 40) \begin{pmatrix} 50 & 60 & 40 \\ 175 & 200 & 125 \\ 40 & 25 & 30 \\ 205 & 235 & 275 \end{pmatrix}$$

$$CG = (12,150, 13,350, 13,950)$$

Inventory has a value of \$12,150 at outlet 1, \$13,350 at outlet 2, and \$13,950 at outlet 3.

Example. For $A = (6, 2, 3)$ and $B = \begin{pmatrix} 2 \\ 5 \\ 4 \end{pmatrix}$, find AB and BA.

The row vector A is a 1 by 3 matrix and the column vector B is a 3 by 1 matrix. The inner product AB gives the 1 by 1 matrix or scalar, $AB = 34$.

$$AB = (6, 2, 3) \begin{pmatrix} 2 \\ 5 \\ 4 \end{pmatrix} = 12 + 10 + 12 = 34$$

Multiplying B by A is equivalent to multiplying a 3 by 1 matrix by a 1 by 3 matrix. Since the number of columns in the first matrix equals the number of rows in the second matrix, the product is defined. The product of an n by 1 vector and a 1 by n vector is termed the *outer product* and has dimension n by n. To illustrate,

$$BA = \begin{pmatrix} 2 \\ 5 \\ 4 \end{pmatrix} (6, 2, 3) = \begin{pmatrix} 12 & 4 & 6 \\ 30 & 10 & 15 \\ 24 & 8 & 12 \end{pmatrix}$$

The outer product of the vectors B and A is the 3 by 3 matrix BA.

Example. For

$$A = \begin{pmatrix} 3 & -1 & 4 \\ 6 & 8 & 2 \\ 1 & -5 & 4 \end{pmatrix} \text{ and } B = \begin{pmatrix} 2 & 6 & -1 \\ 3 & -2 & 4 \\ 5 & 3 & -3 \end{pmatrix}$$

find AB and BA.

$$AB = \begin{pmatrix} 3 & -1 & 4 \\ 6 & 8 & 2 \\ 1 & -5 & 4 \end{pmatrix} \begin{pmatrix} 2 & 6 & -1 \\ 3 & -2 & 4 \\ 5 & 3 & -3 \end{pmatrix} = \begin{pmatrix} 23 & 32 & -19 \\ 46 & 26 & 20 \\ 7 & 28 & -33 \end{pmatrix}$$

$$BA = \begin{pmatrix} 2 & 6 & -1 \\ 3 & -2 & 4 \\ 5 & 3 & -3 \end{pmatrix} \begin{pmatrix} 3 & -1 & 4 \\ 6 & 8 & 2 \\ 1 & -5 & 4 \end{pmatrix} = \begin{pmatrix} 41 & 51 & 16 \\ 1 & -39 & 24 \\ 30 & 34 & 14 \end{pmatrix}$$

The preceding two examples illustrate that the commutative law of multiplication does not apply to the multiplication of matrices. The commutative law of multiplication states that $a \cdot b = b \cdot a$. Since AB is not equal to BA, the commutative law of multiplication does not apply to matrix multiplication.

The associative and distributive laws of multiplication do, however, apply to matrix multiplication. Given three matrices A, B, and C that are conformable for multiplication, the associative law states that $A(BC) = (AB)C$. The distributive law, again if conformability is assumed, states that $A(B + C) = AB + AC$. These properties are illustrated by the following two examples.

Example. Show that the associative law of multiplication applies to the matrices

$$A = \begin{pmatrix} 2 & 3 \\ 4 & 5 \end{pmatrix}, \qquad B = \begin{pmatrix} 3 & 4 \\ 5 & 6 \end{pmatrix}, \qquad C = \begin{pmatrix} 4 & 5 \\ 6 & 7 \end{pmatrix}$$

The associative law of multiplication states that $A(BC) = (AB)C$. To verify the law, we multiply A by BC.

$$BC = \begin{pmatrix} 3 & 4 \\ 5 & 6 \end{pmatrix}\begin{pmatrix} 4 & 5 \\ 6 & 7 \end{pmatrix} = \begin{pmatrix} 36 & 43 \\ 56 & 67 \end{pmatrix}$$

$$A(BC) = \begin{pmatrix} 2 & 3 \\ 4 & 5 \end{pmatrix}\begin{pmatrix} 36 & 43 \\ 56 & 67 \end{pmatrix} = \begin{pmatrix} 240 & 287 \\ 424 & 507 \end{pmatrix}$$

Next, we multiply AB by the matrix C.

$$AB = \begin{pmatrix} 2 & 3 \\ 4 & 5 \end{pmatrix}\begin{pmatrix} 3 & 4 \\ 5 & 6 \end{pmatrix} = \begin{pmatrix} 21 & 26 \\ 37 & 46 \end{pmatrix}$$

$$(AB)C = \begin{pmatrix} 21 & 26 \\ 37 & 46 \end{pmatrix}\begin{pmatrix} 4 & 5 \\ 6 & 7 \end{pmatrix} = \begin{pmatrix} 240 & 287 \\ 424 & 507 \end{pmatrix}$$

Since $A(BC)$ equals $(AB)C$, we conclude that the associative law applies in matrix multiplication.

Example. Show that the distributive law of algebra applies to the matrices

$$A = \begin{pmatrix} 2 & 3 \\ 4 & 5 \end{pmatrix}, \qquad B = \begin{pmatrix} 2 & 3 & 6 \\ 3 & 5 & 1 \end{pmatrix}, \qquad C = \begin{pmatrix} 1 & 4 & 5 \\ 6 & 3 & 1 \end{pmatrix}$$

The distributive law states that $A(B + C) = AB + AC$, provided that B and C are conformable for addition and that AB and AC are conformable for multiplication. The matrix $A(B + C)$ is

$$A(B + C) = \begin{pmatrix} 2 & 3 \\ 4 & 5 \end{pmatrix}\left[\begin{pmatrix} 2 & 3 & 6 \\ 3 & 5 & 1 \end{pmatrix} + \begin{pmatrix} 1 & 4 & 5 \\ 6 & 3 & 1 \end{pmatrix}\right]$$

$$A(B + C) = \begin{pmatrix} 2 & 3 \\ 4 & 5 \end{pmatrix}\begin{pmatrix} 3 & 7 & 11 \\ 9 & 8 & 2 \end{pmatrix} = \begin{pmatrix} 33 & 38 & 28 \\ 57 & 68 & 54 \end{pmatrix}$$

and the matrix $AB + AC$ is

$$AB + AC = \begin{pmatrix} 2 & 3 \\ 4 & 5 \end{pmatrix} \begin{pmatrix} 2 & 3 & 6 \\ 3 & 5 & 1 \end{pmatrix} + \begin{pmatrix} 2 & 3 \\ 4 & 5 \end{pmatrix} \begin{pmatrix} 1 & 4 & 5 \\ 6 & 3 & 1 \end{pmatrix}$$

$$AB + AC = \begin{pmatrix} 13 & 21 & 15 \\ 23 & 37 & 29 \end{pmatrix} + \begin{pmatrix} 20 & 17 & 13 \\ 34 & 31 & 25 \end{pmatrix} = \begin{pmatrix} 33 & 38 & 28 \\ 57 & 68 & 54 \end{pmatrix}$$

Since $A(B + C)$ equals $AB + AC$, the distributive law applies to matrix algebra.

9.4.3 THE IDENTITY AND NULL MATRICES

The *identity matrix* is a square matrix that has 1's on the upper left to lower right diagonal and 0's elsewhere.[1] The 2 by 2 identity matrix, commonly designated by I, is

$$I = \begin{pmatrix} 1 & 0 \\ 0 & 1 \end{pmatrix}$$

Similarly, the 3 by 3 and 4 by 4 identity matrices, also designated by I, are

$$I = \begin{pmatrix} 1 & 0 & 0 \\ 0 & 1 & 0 \\ 0 & 0 & 1 \end{pmatrix} \quad \text{and} \quad I = \begin{pmatrix} 1 & 0 & 0 & 0 \\ 0 & 1 & 0 & 0 \\ 0 & 0 & 1 & 0 \\ 0 & 0 & 0 & 1 \end{pmatrix}$$

The identity matrix has properties in matrix algebra similar to those of the number 1 in ordinary algebra. Specifically, the identity matrix has the property that the product of any matrix A with a conformable identity matrix I is equal to the matrix A. Symbolically, this means that $AI = A$, or alternatively, $IA = A$. Thus, although we pointed out that the commutative law of multiplication does not apply in matrix multiplication, it does apply in the case of multiplication of a matrix by an identity matrix.

Example. For

$$A = \begin{pmatrix} 2 & 3 \\ 4 & 5 \end{pmatrix}$$

verify that $AI = A = IA$.

$$AI = \begin{pmatrix} 2 & 3 \\ 4 & 5 \end{pmatrix} \begin{pmatrix} 1 & 0 \\ 0 & 1 \end{pmatrix} = \begin{pmatrix} 2 & 3 \\ 4 & 5 \end{pmatrix}$$

$$IA = \begin{pmatrix} 1 & 0 \\ 0 & 1 \end{pmatrix} \begin{pmatrix} 2 & 3 \\ 4 & 5 \end{pmatrix} = \begin{pmatrix} 2 & 3 \\ 4 & 5 \end{pmatrix}$$

Therefore, $AI = A = IA$.

---

[1] A square matrix is a matrix that has the same number of rows and columns, i.e., matrices of dimensions 2 by 2, 3 by 3, 4 by 4, etc.

Example. For

$$A = \begin{pmatrix} 2 & 3 & 4 \\ 5 & 6 & 7 \\ 8 & 9 & 10 \end{pmatrix},$$

verify that $AI = A = IA$.

$$AI = \begin{pmatrix} 2 & 3 & 4 \\ 5 & 6 & 7 \\ 8 & 9 & 10 \end{pmatrix} \begin{pmatrix} 1 & 0 & 0 \\ 0 & 1 & 0 \\ 0 & 0 & 1 \end{pmatrix} = \begin{pmatrix} 2 & 3 & 4 \\ 5 & 6 & 7 \\ 8 & 9 & 10 \end{pmatrix}$$

$$IA = \begin{pmatrix} 1 & 0 & 0 \\ 0 & 1 & 0 \\ 0 & 0 & 1 \end{pmatrix} \begin{pmatrix} 2 & 3 & 4 \\ 5 & 6 & 7 \\ 8 & 9 & 10 \end{pmatrix} = \begin{pmatrix} 2 & 3 & 4 \\ 5 & 6 & 7 \\ 8 & 9 & 10 \end{pmatrix}$$

This again demonstrates that $AI = A = IA$.

The *null matrix* is a matrix in which all elements are 0. Unlike the identity matrix, the null matrix need not be a square matrix. The null matrix is commonly designated by the symbol ϕ (phi). A 2 by 3 null matrix would be

$$\phi = \begin{pmatrix} 0 & 0 & 0 \\ 0 & 0 & 0 \end{pmatrix}$$

Similarly, a 1 by 3 null matrix or null row vector is

$$\phi = (0, 0, 0)$$

and a 3 by 1 null matrix or null column vector is

$$\phi = \begin{pmatrix} 0 \\ 0 \\ 0 \end{pmatrix}$$

The product of any matrix with a conformable null matrix is a null matrix, i.e., $A\phi = \phi$. Thus, if

$$A = \begin{pmatrix} 1 & 2 & 3 \\ 4 & 5 & 6 \end{pmatrix} \quad \text{and} \quad \phi = \begin{pmatrix} 0 & 0 \\ 0 & 0 \\ 0 & 0 \end{pmatrix}$$

$$A\phi = \begin{pmatrix} 1 & 2 & 3 \\ 4 & 5 & 6 \end{pmatrix} \begin{pmatrix} 0 & 0 \\ 0 & 0 \\ 0 & 0 \end{pmatrix} = \begin{pmatrix} 0 & 0 \\ 0 & 0 \end{pmatrix}$$

9.5 Matrix Representation of Systems of Linear Equations

One of the important uses of matrices and matrix algebra is in the representation of systems of linear equations. To illustrate, consider the two linear equations

$$a_{11}x_1 + a_{12}x_2 = b_1$$
$$a_{21}x_1 + a_{22}x_2 = b_2$$

(9.3)

These two equations can be represented by the matrix equation

$$AX = B \tag{9.4}$$

where A is the *coefficient matrix* $\begin{pmatrix} a_{11} & a_{12} \\ a_{21} & a_{22} \end{pmatrix}$, X is the *solution vector* $\begin{pmatrix} x_1 \\ x_2 \end{pmatrix}$, and
B is the *right-hand-side vector* $\begin{pmatrix} b_1 \\ b_2 \end{pmatrix}$.

To show that the matrix equation $AX = B$ is an alternative way of writing the two linear equations, we write

$$AX = B$$

or alternatively,

$$\begin{pmatrix} a_{11} & a_{12} \\ a_{21} & a_{22} \end{pmatrix} \begin{pmatrix} x_1 \\ x_2 \end{pmatrix} = \begin{pmatrix} b_1 \\ b_2 \end{pmatrix} \tag{9.5}$$

Multiplying the coefficient matrix by the solution vector and equating the sum to the right-hand side gives

$$a_{11}x_1 + a_{12}x_2 = b_1$$

$$a_{21}x_1 + a_{22}x_2 = b_2$$

The components of the matrix equation $AX = B$ are the coefficient matrix A, the solution vector X, and the right-hand-side vector B. The coefficient matrix contains the coefficients of the linear equations. The solution vector contains the variables in the system, $x_1, x_2, \ldots, x_n$. The right-hand-side or B vector contains those elements customarily written on the right-hand side of the equal sign.[2]

Example. Express the following system of two equations in matrix form:

$$3x_1 + x_2 = 9$$

$$5x_1 - 3x_2 = 1$$

The two equations can be written as

$$\begin{pmatrix} 3 & 1 \\ 5 & -3 \end{pmatrix} \begin{pmatrix} x_1 \\ x_2 \end{pmatrix} = \begin{pmatrix} 9 \\ 1 \end{pmatrix}$$

or alternatively as

$$AX = B$$

---

[2] The system of equations is termed *homogeneous* in those cases in which $B = \phi$. Systems of equations of the type illustrated in this text in which $B \neq \phi$ are, conversely, *nonhomogeneous*.

where

$$A = \begin{pmatrix} 3 & 1 \\ 5 & -3 \end{pmatrix}, \quad X = \begin{pmatrix} x_1 \\ x_2 \end{pmatrix}, \quad B = \begin{pmatrix} 9 \\ 1 \end{pmatrix}$$

Example. Express the system of three equations in matrix form.

$$2x - 3y + 6z = -18$$
$$6x + 4y - 2z = 44$$
$$5x + 8y + 10z = 56$$

The three equations can be written as

$$\begin{pmatrix} 2 & -3 & 6 \\ 6 & 4 & -2 \\ 5 & 8 & 10 \end{pmatrix} \begin{pmatrix} x \\ y \\ z \end{pmatrix} = \begin{pmatrix} -18 \\ 44 \\ 56 \end{pmatrix}$$

or alternatively as

$$AX = B$$

where

$$A = \begin{pmatrix} 2 & -3 & 6 \\ 6 & 4 & -2 \\ 5 & 8 & 10 \end{pmatrix}, \quad X = \begin{pmatrix} x \\ y \\ z \end{pmatrix}, \quad B = \begin{pmatrix} -18 \\ 44 \\ 56 \end{pmatrix}$$

By applying the rules of matrix multiplication, a matrix equation is easily transformed into a system of equations. Consider the following examples.

Example. Write the matrix equation $AX = B$, where

$$A = \begin{pmatrix} 4 & 2 & 3 \\ 3 & 4 & -1 \\ 5 & 3 & 1 \end{pmatrix}, \quad X = \begin{pmatrix} x_1 \\ x_2 \\ x_3 \end{pmatrix}, \quad B = \begin{pmatrix} 16 \\ 13 \\ 16 \end{pmatrix}$$

as three equations in three variables.

The matrix equation $AX = B$ is written as

$$\begin{pmatrix} 4 & 2 & 3 \\ 3 & 4 & -1 \\ 5 & 3 & 1 \end{pmatrix} \begin{pmatrix} x_1 \\ x_2 \\ x_3 \end{pmatrix} = \begin{pmatrix} 16 \\ 13 \\ 16 \end{pmatrix}$$

Applying the rules of matrix multiplication gives

$$4x_1 + 2x_2 + 3x_3 = 16$$
$$3x_1 + 4x_2 - 1x_3 = 13$$
$$5x_1 + 3x_2 + 1x_3 = 16$$

Example. Express

$$\begin{pmatrix} 3 & 2 \\ 6 & -2 \end{pmatrix} \begin{pmatrix} x_1 \\ x_2 \end{pmatrix} = \begin{pmatrix} 14 \\ 4 \end{pmatrix}$$

as two equations.

Multiplying the coefficient matrix by the solution vector and equating this sum to the right-hand side gives

$$3x_1 + 2x_2 = 14$$

$$6x_1 - 2x_2 = 4$$

Matrix algebra can be used in conjunction with the matrix representation of the system for solution of the system of linear equations. To illustrate, consider the preceding example of two equations with variables x_1 and x_2. The two equations were

$$3x_1 + 2x_2 = 14$$

$$6x_1 - 2x_2 = 4$$

The solution set for these two equations consists of values of x_1 and x_2 that satisfy both equalities. We shall demonstrate that these values are $x_1 = 2$ and $x_2 = 4$.

Techniques for the solution of matrix equations are based upon *row operations*. The term row operations means the application of basic algebraic operations to the rows of a matrix. Three operations are defined. These are

1. Any two rows of a matrix may be interchanged.
2. A row may be multiplied by a nonzero constant.
3. A multiple of one row may be added to another row.

The three row operations are applications of fundamental algebraic operations. To illustrate, consider the two equations

$$3x_1 + 2x_2 = 14$$

$$6x_1 - 2x_2 = 4$$

The first row operation states that the rows may be interchanged, i.e.,

$$6x_1 - 2x_2 = 4$$

$$3x_1 + 2x_2 = 14$$

The second row operation states that a row may be multiplied by a nonzero constant. For example, multiplying $6x_1 - 2x_2 = 4$ by $\frac{1}{2}$ gives

$$3x_1 - 1x_2 = 2$$

The third row operation states that a multiple of one row may be added to another row. Multiplying $3x_1 + 2x_2 = 14$ by -2 and adding the result to the equation $6x_1 - 2x_2 = 4$ gives $0x_1 - 6x_2 = -24$. To demonstrate this result, first multiply $3x_1 + 2x_2 = 14$ by -2.

$$-2(3x_1 + 2x_2 = 14) = -6x_1 - 4x_2 = -28$$

This product is then added to the equation $6x_1 - 2x_2 = 4$, i.e.,

$$
\begin{aligned}
6x_1 - 2x_2 &= 4 \\
-6x_1 - 4x_2 &= -28 \\
\hline
0x_1 - 6x_2 &= -24
\end{aligned}
$$

The resulting equation is $0x_1 - 6x_2 = -24$.

This row operation is based upon the algebraic principle that "equals may be added to equals." For the equation $a = b$ and the equation $c = d$, the principle states that $a + c = b + d$. Since $6x_1 - 2x_2 = 4$ and $-6x_1 - 4x_2 = -28$, we can add $6x_1 - 2x_2$ to $-6x_1 - 4x_2$ provided that we add -28 to 4. The result of the addition is $0x_1 - 6x_2 = -24$.

To apply row operations in the solution of matrix equations, we *augment* the coefficient matrix A with the right-hand-side vector B. The augmented matrix is designated by the symbol $A \mid B$ and is expressed in matrix form as

$$
\begin{pmatrix}
a_{11} & a_{12} & b_1 \\
a_{21} & a_{22} & b_2
\end{pmatrix}
$$

Referring to the preceding example, we write the augmented matrix as

$$
A \mid B = \begin{pmatrix}
3 & 2 & 14 \\
6 & -2 & 4
\end{pmatrix}
$$

The solution to the system of equations is determined by applying row operations to the rows of the augmented matrix. The row operations are applied with the objective of obtaining an identity matrix in the position originally occupied by the coefficient matrix A. Once the identity matrix is obtained, the solution to the system of equations appears in the position originally occupied by the right-hand-side vector.

To demonstrate the solution technique, row operations are performed on both the augmented matrix and the system of two equations in two unknowns. Our purpose is to show that applying row operations to augmented matrices is analogous to the algebraic operations used in the solution of systems of equations and can, therefore, be used to determine the solution for systems of equations. The augmented matrix and system of equations are

$$
\begin{pmatrix}
3 & 2 & 14 \\
6 & -2 & 4
\end{pmatrix}
\qquad
\begin{aligned}
3x_1 + 2x_2 &= 14 \\
6x_1 - 2x_2 &= 4
\end{aligned}
$$

The first row operation is to add one times the second row or second equation to the first row or equation. This gives

$$\begin{pmatrix} 9 & 0 & \vert & 18 \\ 6 & -2 & \vert & 4 \end{pmatrix} \qquad \begin{aligned} 9x_1 + 0x_2 &= 18 \\ 6x_1 - 2x_2 &= 4 \end{aligned}$$

Next, multiply the first row or equation by $\frac{1}{9}$.

$$\begin{pmatrix} 1 & 0 & \vert & 2 \\ 6 & -2 & \vert & 4 \end{pmatrix} \qquad \begin{aligned} 1x_1 + 0x_2 &= 2 \\ 6x_1 - 2x_2 &= 4 \end{aligned}$$

Add -6 times the first row or equation to the second row or equation.

$$\begin{pmatrix} 1 & 0 & \vert & 2 \\ 0 & -2 & \vert & -8 \end{pmatrix} \qquad \begin{aligned} 1x_1 + 0x_2 &= 2 \\ 0x_1 - 2x_2 &= -8 \end{aligned}$$

The last operation is to multiply the second row or equation by $-\frac{1}{2}$.

$$\begin{pmatrix} 1 & 0 & \vert & 2 \\ 0 & 1 & \vert & 4 \end{pmatrix} \qquad \begin{aligned} 1x_1 + 0x_2 &= 2 \\ 0x_1 + 1x_2 &= 4 \end{aligned}$$

The solution to the system of equations is read directly as $x_1 = 2$ and $x_2 = 4$. The top element to the right of the vertical line is the solution value of x_1 and the bottom element the solution value of x_2. This can be demonstrated by writing the augmented matrix as the matrix equation,

$$\begin{pmatrix} 1 & 0 \\ 0 & 1 \end{pmatrix} \begin{pmatrix} x_1 \\ x_2 \end{pmatrix} = \begin{pmatrix} 2 \\ 4 \end{pmatrix}$$

and converting the matrix equation to two equations in the two unknowns,

$$1x_1 + 0x_2 = 2$$

$$0x_1 + 1x_2 = 4$$

As a second example of solving matrix equations by using row operations, consider the following three equations:

$$2x + 1y - 2z = -1$$

$$4x - 2y + 3z = 14$$

$$1x - 1y + 2z = 7$$

Row operations are performed on the augmented matrix and the system of equations. This will again demonstrate the parallel between the algebraic solution and the matrix solution of systems of equations. The augmented matrix and the systems of equations are

$$\begin{pmatrix} 2 & 1 & -2 & \vert & -1 \\ 4 & -2 & 3 & \vert & 14 \\ 1 & -1 & 2 & \vert & 7 \end{pmatrix} \qquad \begin{aligned} 2x + 1y - 2z &= -1 \\ 4x - 2y + 3z &= 14 \\ 1x - 1y + 2z &= 7 \end{aligned}$$

Add 1 times the third row to the first row.

$$\begin{pmatrix} 3 & 0 & 0 \\ 4 & -2 & 3 \\ 1 & -1 & 2 \end{pmatrix} \left| \begin{matrix} 6 \\ 14 \\ 7 \end{matrix} \right) \qquad \begin{matrix} 3x + 0y + 0z = 6 \\ 4x - 2y + 3z = 14 \\ 1x - 1y + 2z = 7 \end{matrix}$$

Multiply the first row by $\frac{1}{3}$.

$$\begin{pmatrix} 1 & 0 & 0 \\ 4 & -2 & 3 \\ 1 & -1 & 2 \end{pmatrix} \left| \begin{matrix} 2 \\ 14 \\ 7 \end{matrix} \right) \qquad \begin{matrix} 1x + 0y + 0z = 2 \\ 4x - 2y + 3z = 14 \\ 1x - 1y + 2z = 7 \end{matrix}$$

Add -4 times the first row to the second row and -1 times the first row to the third row.

$$\begin{pmatrix} 1 & 0 & 0 \\ 0 & -2 & 3 \\ 0 & -1 & 2 \end{pmatrix} \left| \begin{matrix} 2 \\ 6 \\ 5 \end{matrix} \right) \qquad \begin{matrix} 1x + 0y + 0z = 2 \\ 0x - 2y + 3z = 6 \\ 0x - 1y + 2z = 5 \end{matrix}$$

Multiply the second row by $-\frac{1}{2}$.

$$\begin{pmatrix} 1 & 0 & 0 \\ 0 & 1 & -\frac{3}{2} \\ 0 & -1 & 2 \end{pmatrix} \left| \begin{matrix} 2 \\ -3 \\ 5 \end{matrix} \right) \qquad \begin{matrix} 1x + 0y + 0z = 2 \\ 0x + 1y - \frac{3}{2}z = -3 \\ 0x - 1y + 2z = 5 \end{matrix}$$

Add 1 times the second row to the third row.

$$\begin{pmatrix} 1 & 0 & 0 \\ 0 & 1 & -\frac{3}{2} \\ 0 & 0 & \frac{1}{2} \end{pmatrix} \left| \begin{matrix} 2 \\ -3 \\ 2 \end{matrix} \right) \qquad \begin{matrix} 1x + 0y + 0z = 2 \\ 0x + 1y - \frac{3}{2}z = -3 \\ 0x + 0y + \frac{1}{2}z = 2 \end{matrix}$$

Multiply the third row by 2.

$$\begin{pmatrix} 1 & 0 & 0 \\ 0 & 1 & -\frac{3}{2} \\ 0 & 0 & 1 \end{pmatrix} \left| \begin{matrix} 2 \\ -3 \\ 4 \end{matrix} \right) \qquad \begin{matrix} 1x + 0y + 0z = 2 \\ 0x + 1y - \frac{3}{2}z = -3 \\ 0x + 0y + 1z = 4 \end{matrix}$$

Add $\frac{3}{2}$ times the third row to the second row.

$$\begin{pmatrix} 1 & 0 & 0 \\ 0 & 1 & 0 \\ 0 & 0 & 1 \end{pmatrix} \left| \begin{matrix} 2 \\ 3 \\ 4 \end{matrix} \right) \qquad \begin{matrix} 1x + 0y + 0z = 2 \\ 0x + 1y + 0z = 3 \\ 0x + 0y + 1z = 4 \end{matrix}$$

The solution to the three equations is $x = 2$, $y = 3$, and $z = 4$.

In the preceding two examples, row operations are applied to the augmented matrix with the objective of obtaining an identity matrix on the left side of the vertical line. Once the identity matrix is obtained, the solution vector can be read from the right side of the vertical line. This is possible since the row operations on the augmented matrix correspond to algebraic operations on the system of equations.

This method for the solution of matrix equations is termed the *Gaussian elimination method* (named for the German mathematician Karl Friedrich Gauss, 1777–1855) or simply the elimination method. To summarize, the technique involves writing the coefficient matrix and the right-hand-side vector as $A \mid B$. Row operations are applied to transform the coefficient matrix A to an identity matrix I. The solution to the system of equations is then read from the right-hand side of the vertical line.

Any combination of row operations is acceptable in obtaining the identity matrix to the left of the vertical line. The procedure many students find useful is to first obtain a 1 in the a_{11} position and to then use multiples of the first row to obtain zeros elsewhere in the first column. The student next obtains a 1 in the a_{22} position and uses multiples of the second row to obtain zeros elsewhere in the second column. This procedure is followed until the identity matrix is obtained. This method is used in the following examples.

Example. Determine the solution vector for the following two equations by using the Gaussian elimination method:

$$-2x_1 + 4x_2 = 4$$
$$3x_1 - 2x_2 = 10$$

Write the equations in matrix form as $A \mid B$.

$$\left(\begin{array}{cc|c} -2 & 4 & 4 \\ 3 & -2 & 10 \end{array} \right)$$

Multiply the first row by $-\frac{1}{2}$.

$$\left(\begin{array}{cc|c} 1 & -2 & -2 \\ 3 & -2 & 10 \end{array} \right)$$

Add -3 times the first row to the second row.

$$\left(\begin{array}{cc|c} 1 & -2 & -2 \\ 0 & 4 & 16 \end{array} \right)$$

Multiply the second row by $\frac{1}{4}$.

$$\left(\begin{array}{cc|c} 1 & -2 & -2 \\ 0 & 1 & 4 \end{array} \right)$$

Add 2 times the second row to the first row.

$$\left(\begin{array}{cc|c} 1 & 0 & 6 \\ 0 & 1 & 4 \end{array} \right)$$

The solution is $x_1 = 6$ and $x_2 = 4$.

Example. Determine the solution vector to the following system of equations by using the Gaussian elimination method:

$$2x + 3y - 4z = 9$$
$$3x - 2y + 3z = -15$$
$$1x + 4y - 2z = 12$$

Write the equations in matrix form as $A \mid B$.

$$\begin{pmatrix} 2 & 3 & -4 \\ 3 & -2 & 3 \\ 1 & 4 & -2 \end{pmatrix} \begin{array}{c} 9 \\ -15 \\ 12 \end{array}$$

Multiply the first row by $\frac{1}{2}$.

$$\begin{pmatrix} 1 & \frac{3}{2} & -2 \\ 3 & -2 & 3 \\ 1 & 4 & -2 \end{pmatrix} \begin{array}{c} \frac{9}{2} \\ -15 \\ 12 \end{array}$$

Add -3 times the first row to the second row and add -1 times the first row to the third row.

$$\begin{pmatrix} 1 & \frac{3}{2} & -2 \\ 0 & -\frac{13}{2} & 9 \\ 0 & \frac{5}{2} & 0 \end{pmatrix} \begin{array}{c} \frac{9}{2} \\ -\frac{57}{2} \\ \frac{15}{2} \end{array}$$

Interchange the second and third rows.

$$\begin{pmatrix} 1 & \frac{3}{2} & -2 \\ 0 & \frac{5}{2} & 0 \\ 0 & -\frac{13}{2} & 9 \end{pmatrix} \begin{array}{c} \frac{9}{2} \\ \frac{15}{2} \\ -\frac{57}{2} \end{array}$$

Multiply the second row by $\frac{2}{5}$.

$$\begin{pmatrix} 1 & \frac{3}{2} & -2 \\ 0 & 1 & 0 \\ 0 & -\frac{13}{2} & 9 \end{pmatrix} \begin{array}{c} \frac{9}{2} \\ 3 \\ -\frac{57}{2} \end{array}$$

Add $-\frac{3}{2}$ times the second row to the first row and add $\frac{13}{2}$ times the second row to the third row.

$$\begin{pmatrix} 1 & 0 & -2 \\ 0 & 1 & 0 \\ 0 & 0 & 9 \end{pmatrix} \begin{array}{c} 0 \\ 3 \\ -9 \end{array}$$

Multiply the third row by $\frac{1}{9}$.

$$\begin{pmatrix} 1 & 0 & -2 \\ 0 & 1 & 0 \\ 0 & 0 & 1 \end{pmatrix} \begin{array}{c} 0 \\ 3 \\ -1 \end{array}$$

Add 2 times the third row to the first row.

$$\begin{pmatrix} 1 & 0 & 0 & | & -2 \\ 0 & 1 & 0 & | & 3 \\ 0 & 0 & 1 & | & -1 \end{pmatrix}$$

The solution is $x = -2$, $y = 3$, and $z = -1$.

9.6   The Inverse: Gaussian Method

The four basic arithmetic operations of ordinary algebra are addition, subtraction, multiplication, and division. The first three operations apply in matrix algebra and have been discussed. The fourth operation, division, is not defined for matrix algebra. The inverse of a matrix, however, is used in matrix algebra in much the same manner as division in ordinary algebra.

Division is used in ordinary algebra to determine the solution for systems of equations. For instance, in the algebraic equation $ax = b$, x is determined by dividing both sides of the equation by a to give $x = b/a$. An alternative method of solving this algebraic equation is to multiply both sides of the equation $ax = b$ by the reciprocal of a. In ordinary algebra the reciprocal of a is $1/a$. The solution to the equation is thus

$$\frac{1}{a} ax = \frac{1}{a} b$$

or

$$x = \frac{1}{a} b, \qquad \text{provided } a \neq 0$$

An inverse matrix performs a function in matrix algebra similar to that of the reciprocal in ordinary algebra. The product of a number a and its reciprocal $1/a$ in ordinary algebra is 1. Similarly, in matrix algebra the product of a matrix A and its inverse, A^{-1} (read A inverse) is the identity matrix I, i.e.,

$$AA^{-1} = I \tag{9.6}$$

The matrix equation $AX = B$ can be solved for the solution vector X by use of the inverse. The procedure is to multiply the matrix equation by the inverse,

$$A^{-1}AX = A^{-1}B \tag{9.7}$$

Since $A^{-1}A = I$, this reduces to

$$IX = A^{-1}B \tag{9.8}$$

and since $IX = X$, the solution vector is

$$X = A^{-1}B \tag{9.9}$$

In order for a matrix to have an inverse, the matrix must be square.[3] If A is a square matrix, the inverse of A is a square matrix with components such that the matrix product of A^{-1} and A is the identity matrix I, i.e., $AA^{-1} =$ I.

As an example, consider the matrix

$$A = \begin{pmatrix} 3 & 2 \\ 6 & -2 \end{pmatrix}$$

and the matrix

$$A^{-1} = \begin{pmatrix} \frac{1}{9} & \frac{1}{9} \\ \frac{1}{3} & -\frac{1}{6} \end{pmatrix} .$$

A^{-1} is the inverse of A, provided that $AA^{-1} =$ I. Since

$$\begin{pmatrix} 3 & 2 \\ 6 & -2 \end{pmatrix} \begin{pmatrix} \frac{1}{9} & \frac{1}{9} \\ \frac{1}{3} & -\frac{1}{6} \end{pmatrix} = \begin{pmatrix} 1 & 0 \\ 0 & 1 \end{pmatrix}$$

we conclude that A^{-1} is the inverse of A. It also follows that the matrix

$$A = \begin{pmatrix} 3 & 2 \\ 6 & -2 \end{pmatrix}$$

is the inverse of the matrix

$$A^{-1} = \begin{pmatrix} \frac{1}{9} & \frac{1}{9} \\ \frac{1}{3} & -\frac{1}{6} \end{pmatrix}$$

since

$$A^{-1}A = \begin{pmatrix} \frac{1}{9} & \frac{1}{9} \\ \frac{1}{3} & -\frac{1}{6} \end{pmatrix} \begin{pmatrix} 3 & 2 \\ 6 & -2 \end{pmatrix} = \begin{pmatrix} 1 & 0 \\ 0 & 1 \end{pmatrix}$$

In summary, for two square matrices A and A^{-1}, A^{-1} is the inverse of A if $AA^{-1} =$ I. Similarly, since $A^{-1}A =$ I, it follows that A is the inverse of A^{-1}. The commutative law of multiplication thus applies to inverse matrices, i.e., $AA^{-1} =$ I $= A^{-1}A$. The relationship between a matrix and its inverse is illustrated in the following examples.

Example. Verify that

$$A^{-1} = \begin{pmatrix} \frac{2}{28} & \frac{3}{28} \\ \frac{6}{28} & -\frac{5}{28} \end{pmatrix}$$

is the inverse matrix of

$$A = \begin{pmatrix} 5 & 3 \\ 6 & -2 \end{pmatrix}$$

---

[3] Not all square matrices have inverses. See Robert L. Childress, *Mathematics for Managerial Decisions* (Englewood Cliffs, N.J.: Prentice-Hall, Inc., 1974). Chap. 4, p. 125.

The fact that A^{-1} is the inverse matrix of A can be verified by multiplying A by A^{-1}. If the product of the two matrices is the identity matrix, then A^{-1} is the inverse of A.

$$AA^{-1} = \begin{pmatrix} 5 & 3 \\ 6 & -2 \end{pmatrix} \begin{pmatrix} \frac{2}{28} & \frac{3}{28} \\ \frac{6}{28} & -\frac{5}{28} \end{pmatrix} = \begin{pmatrix} 1 & 0 \\ 0 & 1 \end{pmatrix}$$

Example. Verify that the commutative law of multiplication applies for inverse matrices. Use the matrices in the preceding example.

The commutative law of multiplication states that $AB = BA$. We have previously shown that this law does not generally hold in matrix multiplication. It does hold true, however, for inverse matrices. Thus,

$$AA^{-1} = \begin{pmatrix} 5 & 3 \\ 6 & -2 \end{pmatrix} \begin{pmatrix} \frac{2}{28} & \frac{3}{28} \\ \frac{6}{28} & -\frac{5}{28} \end{pmatrix} = \begin{pmatrix} 1 & 0 \\ 0 & 1 \end{pmatrix}$$

and

$$A^{-1}A = \begin{pmatrix} \frac{2}{28} & \frac{3}{28} \\ \frac{6}{28} & -\frac{5}{28} \end{pmatrix} \begin{pmatrix} 5 & 3 \\ 6 & -2 \end{pmatrix} = \begin{pmatrix} 1 & 0 \\ 0 & 1 \end{pmatrix}$$

This demonstrates that $A^{-1}A = I = AA^{-1}$.

Example. Verify that

$$A = \begin{pmatrix} 2 & 4 & -6 \\ 4 & 2 & 2 \\ 3 & -3 & 1 \end{pmatrix} \quad \text{and} \quad A^{-1} = \begin{pmatrix} \frac{4}{66} & \frac{7}{66} & \frac{10}{66} \\ \frac{1}{66} & \frac{10}{66} & -\frac{14}{66} \\ -\frac{9}{66} & \frac{9}{66} & -\frac{6}{66} \end{pmatrix}$$

are inverse matrices.

The two matrices are inverse matrices if $AA^{-1} = I$.

$$AA^{-1} = \begin{pmatrix} 2 & 4 & -6 \\ 4 & 2 & 2 \\ 3 & -3 & 1 \end{pmatrix} \begin{pmatrix} \frac{4}{66} & \frac{7}{66} & \frac{10}{66} \\ \frac{1}{66} & \frac{10}{66} & -\frac{14}{66} \\ -\frac{9}{66} & \frac{9}{66} & -\frac{6}{66} \end{pmatrix} = \begin{pmatrix} 1 & 0 & 0 \\ 0 & 1 & 0 \\ 0 & 0 & 1 \end{pmatrix}$$

Since $AA^{-1} = I$, we conclude that A and A^{-1} are inverse matrices.

9.6.1 CALCULATING THE INVERSE BY GAUSSIAN ELIMINATION

The Gaussian elimination method can be used to calculate an inverse matrix. The procedure involves augmenting the square matrix A with the identity matrix I, i.e.

$$A \mid I \tag{9.10}$$

Row operations are then employed to obtain an identity matrix on the left side of the vertical line. Concurrent with the identity matrix on the left of the vertical line, the inverse matrix is obtained on the right of the vertical line. The Gaussian elimination method involves transforming the augmented matrix $A \mid I$ to the inverse matrix $I \mid A^{-1}$ through row operations.

The justification for this procedure can be demonstrated with matrix algebra. First write the matrix A augmented with I.

$$A \mid I$$

If the inverse matrix A^{-1} were known, we could multiply the matrices on both sides of the vertical line by A^{-1}, i.e.,

$$AA^{-1} \mid IA^{-1}$$

This product would give

$$I \mid A^{-1} \tag{9.11}$$

Instead of using the inverse to obtain the identity matrix on the left side of the vertical line, we use row operations. The row operations applied to obtain the identity matrix concurrently yield the inverse matrix on the right side of the vertical line.

The Gaussian elimination method for obtaining the inverse matrix is illustrated by the following examples.

Example. Determine the inverse of

$$A = \begin{pmatrix} 5 & 3 \\ 6 & -2 \end{pmatrix}$$

To determine the inverse, we augment the matrix A with the identity matrix I and apply row operations to obtain an identity matrix on the left side of the vertical line. The augmented matrix is

$$A \mid I = \begin{pmatrix} 5 & 3 & 1 & 0 \\ 6 & -2 & 0 & 1 \end{pmatrix}$$

Multiply the first row by $\frac{1}{5}$.

$$\begin{pmatrix} 1 & \frac{3}{5} & \frac{1}{5} & 0 \\ 6 & -2 & 0 & 1 \end{pmatrix}$$

Add -6 times the first row to the second row.

$$\begin{pmatrix} 1 & \frac{3}{5} & \frac{1}{5} & 0 \\ 0 & -\frac{28}{5} & -\frac{6}{5} & 1 \end{pmatrix}$$

Multiply the second row by $-\frac{5}{28}$.

$$\begin{pmatrix} 1 & \frac{3}{5} & \frac{1}{5} & 0 \\ 0 & 1 & \frac{6}{28} & -\frac{5}{28} \end{pmatrix}$$

Add $-\frac{3}{5}$ times the second row to the first row.

$$\begin{pmatrix} 1 & 0 & \frac{2}{28} & \frac{3}{28} \\ 0 & 1 & \frac{6}{28} & -\frac{5}{28} \end{pmatrix}$$

This results in an identity matrix on the left side of the vertical line. From Eq. (9.11), we know that the inverse of A is given on the right side of the vertical line. The inverse is

$$A^{-1} = \begin{pmatrix} \frac{2}{28} & \frac{3}{28} \\ \frac{6}{28} & -\frac{5}{28} \end{pmatrix}$$

Example. Determine the inverse of

$$A = \begin{pmatrix} 3 & 2 \\ 6 & -2 \end{pmatrix}$$

Augment the matrix A with the identity matrix I.

$$\begin{pmatrix} 3 & 2 & 1 & 0 \\ 6 & -2 & 0 & 1 \end{pmatrix}$$

Add 1 times the second row to the first row.

$$\begin{pmatrix} 9 & 0 & 1 & 1 \\ 6 & -2 & 0 & 1 \end{pmatrix}$$

Multiply the first row by $\frac{1}{9}$.

$$\begin{pmatrix} 1 & 0 & \frac{1}{9} & \frac{1}{9} \\ 6 & -2 & 0 & 1 \end{pmatrix}$$

Add -6 times the first row to the second row.

$$\begin{pmatrix} 1 & 0 & \frac{1}{9} & \frac{1}{9} \\ 0 & -2 & -\frac{2}{3} & \frac{1}{3} \end{pmatrix}$$

Multiply the second row by $-\frac{1}{2}$.

$$\begin{pmatrix} 1 & 0 & \frac{1}{9} & \frac{1}{9} \\ 0 & 1 & \frac{1}{3} & -\frac{1}{6} \end{pmatrix}$$

The inverse of A is given on the right side of the vertical line and is

$$A^{-1} = \begin{pmatrix} \frac{1}{9} & \frac{1}{9} \\ \frac{1}{3} & -\frac{1}{6} \end{pmatrix}$$

Example. Determine the inverse of

$$A = \begin{pmatrix} 2 & 3 \\ -4 & 6 \end{pmatrix}$$

Augment the matrix A with the identity matrix I.

$$\begin{pmatrix} 2 & 3 & | & 1 & 0 \\ -4 & 6 & | & 0 & 1 \end{pmatrix}$$

Add 2 times the first row to the second row.

$$\begin{pmatrix} 2 & 3 & | & 1 & 0 \\ 0 & 12 & | & 2 & 1 \end{pmatrix}$$

Multiply the first row by $\frac{1}{2}$ and the second row by $\frac{1}{12}$.

$$\begin{pmatrix} 1 & \frac{3}{2} & | & \frac{1}{2} & 0 \\ 0 & 1 & | & \frac{1}{6} & \frac{1}{12} \end{pmatrix}$$

Add $-\frac{3}{2}$ times the second row to the first row.

$$\begin{pmatrix} 1 & 0 & | & \frac{1}{4} & -\frac{3}{24} \\ 0 & 1 & | & \frac{1}{6} & \frac{1}{12} \end{pmatrix}$$

The inverse of A is

$$A^{-1} = \begin{pmatrix} \frac{1}{4} & -\frac{3}{24} \\ \frac{1}{6} & \frac{1}{12} \end{pmatrix}$$

Example. Determine the inverse of

$$A = \begin{pmatrix} 2 & 4 & -6 \\ 4 & 2 & 2 \\ 3 & -3 & 1 \end{pmatrix}$$

Augment the matrix A with I.

$$A \mid I = \begin{pmatrix} 2 & 4 & -6 & | & 1 & 0 & 0 \\ 4 & 2 & 2 & | & 0 & 1 & 0 \\ 3 & -3 & 1 & | & 0 & 0 & 1 \end{pmatrix}$$

Multiply the first row by $\frac{1}{2}$.

$$\begin{pmatrix} 1 & 2 & -3 & | & \frac{1}{2} & 0 & 0 \\ 4 & 2 & 2 & | & 0 & 1 & 0 \\ 3 & -3 & 1 & | & 0 & 0 & 1 \end{pmatrix}$$

Add -4 times the first row to the second row and add -3 times the first row to the third row.

$$\begin{pmatrix} 1 & 2 & -3 & | & \frac{1}{2} & 0 & 0 \\ 0 & -6 & 14 & | & -2 & 1 & 0 \\ 0 & -9 & 10 & | & -\frac{3}{2} & 0 & 1 \end{pmatrix}$$

Multiply the second row by $-\frac{1}{6}$.

$$\left(\begin{array}{ccc|ccc} 1 & 2 & -3 & \frac{1}{2} & 0 & 0 \\ 0 & 1 & -\frac{7}{3} & \frac{1}{3} & -\frac{1}{6} & 0 \\ 0 & -9 & 10 & -\frac{3}{2} & 0 & 1 \end{array}\right)$$

Add -2 times the second row to the first row and add 9 times the second row to the third row. Multiply the new third row by $-\frac{1}{11}$.

$$\left(\begin{array}{ccc|ccc} 1 & 0 & \frac{5}{3} & -\frac{1}{6} & \frac{1}{3} & 0 \\ 0 & 1 & -\frac{7}{3} & \frac{1}{3} & -\frac{1}{6} & 0 \\ 0 & 0 & 1 & -\frac{3}{22} & \frac{3}{22} & -\frac{1}{11} \end{array}\right)$$

Add $\frac{7}{3}$ times the third row to the second row and add $-\frac{5}{3}$ times the third row to the first row.

$$\left(\begin{array}{ccc|ccc} 1 & 0 & 0 & \frac{4}{66} & \frac{7}{66} & \frac{10}{66} \\ 0 & 1 & 0 & \frac{1}{66} & \frac{10}{66} & -\frac{14}{66} \\ 0 & 0 & 1 & -\frac{9}{66} & \frac{9}{66} & -\frac{6}{66} \end{array}\right)$$

The inverse of A is

$$A^{-1} = \left(\begin{array}{ccc} \frac{4}{66} & \frac{7}{66} & \frac{10}{66} \\ \frac{1}{66} & \frac{10}{66} & -\frac{14}{66} \\ -\frac{9}{66} & \frac{9}{66} & -\frac{6}{66} \end{array}\right)$$

9.6.2 USING THE INVERSE TO SOLVE MATRIX EQUATIONS

The solution to a matrix equation is relatively straightforward once the inverse has been determined. Given the matrix equation $AX = B$, the solution vector is $X = A^{-1}B$. This result comes from the fact that both sides of the matrix equation can be multiplied by the inverse. That is,

$$A^{-1}AX = A^{-1}B$$

$$IX = A^{-1}B$$

and

$$X = A^{-1}B$$

To illustrate, consider again the example of two simultaneous equations presented on p. 289. The two equations are written in matrix form as

$$\begin{pmatrix} 3 & 2 \\ 6 & -2 \end{pmatrix}\begin{pmatrix} x_1 \\ x_2 \end{pmatrix} = \begin{pmatrix} 14 \\ 4 \end{pmatrix}$$

The inverse of the coefficient matrix was calculated on p. 299 and is

$$A^{-1} = \begin{pmatrix} \frac{1}{9} & \frac{1}{9} \\ \frac{1}{3} & -\frac{1}{6} \end{pmatrix}$$

Multiplying A^{-1} by B gives

$$X = A^{-1}B = \begin{pmatrix} \frac{1}{9} & \frac{1}{9} \\ \frac{1}{3} & -\frac{1}{6} \end{pmatrix}\begin{pmatrix} 14 \\ 4 \end{pmatrix} = \begin{pmatrix} 2 \\ 4 \end{pmatrix}$$

The solution vector is $x_1 = 2$ and $x_2 = 4$. This is, of course, the solution that was obtained by the Gaussian elimination method.[4]

Example. Determine the solution vector X for the matrix equation

$$\begin{pmatrix} 5 & 3 \\ 6 & -2 \end{pmatrix}\begin{pmatrix} x_1 \\ x_2 \end{pmatrix} = \begin{pmatrix} 6 \\ 10 \end{pmatrix}$$

The inverse of the coefficient matrix was found on p. 298 to be

$$A^{-1} = \begin{pmatrix} \frac{2}{28} & \frac{3}{28} \\ \frac{6}{28} & -\frac{5}{28} \end{pmatrix}$$

The solution vector is

$$X = A^{-1}B = \begin{pmatrix} \frac{2}{28} & \frac{3}{28} \\ \frac{6}{28} & -\frac{5}{28} \end{pmatrix}\begin{pmatrix} 6 \\ 10 \end{pmatrix} = \begin{pmatrix} \frac{3}{2} \\ -\frac{1}{2} \end{pmatrix}$$

Example. Determine the solution vector X for the matrix equation

$$\begin{pmatrix} 2 & 4 & -6 \\ 4 & 2 & 2 \\ 3 & -3 & 1 \end{pmatrix}\begin{pmatrix} x_1 \\ x_2 \\ x_3 \end{pmatrix} = \begin{pmatrix} 6 \\ 10 \\ 4 \end{pmatrix}$$

The inverse matrix was given on p. 300 as

$$A^{-1} = \begin{pmatrix} \frac{4}{66} & \frac{7}{66} & \frac{10}{66} \\ \frac{1}{66} & \frac{10}{66} & -\frac{14}{66} \\ -\frac{9}{66} & \frac{9}{66} & -\frac{6}{66} \end{pmatrix}$$

The solution vector is $X = A^{-1}B$.

$$X = \begin{pmatrix} \frac{4}{66} & \frac{7}{66} & \frac{10}{66} \\ \frac{1}{66} & \frac{10}{66} & -\frac{14}{66} \\ -\frac{9}{66} & \frac{9}{66} & -\frac{6}{66} \end{pmatrix}\begin{pmatrix} 6 \\ 10 \\ 4 \end{pmatrix} = \begin{pmatrix} \frac{67}{33} \\ \frac{25}{33} \\ \frac{6}{33} \end{pmatrix}$$

PROBLEMS

1. For the vectors $A = (10, 6, 12)$ and $B = (14, 8, 5)$, perform the operations indicated or state why the operation cannot be done.
   (a) $A - B$ (b) $B - A$
   (c) AB (d) $A + B$

---

[4] The reader will note that the Gaussian elimination method can be used to determine the inverse or can be applied directly to $A|B$ to obtain the solution vector.

2. Using the vectors $X = (6, 4, 8)$, $Y = (3, 6, -2)$, and $Z = (4, 10, 12)$, verify that the associative law of addition applies to the addition of conformable vectors.

3. For the vectors $A = \begin{pmatrix} 6 \\ 4 \\ 2 \end{pmatrix}$ and $B = \begin{pmatrix} 8 \\ -2 \\ 6 \end{pmatrix}$, determine the following:

   (a) $A + B$ (b) $2A - B$

   (c) $3A - 2B$ (d) $0.5A + 1.5B$

4. Find the inner product of $W = (5, 8, 3)$ and $V = \begin{pmatrix} 2 \\ -2 \\ 4 \end{pmatrix}$.

5. For the vectors $A = (2, 8, 6)$, $B = \begin{pmatrix} 5 \\ 4 \\ 6 \end{pmatrix}$, and $C = \begin{pmatrix} 8 \\ 10 \\ 3 \end{pmatrix}$, verify the distributive law by showing that $A(B + C)$ is equal to $AB + AC$.

6. A television dealer has an inventory of 12 black and white sets, 9 color console sets, and 8 color portable sets. The dealer's cost on the different types of sets is $80, $340, and $260, respectively. Form the inventory and price vectors and determine the value of the inventory by finding the inner product of the two vectors.

7. For the matrices $A = \begin{pmatrix} 3 & 5 \\ 2 & 6 \\ 1 & 4 \end{pmatrix}$ and $B = \begin{pmatrix} 2 & -5 \\ 3 & 6 \\ -5 & 5 \end{pmatrix}$, determine the following:

   (a) $A + B$ (b) $B - A$

   (c) $3A - B$ (d) $A - 2B$

8. Find the product of $A = \begin{pmatrix} 6 & 4 & 8 \\ 3 & -3 & 5 \end{pmatrix}$ and $B = \begin{pmatrix} 8 & -3 \\ -2 & 5 \\ 6 & 4 \end{pmatrix}$.

9. Using the matrices $A = \begin{pmatrix} 5 & 2 \\ 3 & -8 \\ 2 & 6 \end{pmatrix}$, $B = \begin{pmatrix} 8 & 5 \\ 4 & 7 \end{pmatrix}$, and $C = \begin{pmatrix} 4 & 7 \\ 2 & 5 \end{pmatrix}$, verify that $A(B + C) = AB + AC$.

10. For the matrices $A = \begin{pmatrix} 3 & 8 & 2 \\ 5 & -2 & 4 \\ 6 & 1 & -5 \end{pmatrix}$ and $B = \begin{pmatrix} 6 & 13 & 7 \\ 7 & 11 & 5 \\ 8 & -1 & 10 \end{pmatrix}$, find AB and BA.

11. Determine the outer product of the vectors W and V given in Problem 4.

12. The matrices A, B, and C have the following dimensions: A has 3 rows and 4 columns, B has 4 rows and 4 columns, C has 2 rows and 3 columns.

Can the following operations be performed? If so, give the dimensions of the resulting matrix.
(a) AB (b) AC
(c) CA (d) $C(AB)$

13. Although the commutative law of multiplication does not generally apply to the multiplication of matrices, it does apply in the multiplication of a matrix by the identity matrix. Verify this statement by using the matrices

$$A = \begin{pmatrix} 6 & 3 & 8 \\ 4 & 2 & 5 \\ 3 & 1 & 7 \end{pmatrix} \quad \text{and} \quad I = \begin{pmatrix} 1 & 0 & 0 \\ 0 & 1 & 0 \\ 0 & 0 & 1 \end{pmatrix}$$

14. Express the following systems of equations in matrix form:
(a) $2x_1 + 6x_2 = 44$
$\quad 3x_1 + 4x_2 = 36$
(b) $2x_1 + x_2 = 8$
$\quad x_1 + x_2 = 10$
(c) $3x_1 + 5x_2 + 4x_3 = 50$
$\quad 6x_1 + 4x_2 - 3x_3 = 26$
$\quad 2x_1 - 5x_2 + 6x_3 = 5$
(d) $2x_1 - 5x_2 + 3x_3 = 14$
$\quad x_1 + x_2 - x_3 = 6$
$\quad 2x_1 - x_2 + 3x_3 = 30$

15. Determine the solutions to the systems of equations in Problem 14 by the use of the Gaussian elimination method.

16. Verify that the matrices

$$A = \begin{pmatrix} 3 & 2 \\ 1 & 5 \end{pmatrix} \quad \text{and} \quad A^{-1} = \begin{pmatrix} \frac{5}{13} & -\frac{2}{13} \\ -\frac{1}{13} & \frac{3}{13} \end{pmatrix}$$

are inverse by determining AA^{-1}.

17. Use the Gaussian method to determine the inverses of the following matrices:

(a) $A = \begin{pmatrix} 2 & 3 \\ 4 & 7 \end{pmatrix}$

(b) $A = \begin{pmatrix} 2 & 3 \\ -5 & -2 \end{pmatrix}$

(c) $A = \begin{pmatrix} 1 & 2 & 3 \\ 1 & 3 & 5 \\ 2 & 5 & 9 \end{pmatrix}$

(d) $A = \begin{pmatrix} 5 & -6 & 7 \\ -10 & 11 & -13 \\ 1 & -1 & 1 \end{pmatrix}$

18. Use the inverses found in Problem 17 to solve the following systems of equations:
(a) $2x_1 + 3x_2 = 10$
$\quad 4x_1 + 7x_2 = 6$
(b) $\quad 2x_1 + 3x_2 = 6$
$\quad -5x_1 - 2x_2 = 7$
(c) $\quad x_1 + 2x_2 + 3x_3 = 8$
$\quad x_1 + 3x_2 + 5x_3 = 5$
$\quad 2x_1 + 5x_2 + 9x_3 = 10$
(d) $\quad 5x_1 - 6x_2 + 7x_3 = 5$
$\quad -10x_1 + 11x_2 - 13x_3 = -10$
$\quad x_1 - x_2 + x_3 = -12$

SUGGESTED REFERENCES

AYRES, FRANK JR., *Theory and Problems of Matrices*, Schaum's Outline Series (New York, N.Y.: McGraw-Hill Book Company, Inc., 1962).

CHILDRESS, ROBERT L., *Mathematics for Managerial Decisions* (Englewood Cliffs, N.J.: Prentice-Hall, Inc., 1974).

DORFMAN, R., et al., *Linear Programming and Economic Analysis* (New York, N.Y.: McGraw-Hill Book Company, Inc., 1958).

FULLER, LEONARD E., *Basic Matrix Theory* (Englewood Cliffs, N.J.: Prentice-Hall, Inc., 1962).

GRAYBILL, FRANKLIN A., *Introduction to Matrices with Applications in Statistics* (Belmont, Ca.: Wadsworth Publishing Company, Inc., 1969).

HADLEY, G., *Linear Algebra* (Reading, Mass.: Addison-Wesley Publishing Company, Inc., 1961).

KEMENY, JOHN G., et al., *Finite Mathematical Structures* (Englewood Cliffs, N.J.: Prentice-Hall, Inc., 1958), 4.

LIPSCHUTZ, SEYMOUR, *Linear Algebra*, Schaum's Outline Series (New York, N.Y.: McGraw-Hill Book Company, Inc., 1968).

STEIN, F. MAX, *An Introduction to Matrices and Determinants* (Belmont, Ca.: Wadsworth Publishing Company, Inc., 1967).

YAMANE, TARO, *Mathematics for Economists*, 2nd ed. (Englewood Cliffs, N.J.: Prentice-Hall, Inc., 1962), 10–12.

Chapter Ten

Markov Chains

This chapter introduces an important kind of probabilistic model, a model commonly referred to as a *Markov chain*. A Markov chain consists of a sequence of experiments or trials. Each trial in the sequence results with certain probability in one of a finite number of possible outcomes. Rather than immediately attempting to distinguish between a Markov chain and other kinds of probabilistic models, we shall introduce the Markov chain by using a simple example. The reader will find that probability and his newly acquired knowledge of matrix algebra are necessary tools for the development of the particular probabilistic models included under the general category of Markov chains.

10.1   An Example

As an introduction to Markov chains, consider the following model developed by the marketing staff of a large consumer products firm. The model was developed to predict the market share of a particular product of the firm along with the market share of the only two competing products. For simplicity, the three products are designated by the numbers 1, 2, and 3. The act of purchasing a product from firm 1 is represented by a_1. Purchases of the two competing products are represented by a_2 and a_3.

Based on historical sales data and market studies, the probability of purchase of each of the three competing products has been established. These probabilities reflect purchases of the products by a "typical customer." It has been found that the typical customer has a tendency to purchase the same brand of product purchased in the past. This factor, termed *brand loyalty*, is taken into account in the model by using conditional probabilities to describe the probability of purchasing each of the three products. To simplify the model, it is assumed that the purchase of a product is conditional only upon the product last purchased by the customer. In other words, the customer takes into account the last purchase but does not take into account any earlier purchases.[1] The probabilities of purchase are thus of the form $P(a_j \mid a_i)$, i.e., the probability of purchasing product j given that the preceding purchase was product i. These probabilities have been established for our typical customer and are shown by the probability matrix in Table 10.1.

Table 10.1.

Next Purchase (a_j)

		a_1	a_2	a_3
Last	a_1	0.60	0.20	0.20
Purchase	a_2	0.10	0.70	0.20
(a_i)	a_3	0.30	0.20	0.50

The probability matrix in Table 10.1 shows the conditional probability of a_j given a_i. The entries in the first row of the matrix represent the probabilities of purchasing products 1, 2, and 3 given that the typical customer's last purchase was product 1. The second row of the matrix gives the probabilities of purchasing each of the three products given that the last purchase was product 2. Similarly, the third row shows the probabilities given that the last purchase was product 3.

Notice that the probabilities shown in the matrix reflect the brand loyalty mentioned earlier. For instance, the first row of the matrix shows that

P(next purchase is product 1 given last purchase was product 1) = 0.60

whereas

P(next purchase is product 2 given last purchase was product 1) = 0.20

and

P(next purchase is product 3 given last purchase was product 1) = 0.20

Brand loyalty for product 1 is reflected by the fact that the probability of a repeat purchase of product 1 is $P(a_1 \mid a_1) = 0.60$. The probability of switching from product 1 to product 2 is $P(a_2 \mid a_1) = 0.20$. The probability of

---

[1] The assumption that the outcome on the kth trial is influenced by the outcome on the $(k-1)$th trial is termed a *Markov process*. This will be discussed in greater detail in the following section.

switching from product 1 to product 3, $P(a_3 \mid a_1)$, is also 0.20. Brand loyalties for products 2 and 3 are reflected by the conditional probabilities given in the second and third rows of the probability matrix.

The entries in each row of the matrix in Table 10.1 are probabilities. Consequently, these numbers must be assigned according to the three axioms of probability.[2] This means that all entries in the matrix must be non-negative and that the probabilities in each row of the matrix must sum to one.[3]

A matrix, such as that in Table 10.1, forms the basis for a Markov chain. From this matrix it is possible to describe the probabilistic behavior of certain processes over time. Our objective in this chapter is to present these properties. We begin by introducing the notation and terminology used in Markov chains.

10.2   Notation and Terminology

A *Markov chain* consists of a sequence of *experiments* or *trials*. This sequence of experiments has certain characteristic features. First, the outcome of each experiment or trial is one of a finite number of possible outcomes. These outcomes can be designated as $a_1, a_2, \ldots, a_n$ and are referred to as *states*. Second, the probability of outcome a_j on any trial in the sequence depends at most on the outcome of the experiment on the immediately preceding trial. Thus, the outcome on the kth trial in the sequence can depend on the outcome of the $(k - 1)$th trial but not on the outcomes of the first $(k - 2)$ trials. A probabilistic process with these characteristics is termed a *Markov process*. A sequence of such experiments—the outcome of each experiment depending only on the outcome of the immediately preceding experiment—is called a Markov chain.

10.2.1   TRANSITION MATRICES AND DIAGRAMS

In a Markov process, the conditional probability of the jth outcome on the kth trial given the ith outcome on the $(k - 1)$th trial is termed a *transition probability*. This probability reflects the likelihood of a change from state i to state j on any trial of the experiment and is represented by p_{ij}.

Transition probabilities are normally given in either a matrix or by a diagram. A *transition matrix* for a Markov process with states 1, 2, and 3 is

$$P = \begin{array}{c} \\ a_1 \\ a_2 \\ a_3 \end{array} \begin{array}{c} \begin{array}{ccc} a_1 & a_2 & a_3 \end{array} \\ \begin{pmatrix} p_{11} & p_{12} & p_{13} \\ p_{21} & p_{22} & p_{23} \\ p_{31} & p_{32} & p_{33} \end{pmatrix} \end{array}, \tag{10.1}$$

---

[2] The three axioms of probability were discussed on pp. 153–56.

[3] Any matrix that satisfies these two conditions—all entries are non-negative and each row sum is one—is called a *probabilistic* or a *stochastic matrix*.

where the rows represent the possible outcomes on the $(k-1)$th trial, the columns the possible outcomes on the kth trial, and the transition probabilities the entries in the matrix.

A transition diagram for the same process is shown in Fig. 10.1. The arrows from each state indicate the possible transitions and the probabilities reflect the likelihood of these transitions.

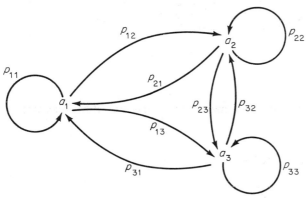

Figure 10.1.

To relate these definitions to our example, the purchase of one of the three products by the hypothetical customer represents an experiment or a trial in the Markov chain. The initial purchase is the first trial, the second purchase the second trial, the kth purchase the kth trial, etc. The outcome of each trial is the purchase of one of the three products. These outcomes are referred to as states. The purchase of product 1 results in state 1, product 2 in state 2, and product 3 in state 3.

We assume in our example that the typical customer's next purchase depends only on the immediately preceding purchase. The corresponding assumption in a Markov process is that the outcome of the kth trial depends only on the current state or, alternatively, the outcome of the $(k-1)$th trial. The probability of our typical customer's next purchasing product j given that the preceding purchase was product i was given in the matrix in Table 10.1. The matrix in Table 10.1 is therefore a transition matrix.

We stated earlier that transition probabilities can be shown by either a transition diagram or a transition matrix. A transition diagram consists of a representation of the possible states along with the probabilities of transitions from one state to another on any trial in the chain. A transition diagram for our typical consumer is shown in Fig. 10.2. The advantage of the diagram is that it provides a pictorial representation of the possible transitions. Once these have been determined, the probabilities can be transferred to a transition matrix.

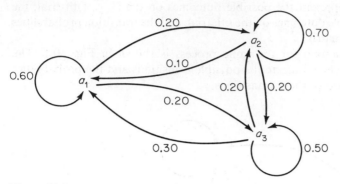

Figure 10.2.

Example. Construct a transition matrix from the transition diagram in Fig. 10.3.

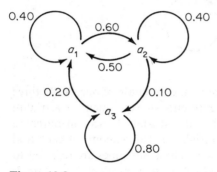

Figure 10.3.

The transition matrix is shown in Table 10.2.

Table 10.2.

State a_j

$$
\begin{array}{c}
State \\
a_i
\end{array}
\begin{array}{c}
 \\
a_1 \\
a_2 \\
a_3
\end{array}
\begin{pmatrix}
a_1 & a_2 & a_3 \\
0.40 & 0.60 & 0 \\
0.50 & 0.40 & 0.10 \\
0.20 & 0 & 0.80
\end{pmatrix}
$$

10.3   *k*-step Transition Matrix

One of the problems of interest in the study of Markov chains is the following. Suppose that the process described by the transition matrix begins in state i. What is the probability that the process is in state j after k trials of the experiment? This probability is represented by $p_{ij}^{(k)}$, where i refers to the initial state, j refers to the terminal state, and k to the intervening number of trials. Note that $p_{ij}^{(k)}$ is *not* the kth power of p_{ij} but rather the probability of going from state i to state j in k trials or steps in the Markov chain. $p_{ij}^{(k)}$ is termed a *k-step transition probability*.

We are normally interested in k-step transition probabilities for all possible initial and terminal states. These probabilities can conveniently be given in a matrix. Such a matrix, termed a k-*step transition matrix*, includes k-step probabilities for transitions from the i initial to the j terminal states. This matrix is designated by $P^{(k)}$ (read "P upper k") and is

$$P^{(k)} = \begin{pmatrix} p_{11}^{(k)} & p_{12}^{(k)} & p_{13}^{(k)} \\ p_{21}^{(k)} & p_{22}^{(k)} & p_{23}^{(k)} \\ p_{31}^{(k)} & p_{32}^{(k)} & p_{33}^{(k)} \end{pmatrix} \tag{10.2}$$

10.3.1 USING TREE DIAGRAMS TO CALCULATE THE k-STEP MATRIX

A k-step transition matrix can easily be calculated by using matrix multiplication. In order to understand this approach, it is important to calculate several k-step transition matrices by using a more elementary technique. This technique requires nothing more than a tree diagram. To illustrate the technique, we shall calculate a two-step transition matrix for our typical customer.

The reader will remember that the probabilities of purchasing three competing products by our typical customer were given in the introductory section. These products were designated as 1, 2, and 3. The transition probabilities for the purchase of product i given a preceding purchase of product j were given by the transition matrix in Table 10.1. This transition matrix is repeated in the following:

$$\begin{array}{c c} & \begin{array}{ccc} a_1 & a_2 & a_3 \end{array} \\ \begin{array}{c} a_1 \\ a_2 \\ a_3 \end{array} & \begin{pmatrix} 0.60 & 0.20 & 0.20 \\ 0.10 & 0.70 & 0.20 \\ 0.30 & 0.20 & 0.50 \end{pmatrix} \end{array}$$

Suppose that we are interested in the probability that the customer purchases product 1, 2, or 3 after an intermediate purchase of one of the three products. In other words, the probability that the process is in each of the three states after two steps or trials. These probabilities are calculated by using the tree diagram shown in Fig. 10.4. The tree diagram has been constructed for a process beginning in state 1, i.e., an initial purchase of product 1.

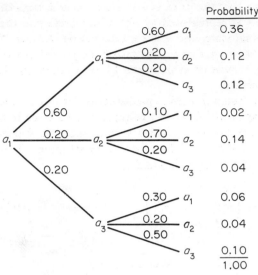

Figure 10.4.

The tree diagram shows a Markov chain that begins in state 1 and terminates after two steps in states 1, 2, or 3. The probabilities associated with each path through the tree are given by the products of the probabilities along the branches. The probability of two successive purchases of product 1 is $(0.60)(0.60) = 0.36$. Similarly, the probability of a purchase of product 1 followed by product 2 is $(0.60)(0.20) = 0.12$. The probability that the process terminates in each of the three states is found by summing the terminal probabilities for each state. These probabilities are entered in the first row of the two-step transition matrix and are

$$(0.44 \quad 0.30 \quad 0.26)$$

The second row of the two-step transition matrix is calculated by using the tree diagram in Fig. 10.5. This row gives the probabilities that the process is in states 1, 2, and 3 after two steps given that the initial state was a_2. These probabilities form the second row of the matrix and are

$$(0.19 \quad 0.55 \quad 0.26)$$

The third row of the matrix gives the two-step transition probabilities for the initial state a_3. The reader should verify by constructing a tree diagram that the probabilities are

$$(0.35 \quad 0.30 \quad 0.35)$$

Probability

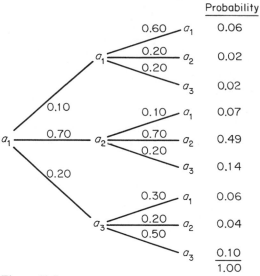

0.60 a_1	0.06	
0.20 a_2	0.02	
0.20		
a_3	0.02	
0.10 a_1	0.07	
0.70 a_2	0.49	
0.20		
a_3	0.14	
0.30 a_1	0.06	
0.20 a_2	0.04	
0.50		
a_3	0.10	

Figure 10.5.

The two-step transition matrix for our typical customer is given in the following:

$$
\begin{array}{cccc}
& a_1 & a_2 & a_3 \\
P^{(2)} = \begin{array}{c} a_1 \\ a_2 \\ a_3 \end{array} & \left(\begin{array}{ccc} 0.44 & 0.30 & 0.26 \\ 0.19 & 0.55 & 0.26 \\ 0.35 & 0.30 & 0.35 \end{array} \right)
\end{array}
$$

We interpret the probabilities in this matrix as follows. Assume that a customer initially purchased product 1. The probabilities of purchasing each of the three products after an intermediate purchase are given in the first row of the two-step transition matrix. The probabilities are $p_{11}^{(2)} = 0.44$, $p_{12}^{(2)} = 0.30$, and $p_{13}^{(2)} = 0.26$. Assuming an initial purchase of product 2, the probability distribution of states 1, 2, and 3 after two trials is given in the second row of the matrix. These probabilities are $p_{21}^{(2)} = 0.19$, $p_{22}^{(2)} = 0.55$, and $p_{23}^{(2)} = 0.26$. Similarly, the probability distribution of the states after two trials assuming the initial state a_3 is $p_{31}^{(2)} = 0.35$, $p_{32}^{(2)} = 0.30$, and $p_{33}^{(2)} = 0.35$. These probabilities are shown in the third row of the matrix.

Example. Determine a two-step transition matrix for

$$
\begin{array}{cccc}
& a_1 & a_2 & a_3 \\
P = \begin{array}{c} a_1 \\ a_2 \\ a_3 \end{array} & \left(\begin{array}{ccc} 0.40 & 0.60 & 0 \\ 0.50 & 0.40 & 0.10 \\ 0.20 & 0 & 0.80 \end{array} \right)
\end{array}
$$

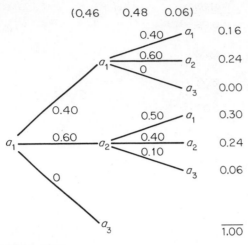

Figure 10.6.

The first row of the two-step transition matrix is calculated by using the tree diagram in Fig. 10.6. From the diagram, the two-step transition probabilities associated with the initial state a_1 are

$$(0.46 \quad 0.48 \quad 0.06)$$

The two-step transition probabilities for the initial states 2 and 3 are calculated in the same manner. The reader should calculate these probabilities and verify that the two-step transition matrix is

$$
P^{(2)} = \begin{array}{c} a_1 \\ a_2 \\ a_3 \end{array}
\begin{array}{c} \quad a_1 \quad\quad a_2 \quad\quad a_3 \\
\begin{pmatrix} 0.46 & 0.48 & 0.06 \\ 0.42 & 0.46 & 0.12 \\ 0.24 & 0.12 & 0.64 \end{pmatrix} \end{array}
$$

10.3.2 MATRIX MULTIPLICATION—AN ALTERNATIVE TO TREE DIAGRAMS

A k-step transition matrix can be easily calculated by using matrix multiplication. The k-step transition matrix, $P^{(k)}$, is equal to the kth power of the transition matrix, P. In mathematical symbols,

$$P^{(k)} = P^k \qquad (10.3)$$

The kth power of a matrix is found by multiplying the matrix by itself k times. For instance, the two-step transition matrix would be $P^{(2)} = P \cdot P$ or alternatively, $P^{(2)} = P^2$. The three-step transition matrix would be $P^{(3)} = P^2 \cdot P$ or P^3. Similarly, the k-step transition matrix would be $P^{(k)} = P^k$.

Instead of proving Eq. (10.3), we shall verify the formula for a two-step transition matrix with states a_1 and a_2. The transition matrix is

$$P = \begin{array}{c} \\ a_1 \\ a_2 \end{array} \begin{array}{cc} a_1 & a_2 \\ \begin{pmatrix} p_{11} & p_{12} \\ p_{21} & p_{22} \end{pmatrix} \end{array}$$

According to Eq. (10.3), the two-step transition matrix is

$$P^{(2)} = \begin{pmatrix} p_{11} & p_{12} \\ p_{21} & p_{22} \end{pmatrix} \begin{pmatrix} p_{11} & p_{12} \\ p_{21} & p_{22} \end{pmatrix}$$

$$P^{(2)} = \begin{pmatrix} p_{11}p_{11} + p_{12}p_{21} & p_{11}p_{12} + p_{12}p_{22} \\ p_{21}p_{11} + p_{22}p_{21} & p_{21}p_{12} + p_{22}p_{22} \end{pmatrix}$$

To verify this formula, consider the tree diagrams in Figs. 10.7 and 10.8. The tree diagram in Fig. 10.7 gives the two-step transition probabilities assuming the initial state is a_1. From this diagram, the probabilities of states a_1 and a_2 are $p_{11}p_{11} + p_{12}p_{21}$ and $p_{11}p_{12} + p_{12}p_{22}$, respectively. The first row of the two-step transition matrix is thus

$$(p_{11}p_{11} + p_{12}p_{21} \quad p_{11}p_{12} + p_{12}p_{22})$$

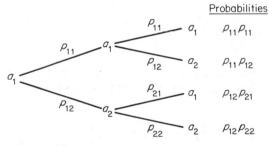

Figure 10.7.

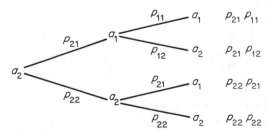

Figure 10.8.

Fig. 10.8 gives the two-step transition probabilities assuming the initial state is a_2. From Fig. 10.8, the second row of the two-step transition matrix is

$$(p_{21}p_{11} + p_{22}p_{21} \quad p_{21}p_{12} + p_{22}p_{22})$$

The two-step transition matrix is

$$P^{(2)} = \begin{pmatrix} p_{11}p_{11} + p_{12}p_{12} & p_{11}p_{12} + p_{12}p_{22} \\ p_{21}p_{11} + p_{22}p_{21} & p_{21}p_{12} + p_{22}p_{22} \end{pmatrix}$$

As expected from Eq. (10.3), the two-step transition matrices given by matrix multiplication and by tree diagrams are the same.

Example. Use Eq. (10.3) to calculate the two-step transition matrix for our typical customer.

The two-step transition matrix is

$$P^{(2)} = \begin{pmatrix} 0.60 & 0.20 & 0.20 \\ 0.10 & 0.70 & 0.20 \\ 0.30 & 0.20 & 0.50 \end{pmatrix} \begin{pmatrix} 0.60 & 0.20 & 0.20 \\ 0.10 & 0.70 & 0.20 \\ 0.30 & 0.20 & 0.50 \end{pmatrix}$$

$$P^{(2)} = \begin{pmatrix} 0.44 & 0.30 & 0.26 \\ 0.19 & 0.55 & 0.26 \\ 0.35 & 0.30 & 0.35 \end{pmatrix}$$

Example. Determine two- and three-step transition matrices for

$$P = \begin{pmatrix} 0.40 & 0.60 & 0 \\ 0.50 & 0.40 & 0.10 \\ 0.20 & 0 & 0.80 \end{pmatrix}$$

A two-step transition matrix was calculated for P by using a tree diagram on p. 313. Using matrix multiplication, we see that the two-step transition matrix is

$$P^{(2)} = \begin{pmatrix} 0.40 & 0.60 & 0 \\ 0.50 & 0.40 & 0.10 \\ 0.20 & 0 & 0.80 \end{pmatrix} \begin{pmatrix} 0.40 & 0.60 & 0 \\ 0.50 & 0.40 & 0.10 \\ 0.20 & 0 & 0.80 \end{pmatrix}$$

$$P^{(2)} = \begin{pmatrix} 0.46 & 0.48 & 0.06 \\ 0.42 & 0.46 & 0.12 \\ 0.24 & 0.12 & 0.64 \end{pmatrix}$$

The three-step transition matrix is given by $P^2 \cdot P$. The three-step transition matrix is

$$P^{(3)} = \begin{pmatrix} 0.436 & 0.468 & 0.096 \\ 0.422 & 0.436 & 0.142 \\ 0.284 & 0.192 & 0.524 \end{pmatrix}$$

10.3.3 PROBABILITY DISTRIBUTION AFTER k-STEPS

The probability of each of the states after k-steps in a Markov process is termed the *state probability distribution*. If the initial state is known, the state probability distribution is given in the appropriate row of the k-step transition matrix. For instance, if the Markov process begins with state 1, the state probability distribution after k-steps is given in the first row of the k-step transition matrix.

Suppose that instead of being given, the initial state is chosen by some chance device. In other words, the Markov process begins initially in state j with probability $p_j^{(0)}$. In this case, the state probability distribution after k-steps is given by the product of the initial probability distribution and the k-step transition matrix. Representing the state probability distribution after k-steps by

$$p^{(k)} = (p_1^{(k)}, p_2^{(k)}, \ldots, p_n^{(k)})$$

we obtain

$$p^{(k)} = p^{(0)} \cdot P^{(k)} \tag{10.4}$$

where

$$p^{(0)} = (p_1^{(0)}, p_2^{(0)}, \ldots, p_n^{(0)})$$

is the *initial probability distribution* and $P^{(k)}$ is the k-step transition matrix.

To illustrate Eq. (10.4), suppose that the initial probability distribution for the purchase of the three competing products by our "typical customer" is given by the probability vector

$$p^{(0)} = (0.50, 0.30, 0.20)$$

This vector shows that the sequence begins in state 1 with probability 0.50, in state 2 with probability 0.30, and in state 3 with probability 0.20. The state probability distribution after k-steps is given by the product of $p^{(0)}$ and $P^{(k)}$. After two steps, the state probability distribution is

$$p^{(2)} = p^{(0)} \cdot P^{(2)}$$

$$p^{(2)} = (0.50, 0.30, 0.20) \begin{pmatrix} 0.44 & 0.30 & 0.26 \\ 0.19 & 0.55 & 0.26 \\ 0.35 & 0.30 & 0.35 \end{pmatrix}$$

$$p^{(2)} = (0.347, 0.375, 0.278)$$

Example. Given the initial probability distribution

$$p^{(0)} = (0.80, 0.10, 0.10)$$

and the transition matrix

$$P = \begin{pmatrix} 0.40 & 0.60 & 0 \\ 0.50 & 0.40 & 0.10 \\ 0.20 & 0 & 0.80 \end{pmatrix}$$

determine the state probability distribution after three steps.

The three-step transition matrix was calculated on p. 316. The distribution is

$$p^{(3)} = p^{(0)} \cdot P^{(3)}$$

$$p^{(3)} = (0.80, 0.10, 0.10) \begin{pmatrix} 0.436 & 0.468 & 0.096 \\ 0.422 & 0.436 & 0.142 \\ 0.284 & 0.192 & 0.524 \end{pmatrix}$$

$$p^{(3)} = (0.4194, 0.4384, 0.1434)$$

10.4   Steady-state Probabilities

Suppose that we consider the state probability distributions for our typical customer after two, four, and eight purchases. For the case in which the initial state is specified, these distributions are given in the appropriate rows of the k-state transition matrix. The state probability distributions after two purchases are given in the rows of the two-step transition matrix, after four purchases in the rows of the four-step transition matrix, etc. The two-step transition matrix is

$$P^{(2)} = \begin{pmatrix} 0.44 & 0.30 & 0.26 \\ 0.19 & 0.55 & 0.26 \\ 0.35 & 0.30 & 0.35 \end{pmatrix}$$

The four- and eight-step transition matrices are given, respectively, by P^4 and P^8 and are

$$P^{(4)} = \begin{pmatrix} 0.3416 & 0.3750 & 0.2834 \\ 0.2791 & 0.4375 & 0.2834 \\ 0.3335 & 0.3750 & 0.2915 \end{pmatrix}$$

$$P^{(8)} = \begin{pmatrix} 0.3158 & 0.3985 & 0.2857 \\ 0.3120 & 0.4023 & 0.2857 \\ 0.3158 & 0.3985 & 0.2857 \end{pmatrix}$$

Consider first the two-step transition matrix. In particular, notice that the state probability distributions (i.e., the probabilities of states 1, 2, and 3) differ considerably from one row to the next. This should, of course, be expected. The first row of this matrix gives the two-step transition probabilities associated with the initial state a_1, the second row with the initial state a_2, and the third row with the initial state a_3. Since these probabilities reflect the likelihood of purchasing each of the three products after only one intervening purchase, these differences are not surprising.

Next consider the four- and eight-step transition matrices. Notice that the differences in the state probability distributions are becoming smaller as k increases. This is particularly noticeable in the eight-step transition matrix.

The differences in the state probability distributions shown in this matrix for each of the three initial states are small. This should also not be surprising. In terms of our example, the probabilities of purchasing each of the three products after seven intervening purchases are not significantly affected by the initial purchase.

The two-, four-, and eight-step transition matrices for our typical customer illustrate an important characteristic of certain kinds of transition matrices. Specifically, for so-called *regular transition matrices*, the state probability distributions approach a common limiting distribution as k increases.[4] This common distribution is termed the *limiting state probability distribution* or the *steady-state probability distribution*. As shown for our typical customer, the steady-state probability distribution does not depend on the initial state of the Markov process. In other words, the long-run probabilities of purchasing each of the three products do not depend on the initial purchase.

This discussion of the steady state can be expressed in more general terms as follows. By definition, a transition matrix is said to be regular if all the entries of $P^{(k)}$ are positive for some integer k. For a regular transition matrix, $P^{(k)}$ approaches a limiting matrix T with rows t as k approaches infinity. All rows of T are equal and are probability vectors. These common probability vectors represent the limiting state or the steady-state probability distribution. The steady-state probability distribution, t, is independent of the initial state of the Markov process.

Example. Determine the steady-state probability distribution for the transition matrix

$$P = \begin{pmatrix} 0.4 & 0.6 \\ 0.2 & 0.8 \end{pmatrix}$$

k-step transition probabilities for two, three, four, and five steps are

$$P^{(2)} = \begin{pmatrix} 0.28 & 0.72 \\ 0.24 & 0.76 \end{pmatrix} \qquad P^{(3)} = \begin{pmatrix} 0.256 & 0.744 \\ 0.248 & 0.752 \end{pmatrix}$$

$$P^{(4)} = \begin{pmatrix} 0.2512 & 0.7488 \\ 0.2496 & 0.7504 \end{pmatrix} \qquad P^{(5)} = \begin{pmatrix} 0.25023 & 0.74977 \\ 0.24992 & 0.75008 \end{pmatrix}$$

As k increases, the k-step transition matrix evidently approaches the matrix

$$T = \begin{pmatrix} 0.25 & 0.75 \\ 0.25 & 0.75 \end{pmatrix}$$

The steady-state probability distribution is $t = (0.25, 0.75)$.

---

[4] A transition matrix is said to be regular if some power of the matrix has only positive components.

Example. Use the steady-state transition matrix from the preceding example to show that the steady-state probability distribution is independent of the initial state of the Markov process.

Assume that the process begins in state 1. The steady-state probability distribution is given by the first row of the steady-state transition matrix T and is $t = (0.25, 0.75)$. Next assume that the process begins in state 2. The steady-state probability distribution is given by the second row of T and is again $t = (0.25, 0.75)$. Finally, assume that the initial state is described by the probability distribution $p^{(0)} = (0.50, 0.50)$. According to Eq. (10.4), the steady-state probabilities are

$$t = (0.50, 0.50) \begin{pmatrix} 0.25 & 0.75 \\ 0.25 & 0.75 \end{pmatrix} = (0.25, 0.75)$$

Since the steady-state probabilities remain constant regardless of the initial state, we conclude that the steady-state probability distribution is independent of the initial state.

Example. Determine the steady-state probability distribution for

$$P = \begin{pmatrix} 0.7 & 0.3 \\ 0.5 & 0.5 \end{pmatrix}$$

The powers of the transition matrix P are

$$P^{(2)} = \begin{pmatrix} 0.64 & 0.36 \\ 0.60 & 0.40 \end{pmatrix} \qquad P^{(3)} = \begin{pmatrix} 0.628 & 0.372 \\ 0.620 & 0.380 \end{pmatrix}$$

$$P^{(4)} = \begin{pmatrix} 0.6256 & 0.3744 \\ 0.6240 & 0.3760 \end{pmatrix} \qquad P^{(5)} = \begin{pmatrix} 0.6251 & 0.3749 \\ 0.6248 & 0.3752 \end{pmatrix}$$

The limiting matrix is evidently

$$T = \begin{pmatrix} 0.625 & 0.375 \\ 0.625 & 0.375 \end{pmatrix}$$

and the steady-state probabilities are

$$t = (0.625, 0.375)$$

Readers who have access to a computer can easily determine the steady-state probabilities. For those who don't, repeated multiplication of transition matrices can prove to be an arduous task. Fortunately, a method is available that eliminates the need for determining ever increasing powers of P. This method is derived from a certain relationship that exists between the transition matrix and the steady-state probability distribution. Formal proof of this method is beyond the scope of this text.

Consider a square matrix A and a row vector X. The row vector X is called a *fixed vector* of the square matrix A if $XA = X$. For instance, the row vector $X = (3, 4)$ is a fixed vector of the matrix

$$A = \begin{pmatrix} \frac{2}{3} & \frac{1}{3} \\ \frac{1}{4} & \frac{3}{4} \end{pmatrix}$$

since $XA = X$, i.e.,

$$(3, 4) \begin{pmatrix} \frac{2}{3} & \frac{1}{3} \\ \frac{1}{4} & \frac{3}{4} \end{pmatrix} = (3, 4)$$

It can be demonstrated that the vector of steady-state probabilities, t, is a fixed vector of a regular transition matrix, P. This useful fact can be expressed symbolically as

$$tP = t \tag{10.5}$$

where t is the unique vector of steady-state probabilities and P is a regular transition matrix. To illustrate Eq. (10.5), consider again the transition matrix

$$P = \begin{pmatrix} 0.4 & 0.6 \\ 0.2 & 0.8 \end{pmatrix}$$

The vector of steady-state probabilities was calculated on p. 319 and was $t = (0.25, 0.75)$. Since t is a fixed vector of P, it follows that $tP = t$, i.e.,

$$(0.25, 0.75) \begin{pmatrix} 0.4 & 0.6 \\ 0.2 & 0.8 \end{pmatrix} = (0.25, 0.75)$$

Example. Verify that $t = (0.625, 0.375)$ is a fixed vector of the transition matrix

$$P = \begin{pmatrix} 0.7 & 0.3 \\ 0.5 & 0.5 \end{pmatrix}$$

According to Eq. (10.5), t is a fixed vector of P if $tP = t$. By multiplication, we see that

$$(0.625, 0.375) \begin{pmatrix} 0.7 & 0.3 \\ 0.5 & 0.5 \end{pmatrix} = (0.625, 0.375)$$

The unique relationship defined by Eq. (10.5) can be used to determine the steady-state probabilities. By way of illustration, suppose that we again calculate the steady-state probabilities for the transition matrix

$$P = \begin{pmatrix} 0.4 & 0.6 \\ 0.2 & 0.8 \end{pmatrix}$$

The steady-state probabilities are represented by the probability vector $t = (p_1, p_2)$. Since t is a fixed vector of P, it follows that

$$(p_1, p_2) \begin{pmatrix} 0.4 & 0.6 \\ 0.2 & 0.8 \end{pmatrix} = (p_1, p_2)$$

Multiplying the vector by the transition matrix gives

$$(0.4p_1 + 0.2p_2, 0.6p_1 + 0.8p_2) = (p_1, p_2)$$

This relationship, together with the fact that t is a probability vector, leads to the following system of equations:

$$p_1 + p_2 = 1$$

$$0.4p_1 + 0.2p_2 = p_1$$

$$0.6p_1 + 0.8p_2 = p_2$$

Discarding either the second or third equation and solving the remaining two equations, we obtain $p_1 = 0.25$ and $p_2 = 0.75$.[5] The steady-state probability vector is thus $t = (0.25, 0.75)$.

Example. Use Eq. (10.5) to determine the steady-state probability distribution of

$$P = \begin{pmatrix} 0.7 & 0.3 \\ 0.5 & 0.5 \end{pmatrix}$$

Multiplying the probability vector $t = (p_1, p_2)$ by P and equating this product with t gives

$$(p_1, p_2) \begin{pmatrix} 0.7 & 0.3 \\ 0.5 & 0.5 \end{pmatrix} = (p_1, p_2)$$

$$(0.7p_1 + 0.5p_2, 0.3p_1 + 0.5p_2) = (p_1, p_2)$$

This, together with the requirement that the probabilities sum to 1, gives the following system of three equations:

$$p_1 + p_2 = 1$$

$$0.7p_1 + 0.5p_2 = p_1$$

$$0.3p_1 + 0.5p_2 = p_2$$

Discarding either the second or third equation and solving the remaining two equations simultaneously result in the steady-state probability vector $t = (0.625, 0.375)$.

10.4.1 INTERPRETING STEADY-STATE PROBABILITIES

Steady-state probabilities can be thought of as the probability of finding a Markov process in each of the several states after the process has reached

---

[5] The second and third equations are not independent. Therefore, one of these equations must be discarded. This leaves a system of two independent equations with two variables.

a point of *equilibrium*. This point occurs when the memory of the initial state is more or less lost. To illustrate, consider the steady-state probability distribution associated with the transition matrix

$$P = \begin{pmatrix} 0.7 & 0.3 \\ 0.5 & 0.5 \end{pmatrix}$$

This distribution was found on p. 322 and is

$$t = (0.625, 0.375)$$

The steady-state probabilities for the process's being in states a_1 and a_2 are $p(a_1) = 0.625$ and $p(a_2) = 0.375$. Since the process has reached a point of equilibrium, these probabilities apply regardless of the initial state in the chain of Markov trials.

There is a second and equally useful interpretation of steady-state probabilities. Instead of predicting the state at some random step in the sequence of trials, we look at the process over a large number of trials. Assuming the Markov process has reached the point of equilibrium, the steady-state probabilities can be viewed as the frequency of occurrence of each of the several states. The process can thus be expected to be in state a_1 with frequency $p(a_1)$, state a_2 with frequency $p(a_2)$, etc.

To illustrate, suppose that we consider 100 trials of a Markov process with steady-state probabilities $p(a_1) = 0.75$ and $p(a_2) = 0.25$. Based on the frequency interpretation of steady-state probabilities, we would expect state a_1 to occur in approximately 75 of the trials and state a_2 to occur in the remaining 25 trials. Since a sufficient number of trials has occurred for the memory of the initial state to be more or less lost, this frequency would be expected regardless of the initial state of the Markov process.

Example. A frustrated commuter has three alternative routes from his home to his office. His commuting habits are as follows: He never takes the same route from his home to his office. If he takes route A one day, he takes route B or route C with probability 0.50 the following day. If he takes route B one day, then he takes route A with probability 0.4 and route C with probability 0.6 the following day. If he takes route C, then he takes route A with probability 0.7 and route B with probability 0.3 the following day. Given that he takes route A on Monday, determine the probability of each of the routes on Wednesday. In addition, determine the long-term frequency of each of the three routes.

The transition matrix is

$$P = \begin{array}{c} A \\ B \\ C \end{array} \begin{pmatrix} 0 & 0.5 & 0.5 \\ 0.4 & 0 & 0.6 \\ 0.7 & 0.3 & 0 \end{pmatrix} \begin{array}{ccc} A & B & C \end{array}$$

The two-step transition matrix is

$$
\begin{array}{c}
\qquad\quad A \quad\ \ B \quad\ \ C \\
P^{(2)} = \begin{array}{c} A \\ B \\ C \end{array}
\begin{pmatrix}
0.55 & 0.15 & 0.30 \\
0.42 & 0.38 & 0.20 \\
0.12 & 0.35 & 0.53
\end{pmatrix}
\end{array}
$$

Given the commuter takes route A on Monday, the probability of each of the routes on Wednesday is

$$(0.55, 0.15, 0.30)$$

The long-term frequency of each route is given by solving the following system of equations:

$$
\begin{aligned}
p_1 + \quad p_2 + \quad p_3 &= 1 \\
0p_1 + 0.4p_2 + 0.7p_3 &= p_1 \\
0.5p_1 + \quad 0p_2 + 0.3p_3 &= p_2 \\
0.5p_1 + 0.6p_2 + \quad 0p_3 &= p_3
\end{aligned}
$$

Solving the first equation and any two of the remaining three equations simultaneously gives $p_1 = 0.342$, $p_2 = 0.283$, and $p_3 = 0.375$. The commuter takes route A with probability 0.342, route B with probability 0.283, and route C with probability 0.375.

10.4.2   A FINAL LOOK AT MARKET SHARES

Let us again turn to the Markov model presented in the introductory section. The reader will remember that the purpose in developing the transition matrix for our typical customer was to determine the long-run market share of the three competing products. Suppose that we consider a large number of purchases. The long-run market share can be viewed as the relative proportion of sales of each of the three products. Once the Markov process reaches equilibrium and the steady-state probabilities apply, the market share of each product is equivalent to the fraction of times the process occupies each of the three possible states. From a frequency interpretation of probability, these fractions are equivalent to the steady-state probabilities of the Markov chain.

The steady-state probabilities are calculated by using Eq. (10.5). This equation, together with the fact that the steady-state probabilities must sum to one, leads to the following system of equations:

$$
\begin{aligned}
p_1 + \quad p_2 + \quad p_3 &= 1 \\
0.6p_1 + 0.1p_2 + 0.3p_3 &= p_1 \\
0.2p_1 + 0.7p_2 + 0.2p_3 &= p_2 \\
0.2p_1 + 0.2p_2 + 0.5p_3 &= p_3
\end{aligned}
$$

Discarding any one of the final three equations and solving the remaining three simultaneously yield $p_1 = 0.3143$, $p_2 = 0.4000$, and $p_3 = 0.2857$. Barring changes in the purchasing habits of consumers, we would thus expect 31.43 percent of the market to go to product 1, 40.00 percent to product 2, and 28.57 percent to product 3. These market shares are expected regardless of the initial state of the system, i.e., regardless of the initial market share of each of the three products.

Example. Suppose that the manufacturer of product 1 is able to increase the probability of a repeat purchase to 0.80. If the increase occurs equally at the expense of products 2 and 3, how would this alter the long-run market shares of each of the three products?

Based upon the change in transition probabilities for product 1, the transition matrix for our typical customer becomes

$$P = \begin{pmatrix} 0.80 & 0.10 & 0.10 \\ 0.10 & 0.70 & 0.20 \\ 0.30 & 0.20 & 0.50 \end{pmatrix}$$

The steady-state probabilities are found by solving the following system of equations:

$$p_1 + \quad p_2 + \quad p_3 = 1$$

$$0.80p_1 + 0.10p_2 + 0.30p_3 = p_1$$

$$0.10p_1 + 0.70p_2 + 0.20p_3 = p_2$$

$$0.10p_1 + 0.20p_2 + 0.50p_3 = p_3$$

Discarding any one of the last three equations and solving the remaining three simultaneously give $p_1 = 0.4778$, $p_2 = 0.3046$, and $p_3 = 0.2176$. The long-term share of product 1 is increased to 47.78 percent. The market shares of products 2 and 3 are 30.46 and 21.76 percent, respectively.

10.5   Absorbing Markov Chains

Suppose that we consider a different kind of Markov process, specifically, a Markov process that contains one or more absorbing states. As an example, consider the following game of chance.

Mr. Murphy and his wealthy friend Mr. Clifford are playing pool at their club. Mr. Murphy has $2 and plans to play until he wins $4 or is broke. The two pool players are equally matched and are playing for $1 per game.

We assume that the probability of either player's winning a single game is

0.50. The states of this Markov process are the amount of money that Murphy has at any time and are $0, $1, $2, $3, and $4. The transition matrix is

$$
P = \begin{array}{c} \\ 0 \\ 1 \\ 2 \\ 3 \\ 4 \end{array} \begin{array}{ccccc} 0 & 1 & 2 & 3 & 4 \\ \begin{pmatrix} 1 & 0 & 0 & 0 & 0 \\ \frac{1}{2} & 0 & \frac{1}{2} & 0 & 0 \\ 0 & \frac{1}{2} & 0 & \frac{1}{2} & 0 \\ 0 & 0 & \frac{1}{2} & 0 & \frac{1}{2} \\ 0 & 0 & 0 & 0 & 1 \end{pmatrix} \end{array}
$$

The transition matrix is interpreted as follows. The states represent Murphy's cash. If Murphy begins with $2, the process begins at state 2. The transition probabilities represent the probabilities of going from state i to state j. For example, beginning with state 2 the process goes to state 1 or to state 3 with probability 0.50. Sooner or later, Murphy either goes broke or wins $4. These events are shown as states 0 and 4, respectively.

Markov chains like this are called *absorbing Markov chains*. An absorbing Markov chain has the following two characteristics. First, the transition matrix contains at least one *absorbing state*, i.e., a state once entered is impossible to leave. Second, given a sufficient number of trials, it is possible to go from every nonabsorbing state to an absorbing state.

To illustrate these characteristics, consider the game of pool between Murphy and Clifford. The game ends when Murphy either goes broke or wins $4. States 0 and 4 are therefore absorbing states. This is reflected by the rows in the transition matrix for these states. These rows contain a single 1 with remaining entries of 0. This means that once the process reaches the absorbing states 0 or 4, the process remains in that state with probability 1.0. Furthermore, it is possible to reach an absorbing state from any initial nonabsorbing state. In other words, Murphy can go broke or win $4 regardless of whether he begins with $1, $2, or $3.

Normally, there are three questions of interest for an absorbing Markov chain.

1. What is the probability that the process will terminate in a given absorbing state?
2. What is the expected number of steps before absorption?
3. What is the expected number of times the process will be in each nonabsorbing state?

The answers to these questions generally depend on the initial state of the Markov process. To illustrate the procedure for determining the answers, we turn again to our hypothetical game of pool.

The first step in the solution procedure is to rearrange the transition matrix so that the absorbing states appear first. The transition matrix for the game of pool becomes

$$
P = \begin{array}{c} \\ 0 \\ 4 \\ 1 \\ 2 \\ 3 \end{array}
\begin{array}{cc}
\begin{array}{cc} 0 & 4 \end{array} & \begin{array}{ccc} 1 & 2 & 3 \end{array} \\
\left(\begin{array}{cc|ccc}
1 & 0 & 0 & 0 & 0 \\
0 & 1 & 0 & 0 & 0 \\
\hline
\frac{1}{2} & 0 & 0 & \frac{1}{2} & 0 \\
0 & 0 & \frac{1}{2} & 0 & \frac{1}{2} \\
0 & \frac{1}{2} & 0 & \frac{1}{2} & 0
\end{array}\right)
\end{array}
$$

The transition matrix rearranged in this fashion is said to be in *canonical* (or *standard*) form. If there are r absorbing states and s nonabsorbing states, the canonical form is

$$
P = \left(\begin{array}{c|c} I & \phi \\ \hline R & Q \end{array}\right) \tag{10.6}
$$

where I is an r by r identity matrix, ϕ is an r by s null matrix, R is an s by r matrix, and Q is an s by s matrix. In our example, the identity matrix is

$$
I = \begin{pmatrix} 1 & 0 \\ 0 & 1 \end{pmatrix}
$$

the null matrix is

$$
\phi = \begin{pmatrix} 0 & 0 & 0 \\ 0 & 0 & 0 \end{pmatrix}
$$

the matrix R is

$$
R = \begin{pmatrix} \frac{1}{2} & 0 \\ 0 & 0 \\ 0 & \frac{1}{2} \end{pmatrix}
$$

and the matrix Q is

$$
Q = \begin{pmatrix} 0 & \frac{1}{2} & 0 \\ \frac{1}{2} & 0 & \frac{1}{2} \\ 0 & \frac{1}{2} & 0 \end{pmatrix}
$$

In order to answer the three questions asked earlier, we calculate the *fundamental matrix* for the absorbing Markov chain. The fundamental matrix is defined as

$$
N = (I - Q)^{-1} \tag{10.7}
$$

where I is an s by s identity matrix, Q is the s by s matrix of transition probabilities for the nonabsorbing states, and N is the inverse of $(I - Q)$.[6] In the

---

[6] Methods for determining the inverse of a matrix are discussed in Chap. 9.

example of the game of pool between Murphy and Clifford,

$$I - Q = \begin{pmatrix} 1 & -\frac{1}{2} & 0 \\ -\frac{1}{2} & 1 & -\frac{1}{2} \\ 0 & -\frac{1}{2} & 1 \end{pmatrix}$$

and

$$N = (I - Q)^{-1} = \begin{pmatrix} \frac{3}{2} & 1 & \frac{1}{2} \\ 1 & 2 & 1 \\ \frac{1}{2} & 1 & \frac{3}{2} \end{pmatrix}$$

We are now able to answer the questions asked on p. 326.

The first question is answered by multiplying the fundamental matrix N by the matrix R. The entries in the matrix product NR represent the probabilities that a Markov process beginning in a nonabsorbing state i will terminate in the absorbing state j. To illustrate, the probabilities that Murphy either goes broke or wins \$4 for each of the possible initial cash positions are given in the matrix

$$NR = \begin{pmatrix} \frac{3}{2} & 1 & \frac{1}{2} \\ 1 & 2 & 1 \\ \frac{1}{2} & 1 & \frac{3}{2} \end{pmatrix} \begin{pmatrix} \frac{1}{2} & 0 \\ 0 & 0 \\ 0 & \frac{1}{2} \end{pmatrix} = \begin{pmatrix} \frac{3}{4} & \frac{1}{4} \\ \frac{1}{2} & \frac{1}{2} \\ \frac{1}{4} & \frac{3}{4} \end{pmatrix}$$

The three rows of NR represent the three nonabsorbing initial states (i.e., states 1, 2, and 3) and the two columns represent the two absorbing terminal states (i.e., states 0 and 4). If Murphy begins with \$1, the probability of going broke is $\frac{3}{4}$ and the probability of winning \$4 is $\frac{1}{4}$. Should Murphy begin with \$2, the probabilities of going broke or winning \$4 are each $\frac{1}{2}$. If Murphy begins with \$3, the probabilities of going broke and winning \$4 are $\frac{1}{4}$ and $\frac{3}{4}$, respectively.

The second question concerned the expected number of steps before absorption. This question is also answered by using the fundamental matrix. To determine the expected number of steps before absorption for a given initial state, we merely sum the entries in the row of the fundamental matrix for that state. In other words, adding all the entries in the ith row of the fundamental matrix gives the expected number of steps before absorption for the ith initial state.

The fundamental matrix for the game of pool between Murphy and Clifford was

$$N = \begin{pmatrix} \frac{3}{2} & 1 & \frac{1}{2} \\ 1 & 2 & 1 \\ \frac{1}{2} & 1 & \frac{3}{2} \end{pmatrix}$$

The sum of the entries in the first row of this matrix is 3, the second row is 4, and the third row is 3. Since the sum of the entries of the ith row gives the expected number of steps before absorption for the ith initial state, the expected number of steps to absorption starting in state 1 is 3, starting in state 2 is 4, and starting in state 3 is 3. For instance, if Murphy begins playing pool with $2, he could expect on the average to play four games before he either goes broke or wins $4.

The third question concerned the expected number of times that the process is in each nonabsorbing state. This is again given by the entries in the fundamental matrix. For each possible nonabsorbing starting state i, the a_{ij} entries for that state represent the expected number of times that the process is in the nonabsorbing state j. In our example, the fundamental matrix is

$$N = \begin{array}{c} \\ 1 \\ 2 \\ 3 \end{array} \begin{array}{ccc} 1 & 2 & 3 \\ \begin{pmatrix} \frac{3}{2} & 1 & \frac{1}{2} \\ 1 & 2 & 1 \\ \frac{1}{2} & 1 & \frac{3}{2} \end{pmatrix} \end{array}$$

Thus, starting in state 1, the expected number of times in state 1 before absorption is $\frac{3}{2}$, in state 2 is 1, and in state 3 is $\frac{1}{2}$. If Murphy begins to play with $2, the expected number of times in state 1 before absorption is 1, in state 2 is 2, and in state 3 is 1. If Murphy starts with $3, the expected number of times in each of the three states is $\frac{1}{2}$, 1, and $\frac{3}{2}$, respectively.

We can summarize our results as follows. The answers to the three questions can all be found by using the fundamental matrix N. The matrix product NR gives the probability of absorption in each of the absorbing states for each starting state. The sum of the entries in each row of N gives the expected number of steps before absorption for each particular starting state. Finally, the matrix N itself gives us the expected number of times in each state before absorption.

Example. An absorbing Markov process has the following transition matrix:

$$P = \begin{array}{c} \\ 1 \\ 2 \\ 3 \end{array} \begin{array}{ccc} 1 & 2 & 3 \\ \begin{pmatrix} 0.4 & 0.6 & 0 \\ 0.5 & 0.3 & 0.2 \\ 0 & 0 & 1 \end{pmatrix} \end{array}$$

Determine the fundamental matrix.

Rewriting the transition matrix with the absorbing state listed first gives

$$\begin{array}{c} \\ 3 \\ 1 \\ 2 \end{array} \begin{array}{c|cc} 3 & 1 & 2 \\ \hline 1 & 0 & 0 \\ 0 & 0.4 & 0.6 \\ 0.2 & 0.5 & 0.3 \end{array}$$

The matrix Q is

$$Q = \begin{pmatrix} 0.4 & 0.6 \\ 0.5 & 0.3 \end{pmatrix}$$

and $(I - Q)$ is

$$(I - Q) = \begin{pmatrix} 1 & 0 \\ 0 & 1 \end{pmatrix} - \begin{pmatrix} 0.4 & 0.6 \\ 0.5 & 0.3 \end{pmatrix}$$

$$(I - Q) = \begin{pmatrix} 0.6 & -0.6 \\ -0.5 & 0.7 \end{pmatrix}$$

The inverse of $(I - Q)$ is

$$N = (I - Q)^{-1} = \begin{pmatrix} \frac{35}{6} & \frac{30}{6} \\ \frac{25}{6} & \frac{30}{6} \end{pmatrix}$$

State 3 is the only absorbing state in this example. Since it is possible to reach state 3 from state 1 via state 2 or state 2 directly, the process must terminate in state 3. Verify this by calculating NR.

$$NR = \begin{pmatrix} \frac{35}{6} & \frac{30}{6} \\ \frac{25}{6} & \frac{30}{6} \end{pmatrix} \begin{pmatrix} 0 \\ 0.2 \end{pmatrix} = \begin{pmatrix} 1.0 \\ 1.0 \end{pmatrix}$$

The probability of terminating in state 3 given initial state 1 is shown by the first row of the column vector. Similarly, the probability of terminating in state 3 given initial state 2 is shown by the second row. As expected, these entries are both 1.0.

For each of the initial states, calculate the expected number of steps before absorption.

Summing the entries in the first row of the fundamental matrix N gives the expected number of steps before absorption for initial state 1. This sum is $\frac{65}{6}$. The sum of the entries in the second row gives the expected number of steps before absorption for initial state 2. This sum is $\frac{55}{6}$.

Calculate the expected number of times that the process will be in each nonabsorbing state.

The expected number of times that the process will be in each nonabsorbing state is given by the fundamental matrix. If the process begins in state 1, the expected number of times that the process will be in state 1 is $\frac{35}{6}$ and the expected number of times in state 2 is $\frac{30}{6}$. If the process begins in state 2, the numbers are $\frac{25}{6}$ and $\frac{30}{6}$.

Example. The producer of a television quiz show has stated that the questions asked on the show are such that the probability is 0.80 that a contestant can correctly answer the question. A contestant remains on the show until she misses two consecutive questions. Determine the transition matrix that describes the probabilities of correctly answering the questions and calculate the expected number of questions asked per contestant.

The transition matrix is

$$
\begin{array}{c}
 & \begin{array}{ccc} C & I & T \end{array} \\
\begin{array}{c} C \\ I \\ T \end{array} &
\left(\begin{array}{ccc}
0.8 & 0.2 & 0 \\
0.8 & 0 & 0.2 \\
0 & 0 & 1
\end{array} \right)
\end{array}
$$

where C represents a correct answer, I represents an incorrect answer, and T represents two consecutive incorrect answers.

To calculate the expected number of questions asked per contestant, we must determine the fundamental matrix. The steps are as follows:

$$
\begin{array}{c}
 & \begin{array}{ccc} T & C & I \end{array} \\
\begin{array}{c} T \\ C \\ I \end{array} &
\left(\begin{array}{c|cc}
1 & 0 & 0 \\
\hline
0 & 0.8 & 0.2 \\
0.2 & 0.8 & 0
\end{array} \right)
\end{array}
$$

$$
I - Q = \begin{pmatrix} 1 & 0 \\ 0 & 1 \end{pmatrix} - \begin{pmatrix} 0.8 & 0.2 \\ 0.8 & 0 \end{pmatrix} = \begin{pmatrix} 0.2 & -0.2 \\ -0.8 & 1 \end{pmatrix}
$$

$$
N = (I - Q)^{-1} = \begin{pmatrix} 25 & 5 \\ 20 & 5 \end{pmatrix}
$$

The game begins in state C. Thus, the expected number of questions is 30 per contestant.

PROBLEMS

1. Construct transition matrices for the transition diagrams in Fig. 10.9.

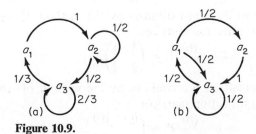

Figure 10.9.

2. Construct transition matrices for the transition diagrams in Fig. 10.10.

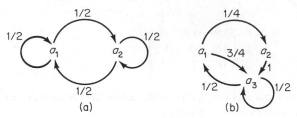

Figure 10.10.

3. Determine the missing transition probabilities in the transition matrices in Fig. 10.11.

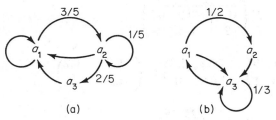

(a)   (b)

Figure 10.11.

4. Draw a transition diagram for the Markov process with transition probabilities given by the following matrices:
   (a)
   $$P = \begin{pmatrix} \frac{1}{3} & 0 & \frac{2}{3} \\ \frac{1}{2} & \frac{1}{2} & 0 \\ \frac{1}{4} & \frac{1}{2} & \frac{1}{4} \end{pmatrix}$$
   (b)
   $$P = \begin{pmatrix} \frac{1}{4} & \frac{3}{4} \\ \frac{1}{2} & \frac{1}{2} \end{pmatrix}$$

5. Draw a transition diagram for the Markov process with transition probabilities given by the following matrices:
   (a)
   $$P = \begin{pmatrix} 0 & 1 & 0 \\ \frac{1}{2} & 0 & \frac{1}{2} \\ 0 & \frac{3}{4} & \frac{1}{4} \end{pmatrix}$$
   (b)
   $$P = \begin{pmatrix} \frac{1}{2} & \frac{1}{2} & 0 \\ 0 & \frac{1}{2} & \frac{1}{2} \\ \frac{1}{2} & 0 & \frac{1}{2} \end{pmatrix}$$

6. Determine the two-step transition probabilities by the method of tree diagrams for the following transition matrices:
   (a) $P = \begin{pmatrix} 0 & 1 \\ \frac{1}{2} & \frac{1}{2} \end{pmatrix}$
   (b) $P = \begin{pmatrix} \frac{1}{2} & \frac{1}{2} \\ \frac{1}{4} & \frac{3}{4} \end{pmatrix}$

7. Determine the two-step transition probabilities by the method of tree diagrams for the following transition matrices:
   (a) $P = \begin{pmatrix} 0.3 & 0.7 \\ 0.6 & 0.4 \end{pmatrix}$
   (b) $P = \begin{pmatrix} 0.1 & 0.9 \\ 0.5 & 0.5 \end{pmatrix}$

8. Find the matrices $P^{(2)}$ and $P^{(3)}$ for the following matrix of transition probabilities:

$$P = \begin{pmatrix} \frac{1}{2} & \frac{1}{2} \\ \frac{3}{4} & \frac{1}{4} \end{pmatrix}$$

9. Determine the two- and three-step transition matrices for

$$P = \begin{pmatrix} \frac{1}{2} & \frac{1}{2} & 0 \\ 0 & 0 & 1 \\ \frac{1}{2} & 0 & \frac{1}{2} \end{pmatrix}$$

10. Suppose that a Markov chain has two states, a_1 and a_2, with transition probabilities given by

$$P = \begin{pmatrix} \frac{1}{4} & \frac{3}{4} \\ \frac{1}{2} & \frac{1}{2} \end{pmatrix}$$

Assume that the initial state is selected by a chance device that chooses a_1 with probability $\frac{3}{4}$ and a_2 with probability $\frac{1}{4}$. Find the state probability distribution after two steps.

11. Determine the steady-state probability distribution for the Markov chain with transition probabilities

$$P = \begin{pmatrix} 0.4 & 0.6 \\ 0.5 & 0.5 \end{pmatrix}$$

12. A professor gives either a true–false test or a multiple-choice test. If he gives a true–false test, the next test will be a true–false test with probability 0.4. If he gives a multiple-choice test, the next test will be a multiple-choice test with probability 0.5. Construct a transition matrix that describes this Markov process and determine the long-run proportions of each kind of test.

13. A study of the weather in Los Angeles shows that if it is smoggy, then 80 percent of the time the next day will also be smoggy. A clear day and rain are equally likely to follow a smoggy day. A clear day is followed 60 percent of the time by a clear day, 30 percent of the time by a smoggy day, and 10 percent of the time by a rainy day. A rainy day is followed 40 percent of the time by a clear day, 20 percent of the time by a smoggy day, and 40 percent of the time by more rain. Assuming rain on Friday, what is the probability of a clear day on Sunday?

14. A small Kansas town has three gasoline stations. Over the past year, station A retained 90 percent of its customers but lost 5 percent to station B and 5 percent to station C. Station B retained 80 percent of its customers but lost 10 percent to A and 10 percent to C. Station C retained 60 percent of its customers but lost 30 percent to A and 10 percent to B. On January 1, 1976, station A had 20 percent of the market, station

B had 30 percent, and station C had 50 percent. If the same pattern of customer gains and losses continues, determine the percent of the market for each station on January 1, 1978.

15. The price of stock in Acme Corporation has been observed over a period of days. Each day the stock either goes up by one point, remains unchanged, or goes down by one point. The probabilities are given in the following table:

From \ To	Up One	Unchanged	Down One
Up one	0.6	0.3	0.1
Unchanged	0.4	0.3	0.3
Down one	0.2	0.3	0.5

If the stock goes down on Monday, what is the probability that it also goes down on Wednesday?

16. Suppose that an individual's work is classified as professional, semiprofessional, or laborer. Assume that of the sons and daughters of professional individuals, 80 percent are professional, 10 percent are semiprofessional, and 10 percent are laborers. In the case of the offspring of semiprofessionals, 20 percent are professional, 60 percent are semiprofessional, and 20 percent are laborers. Of the sons and daughters of laborers, 40 percent are professionals, 20 percent are semiprofessionals, and 40 percent are laborers. Determine the probability that the grandson of a laborer is also a laborer.

17. A certain part operates properly with a probability of 0.9 and fails with a probability of 0.1. Once the part has failed, it must be withdrawn from service. Treating the operation of this part as a Markov process, determine the probability that the part is still operating properly after four trials.

18. Tom and Harry have a standing tennis match each Saturday. If Tom wins the match on any given Saturday, the probability is 0.60 that he will also win the following Saturday. If Harry wins, he wins the following Saturday with probability 0.50. Tom won the last match. What is the probability that Tom wins two Saturdays hence?

19. Determine the steady-state probability distribution for the following transition matrices:

(a) $P = \begin{pmatrix} 0.8 & 0.2 \\ 0.5 & 0.5 \end{pmatrix}$ (b) $P = \begin{pmatrix} 0.1 & 0.9 \\ 1 & 0 \end{pmatrix}$

20. Determine the steady-state probability distribution for the following transition matrices:

(a)
$$P = \begin{pmatrix} 0.8 & 0.2 \\ 0.2 & 0.8 \end{pmatrix}$$

(b)
$$P = \begin{pmatrix} 0 & 1 & 0 \\ \frac{1}{2} & 0 & \frac{1}{2} \\ \frac{1}{2} & \frac{1}{2} & 0 \end{pmatrix}$$

21. Determine the steady-state probability distributions for the Markov processes described by the transition diagrams given in Problem 1.

22. Using the description of Los Angeles weather given in Problem 13, determine the long-run proportion of smoggy days.

23. Using the sales data in Problem 14, what long-run proportion of customers can be expected by each service station?

24. Using the data from Problem 18, determine the long-run proportion of tennis matches won by Tom and Harry.

25. Which of the following transition matrices contain at least one absorbing state?

(a)
$$P = \begin{pmatrix} 0 & 1 \\ 1 & 0 \end{pmatrix}$$

(b)
$$P = \begin{pmatrix} \frac{1}{2} & \frac{1}{2} \\ 0 & 1 \end{pmatrix}$$

(c)
$$P = \begin{pmatrix} 0 & \frac{1}{2} & \frac{1}{2} \\ 0 & 0 & 1 \\ \frac{1}{2} & \frac{1}{2} & 0 \end{pmatrix}$$

(d)
$$P = \begin{pmatrix} 0 & \frac{1}{2} & \frac{1}{2} & 0 \\ 1 & 0 & 0 & 0 \\ 0 & 0 & 1 & 0 \\ 0 & \frac{1}{2} & \frac{1}{2} & 0 \end{pmatrix}$$

26. Calculate the fundamental matrix for the Markov processes described by the following transition matrices:

(a)
$$P = \begin{pmatrix} 1 & 0 & 0 \\ 0 & \frac{1}{2} & \frac{1}{2} \\ \frac{1}{3} & 0 & \frac{2}{3} \end{pmatrix}$$

(b)
$$P = \begin{pmatrix} \frac{1}{2} & \frac{1}{2} & 0 \\ 0 & 1 & 0 \\ 1 & 0 & 0 \end{pmatrix}$$

27. For the Markov process described by the transition matrix

$$P = \begin{pmatrix} 1 & 0 & 0 & 0 \\ 0 & \frac{1}{4} & \frac{1}{4} & \frac{1}{2} \\ 0 & 0 & 1 & 0 \\ \frac{1}{2} & 0 & \frac{1}{4} & \frac{1}{4} \end{pmatrix}$$

(a) Determine the probability that the process will terminate in each absorbing state.

(b) Determine the expected number of steps before absorption.

(c) Determine the expected number of times that the process will be in each nonabsorbing state.

28. For the Markov process described by the transition matrix

$$P = \begin{pmatrix} 0 & \frac{1}{2} & \frac{1}{2} & 0 \\ \frac{1}{4} & 0 & \frac{1}{2} & \frac{1}{4} \\ 0 & 0 & 1 & 0 \\ 0 & 0 & 0 & 1 \end{pmatrix}$$

   (a) Determine the probability that the process will terminate in each absorbing state.

   (b) Determine the expected number of steps before absorption.

   (c) Determine the expected number of times that the process will be in each nonabsorbing state.

29. Mike and Joe are matching quarters. Mike initially has two quarters and Joe has three quarters. The game ends when one of them wins all five quarters. Let the states be labeled with the number of quarters Joe has. Determine the probability of Mike's winning all of the quarters. Of Joe's winning the quarters.

30. A fair coin is tossed until heads occur three times in a row. Determine the expected number of tosses. (Hint: Let the states be a_i for $i = 0, 1, 2, 3$, where i represents the number of successive occurrences of a head.)

SUGGESTED REFERENCES

GOODMAN, A. W., and J. S. RATTI, *Finite Mathematics with Applications* (New York, N.Y.: The Macmillan Company, Inc., 1971).

KEMENY, JOHN G., et al., *Finite Mathematics with Business Applications* (Englewood Cliffs, N.J.: Prentice-Hall, Inc., 1972).

LEVIN, R. I., and C. A. KIRKPATRICK, *Quantitative Approaches to Management* (New York, N.Y.: McGraw-Hill Book Company, Inc., 1975).

LIPSCHUTZ, SEYMOUR, *Finite Mathematics*, Schaum's Outline Series (New York, N.Y.: McGraw-Hill Book Company, Inc., 1966).

Chapter Eleven

Linear Programming

Linear programming is one of the most frequently and successfully applied mathematical approaches to managerial decisions. The objective in using linear programming is to develop a model to aid the decision maker in determining the optimal allocation of the firm's resources among the alternative products of the firm. Since the resources employed in a firm have an economic value and the products of the firm lead to profits and costs, the linear programming problem becomes that of allocating the scarce resources to the products in a manner such that profits are a maximum, or alternatively, costs are a minimum.

The popularity of linear programming stems directly from its usefulness to the business firm. One of the fundamental tasks of management involves the allocation of resources. For instance, the petroleum firm must allocate sources of crude oil for the production of various petroleum products. Similarly, the wood products firm must allocate timber resources for the production of different lumber and paper products. In both examples, the allocation must be made subject to constraints on the availability of resources and the demand for the products. The objective in allocating the resources to the products is to maximize the profit of the firm.

Our objective in this and the following two chapters is twofold. First, we shall describe the linear programming problem and the solution techniques available for this problem. Our approach will be to introduce the important characteristics of linear programming through the use of very simple two-product linear programming problems and to solve these problems by using a graphical solution technique. This, in turn, will lead to a discussion in Chap. 12 of a more general solution technique. This technique, termed the *simplex algorithm*, relies largely on the use of the elementary row operations of matrix algebra to obtain the solution of systems of simultaneous equations. A second, and equally important objective, will be to explain and offer examples of important applications of linear programming. These examples are in sufficient detail for the student to understand industrial applications of linear programming and, hopefully, to use the technique in his job.

The importance of the computer in the solution of linear programming problems should be emphasized. Although "linear programming" and "computer programming" are not related, the computer is extremely valuable in obtaining the solution to linear programming problems. Many industrial applications of linear programming require more than 100 equations and variables. Problems of this size can be solved only through the use of modern electronic computers.

11.1   The Linear Programming Problem

The linear programming problem consists of allocating scarce resources among competing products or activities. Two factors give rise to the allocation problem. First, resources available to management have a cost and are limited in supply. Therefore, management must determine how these limited resources will be used. Second, the allocation of these resources must be made in accordance with some overall objective. In the business firm, this objective is normally the maximization of profit or the minimization of cost.

As an illustration of the allocation of resources using linear programming, consider the following problem concerning the scheduling of production in a machine shop. Production must be scheduled for two kinds of machines, machine 1 and machine 2. One hundred twenty hours of time can be scheduled for machine 1, and 80 hours can be scheduled for machine 2. Production during the scheduling period is limited to two products, A and B. Each unit of product A requires 2 hours of process time on each machine. Each unit of product B requires 3 hours on machine 1 and 1.5 hours on machine 2. The profit is $4.00 per unit of product A and $5.00 per unit of product B. Both products can be readily marketed; consequently, production should be scheduled with the objective of maximizing profit. The linear programming problem can be formulated in the following manner.

Let

x_1 = the number of units of product A scheduled for production

x_2 = the number of units of product B scheduled for production

P = contribution to profit

The linear programming problem is

Maximize: $P = 4x_1 + 5x_2$ (profit)

Subject to: $2x_1 + 3x_2 \leq 120$ (machine 1 resource)
 $2x_1 + 1.5x_2 \leq 80$ (machine 2 resource)
 $x_1, x_2 \geq 0$ (non-negativity)

The mathematical formulation states that production of products A and B should be scheduled with the objective of maximizing profit. Production is limited, however, by the availability of resources, i.e., 120 hours of machine 1 time and 80 hours of machine 2 time. An additional limitation, shown by $x_1, x_2 \geq 0$, is that negative production is not allowed, i.e., the number of units of either product must be zero or positive.

The solution in terms of the production quantities of products A and B that maximizes profit can be determined in this problem by either a graphical approach or the simplex method. The graphical approach for solving two variable problems provides important insights into the nature of linear programming problems. This approach is shown in Sec. 11.2. The simplex method is presented in Chap. 12.

To illustrate a linear programming problem in which costs are minimized, consider the problem facing a specialty metals manufacturer. The firm produces an alloy that is made from steel and scrap metal. The cost per ton of steel is $50 and the cost per ton of scrap is $20. The technological requirements for the alloy are (1) a minimum of 1 ton of steel is required for every 2 tons of scrap; (2) 1 hour of processing time is required for each ton of steel, and 4 hours of processing time are required for each ton of scrap; (3) the steel and scrap combine linearly to make the alloy. The process loss from the steel is 10 percent and the loss from scrap is 20 percent. Although production may exceed demand, a minimum of 40 tons of the alloy must be manufactured. To maintain efficient plant operation, a minimum of 80 hours of processing time must be used. The supply of both scrap and steel is adequate for production of the alloy.

The objective of the manufacturer is to produce the alloy at a minimum cost. The problem can be formulated as follows:

Let $x_1 = $ the number of tons of steel

 $x_2 = $ the number of tons of scrap

 $C = $ the cost of the alloy

The linear programming problem is

Minimize: $C = 50x_1 + 20x_2$ (cost)

Subject to: $2x_1 - 1x_2 \geq 0$ (tech. req'm't.)
                               $1x_1 + 4x_2 \geq 80$ (process)
                        $0.90x_1 + 0.80x_2 \geq 40$ (demand)
                                    $x_1, x_2 \geq 0$ (non-negativity)

The problem is to minimize the cost of producing the alloy. The cost function is equal to \$50 per ton of steel multiplied by the number of tons of steel plus \$20 per ton of scrap multiplied by the number of tons of scrap. The first constraint states that a minimum of 1 ton of steel must be used for every 2 tons of scrap. The second constraint concerns process time, and the third constraint shows that the steel and scrap combine linearly to make the alloy with a 10 percent and 20 percent loss, respectively. The solution to the problem is determined by using the graphical method in Sec. 11.2.

11.2   Graphical Solution

The graphical solution technique provides a convenient method for solving simple two-variable linear programming problems. It is also useful in illustrating basic concepts in linear programming. For problems in which more than two variables are required, the simplex algorithm is used to determine the solution.

11.2.1   MAXIMIZATION

The problem concerning the scheduling of production in a machine shop was expressed mathematically as

Maximize: $P = 4x_1 + 5x_2$

Subject to: $2x_1 + 3x_2 \leq 120$
                               $2x_1 + 1.5x_2 \leq 80$
                                    $x_1, x_2 \geq 0$

The function to be optimized, $P = 4x_1 + 5x_2$, is termed the *objective function*. If x_1 and x_2 were unrestricted, there would be no limit on the value of the objective function. There are, however, *constraints* on the values of the variables.[1] The constraints are expressed as inequalities and are

$$2x_1 + 3x_2 \leq 120$$

$$2x_1 + 1.5x_2 \leq 80$$

$$x_1, x_2 \geq 0$$

The first two constraints come from the technological requirements specified in the problem. The remaining constraints, $x_1 \geq 0$ and $x_2 \geq 0$, are termed *non-negativity* constraints. These constraints are necessary to assure that the variables take on only zero or positive values.

The constraints are graphed in Fig. 11.1 by plotting the equations. Since there are four constraints in the problem, four equations must be plotted. These are

$$2x_1 + 3x_2 = 120$$

$$2x_1 + 1.5x_2 = 80$$

$$x_1 = 0$$

$$x_2 = 0$$

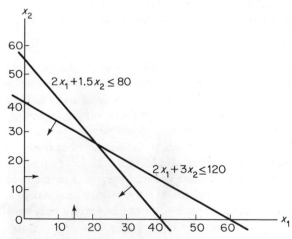

Figure 11.1.

Each of the equations is linear; consequently, each equation can be plotted from two points. The intercept points ($x_1 = 0$, $x_2 = 40$) and ($x_1 = 60$, $x_2 = 0$) are used to plot the first equation. Similarly, the second equation is plotted

---

[1] Constraints can take the form $ax_1 + bx_2 \leq r$, $ax_1 + bx_2 = r$, or $ax_1 + bx_2 \geq r$.

by using ($x_1 = 0$, $x_2 = 53.3$) and ($x_1 = 40$, $x_2 = 0$). The remaining equations, $x_1 = 0$ and $x_2 = 0$, represent the vertical and horizontal axes of the graph.[2]

The constraints in this problem are expressed as inequalities. It was shown in Chap. 2 that the solution to an inequality consisted of the set $\{(x_1, x_2) \mid ax_1 + bx_2 \leq r\}$. The direction of each inequality is shown in Fig. 11.1 by an arrow. For instance, the inequality $2x_1 + 1.5x_2 \leq 80$ can be thought of as consisting of all ordered pairs for which $2x_1 + 1.5x_2 = 80$ and for which $2x_1 + 1.5x_2 < 80$. The line shows the ordered pairs for which $2x_1 + 1.5x_2 = 80$, and the arrow pointing downward and to the left shows the ordered pairs for which $2x_1 + 1.5x_2 < 80$.

The linear programming problem consists of maximizing the objective function, $P = 4x_1 + 5x_2$, subject to the constraints. To define feasible values of (x_1, x_2), we denote the four constraints by c_1, c_2, c_3, and c_4.

$$c_1: \quad 2x_1 + 3x_2 \leq 120$$

$$c_2: \quad 2x_1 + 1.5x_2 \leq 80$$

$$c_3: \quad x_1 \geq 0$$

$$c_4: \quad x_2 \geq 0$$

The values of x_1 and x_2 are limited to the set of ordered pairs that are solutions to the constraints. Using set notation, we express this set of ordered pairs as

$$F = \{(x_1, x_2) \mid c_1 \cap c_2 \cap c_3 \cap c_4\}$$

The set F contains all ordered pairs that are *feasible* solutions to the linear programming problem. These ordered pairs are shown in Fig. 11.2 by the shaded area. This area is termed the *feasible region*. The ordered pairs defined by F are members of the *solution set* of the linear programming problem.

The optimal solution to the linear programming problem is the member of F that maximizes the objective function. Even in this simplified example, it is obvious that there are an infinite number of ordered pairs in the feasible region (i.e., shaded area). Since each of these ordered pairs is a member of the solution set, it would not be practical to investigate each ordered pair for optimality. Instead, we can make use of one of the important theorems in linear programming. This theorem, termed the *extreme point theorem*, states that the *optimal value of the objective function occurs at one of the extreme points of the feasible region*.[3] An extreme point in two-dimensional space is

---

[2] Some readers may find it helpful to review the method of plotting equations discussed in Chap. 2.

[3] This theorem is true provided that a unique, finite, optimal solution exists for the linear programming problem. There are three exceptions to this theorem. These are discussed in Sec. 11.3.3.

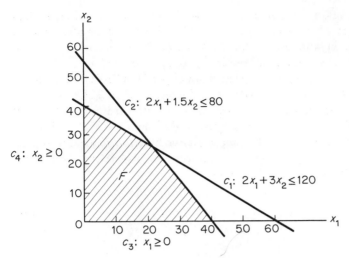

Figure 11.2.

defined by the intersection of two equations. By referring to Fig. 11.2, we see
that there are four extreme points. The values of x_1 and x_2 for each of these
extreme points are determined by solving the appropriate pairs of intersecting
equations. The solutions are

$$E = \{(0, 0), (40, 0), (20, 26.67), (0, 40)\}$$

According to the theorem, the objective function will be a maximum at one of
these four ordered pairs. To determine which ordered pair maximizes the
objective function, the objective function is evaluated for each extreme point.
Table 11.1 shows that the objective function is a maximum when $x_1 = 20$ and
$x_2 = 26.67$.

Table 11.1.

(x_1, x_2)	P
(0, 0)	\$0
(40, 0)	160.00
(20, 26.67)	213.33
(0, 40)	200.00

The validity of the extreme point theorem can be demonstrated for the
two-variable linear programming problem by superimposing the objective
function on a graph of the feasible region. The objective function is

$$P = 4x_1 + 5x_2$$

which by rearranging terms can also be written as

$$x_2 = 0.2P - 0.8x_1$$

The objective function, superimposed on the feasible region in Fig. 11.3, is shown as a series of parallel lines. Each line shows the combination of ordered pairs (x_1, x_2) that gives a specific value of P. For instance, the set of ordered pairs that are solutions to the equation $x_2 = 0.2(\$100) - 0.8x_1$ represents all combinations of (x_1, x_2) that result in profit of \$100. The objective function is also plotted for $P = \$150$, $P = \$200$, and $P = \$213.33$.

Profit increases as the objective function moves upward and to the right in the direction perpendicular to the slope of the function. Fig. 11.3 shows that profit will be a maximum when the objective function passes through an extreme point of the feasible region. In this particular problem, the extreme point is (20, 26.67), and profit is \$213.33.

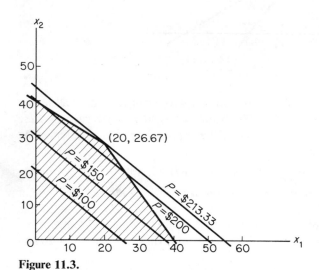

Figure 11.3.

Example. Use the graphical solution technique to determine the solution to the following linear programming problem:

$$\text{Maximize:} \quad P = 4x_1 + 3x_2$$

$$\text{Subject to:} \quad
\begin{aligned}
21x_1 + 16x_2 &\le 336 \\
13x_1 + 25x_2 &\le 325 \\
15x_1 + 18x_2 &\le 270 \\
x_1, x_2 &\ge 0
\end{aligned}$$

The feasible region is defined by the intersection of the five linear inequalities. The inequalities $x_1 \ge 0$ and $x_2 \ge 0$ limit the solution to the first quadrant, i.e., non-negative values of x_1 and x_2. The remaining three inequalities act as an upper bound on the values of x_1 and x_2. The feasible region is shown as the shaded area of Fig. 11.4.

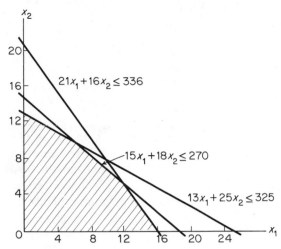

Figure 11.4.

The objective function has an optimal value at an extreme point of the feasible region. If the ordered pair $(0, 0)$ is included as an extreme point, a total of five extreme points must be evaluated. In two-dimensional space, an extreme point is defined by the intersection of two equations. The five extreme points can thus be determined by specifying the five pairs of intersecting equations and solving each pair of equations simultaneously. The sets of ordered pairs specified by the five pairs of intersecting equations are

1. $\{(x_1, x_2) \mid x_1 = 0 \cap x_2 = 0\}$
2. $\{(x_1, x_2) \mid x_1 = 0 \cap 13x_1 + 25x_2 = 325\}$
3. $\{(x_1, x_2) \mid 15x_1 + 18x_2 = 270 \cap 13x_1 + 25x_2 = 325\}$
4. $\{(x_1, x_2) \mid 21x_1 + 16x_2 = 336 \cap 15x_1 + 18x_2 = 270\}$
5. $\{(x_1, x_2) \mid 21x_1 + 16x_2 = 336 \cap x_2 = 0\}$

Solving each pair of equations simultaneously gives the set of extreme points, $E = \{(0, 0), (0, 13), (6.38, 9.68), (12.45, 4.57), (16, 0)\}$.

Not all pairs of intersecting equations in a linear programming problem specify an extreme point. For instance, the equations $21x_1 + 16x_2 = 336$ and $13x_1 + 25x_2 = 325$ intersect at $(9.7, 8.2)$. This point is not in the feasible region. Consequently, the ordered pair $(9.7, 8.2)$ is *nonfeasible* and is not considered in evaluating the objective function. Table 11.2 shows that the objective function is a maximum when $x_1 = 16$ and $x_2 = 0$.

11.2.2   MINIMIZATION

The graphical solution for a minimization problem is found by using the technique introduced in the preceding section. The difference between the mini-

Table 11.2.

(x_1, x_2)	P
(0, 0)	$0
(0, 13)	39.00
(6.38, 9.68)	54.56
(12.45, 4.57)	63.51
(16, 0)	64.00

mization and maximization problems is that the extreme point that results in a minimum value of the objective function is the solution to the minimization problem rather than that which results in a maximum value of the function. To illustrate, consider the problem of the specialty metals manufacturer previously described in Sec. 11.1. The mathematical formulation of the problem was

$$\text{Minimize:} \quad C = 50x_1 + 20x_2$$

$$\text{Subject to:} \quad \begin{aligned} 2x_1 - 1x_2 &\geq 0 \\ 1x_1 + 4x_2 &\geq 80 \\ 0.90x_1 + 0.80x_2 &\geq 40 \\ x_1, x_2 &\geq 0 \end{aligned}$$

All ordered pairs (x_1, x_2) that satisfy the five inequalities are feasible solutions. These are shown by the shaded area in Fig. 11.5.

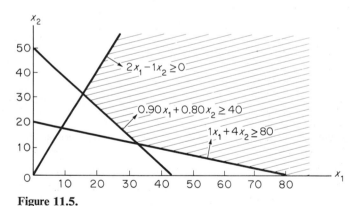

Figure 11.5.

The objective function is minimized at an extreme point of the feasible region. Figure 11.5 shows that three extreme points must be evaluated. These points are described by the sets

1. $\{(x_1, x_2) \mid 2x_1 - 1x_2 = 0 \cap 0.90x_1 + 0.80x_2 = 40\}$
2. $\{(x_1, x_2) \mid 0.90x_1 + 0.80x_2 = 40 \cap 1x_1 + 4x_2 = 80\}$
3. $\{(x_1, x_2) \mid 1x_1 + 4x_2 = 80 \cap x_2 = 0\}$

The solution of each of the three pairs of simultaneous equations gives the set of extreme points, $E = \{(16, 32), (34.32, 11.42), (80, 0)\}$. The value of the objective function at each extreme point is shown in Table 11.3. The objective

Table 11.3.

(x_1, x_2)	C
(16, 32)	$1440
(34.32, 11.42)	$1944.40
(80, 0)	$4000

function is minimized for $x_1 = 16$ and $x_2 = 32$.

Example. Use the graphical solution technique to determine the solution to the following linear programming problem:

$$\text{Minimize:} \quad C = 3x_1 + 2x_2$$

$$\text{Subject to:} \quad \begin{aligned} 2x_1 + x_2 &\geq 5 \\ x_1 + 3x_2 &\geq 6 \\ x_1 + x_2 &\geq 4 \\ x_1, x_2 &\geq 0 \end{aligned}$$

The feasible region for the linear programming problem is shown in Fig. 11.6. From the extreme point theorem, we know that the objective function is a minimum at one of the extreme points. The extreme points are determined

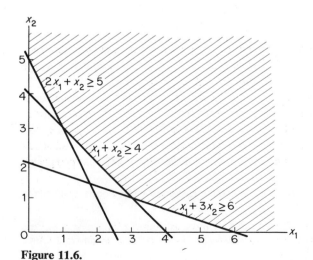

Figure 11.6.

by selecting the pairs of equations that define the extreme points and solving each pair of equations simultaneously. This results in the set of extreme points

$$E = \{(0, 5), (1, 3), (3, 1), (6, 0)\}$$

By evaluating the objective function at each of these extreme points, we find that the objective function has a minimum value of $C = 9$ at the extreme point $x_1 = 1$, $x_2 = 3$.

11.3   Assumptions, Terminology, and Special Cases

Certain assumptions and terms are required in the linear programming model. Several of these have been introduced. The purpose of this section is to introduce the additional assumptions of linearity, continuity, and convexity, to show that linear programming implies constant returns to scale, and to illustrate exceptions to the extreme point theorem of linear programming.

11.3.1   LINEARITY, CONTINUITY, AND RETURNS TO SCALE

One of the assumptions in linear programming is that the objective function and constraints are linear. The problem of scheduling production in a machine shop illustrates this assumption. The problem was

$$\text{Maximize:} \quad P = 4x_1 + 5x_2$$

$$\text{Subject to:} \quad 2x_1 + 3x_2 \leq 120$$
$$2x_1 + 1.5x_2 \leq 80$$
$$x_1, x_2 \geq 0$$

The assumption of linearity imposes an important restriction on the structure of the problem. This restriction is that both profits and resource usage are a linear combination of the number of units of each product. In this problem, profit is given by $P = 4x_1 + 5x_2$. The linear form of the objective function thus prohibits the competitive and substitutive effects that sometimes exist between products or activities. In certain industries, for instance, the manufacture of one product enables the manufacture of a second product at lesser cost. Such "second-order effects" are not included in the linear programming model.[4]

The result of the assumption of linearity involves what economists term

---

[4] These effects can, however, be included in more advanced programming models. See Harvey M. Wagner, *Principles of Operations Research*, 2nd ed. (Englewood Cliffs, N.J.: Prentice-Hall, Inc., 1975), Chap. 14.

"constant returns to scale." In the example, two units of both resource one and resource two are required to manufacture one unit of product one. This requirement remains regardless of the number of units of product one manufactured. The assumption of linearity, therefore, implies that the productivity of the resource is independent of the number of units of the product manufactured; i.e., the returns from the manufacturing process are constant per unit of product regardless of the scale of production.

An additional restriction in the linear programming model comes from the assumption of continuity. The solution to the linear programming problem occurs at an extreme point. Since the equations that define extreme points are continuous, the solutions need not have integer values. This implies that fractional units of products are possible. For instance, the optimal solution to the problem of scheduling production in the machine shop was $x_1 = 20$ and $x_2 = 26.67$.

The restrictions on the linear programming model due to the assumptions of linearity and continuity are not as severe as one might expect. Profit (or cost) and resource usage are often described satisfactorily by linear relationships, especially in the relevant range of production. Fractional solutions may present no problem. For cases in which fractional solutions are impossible, judicial rounding sometimes provides a practical suboptimal solution. If rounding is unsatisfactory, *integer programming* can be employed.[5]

11.3.2 CONVEX SETS

The solution set or feasible region in a linear programming problem is sometimes referred to as a *convex set*. By definition, a set is *convex* if each point on a straight line that joins two points in the set is also in the set. To illustrate this definition, consider the two sets in Fig. 11.7. The set in Fig. 11.7(a) represents a typical solution set for a two-variable linear programming problem and is convex. It is impossible to select two points in this set such that all points on a straight line that joins the two points are not also in the set. This convex solution set was formed by the intersection of a system of linear inequalities.

The set in Fig. 11.7(b) does not have the property of convexity. This is shown by the fact that points on the line connecting the points (1, 4) and (3, 3) are not in the set.

It can be seen that the set in Fig. 11.7(b) is not formed by the intersection of a system of linear inequalities. If the lines that form the boundaries of the set are extended, we see that there is some area on both sides of at least one

[5] A good discussion of integer programming is provided by Donald R. Plane and Claude McMillan, Jr., *Discrete Optimization* (Englewood Cliffs, N.J.: Prentice-Hall, Inc., 1971).

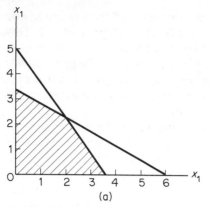 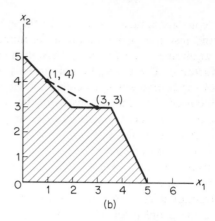

Figure 11.7.

line. This means, of course, that at least one of the inequalities that forms the set in Fig. 11.7(b) is not satisfied.

Since the intersection of a system of linear inequalities defines the solution set for a linear programming problem and the set of points generated by this intersection has the property of convexity, one may ask the reason for introducing the concept of convex sets. The answer, aside from the theoretical importance of convexity, is that the analyst need be aware that the inequalities in a linear programming problem must be satisfied for all possible values of the variables rather than for only a selected subset of values.

11.3.3 NONFEASIBLE, MULTIPLE OPTIMAL, AND UNBOUNDED SOLUTIONS

The linear programming problems discussed in the preceding sections have had a single optimal solution. Although problems of this type are common, other possibilities exist. These include (1) no feasible solution; (2) multiple optimal solutions; and (3) unbounded optimal solutions. These are the exceptions referred to in the footnote on p. 343 to the extreme point theorem.

No Feasible Solution. The solution set for a linear programming problem is formed by the intersection of the system of constraining inequalities. If this intersection is empty, no feasible solution exists for the problem. As an example, assume that the machine shop problem is modified by adding the constraint, $x_1 + x_2 \geq 50$. The problem becomes

$$\text{Maximize:} \quad P = 4x_1 + 5x_2$$

$$\text{Subject to:} \quad
\begin{aligned}
2x_1 + 3x_2 &\leq 120 \\
2x_1 + 1.5x_2 &\leq 80 \\
x_1 + x_2 &\geq 50 \\
x_1, x_2 &\geq 0
\end{aligned}$$

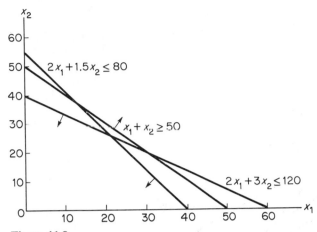

Figure 11.8.

The constraints are shown in Fig. 11.8. It can be seen from this figure that there is no set of points that simultaneously satisfies all the inequalities. Consequently, there is no feasible solution to the linear programming problem.

Multiple Optimal Solutions. We have stressed that an optimal solution occurs at an extreme point of the feasible region. If the objective function is parallel to one of the constraints, however, there will be multiple optimal solutions rather than a unique optimal solution. To illustrate, assume that the objective function for the machine shop problem is changed to $P = 4x_1 + 6x_2$. The problem becomes

$$\text{Maximize:} \quad P = 4x_1 + 6x_2$$

$$\text{Subject to:} \quad 2x_1 + 3x_2 \leq 120$$
$$2x_1 + 1.5x_2 \leq 80$$
$$x_1, x_2 \geq 0$$

The constraints and the objective function are shown in Fig. 11.9.

The figure illustrates that the objective function is parallel to the constraint $2x_1 + 3x_2 \leq 120$. Increases in profit are shown by moving the objective function upward and to the right. Profit is a maximum when the objective function and the constraint are coincident. All ordered pairs described by the constraint between the extreme points (0, 40) and (20, 26.67) are optimal solutions. Profit is $240 at either extreme point or at any linear combination of these extreme points.

Unbounded Optimal Solution. In problems with multiple optimal solutions, the objective function has the same value for all optimal solutions. In

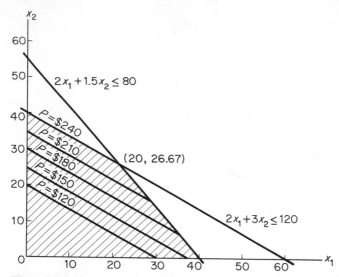

Figure 11.9.

the preceding example this value was $P = \$240$. Based upon the constraints, the upper *bound* on the objective function was $\$240$.

Certain linear programming problems do not have a bound on the value of the objective function. In these cases the objective function is *unbounded*. To illustrate, consider the problem

$$\text{Maximize:} \quad P = \quad x_1 + 2x_2$$
$$\text{Subject to:} \quad -x_1 + \quad x_2 \leq 2$$
$$x_1 + \quad x_2 \geq 4$$
$$x_1, x_2 \geq 0$$

The problem is shown in Fig. 11.10.

The feasible region for this problem is convex. The extreme points of the feasible region are $(4, 0)$ and $(1, 3)$. Neither of these extreme points, however, provides an optimal solution to the problem. This is shown in Fig. 11.10 by the fact that the objective function can increase indefinitely without reaching an upper bound. In fact, given any value of the objective function, there is always a solution that gives a greater value of the objective function. Consequently, the objective function for this problem is unbounded.

These examples illustrate the exception given in the footnote on p. 343 to the extreme point theorem of linear programming. The theorem stated that the optimal value of the objective function occurs at an extreme point of the feasible region. The footnote added that the theorem is true provided that a unique finite optimal solution exists. In the first example, the solution set was empty. There was no feasible solution for the problem. The second example

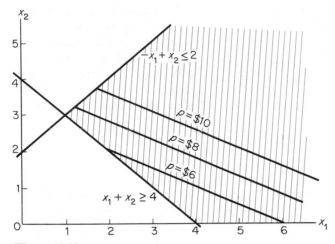

Figure 11.10.

illustrated the case of multiple optimal solutions. Instead of a unique optimal solution, there were an infinite number of optimal solutions, each having the same value. In the final example, the linear programming problem had no finite optimal solution. For any value of the objective function, it was possible to find a solution for which the objective function had a greater value. This third and final exception to the extreme point theorem occurred when the objective function was unbounded.

11.4   Applications of Linear Programming[6]

One of the objectives of this chapter is to provide examples of important applications of linear programming. On the basis of these examples, the reader can begin to express problems in the framework of the linear programming model. These examples involve problems of allocation, such as product mix, feed mix, fluid blending, and portfolio selection. The analytical technique needed to solve these problems is presented in Chap. 12.[7]

11.4.1   PRODUCT MIX

In product mix problems, management must determine the quantities of products to be manufactured during a specific time period. To illustrate, consider

---

[6] This section may be omitted without loss of continuity.

[7] Only the more elementary linear programming problems should be solved by using manual calculations. The more complicated problems, including some in this section, should be solved with the aid of a computer.

the case of the Hall Manufacturing Company. Hall manufactures four types of electronic subassemblies for use in aircraft avionic equipment. Each subassembly requires assembly labor, test labor, resistors, and capacitors. The requirements, availability of parts and labor, and the profit per subassembly are projected for the month of September in Table 11.4. Determine the quantities of the four products that should be manufactured during September in order to maximize profit. Assume that Hall is contractually required to supply 40 type 1 subassemblies.

Table 11.4.

Resource	Electronic Subassembly				Resource Availabilities
	Type 1	Type 2	Type 3	Type 4	
Assembly labor	6.0	5.0	3.5	4.0	600 hours
Test labor	1.0	1.5	1.2	1.2	120 hours
Resistors	4	3	3	3	400 units
Capacitors	2	2	2	3	300 units
Profit	$4.25	$6.25	$5.00	$4.50	

To formulate the linear programming problem, we let x_j, for $j = 1, 2, 3, 4$, represent the quantity of product j manufactured. The problem can be expressed as

Maximize: $P = 4.25x_1 + 6.25x_2 + 5.00x_3 + 4.50x_4$

Subject to:
$$6.0x_1 + 5.0x_2 + 3.5x_3 + 4.0x_4 \leq 600$$
$$1.0x_1 + 1.5x_2 + 1.2x_3 + 1.2x_4 \leq 120$$
$$4x_1 + 3x_2 + 3x_3 + 3x_4 \leq 400$$
$$2x_1 + 2x_2 + 2x_3 + 3x_4 \leq 300$$
$$x_1 \geq 40$$
$$x_2, x_3, x_4 \geq 0$$

Because of the number of variables, the problem cannot be solved graphically. The solution is instead obtained by using the procedure discussed in Chap. 12. The optimal solution is $x_1 = 76.9$ type 1 subassemblies, $x_2 = 20.5$ type 2 subassemblies, $x_3 = 10.3$ type 3 subassemblies, and $x_4 = 0$ type 4 subassemblies. Profit for the month is $506.40.

The following two examples provide additional illustrations of formulations of the product mix problem.

Example. A manufacturer can produce three different products during the month of October. Each of these products requires casting, grinding, assembly, and testing. The maximum available hours of capacity of each of the processes during the month, the requirements for each unit of product, and the profit per unit of product are given in Table 11.5. Formulate the linear programming problem. Assume that all products produced during the month can be sold.

Table 11.5.

	Product			Available Resource Hours
	A	B	C	
Casting	1.0	1.5	2.0	200
Grinding	0.6	1.0	0.8	120
Assembly	1.5	2.0	1.4	240
Testing	0.3	0.2	0.2	40
Profit	$12.00	$18.00	$16.00	

Let x_j, for $j = 1, 2, 3$, represent quantities of products A, B, and C. The linear programming problem is

$$\text{Maximize:} \quad P = 12x_1 + 18x_2 + 16x_3$$

$$\text{Subject to:} \quad 1.0x_1 + 1.5x_2 + 2.0x_3 \leq 200$$
$$0.6x_1 + 1.0x_2 + 0.8x_3 \leq 120$$
$$1.5x_1 + 2.0x_2 + 1.4x_3 \leq 240$$
$$0.3x_1 + 0.2x_2 + 0.2x_3 \leq 40$$
$$x_j \geq 0 \quad \text{for } j = 1, 2, 3$$

Example. McGraw Chemical Company uses nitrates, phosphates, potash, and an inert filler material in the manufacture of chemical fertilizers. The firm mixes these ingredients to make three basic fertilizers: 5-10-5, 5-8-8, and 8-12-12 (numbers represent percent by weight of nitrates, phosphates, and potash in each ton of fertilizer). McGraw receives $70, $75, and $80, respectively, for each ton of the 5-10-5, 5-8-8, and 8-12-12 fertilizer. The cost of the ingredients is $200 per ton of nitrates, $60 per ton of phosphates, $100 per ton of potash, and $20 per ton of the inert filler material. The direct costs of mixing, packing, and selling the fertilizer are $25 per ton.

McGraw has 1200 tons of nitrates, 2000 tons of phosphates, 1500 tons of potash, and an unlimited supply of the inert filler on hand. It will not receive additional chemicals until next month. McGraw has a delivery contract for 8000 tons of 5-8-8 during the month. It believes that it can sell or store at negligible cost all fertilizer produced during the month. Formulate the linear programming problem to maximize profits.

Let x_j, for $j = 1, 2, 3$, represent tons of 5-10-5, 5-8-8, and 8-12-12 fertilizers. The coefficients of the objective function are calculated by subtracting the cost of the ingredients and the $25 mixing cost from the sales price. To illustrate, consider the objective function coefficient of x_1 (i.e., 5-10-5 fertilizer). The cost of the nitrates in one ton of 5-10-5 fertilizer is 0.05($200) or $10. The cost of the phosphates is 0.10($60) or $6. Similarly, the cost of the potash is 0.05($100) = $5, and the cost of the inert filler is 0.80($20) = $16. Subtracting these costs plus the cost of mixing, packing, and selling the 5-10-5 fertilizer from the sales price gives an objective function coefficient of

$$c_1 = \$70 - 0.05(\$200) - 0.10(\$60) - 0.05(\$100) - 0.80(\$20) - \$25 = \$8.00$$

The remaining objective function coefficients arc calculated in the same manner.

The constraints shown below describe the resource limitations and the contractual requirements. The linear programming problem is

$$\text{Maximize:} \quad P = 8.00x_1 + 11.40x_2 + 6.20x_3$$

$$\begin{array}{llll}
\text{Subject to:} & 0.05x_1 + & 0.05x_2 + 0.08x_3 \leq 1200 \\
& 0.10x_1 + & 0.08x_2 + 0.12x_3 \leq 2000 \\
& 0.05x_1 + & 0.08x_2 + 0.12x_3 \leq 1500 \\
& & x_2 \geq 8000 \\
& & x_1, x_3 \geq 0
\end{array}$$

11.4.2   FEED MIX

The feed mix problem is another form of the allocation problem. This problem involves determining the proportion of several ingredients that, when mixed, satisfy certain criteria and give a minimum cost. To illustrate, consider the problem of mixing feed for a dairy herd. The dairyman may purchase and mix quantities of three kinds of grain, each containing differing amounts of four nutritional elements. The dairyman has developed nutritional requirements for the feed mixture. His objective is to minimize the cost of the mixture while meeting the nutritional requirements. The data are given in Table 11.6.

Let x_j, for $j = 1, 2, 3$, represent pounds of grain j in the feed mix. The linear programming problem to minimize cost is

$$\text{Minimize:} \quad C = 0.025x_1 + 0.020x_2 + 0.017x_3$$

$$\begin{array}{llll}
\text{Subject to:} & 4x_1 + 6x_2 + & 8x_3 \geq 2500 \\
& 2x_1 + 3x_2 + & 1x_3 \geq 750 \\
& 4x_1 + 3x_2 + & 1x_3 \geq 900 \\
& 0.7x_1 + 0.3x_2 + 0.8x_3 \geq 250 \\
& & x_2 \leq 1000 \\
& & x_1, x_2, x_3 \geq 0
\end{array}$$

Table 11.6.

Nutritional Elements	Units of Nutritional Elements Contained in One Pound of			Minimum Number of Units Required
	Grain 1	Grain 2	Grain 3	
1	4	6	8	2500
2	2	3	1	750
3	4	3	1	900
4	0.7	0.3	0.8	250
Cost per pound	$0.025	$0.020	$0.017	
Available	Unlimited	1000 pounds	Unlimited	

The optimal solution is found by using the simplex algorithm. The solution is $x_1 = 75.0$ pounds of grain 1, $x_2 = 134.4$ pounds of grain 2, and $x_3 = 196.4$ pounds of grain 3. The cost of this mix is $7.90.

Example. American Foods, Inc., is developing a low-calorie, high-protein diet supplement called Hi-Pro. The specifications for Hi-Pro have been established by a panel of medical experts. These specifications, along with the calorie, protein, and vitamin content of three basic foods, are given in Table 11.7. Formulate the linear programming model to minimize cost.

Table 11.7.

Nutritional Elements	Units of Nutritional Elements per 8-Ounce Serving of Basic Foods			
	Basic Foods			Hi-Pro Specifications
	No. 1	No. 2	No. 3	
Calories	350	250	200	≤ 300
Protein	250	300	150	≥ 200
Vitamin A	100	150	75	≥ 100
Vitamin C	75	125	150	≥ 100
Cost per serving	$0.15	$0.20	$0.12	

Let x_j, for $j = 1, 2, 3$, represent the proportion of basic foods 1, 2, and 3 in an 8-ounce serving of Hi-Pro. The linear programming problem is

$$\text{Minimize:} \quad C = 0.15x_1 + 0.20x_2 + 0.12x_3$$

$$\text{Subject to:} \quad
\begin{aligned}
350x_1 + 250x_2 + 200x_3 &\leq 300 \\
250x_1 + 300x_2 + 150x_3 &\geq 200 \\
100x_1 + 150x_2 + 75x_3 &\geq 100 \\
75x_1 + 125x_2 + 150x_3 &\geq 100 \\
x_1 + x_2 + x_3 &= 1 \\
x_j \geq 0 \quad \text{for } j &= 1, 2, 3
\end{aligned}$$

The first four constraints describe the Hi-Pro specifications. The fifth constraint is necessary to ensure that the proportions total 1.

Example. Chico Candy Company, Inc., mixes three kinds of candies to obtain a one-pound box of candy. The box of candy sells for $0.85 per pound, and the three-ingredient candies cost $1.00, $0.50, and $0.25 per pound, respectively. The mixture must contain at least 0.3 pounds of the first kind of candy, and the weight of the first two candies must at least equal the weight of the third. Formulate the linear programming problem to maximize profit.

Let x_j, for $j = 1, 2, 3$, represent the quantity in pounds of each of the three-ingredient candies. The objective function coefficients are found by subtracting the cost of each kind of candy from the sales price of the box. For instance, the coefficient of x_1 is $0.85 - $1.00 = -$0.15$. The linear programming problem is

$$\text{Maximize:} \quad P = -0.15x_1 + 0.35x_2 + 0.60x_3$$

$$\text{Subject to:} \quad
\begin{aligned}
x_1 \qquad\qquad &\geq 0.3 \\
x_1 + x_2 - x_3 &\geq 0 \\
x_1 + x_2 + x_3 &= 1 \\
x_2, x_3 &\geq 0
\end{aligned}$$

11.4.3 FLUID BLENDING

Another variation of the allocation problem occurs when fluids, such as chemicals, plastics, molten metals, and oils, must be blended to form a product. Each of the resources included in the blend has certain properties and costs. The fluid blending problem involves blending the different resources to form a final product that meets certain criteria and has a minimum cost or a maximum profit. To illustrate, consider a simplified example of gasoline blending.

The Sigma Oil Company markets three brands of gasoline: Sigma Supreme, Sigma Plus, and Sigma. The three gasolines are made by blending two grades of gasoline, each with a different octane rating, and a special high-octane lead additive. In the gasoline blends, the octane rating of the blend is a linear combination of the octane ratings of the component gasolines and additives. The relevant data are shown in Table 11.8.

Table 11.8.

Brands	Minimum Octane	Sales Price to Retailer (per Gallon)
Sigma Supreme	100	$0.20
Sigma Plus	94	0.16
Sigma	86	0.14

Blending Component	Octane	Cost (per Gallon)	Supply (Gallons)
1	110	$0.10	20,000
2	80	0.07	12,000
3	1500	1.00	500

It is assumed that all gasoline blended during the period can be sold or stored at negligible cost. An alternative assumption, not included in this problem, is that forecasts of demand for the three brands of gasoline are available and that gasoline unsold at the end of the period is stored at a cost and sold during the following period.

To formulate the linear programming model, let x_{ij}, for $i = 1, 2, 3$ and $j = 1, 2, 3$, represent the gallons of blending component i used in gasoline blend j. The problem can be expressed as

Maximize: $P = 0.10x_{11} + 0.13x_{21} - 0.80x_{31} + 0.06x_{12} + 0.09x_{22}$
$$- 0.84x_{32} + 0.04x_{13} + 0.07x_{23} - 0.86x_{33}$$

Subject to:
$$110x_{11} + 80x_{21} + 1500x_{31} \geq 100(x_{11} + x_{21} + x_{31})$$
$$110x_{12} + 80x_{22} + 1500x_{32} \geq 94(x_{12} + x_{22} + x_{32})$$
$$110x_{13} + 80x_{23} + 1500x_{33} \geq 86(x_{13} + x_{23} + x_{33})$$
$$x_{11} + x_{12} + x_{13} \leq 20{,}000$$
$$x_{21} + x_{22} + x_{23} \leq 12{,}000$$
$$x_{31} + x_{32} + x_{33} \leq 500$$
$$x_{ij} \geq 0 \quad \text{for } i = 1, 2, 3 \text{ and } j = 1, 2, 3$$

The objective function shows that there is a $0.10 profit for each gallon of blending component 1 used in Sigma Supreme, a $0.13 profit for each gallon of blending component 2 used in Sigma Supreme, an $0.80 loss for each gallon of blending component 3 used in Sigma Supreme, etc. These numbers are calculated by subtracting the cost of the blending component from the sales price of the gasoline. For instance, each gallon of blending component 1 used in Sigma Supreme costs $0.10 and is sold for $0.20, giving a profit of $0.10.

The first three constraints specify that the blends must have minimum octane ratings. The next three constraints restrict the quantity of each component used for blending to the supply available during the period. The solution to the problem is given in Table 11.9. Profit for the period is $3238.

Table 11.9. Gallons of Blend i Used in Gasoline j

Blend	Supreme	Plus	Sigma	Total
1	20,000	0	0	20,000
2	12,000	0	0	12,000
3	403	0	0	403
Total	32,403	0	0	32,403

Example. American whiskeys are made by blending different kinds of bourbon whiskeys. Assume that Kentucky Whiskeys, Inc. mixes three bourbon whiskeys, A, B, and C, to obtain two blends, Kentucky Premium and Kentucky Smooth. The recipes used by Kentucky for their two blends are given in Table 11.10.

Table 11.10.

Blend	Ingredients	Price (*per Gallon*)
Kentucky Premium	Not less than 60% A Not more than 20% C	$7.00
Kentucky Smooth	Not less than 30% A Not more than 60% C	$5.50

Table 11.11.

Bourbon	Quantity Available (*Gallons*)	Cost (*per Gallon*)
A	1500	$6.75
B	2000	$6.00
C	1000	$4.00

Supplies of the basic bourbon whiskeys and their costs are shown in Table 11.11. Using this data, formulate the linear programming problem to maximize profits.

Let x_{ij}, for $i = 1, 2, 3$ and $j = 1, 2$ represent the quantity of bourbon i used in one gallon of blend j. The linear programming problem is

$$\text{Maximize:} \quad P = 0.25x_{11} - 1.25x_{12} + 1.00x_{21} - 0.50x_{22}$$
$$+ 3.00x_{31} + 1.50x_{32}$$

$$\text{Subject to:} \quad x_{11} \geq 0.60(x_{11} + x_{21} + x_{31})$$
$$x_{31} \leq 0.20(x_{11} + x_{21} + x_{31})$$
$$x_{12} \geq 0.30(x_{12} + x_{22} + x_{32})$$
$$x_{32} \leq 0.60(x_{12} + x_{22} + x_{32})$$
$$x_{11} + x_{12} \leq 1500$$
$$x_{21} + x_{22} \leq 2000$$
$$x_{31} + x_{32} \leq 1000$$
$$x_{ij} \geq 0$$

The objective function shows that there is a \$0.25 profit on each gallon of bourbon A used in Kentucky Premium. This is determined by subtracting the cost per gallon of bourbon A from the price per gallon of Kentucky Premium. Similarly, there is a loss of \$1.25 per gallon for each gallon of bourbon A used in Kentucky Smooth.

The first four constraints specify the maximum and minimum quantities of bourbons used in the blended whiskeys. For example, the first constraint shows that the proportion of blend A in Kentucky Premium must be not less than 60 percent. Similarly, the second constraint shows that the proportion of blend C in Kentucky Premium must not exceed 20 percent. The last three constraints limit the supplies of the bourbon whiskeys available for blending during the period.

Example. Plains Coffee Company mixes Brazilian, Colombian, and Mexican coffees to make two brands of coffee, Plains X and Plains XX. The characteristics used in blending the coffees include strength, acidity, and caffeine. Test results of the available supplies of Brazilian, Colombian, and Mexican coffees are shown in Table 11.12.

Table 11.12.

Imported Coffee	Price per Pound	Strength Index	Acidity Index	Percent Caffeine	Supply Available
Brazilian	\$0.30	6	4.0	2.0	40,000 pounds
Colombian	0.40	8	3.0	2.5	20,000 pounds
Mexican	0.35	5	3.5	1.5	15,000 pounds

The requirements for Plains X and Plains XX coffees are given in Table 11.13. Assume that 35,000 pounds of Plains X and 25,000 pounds of Plains XX are to be sold. Formulate the linear programming problem to maximize profits.

Table 11.13.

Plains Coffee	Price per Pound	Minimum Strength	Maximum Acidity	Maximum Percent Caffeine	Quantity Demanded
X	$0.45	6.5	3.8	2.2	35,000 pounds
XX	0.55	6.0	3.5	2.0	25,000 pounds

Let x_{ij}, for $i = 1, 2, 3$ and $j = 1, 2$ represent the pounds of imported coffees used in Plains coffees. The linear programming problem is

Maximize: $P = 0.15x_{11} + 0.25x_{12} + 0.05x_{21} + 0.15x_{22}$
$+ 0.10x_{31} + 0.20x_{32}$

Subject to:

$$6x_{11} + \quad 8x_{21} + \quad 5x_{31} \geq 6.5(x_{11} + x_{21} + x_{31})$$
$$6x_{12} + \quad 8x_{22} + \quad 5x_{32} \geq 6.0(x_{12} + x_{22} + x_{32})$$
$$4.0x_{11} + \quad 3.0x_{21} + \quad 3.5x_{31} \leq 3.8(x_{11} + x_{21} + x_{31})$$
$$4.0x_{12} + \quad 3.0x_{22} + \quad 3.5x_{32} \leq 3.5(x_{12} + x_{22} + x_{32})$$
$$2.0x_{11} + \quad 2.5x_{21} + \quad 1.5x_{31} \leq 2.2(x_{11} + x_{21} + x_{31})$$
$$2.0x_{12} + \quad 2.5x_{22} + \quad 1.5x_{32} \leq 2.0(x_{12} + x_{22} + x_{32})$$
$$x_{11} + x_{12} \leq 40,000$$
$$x_{21} + x_{22} \leq 20,000$$
$$x_{31} + x_{32} \leq 15,000$$
$$x_{11} + x_{21} + x_{31} = 35,000$$
$$x_{12} + x_{22} + x_{32} = 25,000$$
$$x_{ij} \geq 0$$

The coefficients of the objective function represent profits from imported coffee i in mix j. The first six constraints establish restrictions on strength, acidity, and caffeine. Limitations on the supply of imported coffees are given by the next three constraints and the quantity of Plains X and Plains XX coffees demanded during the period is described by the final two constraints.

11.4.4   PORTFOLIO AND MEDIA SELECTION

Another example of the allocation problem occurs in portfolio selection. Various financial institutions such as pension funds, insurance companies, and mutual funds, along with private individuals, are frequently faced with the problem of investing funds among the alternatives available. These alternatives vary in terms of safety, liquidity, income, growth, and other factors. Linear programming can be used to suggest the allocation of funds among the competing alternatives.

To illustrate the use of linear programming in portfolio selection, consider the case of the ABC Mutual Fund. ABC is an income-oriented mutual fund. To aid in its investment decisions, ABC has developed the investment alternatives given in Table 11.14. The return on investment is expressed as an annual rate of return on the invested capital. The risk is a subjective estimate on a scale from 0 to 10 made by the portfolio manager of the safety of the investment. The term of investment is the average length of time required to realize the return on investment indicated in Table 11.14.

ABC's objective is to maximize the return on investment. The guidelines for selecting the portfolio are (1) the average risk should not exceed 2.5; (2) the average term of investment should not exceed 6 years; and (3) at least 15 percent of the funds should be retained in the form of cash. To maximize the return on investment, let x_j, for $j = 1, 2, 3, 4, 5$ represent the proportion of funds to be invested in the jth alternative. The objective function and constraints are

$$\text{Maximize:} \quad P = 12x_1 + 10x_2 + 15x_3 + 25x_4 + 0x_5$$

$$
\begin{array}{llllll}
\text{Subject to:} & x_1 + & x_2 + & x_3 + & x_4 + & x_5 = 1 \\
& 2x_1 + & 1x_2 + & 3x_3 + & 4x_4 & \leq 2.5 \\
& 4x_1 + & 8x_2 + & 2x_3 + & 10x_4 & \leq 6 \\
& & & & & x_5 \geq 0.15
\end{array}
$$

$$x_j \geq 0 \quad \text{for } j = 1, 2, 3, 4, 5$$

The first constraint states that the proportions of funds invested in the various alternatives must total 1. The remainder of the constraints describe the risk, term of investment, and cash requirements. The solution to the problem is $x_1 = 40.7$ percent, $x_2 = 2.85$ percent, $x_3 = 0$, $x_4 = 41.4$ percent, and $x_5 = 15.0$ percent. The overall rate of return on the portfolio is 15.5 percent.

Table 11.14.

Investment Alternative	Yearly Return on Investment	Risk	Term of Investment (Years)
1. Blue chip stock	12%	2	4
2. Bonds	10	1	8
3. Growth stock	15	3	2
4. Speculation	25	4	10
5. Cash	0	0	0

The portfolio selection model is similar to the media selection model. Media selection involves the allocation of the advertising budget of a firm among various communication media in an effort to reach appropriate audiences. The objective is to maximize total effective exposures, subject to constraints on the advertising budget and the number of exposures per media.

To illustrate a simplified media selection model, consider the case of the Edwards Advertising Agency. Edwards is formulating an advertising campaign for one of its customers. A careful analysis of the available media has narrowed the allocation problem to that of determining the number of messages to appear in each of three magazines. The cost of advertisements and other relevant data are given in Table 11.15.

Table 11.15.

	Media		
	1	*2*	*3*
Cost per advertisement	$500	$750	$1,000
Maximum number of ads	24	52	24
Minimum number of ads	6	12	0
Reader characteristics			
Age: 21–40	50%	60%	65%
Income: $10,000 or more	40%	50%	80%
Education: college	40%	40%	60%
Audience size	500,000	800,000	1,200,000

The cost per advertisement, characteristics of the media, and the audience size are available from industry sources. Edwards has established the maximum and minimum number of advertisements per medium and the relative importance of the characteristics. The relative importance of the characteristics are: age, 0.5; income, 0.3; education, 0.2.

The linear programming problem is to maximize total effective exposures. The coefficients of the objective function are the product of the audience size multiplied by the "effectiveness coefficient" for each medium. The effectiveness coefficient is defined as equaling the sum of the products of the characteristics and their relative importance. The effectiveness coefficient for media 1 is

$$\text{effectiveness coefficient} = 0.50(0.5) + 0.40(0.3) + 0.40(0.2) = 0.450$$

Similarly, the effectiveness coefficients for media 2 and 3 are 0.530 and 0.685, respectively. The product of the effectiveness coefficients and the audience size gives the coefficients of the objective function. To illustrate, the coefficient of the objective function for media 1 is 0.450(500,000) or 225,000.

If it is assumed that an advertising budget of $35,000 has been established, the linear problem to maximize total effective exposure is

Maximize: $E = 225{,}000x_1 + 424{,}000x_2 + 822{,}000x_3$

Subject to:

$$500x_1 + 750x_2 + 1000x_3 \le 35{,}000$$
$$x_1 \ge 6$$
$$x_1 \le 24$$
$$x_2 \ge 12$$
$$x_2 \le 52$$
$$x_3 \le 24$$
$$x_3 \ge 0$$

where x_j, for $j = 1, 2, 3$, represents the number of advertisements in the jth media.

The solution to the problem is $x_1 = 6$, $x_2 = 12$, and $x_3 = 23$. The total number of effective exposures is 25,344,000.

PROBLEMS

1. Determine the solutions to the following linear programming problems by graphing the linear inequalities and evaluating the objective function at each extreme point of the feasible region:

(a)   Maximize: $Z = 5x_1 + 6x_2$

Subject to:
$$2x_1 + 3x_2 \le 30$$
$$2x_1 + x_2 \le 18$$
$$x_1, x_2 \ge 0$$

(b)   Maximize: $Z = 3.0x_1 + 1.0x_2$

Subject to:
$$6.0x_1 + 4.0x_2 \le 48$$
$$3.0x_1 + 6.0x_2 \le 42$$
$$x_1, x_2 \ge 0$$

(c)   Minimize: $C = 10x_1 + 8x_2$

Subject to:
$$6x_1 + 2x_2 \ge 12$$
$$2x_1 + 2x_2 \ge 8$$
$$4x_1 + 12x_2 \ge 24$$
$$x_1, x_2 \ge 0$$

(d)   Minimize: $C = 3x_1 + 4x_2$

Subject to:
$$x_1 + x_2 \ge 8$$
$$x_1 + 2x_2 \ge 12$$
$$x_2 \ge 2$$
$$x_1 \ge 0$$

2. Solve the following linear programming problems:

(a) Maximize: $Z = 30x_1 + 20x_2$

Subject to:
$$5x_1 + 2x_2 \leq 180$$
$$3x_1 + 3x_2 \leq 135$$
$$x_1, x_2 \geq 0$$

(b) Maximize: $Z = 3x_1 + 2x_2$

Subject to:
$$12x_1 + 4x_2 \leq 60$$
$$4x_1 + 8x_2 \leq 40$$
$$x_1, x_2 \geq 0$$

(c) Minimize: $C = 6x_1 + 4x_2$

Subject to:
$$5x_1 + 10x_2 \geq 120$$
$$25x_1 + 5x_2 \geq 150$$
$$x_1, x_2 \geq 0$$

(d) Minimize: $C = 2x_1 + 3x_2$

Subject to:
$$6x_1 + 2x_2 \geq 18$$
$$2x_1 + 4x_2 \geq 16$$
$$x_1, x_2 \geq 0$$

3. Solve the following linear programming problems:

(a) Maximize: $Z = 3x_1 + 5x_2$

Subject to:
$$13x_1 + 9x_2 \leq 234$$
$$9x_1 + 15x_2 \leq 270$$
$$x_1, x_2 \geq 0$$

(b) Maximize: $Z = 2x_1 + 5x_2$

Subject to:
$$4x_1 + 3x_2 \leq 24$$
$$3x_1 + 5x_2 \leq 30$$
$$x_1 + x_2 \geq 8$$
$$x_1, x_2 \geq 0$$

(c) Minimize: $C = x_1 + x_2$

Subject to:
$$4x_1 + 3x_2 \geq 12$$
$$-3x_1 + 4x_2 \leq 12$$
$$x_1, x_2 \geq 0$$

(d) Maximize: $Z = x_1 + x_2$

Subject to:
$$8x_1 + 5x_2 \geq 40$$
$$-4x_1 + 6x_2 \geq 24$$
$$x_1, x_2 \geq 0$$

4. Piper Farms plans to introduce two new gift packages of fruit for the Christmas market. Box A will contain 20 apples and 15 pears. Box B will contain 40 apples and 20 pears. Piper has 18,000 apples and 12,000 pears available for packaging. They believe that all fruit packaged can be sold. Profits are estimated as $0.60 for A and $1.00 for B. Determine the number of boxes of A and B that should be prepared to maximize profits.

5. Radio Manufacturing Company must determine production quantities for this month for two different models, A and B. Data per unit are given in the following table:

Model	Revenue	Subassembly Time (hours)	Final Assembly Time (hours)	Quality Inspection (hours)
A	$10	1.0	0.8	0.5
B	$20	1.2	2.0	0

The maximum time available for these products is 1200 hours for subassembly, 1600 hours for final assembly, and 500 hours for quality inspection. Orders outstanding require that at least 200 units of A and 100 units of B be produced. Determine the quantities of A and B that maximize total revenue.

6. A baseball manufacturer makes two kinds of baseballs, Major League and Minor League. He has four manufacturing processes that are used in making each baseball. These processes are designated as J, K, L, and M. Available time and time required per baseball for each process are listed below.

Process	Major League (time/baseball)	Minor League (time/baseball)	Total Time Available
J	2	3	6000
K	8	5	20,000
L	1	$3\frac{1}{3}$	5000
M	3	1	6000

The contribution to profit is $2 for Major and $1 for Minor League baseballs. Determine the product mix that leads to maximum profit.

7. A company makes two types of leather wallets. Type A is a high-quality wallet and type B is a medium-quality wallet. The profits on the two wallets are $0.80 for type A and $0.60 for type B. The type A wallet requires twice as much time to manufacture as the type B. If all wallets were type B, the company could make 1000 per day. The supply of leather

is sufficient for a maximum of 800 wallets per day (both A and B combined). A special process further limits production to a maximum of 450 wallets of type A and 700 wallets of type B. Assuming that all wallets manufactured can be sold, determine the number of each type to maximize profits.

8. Kawar Corporation produces two types of tract homes, the Tempo model and the Trend model. Each type is built on the same size lot and the tract will allow for 120 homes. The building materials used for both types of home are the same except for the framing materials. The Tempo model uses lumber for framing and the Trend model uses aluminum. Sixty thousand board feet of lumber and 72,000 running feet of standard aluminum framing are available during the construction period. Because of its architectural styling, the Tempo model requires 600 board feet of lumber and 3000 man-hours, while the Trend model requires 800 running feet of standard aluminum frame and 6000 man-hours. Total labor available during the construction period is 480,000 man-hours. The profit margin is $5000 for each Tempo model and $7000 for each Trend model. Determine the product mix for maximum profit.

9. A poultry farmer must supplement the vitamins in the feed he buys. He is considering two supplements, each of which contains the three vitamins required but in differing amounts. He must meet or exceed the minimum vitamin requirements. The vitamin content per ounce of the supplements is given in the following table:

Vitamin	Supplement 1	Supplement 2
1	6 units	12 units
2	20 units	10 units
3	10 units	10 units

Supplement 1 costs $0.03 per ounce and supplement 2 costs $0.04 per ounce. The feed must contain at least 48 units of vitamin 1, 80 units of vitamin 2, and 60 units of vitamin 3. Determine the mixture that has the minimum cost.

10. The Whoop-Bang Novelty Co. makes three basic types of noisemakers: Toot, Wheet, and Honk. A Toot can be made in 30 minutes and has a feather attached to it. A Wheet requires 20 minutes, has two feathers, and is sprinkled with 0.5 ounce of sequin powder. The Honk requires 30 minutes, has three feathers, and 1 ounce of sequin powder. The net profit is $0.45 per Toot, $0.55 per Wheet, and $1.00 per Honk. The following resources are available: 80 hours of labor, 90 ounces of sequin powder, and 360 feathers. Set up the linear programming problem to maximize profits.

11. A company manufactures three products: A, B, and C. The products require four operations: grinding, turning, assembling, and testing. The requirements per unit of product in hours for each operation are as follows:

Product	Grinding	Turning	Assembling	Testing
A	0.03	0.11	0.30	0.08
B	0.02	0.14	0.20	0.07
C	0.03	0.20	0.26	0.08
Capacity (hours)	1000	4000	9000	3000

The minimum monthly sales requirements are

A	9000 units
B	9000 units
C	6000 units

The profit per unit sold of each product is $0.15 for A, $0.12 for B, and $0.09 for C. All units produced above the minimum can be sold. What quantities of each product should be produced next month for maximum profit? Set up only.

12. A certain firm has two plants. Orders from four customers have been received. The number of units ordered by each customer and the shipping costs from each plant are shown in the following table:

Customer	Units Ordered	Shipping Cost/Unit	
		From Plant 1	From Plant 2
A	500	$1.50	$4.00
B	300	2.00	3.00
C	1000	3.00	2.50
D	200	3.50	2.00

Each unit of the product must be machined and assembled. These costs, together with the capacities at each plant, are shown below:

	Hours/Unit	Cost/Hour	Hours Available
Plant No. 1:			
Machining	0.10	$4.00	120
Assembling	0.20	3.00	260
Plant No. 2:			
Machining	0.11	4.00	140
Assembling	0.22	3.00	250

Formulate the linear programming problem to minimize cost. Set up only.

13. Crane Feed Company markets two feed mixes for rabbits. The first mix, Fertilex, requires at least twice as much wheat as barley. The second mix, Multiplex, requires at least twice as much barley as wheat. Wheat costs $0.49 per pound, and only 1000 pounds are available this month. Barley costs $0.36 per pound, and 1200 pounds are available. Fertilex sells for $1.59 per pound up to 99 pounds, and each additional pound over 99 sells for $1.43. Multiplex sells at $1.35 per pound up to 99 pounds, and each additional pound over 99 sells for $1.18. Rancho Farms will buy any and all amounts of both mixes Crane Feed Company will mix. Set up the linear programming problem to determine the product mix that results in maximum profits.

14. A farmer must decide how many pounds of each of several types of grain he should purchase in order that his livestock receive the minimum nutrient requirement at the lowest cost possible. The relevant information is as follows:

	One Pound of				Minimum Nutrient Requirement
	Grain 1	Grain 2	Grain 3	Grain 4	
Nutrient 1	4	5	6	3	1000
Nutrient 2	2	1	0	3	850
Nutrient 3	1	2	3	1	700
Nutrient 4	2	3	1	2	1320
Nutrient 5	0	2	1	1	550
Cost/pound	$0.35	$0.42	$0.45	$0.37	

Set up the linear programming problem to determine the mix of grains that minimizes cost.

15. The Franklin Oil Company produces three oils of different viscosities. Franklin Oil makes its products by blending two grades of oil, each with a different viscosity. The final viscosity is linearly proportional to the blending viscosities. There is an unlimited demand for the three oils.

Blending Component	Viscosity	Cost/Quart	Supply
A	10	$0.20	4000 quart/week
B	50	$0.30	2000 quart/week

Brand	Minimum Viscosity	Sales Price/Quart
S	20	$0.43
SS	30	0.48
SSS	40	0.53

Formulate the linear programming problem to maximize profits.

16. Western Oil Company makes three brands of gasoline: Super, Low Lead, and Regular. Western makes its gasolines by blending two grades of gasolines and a high-octane lead additive. Each brand of gasoline must have an octane rating of at least the stated minimum. The brands are made by blending the two grades of gasoline and the high-octane lead additive. The relevant data are given below:

Blending Component	Octane	Cost per Gallon	Supply from Refinery (per week)
A	120	$0.12	18,000 gallons
B	80	0.09	26,000 gallons
L	1200	1.40	Unlimited

Gasoline Brand	Minimum Octane	Sales Price per Gallon
Super	105	$0.18
Low Lead*	94	0.16
Regular	96	0.14

* Low Lead can contain no more than 0.1 percent by volume of blending component L.

Assuming that Western Oil can sell all gasoline produced, set up the linear programming problem to maximize profit.

17. The Hamilton Data Processing Company performs three types of data-processing activities: payrolls, accounts receivable, and inventories. The profit and time requirements for key punch, computation, and off-line printing for a "standard job" are shown in the following table:

Job	Profit per Standard Job	Time Requirements (minutes)		
		Key Punch	Computation	Print
Payroll	$275	1200	20	100
Accounts receivable	125	1400	15	60
Inventory	225	800	35	80

Hamilton guarantees overnight completion of the job. Any job scheduled during the day can be completed during the day or during the night. Any job scheduled during the night, however, must be completed during the night. The capacity for both day and night is shown in the following table:

Capacity (minutes)	Key Punch	Operation Computation	Print
Day	4200	150	400
Night	9200	250	650

Set up the linear programming problem to determine the mixture of "standard jobs" that should be accepted during the day and during the night in order to maximize profit.

18. The Trust Department of the Barclay Bank is preparing an investment portfolio for a wealthy customer. The investment alternatives along with their current yields and risk factors are given in the following table:

Investment	Yield	Risk
AAA corporate bonds	6.5%	1.0
Convertible debentures	7.2	1.5
Selected high-grade common stock	8.3	2.0
Preferred stock	6.3	1.8
Growth stock	9.0	3.2

The risk factor is a subjective estimate on a scale from 0 to 10, the lower numbers representing the lower risk investments. On the assumption that the current yields will continue, the Trust Department wishes to design a portfolio with the maximum yield. However, the average risk must not exceed 2.4. In addition, the customer has specified that the investment in growth stock cannot exceed the combined total investment in corporate bonds and convertible debentures. Set up the linear programming problem to determine the proportion of funds in each investment alternative.

19. G and H Advertising, Inc., is preparing a proposal for an advertising campaign for White Chemical Company, manufacturers of Gro-More Lawn products. The advertising copy has been written and approved. The final step in the proposal, therefore, is to recommend an allocation of advertising funds in order to maximize the total number of effective ex-

posures. The characteristics of three alternative publications are shown in the following table:

	Publication		
	House Beautiful	Home & Garden	Lawn Care
Cost per advertisement	$600	$800	$450
Maximum number of advertisements	12	24	12
Minimum number of advertisements	3	6	2
Characteristics:			
Homeowner	80%	70%	20%
Income: $10,000 or more	70%	80%	60%
Occupation: gardener	15%	20%	40%
Audience size	600,000	800,000	300,000

The relative importance of the three characteristics are: homeowner, 0.4; income, 0.2; gardener, 0.4. The advertising budget is $20,000. Formulate the linear programming problem to determine the most effective number of exposures in each magazine.

20. Fast Eddie is a pool shark. To keep sharp he must play at least 7 hours a day and not more than 12 hours. Of his playing hours, at least 4 hours must be straight practicing. Eddie's average income from his three games is: snooker, $30 per hour; pocket billiards, $35 per hour; three-rail billiards, $55 per hour.

Eddie has the following problem: For every hour of "hustling" a given game, he must practice the other two games to remain in top hustling condition. The following table shows the relationships:

Game (1 hour)	Minimum Practice Required	
Snooker	$\frac{1}{4}$ hour Pocket billiards	$\frac{1}{6}$ hour Three-rail
Pocket billiards	$\frac{1}{6}$ hour Snooker	$\frac{1}{4}$ hour Three-rail
Three-rail billiards	$\frac{1}{2}$ hour Pocket billiards	$\frac{1}{3}$ hour Snooker

Eddie also feels that he must practice each of the three games at least 1 hour apiece. Set up the linear programming problem to determine the mix of hustling and practicing that will maximize Eddie's profit.

21. Next Monday, Gordon Shipley has a term paper due in Marketing and exams in Quantitative Methods and Accounting. Shipley has at most 20 hours of study time available. Shipley has completed the rough draft of the term paper, but he has not started the final draft. He estimates that he could turn in the rough draft and receive a grade of 60. With 10 hours of effort on the final draft he could get an 80 and with 20 hours of study effort he could get 100. Without studying he could get a 60 on the Quantitative Methods exam. With 4 hours of study, however, he could get an 80 and 8 hours of study should result in a perfect score. Shipley is a little behind in Accounting. Without any study he would score 40. With 5 hours of study he would score 70, and with 10 hours of study he would expect to score 100 on the exam.

The term paper is 50 percent of the Marketing course grade, the Quantitative Methods exam is 25 percent, and the Accounting exam is 20 percent. Taking into account that Accounting and Quantitative Methods are four-unit courses and Marketing is a two-unit course, how should Shipley allocate his time to maximize his overall GPA (grade point average) and still not receive a grade of less than 70 in any subject? Set up the problem only.

SELECTED REFERENCES

BAUMOL, WILLIAM J., *Economic Theory and Operations Analysis*, 2nd ed. (Englewood Cliffs, N.J.: Prentice-Hall, Inc., 1965).

BOULDING, K. E., and W. A. SPIVEY, *Linear Programming and the Theory of the Firm* (New York, N.Y.: The Macmillan Company, Inc., 1960).

CHARNES, A., and W. W. COOPER, *Management Models and Industrial Applications of Linear Programming* (New York, N.Y.: John Wiley and Sons, Inc., 1961).

GARVIN, WALTER W., *Introduction to Linear Programming* (New York, N.Y.: The McGraw-Hill Book Company, Inc., 1960).

GASS, SAUL I., *Linear Programming: Methods and Applications* (New York, N.Y.: The McGraw-Hill Book Company, Inc., 1964).

HADLEY, GEORGE, *Linear Programming* (Reading, Mass.: Addison-Wesley Publishing Company, Inc., 1962).

KIM, CHAIHO, *Introduction to Linear Programming* (New York, N.Y.: Holt, Rinehart, and Winston, Inc., 1971).

LEVIN, R. I., and RUDY P. LAMONE, *Linear Programming for Management Decisions* (Homewood, Ill.: Richard D. Irwin, Inc., 1969).

NAYLOR, T. H., E. T. BRYNE, and J. M. VERNON, *Introduction to Linear Programming: Methods and Cases* (Belmont, Ca.: Wadsworth Publishing Company, Inc., 1971).

SPIVEY, W. A., and R. M. THRALL, *Linear Optimization* (New York, N.Y.: Holt, Rinehart, and Winston, Inc., 1970).

STRUM, JAY E., *Introduction to Linear Programming* (San Francisco, Ca.: Holden-Day, Inc., 1972).

VANDERMEULEN, DANIEL C., *Linear Economic Theory* (Englewood Cliffs, N.J.: Prentice-Hall Inc., 1971).

Chapter Twelve

The Simplex Method

The simplex method is a solution algorithm for solving linear programming problems.[1] The simplex algorithm was developed by George Dantzig in 1947 and made generally available in 1951. The importance of the algorithm is demonstrated by the fact that linear programming has become one of the most widely used quantitative approaches to decision making.

12.1 System of Equations

The simplex algorithm utilizes the *extreme point theorem* of linear programming. This theorem states that *an optimal solution to a linear programming problem occurs at one of the extreme points of the feasible region.* The graphical solution technique was used to show that extreme points are defined by the intersection of systems of equations. The first step in the simplex algorithm, therefore, is to convert the system of linear inequalities to linear equations. These equations are then used to determine the extreme points.

12.1.1 CONVERTING INEQUALITIES TO EQUATIONS

A straightforward technique for converting an inequality to an equation involves adding a *slack variable* to the left side of inequalities of the form $ax_1 + bx_2 \leq c$ and subtracting a *surplus variable* from the left side of

---

[1] The term *algorithm* refers to a systematic procedure or series of rules for solving a problem.

376

inequalities of the form $ax_1 + bx_2 \geq c$. To illustrate this technique, consider the system of inequalities

$$3x_1 + 2x_2 \leq 40$$

$$2x_1 + x_2 \geq 10$$

The first inequality is converted to an equation by adding the slack variable x_3. The resulting equation is

$$3x_1 + 2x_2 + x_3 = 40$$

The slack variable x_3 represents the difference between $3x_1 + 2x_2$ and 40. If $3x_1 + 2x_2$ equals 40, then x_3 has the value 0. If, however, $3x_1 + 2x_2$ is less than 40, x_3 assumes the positive value equal to the difference between $3x_1 + 2x_2$ and 40.

The second inequality is converted to an equation by subtracting the surplus variable x_4. The resulting equation is

$$2x_1 + x_2 - x_4 = 10$$

The surplus variable x_4 represents the difference between the expression $2x_1 + x_2$ and 10. As in the case of slack variables, if $2x_1 + x_2$ equals 10, then the surplus variable x_4 has the value 0. If, however, $2x_1 + x_2$ is greater than 10, then x_4 has a positive value that is equal to the difference between $2x_1 + x_2$ and 10. Since $2x_1 + x_2 \geq 10$, the value of the surplus variable subtracted from the left side of the inequality must be greater than or equal to 0.

Several important characteristics of the new system of equations should be noted. First, both the slack and surplus variables are non-negative. Second, a slack variable must be added for each "less than or equal to" inequality and a surplus variable subtracted for each "greater than or equal to" inequality. Finally, slack and surplus variables would only coincidently have equivalent values. Consequently, different variables must be used for each inequality.

Example. A manufacturer makes two kinds of bookcases, A and B. Type A requires 2 hours on machine 1 and 4 hours on machine 2. Type B requires 3 hours on machine 1 and 2 hours on machine 2. The machines work no more than 16 hours per day. The profit is $3 per bookcase A and $4 per bookcase B. Formulate the linear programming problem, convert the constraining inequalities to equations, and interpret the meaning of the slack variables.

Let x_1 and x_2 represent the number of bookcases of type A and B, respectively. The linear programming problem is

$$\text{Maximize:} \quad P = 3x_1 + 4x_2$$

$$\text{Subject to:} \quad 2x_1 + 3x_2 \leq 16$$
$$4x_1 + 2x_2 \leq 16$$
$$x_1, x_2 \geq 0$$

The constraining inequalities are converted to equations by adding slack variables. The problem becomes

$$\text{Maximize:} \quad P = 3x_1 + 4x_2$$

$$\text{Subject to:} \quad \begin{aligned} 2x_1 + 3x_2 + x_3 &= 16 \\ 4x_1 + 2x_2 + x_4 &= 16 \\ x_1, x_2, x_3, x_4 &\geq 0 \end{aligned}$$

The slack variable x_3 represents unused machine 1 time and the slack variable x_4 represents unused machine 2 time. If both machines are fully utilized, x_3 and x_4 would have values of zero.

Example. American Foods, Inc., is developing a low-calorie, high-protein diet supplement. The linear programming formulation of this problem along with the data is given in Chap. 11, p. 357. Convert the system of inequalities to equations by adding slack and subtracting surplus variables. Interpret the meaning of the slack and surplus variables.

After adding the slack and subtracting the surplus variables, the problem becomes

$$\text{Minimize:} \quad C = 0.15x_1 + 0.20x_2 + 0.12x_3$$

$$\text{Subject to:} \quad \begin{aligned} 350x_1 + 250x_2 + 200x_3 + x_4 &= 300 \\ 250x_1 + 300x_2 + 150x_3 - x_5 &= 200 \\ 100x_1 + 150x_2 + 75x_3 - x_6 &= 100 \\ 75x_1 + 125x_2 + 150x_3 - x_7 &= 100 \\ x_j \geq 0 \quad \text{for } j = 1, 2, \ldots, 7 \end{aligned}$$

The slack variable x_4 represents the difference between the calories in the diet supplement and the maximum number of calories allowed by the specifications. The surplus variables x_5, x_6, and x_7 represent, respectively, the number of units of protein, vitamin A, and vitamin C in excess of that required by the specifications.

12.1.2 BASIC FEASIBLE SOLUTIONS

The addition of slack variables and subtraction of surplus variables enable linear inequalities to be expressed in the form of linear equations. As stated earlier, the reason that this transformation from inequalities to equations is important in the simplex algorithm is that it allows one to solve the resulting systems of equations for extreme points. The objective function in a linear programming problem, according to the extreme point theorem of linear programming, reaches an optimal value at one of the extreme points. Since the extreme points are defined by the intersection of sets of linear equations, the transformation from inequalities to equations is necessary.

To illustrate, consider the linear programming problem

$$\text{Maximize:} \quad P = x_1 + x_2$$
$$\text{Subject to:} \quad 3x_1 + 2x_2 \leq 40$$
$$2x_1 + x_2 \geq 10$$
$$x_1, x_2 \geq 0$$

The system of linear inequalities is transformed to a system of equations by adding a slack variable x_3 and subtracting a surplus variable x_4. This gives the system of linear equations

$$3x_1 + 2x_2 + x_3 \qquad = 40$$
$$2x_1 + x_2 \qquad - x_4 = 10$$

This system of two equations and four variables does not have a unique solution. In Chap. 2 it is shown that a system of m equations and n variables, where $n > m$, has an infinite number of solutions, provided the system of equations is consistent. Since this system of equations is consistent, it has an infinite number of solutions rather than a unique solution.[2]

Although the system of equations has an infinite number of solutions, it does not have an infinite number of extreme points. In fact, in this system of two equations and four variables there are only six extreme points. The extreme points can be determined by applying the *basis theorem* of linear programming. The *basis theorem* states that *for a system of* m *equations and* n *variables, where* n $>$ m, *a solution in which at least* n $-$ m *of the variables have values of zero is an extreme point.* Any solution found by setting $n - m$ of the variables equal to zero and solving the resulting system of m equations for the m remaining variables gives an extreme point. This solution is termed a *basic solution.*

The basic solutions for the preceding system of equations are determined by selecting two of the four variables and equating these variables with zero. The resulting system of two equations and two basic variables is solved simultaneously. For instance, if $x_3 = 0$ and $x_4 = 0$, the system of equations is

$$3x_1 + 2x_2 = 40$$
$$2x_1 + x_2 = 10$$

and the basic solution is $x_1 = -20$ and $x_2 = 50$. Similarly, if $x_1 = 0$ and $x_2 = 0$, the resulting basic solution is $x_3 = 40$ and $x_4 = -10$. The six basic solutions to the system of equations are shown in Table 12.1.[3]

---

[2] Consistent systems of equations are discussed in Chap. 2, p. 56.

[3] The number of basic solutions is given by

$$\frac{n!}{m!(n - m)!}$$

where n represents the number of variables and m the number of equations. To illustrate, the number of basic solutions when $n = 4$ and $m = 2$ is 6.

Table 12.1.

x_1	x_2	x_3	x_4	Objective Function
0	0	40	−10	Nonfeasible
0	20	0	10	20
0	10	20	0	10
13.3	0	0	16.6	13.3
5	0	25	0	5
−20	50	0	0	Nonfeasible

The extreme point theorem of linear programming can be extended to state that *the objective function is optimal at at least one of the basic solutions.* In the example given in Table 12.1 there are only six extreme points or, alternatively, only six basic solutions. Two of the solutions have negative values for variables and are, therefore, *nonfeasible.* The optimal value of the objective function, therefore, occurs at one of the four *basic feasible solutions.* Table 12.1 shows that the maximum value of the objective function occurs when $x_1 = 0$, $x_2 = 20$, $x_3 = 0$, and $x_4 = 10$. The value of the objective function at this basic solution is $P = 20$.

It can be shown that the extreme point and basis theorems of linear programming are related. This relationship is demonstrated through the use of the simple two-variable problem,

$$\text{Maximize:} \quad P = 2x_1 + 3x_2$$

$$\text{Subject to:} \quad 3x_1 + 2x_2 \leq 12$$
$$2x_1 + 4x_2 \leq 16$$
$$x_1, x_2 \geq 0$$

The feasible region for this problem is shown in Fig. 12.1.

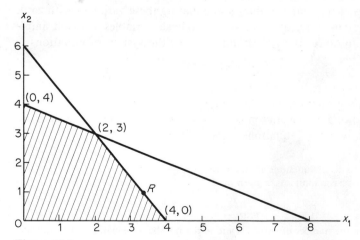

Figure 12.1.

The objective function is a maximum at an extreme point. The four extreme points are $E = \{(0, 0), (0, 4), (2, 3), (4, 0)\}$.

This same set of extreme points is generated from the basis theorem. To apply the basis theorem, the inequalities are first converted to equations by adding the slack variables x_3 and x_4, i.e.,

$$3x_1 + 2x_2 + x_3 \qquad = 12$$
$$2x_1 + 4x_2 \qquad + x_4 = 16$$

With four variables and two equations, the basic solutions are found by equating two of the four variables to zero. The four basic feasible solutions are given in Table 12.2. These were found by solving the four sets of two simultaneous equations. The basic solutions A, B, C, and D correspond, respectively, to the extreme points $(0, 0)$, $(0, 4)$, $(2, 3)$, and $(4, 0)$.

Table 12.2.

Solution	x_1	x_2	x_3	x_4
A	0	0	12	16
B	0	4	4	0
C	2	3	0	0
D	4	0	0	8

Nonbasic solutions correspond to points on the boundary of the constraining inequalities. The point R in Fig. 12.1 is an example of a nonbasic solution. The coordinates of R are $x_1 = 3\frac{1}{3}$ and $x_2 = 1$. Substituting these coordinates into the system of equations gives $x_1 = 3\frac{1}{3}, x_2 = 1, x_3 = 0$, and $x_4 = 5\frac{1}{3}$. This solution is nonbasic, since less than $n - m$ of the variables have values of zero.

This example illustrates the relationship among the basis theorem, the extreme point theorem, and the optimal solution. The extreme point theorem states that the optimal value of the objective function occurs at one of the extreme points of the feasible region. The example shows that basic solutions are extreme points. Consequently, the relationship between the basis theorem, the extreme point theorem, and the optimal solution is that (1) a basic solution occurs for a system of m equations and n variables when at least $n - m$ of the variables have values of zero and (2) the objective function is optimal at one of the basic solutions. We can thus determine the optimal solution by finding all basic solutions and evaluating the objective function at each of these basic solutions.

Example. Determine all basic solutions for the system of equations,

$$2x_1 + x_2 - 2x_3 + x_4 = 300$$
$$3x_1 - 2x_2 + 2x_3 + x_5 = 200$$

Since $m = 2$ equations and $n = 5$ variables, there are

$$\frac{n!}{m!\,(n-m)!} = \frac{5!}{2!\,3!} = 10 \text{ basic solutions}$$

The ten basic solutions are found by equating all possible combinations of three of the five variables to zero and solving the resulting systems of two equations and two basic variables. The basic solutions are given in Table 12.3. The solutions that are feasible are also indicated in the table.

Table 12.3.

x_1	x_2	x_3	x_4	x_5	Feasible/ Nonfeasible
0	0	0	300	200	Feasible
0	0	−150	0	500	Nonfeasible
0	0	100	500	0	Feasible
0	300	0	0	800	Feasible
0	−100	0	400	0	Nonfeasible
0	−500	−400	0	0	Nonfeasible
150	0	0	0	−250	Nonfeasible
66.7	0	0	167.3	0	Feasible
100	0	−50	0	0	Nonfeasible
114.3	71.4	0	0	0	Feasible

12.1.3 ALGEBRAIC SOLUTIONS

Before we introduce the simplex algorithm, it will be useful formally to introduce the algebraic solution technique. This technique, although inefficient, shares several characteristics with the simplex algorithm. The most important of these is that it uses the same algebraic procedure for determining basic solutions.

The algebraic solution technique requires the following steps: (1) determine all basic solutions; (2) evaluate each basic feasible solution for optimality. To illustrate, consider the linear programming problem

$$\text{Maximize:} \quad P = 3x_1 + 4x_2$$
$$\text{Subject to:} \quad 2x_1 + 3x_2 \leq 16$$
$$4x_1 + 2x_2 \leq 16$$
$$x_1, x_2 \geq 0$$

The inequalities are converted to equalities by adding slack variables. The linear programming problem becomes

$$\text{Maximize:} \quad P = 3x_1 + 4x_2 + 0x_3 + 0x_4$$
$$\text{Subject to:} \quad 2x_1 + 3x_2 + 1x_3 + 0x_4 = 16$$
$$4x_1 + 2x_2 + 0x_3 + 1x_4 = 16$$
$$x_1, x_2, x_3, x_4 \geq 0$$

The slack variables are included in the objective function with zero coefficients. The zero coefficients show that unused resource, or slack, does not contribute to profits.

The basic solutions are determined by selecting all possible combinations of two of the four variables and equating these variables with zero. The solutions of the resulting systems of two equations and two variables are basic solutions. For each basic solution, the variables that are equated to zero are termed *not in solution*, or alternatively, *not in the basis*. Conversely, those variables that are not equated to zero are said to be *in solution, in the basis*, or alternatively, *basic variables*.[4]

As a starting point, assume that the slack variables are in the basis. If the slack variables are in the basis, the variables x_1 and x_2 are equated with zero. Since the coefficients of the slack variables form an identity matrix, the initial basic solution can be read directly from the system of equations. The initial basic solution is $x_1 = 0$, $x_2 = 0$, $x_3 = 16$, and $x_4 = 16$. Since x_1, x_2, x_3, and x_4 are greater than or equal to zero, the solution is feasible. The value of the objective function at this initial basic feasible solution is 0, i.e.,

$$P = 3(0) + 4(0) + 0(16) + 0(16) = 0$$

For a second solution, let x_1 and x_4 be in the basis. According to the basis theorem of linear programming, x_2 and x_3 are not in the basis and are valued at zero. The solution to the system of two equations and two basic variables can be obtained by using the row operations of matrix algebra.[5]

The objective is to solve the system of equations for x_1 and x_4. The system could be written as

$$2x_1 + 0x_4 = 16$$

$$4x_1 + 1x_4 = 16$$

Alternatively, the system could be written in matrix form as

$$\begin{pmatrix} 2 & 0 \\ 4 & 1 \end{pmatrix} \begin{pmatrix} x_1 \\ x_4 \end{pmatrix} = \begin{pmatrix} 16 \\ 16 \end{pmatrix}$$

Since x_2 and x_3 have values of zero, both formulations are equivalent to the original system of equations,

$$2x_1 + 3x_2 + 1x_3 + 0x_4 = 16$$

$$4x_1 + 2x_2 + 0x_3 + 1x_4 = 16$$

---

[4] It is possible for the solution value of a basic variable to be zero. When this occurs, the solution is termed *degenerate*. A degenerate solution is treated the same as any other basic solution in the algebraic or simplex algorithms.

[5] Chap. 9, pp. 289–95.

In solving the system of equations, it is customary to apply row operations to the entire system of equations rather than to only the coefficients of the basic variables. This has the advantage of reducing the number of arithmetic calculations required in moving from one basic solution to another.

The system of equations can be solved for x_1 and x_4 by using row operations to obtain an identity matrix as the coefficient matrix of x_1 and x_4. Multiplying the first equation by $\frac{1}{2}$ gives

$$1x_1 + \tfrac{3}{2}x_2 + \tfrac{1}{2}x_3 + 0x_4 = 8$$

By multiplying the new equation by -4 and adding this product to the second equation, the second equation becomes

$$0x_1 - 4x_2 - 2x_3 + 1x_4 = -16$$

and the system of equations is

$$1x_1 + \tfrac{3}{2}x_2 + \tfrac{1}{2}x_3 + 0x_4 = \quad 8$$
$$0x_1 - 4x_2 - 2x_3 + 1x_4 = -16$$

The coefficients of the basic variables form an identity matrix. The nonbasic variables are valued at zero, and the basic solution can be read directly as $x_1 = 8$, $x_2 = 0$, $x_3 = 0$, and $x_4 = -16$. This solution is nonfeasible.

To determine a third solution, let x_1 and x_2 be basic variables and x_3 and x_4 be nonbasic variables. The values of x_1 and x_2 can be found by solving the original two equations simultaneously. An alternative approach, used in the simplex method, is to use the equations from the preceding iteration in solving for x_1 and x_2. The equations were

$$1x_1 + \tfrac{3}{2}x_2 + \tfrac{1}{2}x_3 + 0x_4 = \quad 8$$
$$0x_1 - 4x_2 - 2x_3 + 1x_4 = -16$$

Multiplying the second equation by $-\frac{1}{4}$ gives

$$0x_1 + 1x_2 + \tfrac{1}{2}x_3 - \tfrac{1}{4}x_4 = 4$$

Multiplying the new second equation by $\frac{3}{2}$ and subtracting it from the first equation gives

$$1x_1 + 0x_2 - \tfrac{1}{4}x_3 + \tfrac{3}{8}x_4 = 2$$

The system of equations is

$$1x_1 + 0x_2 - \tfrac{1}{4}x_3 + \tfrac{3}{8}x_4 = 2$$
$$0x_1 + 1x_2 + \tfrac{1}{2}x_3 - \tfrac{1}{4}x_4 = 4$$

The coefficients of x_1 and x_2 form an identity matrix and x_3 and x_4 are not in the basis. The solution can be read directly as $x_1 = 2$, $x_2 = 4$, $x_3 = 0$, and $x_4 = 0$. The solution is feasible. The value of the objective function is 22, i.e.,

$$P = 3(2) + 4(4) + 0(0) + 0(0) = 22$$

A fourth solution can be found by assuming that x_1 and x_3 are in the basis and x_2 and x_4 are not. Instead of solving the original system of equations, we use the equations from the preceding iteration. The system of equations

$$1x_1 + 0x_2 - \tfrac{1}{4}x_3 + \tfrac{3}{8}x_4 = 2$$

$$0x_1 + 1x_2 + \tfrac{1}{2}x_3 - \tfrac{1}{4}x_4 = 4$$

with basic variables x_1 and x_3 can be solved by multiplying the second equation by 2 and adding the product of $\tfrac{1}{4}$ the new second equation to the first equation. This results in an identity matrix as the coefficient matrix for the basic variables x_1 and x_3.

$$1x_1 + \tfrac{1}{2}x_2 + 0x_3 + \tfrac{1}{4}x_4 = 4$$

$$0x_1 + 2x_2 + 1x_3 - \tfrac{1}{2}x_4 = 8$$

The solution is $x_1 = 4$, $x_2 = 0$, $x_3 = 8$, $x_4 = 0$. The solution is feasible, and the value of the objective function is $P = 12$.

For a fifth solution, let x_2 and x_3 be the basic variables. The system of equations from the preceding iteration,

$$1x_1 + \tfrac{1}{2}x_2 + 0x_3 + \tfrac{1}{4}x_4 = 4$$

$$0x_1 + 2x_2 + 1x_3 - \tfrac{1}{2}x_4 = 8$$

can be solved for x_2 and x_3 by multiplying the first equation by 2 and subtracting the product of 2 times the new first equation from the second equation. This gives

$$2x_1 + 1x_2 + 0x_3 + \tfrac{1}{2}x_4 = 8$$

$$-4x_1 + 0x_2 + 1x_3 - \tfrac{3}{2}x_4 = -8$$

The solution, $x_1 = 0$, $x_2 = 8$, $x_3 = -8$, $x_4 = 0$, is not feasible.

The final solution occurs when x_2 and x_4 are in the basis. The system of equations

$$2x_1 + 1x_2 + 0x_3 + \tfrac{1}{2}x_4 = 8$$

$$-4x_1 + 0x_2 + 1x_3 - \tfrac{3}{2}x_4 = -8$$

can be solved for x_2 and x_4 by multiplying the second equation by $-\tfrac{2}{3}$ and subtracting the product of $\tfrac{1}{2}$ times this equation from the first equation. This gives

$$\tfrac{2}{3}x_1 + 1x_2 + \tfrac{1}{3}x_3 + 0x_4 = \tfrac{16}{3}$$

$$\tfrac{8}{3}x_1 + 0x_2 - \tfrac{2}{3}x_3 + 1x_4 = \tfrac{16}{3}$$

The solution is $x_1 = 0$, $x_2 = \tfrac{16}{3}$, $x_3 = 0$, $x_4 = \tfrac{16}{3}$. The value of the objective function is $P = 16$.

Table 12.4.

Solution	x_1	x_2	x_3	x_4	Objective Function
1	0	0	16	16	0
2	8	0	0	-16	Nonfeasible
3	2	4	0	0	22
4	4	0	8	0	12
5	0	8	-8	0	Nonfeasible
6	0	$\frac{16}{3}$	0	$\frac{16}{3}$	16

The six solutions to the linear programming problem are summarized in Table 12.4. The optimal value of the objective function, $P = 22$, occurs when $x_1 = 2$, $x_2 = 4$, $x_3 = 0$, and $x_4 = 0$.

The algebraic and simplex algorithms have several common characteristics. The most important of these is that both algorithms utilize the extreme point theorem of linear programming. This theorem states that an optimal value of the objective functions occurs at one of the basic solutions to the system of equations. Basic solutions to a system of m equations and n variables are found in both algorithms by equating $n - m$ of the variables to zero and solving the resulting system of m equations and m basic variables. Consequently, much of the arithmetic required for the algebraic solution to a linear programming problem is also required for a simplex solution.

12.2   The Simplex Algorithm

The simplex algorithm is an iterative procedure for determining basic feasible solutions to a system of equations and testing each solution for optimality. The algorithm involves moving from one basic feasible solution to another, always maintaining or improving the value of the objective function, until an optimal solution is reached.

Several of the inefficiencies of the algebraic method are eliminated by the simplex algorithm. First, feasible solutions found by using the algebraic technique cannot be distinguished from nonfeasible solutions until after the solution has been determined. In contrast, all solutions found by using the simplex algorithm are feasible. Since only feasible solutions are considered, the computational requirements are reduced by the simplex algorithm. Second, the algebraic technique requires that all basic feasible solutions be evaluated. In comparison, each iteration of the simplex algorithm gives a solution that improves or maintains the value of the objective function. This means that only a subset of the set of extreme points need be evaluated. This results in a further reduction in the computational requirements.

12.2.1 SIMPLEX TABLEAU

The first step in applying the simplex algorithm is to transform the inequalities of the linear programming problem to equations by adding slack and subtracting surplus variables. The coefficients of the objective function and constraining equations along with the right-hand-side values are then transferred to the simplex tableau. As an example, consider the linear programming problem

$$\text{Maximize:} \quad P = 3x_1 + 4x_2 + 5x_3 + 4x_4$$

$$\text{Subject to:} \quad \begin{aligned} 2x_1 + 5x_2 + 4x_3 + 3x_4 &\leq 224 \\ 5x_1 + 4x_2 - 5x_3 + 10x_4 &\leq 280 \\ 2x_1 + 4x_2 + 4x_3 - 2x_4 &\leq 184 \end{aligned}$$

$$x_j \geq 0 \quad \text{for } j = 1, 2, 3, 4$$

By adding slack variables, the problem becomes

$$\text{Maximize:} \quad P = 3x_1 + 4x_2 + 5x_3 + 4x_4 + 0x_5 + 0x_6 + 0x_7$$

$$\text{Subject to:} \quad \begin{aligned} 2x_1 + 5x_2 + 4x_3 + 3x_4 + 1x_5 + 0x_6 + 0x_7 &= 224 \\ 5x_1 + 4x_2 - 5x_3 + 10x_4 + 0x_5 + 1x_6 + 0x_7 &= 280 \\ 2x_1 + 4x_2 + 4x_3 - 2x_4 + 0x_5 + 0x_6 + 1x_7 &= 184 \end{aligned}$$

$$x_j \geq 0 \quad \text{for } j = 1, 2, \ldots, 7$$

The coefficients of the objective function and equations along with the right-hand-side values can now be transferred to the simplex tableau. The initial simplex tableau is shown in Table 12.5.

The first two rows of the tableau give the coefficients of the objective function and the column headings. The coefficients of the objective function are copied directly from the objective function. *Both slack and surplus variables have zero coefficients in the objective function.*

Table 12.5. Initial Simplex Tableau

	c_j	3	4	5	4	0	0	0	
c_b	Basis	x_1	x_2	x_3	x_4	x_5	x_6	x_7	Solution
0	x_5	2	5	4	3	1	0	0	224
0	x_6	5	4	−5	10	0	1	0	280
0	x_7	2	4	4	−2	0	0	1	184
	z_j	0	0	0	0	0	0	0	0
	$c_j - z_j$	3	4	5	4	0	0	0	

The coefficients of the constraining equations are shown in the tableau under the appropriate column heading. The coefficients of the variable x_1 in the constraining equations are given in the column labeled x_1, the coefficients of x_2 are given in the column labeled x_2, etc. The right-hand-side values are listed under the column labeled *Solution*.

The column labeled *Basis* contains the basic variables, i.e., the variables that are "in the basis." In the initial tableau, the variables x_5, x_6, and x_7 are in the basis. This means that the variables x_1, x_2, x_3, and x_4 are not in the basis and, according to the basis theorem of linear programming, are valued at zero. By referring to the system of linear equations, we can see that if x_1, x_2, x_3, and x_4 equal zero, the system reduces to

$$2(0) + 5(0) + 4(0) + 3(0) + 1x_5 + 0x_6 + 0x_7 = 224$$
$$5(0) + 4(0) - 5(0) + 10(0) + 0x_5 + 1x_6 + 0x_7 = 280$$
$$2(0) + 4(0) + 4(0) - 2(0) + 0x_5 + 0x_6 + 1x_7 = 184$$

The solution to this system of three equations and three basic variables can be read from the tableau. The variables x_1, x_2, x_3, and x_4 are not in the basis and have values of zero. Furthermore, the coefficients of the basic variables x_5, x_6, and x_7 form an identity matrix. The values of the variables in the basis can thus be read directly from the Solution column as $x_5 = 224$, $x_6 = 280$, and $x_7 = 184$.

The column labeled c_b contains the objective function coefficients of the basic variables. Since the basic variables in the initial tableau of our example problem are slack variables, the column labeled c_b contains zeros in the initial tableau.

The row labeled z_j contains the sum of the numbers in the c_b column multiplied by the coefficients in the x_j column. In the example,

$$z_1 = 0(2) + 0(5) + 0(2) = 0$$
$$z_2 = 0(5) + 0(4) + 0(4) = 0$$
$$z_3 = 0(4) + 0(-5) + 0(4) = 0, \text{ etc.}$$

The values of z_j in this example are zero for each of the j columns in the initial tableau. This, of course, is because of the fact that the objective function coefficients of the slack variables in the c_b column are zero. Many of the z_j values will be nonzero after the initial tableau.

The $c_j - z_j$ row is determined by subtracting the z_j value for the jth column from c_j, the objective function coefficient for that column. Since the z_j values are zero in the initial tableau, the $c_j - z_j$ values in this example are the same as the values in the c_j row.

The value of the objective function for the initial basic feasible solution is zero. This value is calculated by summing the products of the objective function coefficients of the basic variables and the solution values of the basic variables. This value is shown in the tableau as the last entry in the Solution column.

12.2.2 CHANGE OF BASIS

The value of the objective function can be increased by including one of the nonbasic variables in the basis. In Table 12.5, the nonbasic variables are x_1, x_2, x_3, and x_4. Since the number of variables in the basis remains constant, adding a nonbasic variable means that one of the current basic variables must be removed. Thus, x_5, x_6, or x_7 must leave the basis. The process of adding a variable and, concurrently, removing a variable is termed a *change of basis*.

The $c_j - z_j$ row is used to determine the variable to include in the basis. The numbers in the $c_j - z_j$ row give the change in the objective function resulting from including one unit of variable x_j in the basis. In the initial tableau, the $c_j - z_j$ row shows that the objective function will increase by 3 for each unit of x_1, by 4 for each unit of x_2, by 5 for each unit of x_3, etc. Since the objective is to maximize P, the variable that results in the largest per unit increase in the objective function should be placed in the basis. In this example, the objective function increases by 5 for each unit of x_3. Consequently, variable x_3 enters the basis. The rule for deciding the variable that enters the basis is summarized by Simplex Rule 1.

> Simplex Rule 1. *The selection of the variable to enter the basis is based upon the value of* $c_j - z_j$. *For maximization, the variable selected should have the largest value of* $c_j - z_j$. *If all values of* $c_j - z_j$ *are zero or negative, the current basic solution is optimal. For minimization, the variable selected should have the smallest (most negative) value of* $c_j - z_j$. *If all values of* $c_j - z_j$ *are zero or positive, the objective function is optimal.*

On the basis of Simplex Rule 1, the variable x_3 is included in the basis.

In certain cases it is possible that there is no single largest $c_j - z_j$ value. For instance, if the $c_j - z_j$ value for x_3 were changed from 5 to 4, the $c_j - z_j$ values for x_2, x_3, and x_4 would be the same. In these cases of ties, it is acceptable arbitrarily to select one of the variables for entry into the basis.

For x_3 to enter the basis, one of the variables currently included in the basis must be removed. Simplex Rule 2 is used to determine this variable.

> Simplex Rule 2. *The selection of the variable to leave the basis is made by dividing the numbers in the Solution column by the coefficients in the* x_j *column (i.e., the coefficients of the variable entering the basis). Select the row with the minimum ratio (ignore ratios with zero or negative numbers in the denominator). The variable associated with this row leaves the basis.*[6]

This rule applies for both maximization and minimization problems.

---

[6] In case of ties among the ratios, it is acceptable arbitrarily to select one of the tied variables to leave the basis.

Table 12.6.

Basis Variables	Current Solution	÷	Coefficients of Entering Variable		Ratios
x_5	224	÷	4	=	56
x_6	280	÷	−5	=	—
x_7	184	÷	4	=	46

Simplex Rule 2 is illustrated for the example problem in Table 12.6. This table shows the ratios of the Solution column to the coefficients of the variable entering the basis. The row associated with variable x_7 has the minimum ratio, and x_3 therefore replaces x_7 in the basis.

The reason underlying Simplex Rule 2 can be explained by referring to the example problem. This problem was

$$\text{Maximize:} \quad P = 3x_1 + 4x_2 + 5x_3 + 4x_4$$

$$\text{Subject to:} \quad \begin{aligned} 2x_1 + 5x_2 + 4x_3 + 3x_4 &\le 224 \\ 5x_1 + 4x_2 - 5x_3 + 10x_4 &\le 280 \\ 2x_1 + 4x_2 + 4x_3 - 2x_4 &\le 184 \end{aligned}$$

$$x_j \ge 0 \quad \text{for } j = 1, 2, 3, 4$$

To illustrate the rule, assume that this is a product mix problem and that the variables represent quantities of product. By applying Simplex Rule 1 to the initial tableau, variable x_3 is selected to enter the basis. The limitation on the maximum amount of x_3 that can be produced comes from the third constraint. That is, $\frac{184}{4}$ or 46 units of product 3 can be produced without exceeding the third constraint, while $\frac{224}{4}$ or 56 units can be produced without exceeding the first constraint. Notice that 46 units of x_3 is a feasible solution, while 56 units is not (i.e., $x_3 = 56$ violates the third constraint). If it is assumed that 46 units of x_3 are produced, the values of the four product variables are $x_1 = 0$, $x_2 = 0$, $x_3 = 46$, and $x_4 = 0$. The values of the slack variables are $x_5 = 40$, $x_6 = 510$, and $x_7 = 0$. Since $x_7 = 0$, it is removed from the basis and is replaced by x_3.

The second tableau has the variables x_5, x_6, and x_3 in the basis. The system of equations is

$$2x_1 + 5x_2 + 4x_3 + 3x_4 + 1x_5 + 0x_6 + 0x_7 = 224$$

$$5x_1 + 4x_2 - 5x_3 + 10x_4 + 0x_5 + 1x_6 + 0x_7 = 280$$

$$2x_1 + 4x_2 + 4x_3 - 2x_4 + 0x_5 + 0x_6 + 1x_7 = 184$$

The system of three equations is solved for the basic variables x_5, x_6, and x_3 by using row operations. Multiply the third row by $\frac{1}{4}$:

$$\tfrac{1}{2}x_1 + 1x_2 + 1x_3 - \tfrac{1}{2}x_4 + 0x_5 + 0x_6 + \tfrac{1}{4}x_7 = 46$$

Next, subtract four times the new third row from the first row and add five times the new third row to the second row. The resulting system of equations is

$$0x_1 + 1x_2 + 0x_3 + 5x_4 + 1x_5 + 0x_6 - 1x_7 = 40$$

$$\tfrac{15}{2}x_1 + 9x_2 + 0x_3 + \tfrac{15}{2}x_4 + 0x_5 + 1x_6 + \tfrac{5}{4}x_7 = 510$$

$$\tfrac{1}{2}x_1 + 1x_2 + 1x_3 - \tfrac{1}{2}x_4 + 0x_5 + 0x_6 + \tfrac{1}{4}x_7 = 46$$

Since the coefficient matrix of the basic variables x_5, x_6, and x_3 is an identity matrix and the nonbasic variables have values of zero, the basic solution can be read directly from the system of equations. The coefficients of this system of equations and the solution values of the basic variables are entered in the second tableau. This tableau is shown in Table 12.7.

The use of row operations to solve the systems of equations for the new basic variables is termed *pivoting*. The *pivot element* is the element at the intersection of the column headed by the variable entering the basis and the row associated with the variable leaving the basis. In the example, the pivot element in the initial tableau (Table 12.5) is the element at the intersection of the column labeled x_3 and the row labeled x_7. The pivot element in the second simplex tableau (Table 12.7) is circled and is at the intersection of the column labeled x_4 and the row labeled x_5.

The pivoting process involves obtaining an identity matrix as the coefficient matrix of the basic variables. Since only one new basic variable is entered in the tableau at any iteration, the identity matrix is completed by using row operations to obtain a one as the pivot element and zeros elsewhere in the *pivot column*. The mechanics of pivoting are straightforward. The first row operation is to multiply the *pivot row* by the reciprocal of the pivot element. In the example, the pivot row in the initial tableau is multiplied by $\tfrac{1}{4}$. This new row is entered in the second tableau, Table 12.7. The remainder of the

Table 12.7. Second Simplex Tableau

	c_j	3	4	5	4	0	0	0	
c_b	*Basis*	x_1	x_2	x_3	x_4	x_5	x_6	x_7	*Solution*
0	x_5	0	1	0	⑤	1	0	-1	40
0	x_6	$\tfrac{15}{2}$	9	0	$\tfrac{15}{2}$	0	1	$\tfrac{5}{4}$	510
5	x_3	$\tfrac{1}{2}$	1	1	$-\tfrac{1}{2}$	0	0	$\tfrac{1}{4}$	46
	z_j	$\tfrac{5}{2}$	5	5	$-\tfrac{5}{2}$	0	0	$\tfrac{5}{4}$	230
	$c_j - z_j$	$\tfrac{1}{2}$	-1	0	$\tfrac{13}{2}$	0	0	$-\tfrac{5}{4}$	

row operations are made to obtain zeros in the pivot column. The new pivot row is multiplied by -4 and added to the first equation. The resulting equation is entered in the second tableau. The new pivot row is then multiplied by 5 and added to the second equation. This equation is also entered in the second tableau. These operations result in values of 1 for the pivot element and 0 for the remaining elements in the pivot column.

The pivoting process is equivalent to solving the system of equations simultaneously for the new basic variables. Since $x_1 = 0$, $x_2 = 0$, $x_4 = 0$, and $x_7 = 0$, the system of equations shown in the second tableau can be written as

$$0(x_3) + 1(x_5) + 0(x_6) = 40$$

$$0(x_3) + 0(x_5) + 1(x_6) = 510$$

$$1(x_3) + 0(x_5) + 0(x_6) = 46$$

The values of the variables at the new extreme point can be read directly as $x_1 = 0$, $x_2 = 0$, $x_3 = 46$, $x_4 = 0$, $x_5 = 40$, $x_6 = 510$, and $x_7 = 0$.

The solution in the second tableau is optimal if all entries in the $c_j - z_j$ row are less than or equal to zero. The z_j values in Table 12.7 are calculated by summing the products of the entries in the c_b column and the coefficients in the x_j column. In the second tableau the values are

$$z_1 = 0(0) + 0(\tfrac{15}{2}) + 5(\tfrac{1}{2}) \quad = \tfrac{5}{2}$$

$$z_2 = 0(1) + 0(9) + 5(1) \quad = 5$$

$$z_3 = 0(0) + 0(0) + 5(1) \quad = 5$$

$$z_4 = 0(5) + 0(\tfrac{15}{2}) + 5(-\tfrac{1}{2}) = -\tfrac{5}{2}, \text{etc.}$$

The $c_j - z_j$ values are found by subtracting the z_j values from c_j. The $c_j - z_j$ values for x_1 and x_4 are positive; consequently, at least one more iteration is required.

The selection of the new variable to enter the basis in the third tableau is made by using Simplex Rule 1. For maximization, the variable selected should be that variable that has the largest value of $c_j - z_j$. Since the $c_j - z_j$ value for x_4 is $\tfrac{13}{2}$ and the $c_j - z_j$ value for x_1 is $\tfrac{1}{2}$, variable x_4 is selected to enter the basis.

Simplex Rule 2 is used to determine the variable to be removed from the basis. The ratios of the values in the Solution column to the positive entries in the x_4 column are $\tfrac{40}{5} = 8$ and $510/(15/2) = 68$. For both maximization and minimization problems, the variable leaving the basis is the variable associated with the row that has the smallest ratio. Since the ratio associated with the x_5 row is the smallest, x_5 leaves the basis and is replaced by x_4.

The values of the new basic variables are determined by the pivoting calculations. These calculations again involve completing the identity matrix by obtaining a one as the pivot element and zeros elsewhere in the pivot column. The pivot element is circled in Table 12.7. The pivot row is multiplied by the reciprocal of the pivot element and entered in the new tableau. The product of $\frac{1}{5}$ times the pivot row is shown as the x_4 row in Table 12.8.

Table 12.8.   Third Simplex Tableau

c_b	c_j Basis	3 x_1	4 x_2	5 x_3	4 x_4	0 x_5	0 x_6	0 x_7	Solution
4	x_4	0	$\frac{1}{5}$	0	1	$\frac{1}{5}$	0	$-\frac{1}{5}$	8
0	x_6	$\left(\frac{15}{2}\right)$	$\frac{15}{2}$	0	0	$-\frac{3}{2}$	1	$\frac{11}{4}$	450
5	x_3	$\frac{1}{2}$	$\frac{11}{10}$	1	0	$\frac{1}{10}$	0	$-\frac{3}{20}$	50
	z_j	$\frac{5}{2}$	$\frac{63}{10}$	5	4	$\frac{13}{10}$	0	$-\frac{1}{20}$	282
	$c_j - z_j$	$\frac{1}{2}$	$-\frac{23}{10}$	0	0	$-\frac{13}{10}$	0	$\frac{1}{20}$	

The remaining pivot calculations are made to obtain zeros in the pivot column. The new pivot row is multiplied by $-\frac{15}{2}$ and added to the x_6 row. It is then multiplied by $\frac{1}{2}$ and added to the x_3 row. This results in a one as the pivot element and zeros elsewhere in the column. The identity matrix associated with the basic variables is now complete. The values of the basic variables at the new extreme point can be read directly from the Solution column in the table. The solution is $x_1 = 0$, $x_2 = 0$, $x_3 = 50$, $x_4 = 8$, $x_5 = 0$, $x_6 = 450$, and $x_7 = 0$.

The $c_j - z_j$ values in Table 12.8 are positive for variables x_1 and x_7. Since the $c_j - z_j$ value for x_1 is the largest, x_1 is entered in the basis. The variable leaving the basis is determined by calculating the ratios of the numbers in the Solution column to the positive numbers in the x_1 column. The ratio for the x_6 row is the smallest; consequently, x_6 is removed from the basis and replaced by x_1.

The pivot element is circled in Table 12.8. This element is converted to a one by multiplying the pivot row by $\frac{2}{15}$. The new pivot row is then used to obtain zeros elsewhere in the pivot column. The results of the pivoting calculations are shown in Table 12.9.

The $c_j - z_j$ values in Table 12.9 are all zero or negative. This indicates that the solution shown in the tableau is optimal. The solution is $x_1 = 60$, $x_2 = 0$, $x_3 = 20$, $x_4 = 8$, $x_5 = 0$, $x_6 = 0$, and $x_7 = 0$. The value of the objective function is $P = 312$.

Table 12.9. Fourth Simplex Tableau

	c_j	3	4	5	4	0	0	0	
c_b	Basis	x_1	x_2	x_3	x_4	x_5	x_6	x_7	Solution
4	x_4	0	$\frac{1}{5}$	0	1	$\frac{1}{5}$	0	$-\frac{1}{5}$	8
3	x_1	1	1	0	0	$-\frac{1}{5}$	$\frac{2}{15}$	$\frac{11}{30}$	60
5	x_3	0	$\frac{3}{5}$	1	0	$\frac{1}{5}$	$-\frac{1}{15}$	$-\frac{1}{30}$	20
	z_j	3	$\frac{34}{5}$	5	4	$\frac{6}{5}$	$\frac{1}{15}$	$\frac{2}{15}$	312
	$c_j - z_j$	0	$-\frac{14}{5}$	0	0	$-\frac{6}{5}$	$-\frac{1}{15}$	$-\frac{2}{15}$	

12.2.3 OPTIMALITY

According to Simplex Rule 1, the solution to a linear maximization problem is optimal if all values of $c_j - z_j$ are zero or negative. Conversely, the solution to the minimization problem is optimal if all values of $c_j - z_j$ are zero or positive. To understand this rule, consider the example problem just completed.

In the initial tableau of the example problem, Table 12.5, slack variables are in the basis. According to Simplex Rule 1, a change of basis should be made if one or more of the $c_j - z_j$ values are positive. Since the $c_j - z_j$ value for x_3 in the initial tableau is positive and is greater than the $c_j - z_j$ values for the other variables, x_3 enters the basis. On the basis of Simplex Rule 2, variable x_7 leaves the basis.

The net contribution from including x_3 in the basis and removing x_7 is given by the $c_j - z_j$ row. The $c_j - z_j$ value for x_3 is 5 in the initial tableau. This means that each unit of x_3 will add 5 to the objective function. Since $x_3 = 0$ before the iteration (Table 12.5), and $x_3 = 46$ after the iteration (Table 12.7), a total of 46 units of x_3 has been included in the current solution. The change in the objective function is equal to the net contribution from including x_3 in the basis multiplied by the number of units in the solution, i.e., 5(46) or 230.

The $c_j - z_j$ values in the second tableau, Table 12.7, are positive for x_1 and x_4. Applying the simplex rules, x_4 is entered in the basis and x_5 removed from the basis. The $c_j - z_j$ value for x_4 shows that the net contribution per unit from x_4 is $\frac{13}{2}$. After the pivoting process, the solution is $x_1 = 0, x_2 = 0$, $x_3 = 50, x_4 = 8, x_5 = 0, x_6 = 450, x_7 = 0$, and $P = 282$. Notice that the increase in the objective function from $P = 230$ to $P = 282$ is equal to the $c_j - z_j$ value of x_4 in the second tableau multiplied by the solution value of x_4 in the third tableau, that is, $\frac{13}{2}(8) = 52$.

The $c_j - z_j$ values for x_1 and x_7 are positive in the third tableau. Since the $c_j - z_j$ value of $\frac{1}{2}$ for x_1 is the largest, x_1 is entered in the basis. The solution for the new basic variables is given in Table 12.9 as $x_1 = 60$, $x_2 = 0$, $x_3 = 20$, $x_4 = 8$, $x_5 = 0$, $x_6 = 0$, $x_7 = 0$, and $P = 312$. The increase in the objective function from $P = 282$ to $P = 312$ is again equal to the $c_j - z_j$ value for x_1 multiplied by the solution value of x_1, that is, $\frac{1}{2}(60) = 30$.

The $c_j - z_j$ values in Table 12.9 are less than or equal to zero. This means that introducing any of the nonbasic variables in this tableau into the basis would lead to a decrease in the value of the objective function. For instance, if x_2 is entered in the basis, the objective function will decrease by $\frac{14}{5}$ times the solution value of x_2. Since the objective is to maximize the objective function, and since any change of the basic variables will decrease the objective function, the current solution is optimal.

Example. A firm uses manufacturing labor and assembly labor to produce three different products. There are 120 hours of manufacturing labor and 260 hours of assembly labor available for scheduling. One unit of product 1 requires 0.10 hour of manufacturing labor and 0.20 hour of assembly labor. Product 2 requires 0.25 hour of manufacturing labor and 0.30 hour of assembly labor for each unit produced. One unit of product 3 requires 0.40 hour of assembly labor but no manufacturing labor. The contribution to profit from products 1, 2, and 3 is \$3.00, \$4.00, and \$5.00, respectively. Formulate the linear programming problem and solve by using the simplex algorithm.

Let x_j, for $j = 1, 2, 3$, represent the number of units of products 1, 2, and 3. The linear programming problem is

$$\text{Maximize:} \quad P = \quad 3x_1 + \quad 4x_2 + \quad 5x_3$$

$$\text{Subject to:} \quad 0.10x_1 + 0.25x_2 + \quad 0x_3 \leq 120$$
$$0.20x_1 + 0.30x_2 + 0.40x_3 \leq 260$$
$$x_1, x_2, x_3 \geq 0$$

The inequalities are converted to equalities by the addition of slack variables. This gives

$$\text{Maximize:} \quad P = \quad 3x_1 + \quad 4x_2 + \quad 5x_3 + 0x_4 + 0x_5$$

$$\text{Subject to:} \quad 0.10x_1 + 0.25x_2 + \quad 0x_3 + 1x_4 + 0x_5 = 120$$
$$0.20x_1 + 0.30x_2 + 0.40x_3 + 0x_4 + 1x_5 = 260$$
$$x_j \geq 0, \quad j = 1, 2, 3, 4, 5$$

The slack variables x_4 and x_5 represent unused manufacturing and assembly labor. Since idle labor hours do not contribute to profit, the coefficients of the slack variables in the objective function are zero.

The initial tableau is shown in Table 12.10. The initial basic feasible

Table 12.10.

Tableau		c_j	3	4	5	0	0	
	c_b	Basis	x_1	x_2	x_3	x_4	x_5	Solution
Initial	0	x_4	0.10	0.25	0	1	0	120
	0	x_5	0.20	0.30	0.40	0	1	260
		z_j	0	0	0	0	0	0
		$c_j - z_j$	3	4	5	0	0	
					↑			
Second	0	x_4	0.10	0.25	0	1	0	120
	5	x_3	0.50	0.75	1	0	2.50	650
		z_j	2.50	3.75	5	0	12.50	3250
		$c_j - z_j$	0.50	0.25	0	0	−12.50	
			↑					
Third	3	x_1	1	2.50	0	10	0	1200
	5	x_3	0	−0.50	1	−5	2.50	50
		z_j	3	5	5	5	12.50	3850
		$c_j - z_j$	0	−1	0	−5	−12.50	

solution is $x_1 = 0$, $x_2 = 0$, $x_3 = 0$, $x_4 = 120$, $x_5 = 260$, and $P = 0$. By Simplex Rules 1 and 2, variable x_3 enters the basis and x_5 leaves. The solution for the new basic variables is given in the second tableau as $x_1 = 0$, $x_2 = 0$, $x_3 = 650$, $x_4 = 120$, $x_5 = 0$, and $P = \$3250$. Notice that the change in the objective function from $P = 0$ to $P = 3250$ is equal to the $c_j - z_j$ value for x_3 in the initial tableau times the solution value for x_3 after the iteration, that is, $\$5(650) = \$3250$.

We have shown that the $c_j - z_j$ value of variable x_j gives the net contribution to the objective function for each unit of x_j included in the solution. The reason for this relationship can perhaps be better understood by examining the $c_j - z_j$ values for the variables in the second tableau. The $c_j - z_j$ value for x_1 and x_2 in the second tableau are positive. The $c_j - z_j$ value of $\$0.50$ for x_1 is the larger of the two values; therefore, x_1 enters the basis. Notice in the original problem that x_1 and x_3 use assembly hours in the ratio of 1 to 2. Since all assembly hours are being utilized in the second tableau (i.e., the value of the slack variable in the second tableau for assembly hours is $x_5 = 0$), each unit of x_1 included in the solution in the third tableau reduces the production of x_3 by one-half unit. This, of course, results in a decrease in the objective function of $2.50 for each one-half hour of assembly labor diverted from x_3 to x_1. This effect on the objective function of diverting assembly labor from x_3 to x_1 is shown by z_1. The z_1 value in the second tableau of $2.50 means that the production of each unit of x_1 will reduce the contribution to the objective function from the variables currently in the basis by $2.50. Since, however, the per unit contribution of x_1 is $3.00, the net change in the objective function for each unit of x_1 included in the basis in the third tableau is $3.00 − $2.50 or $0.50. This value is shown in the $c_j − z_j$ row in the second tableau for x_1.

The final tableau contains basic variables x_1 and x_3. All $c_j − z_j$ values are zero or negative; consequently, the solution is optimal. As expected, the increase in the objective function from $P = \$3250$ to $P = \$3850$ is equal to the $c_j − z_j$ value of $0.50 for x_1 times the solution value of x_1, that is, $0.50(1200) = \$600$. The solution to the problem is $x_1 = 1200$, $x_2 = 0$, $x_3 = 50$, $x_4 = 0$, $x_5 = 0$, and $P = \$3850$.

12.2.4 MARGINAL VALUE OF A RESOURCE

The linear programming problem has been described as that of allocating scarce resources among competing products or activities. Since these resources are combined to produce a salable product, the resources have a value to the firm. One measure of this value is termed the *marginal value* of the resource.[7]

The marginal value of a resource is given by the change in the objective function resulting from employing one additional unit of the resource. To illustrate, consider the preceding example. In this example the marginal value of manufacturing labor is equal to the change in the objective function resulting from employing an additional hour of manufacturing labor. Similarly, the marginal value of assembly labor is equal to the change in the objective function resulting from employing one additional hour of assembly labor.

The marginal value of a resource can be determined directly from the $c_j − z_j$ row of the final tableau. In the final tableau (Table 12.10) the $c_j − z_j$ values for x_4 and x_5 are $−5$ and $−12.50$, respectively. Including x_4 in the basis would, therefore, decrease the objective function by $5.00 for each unit of x_4 in the solution. Similarly, including x_5 in the basis would decrease the objective function by $12.50 for each unit of x_5. Since x_4 represents unused manufacturing labor hours and x_5 represents unused assembly labor hours,

---

[7] The marginal value is sometimes referred to as the *shadow price* of the resource.

the objective function will decrease by $5.00 for each hour of manufacturing labor withheld from production and by $12.50 for each hour of assembly labor withheld from production. Conversely, as long as the basic variables do not change, each additional hour of manufacturing labor included in production results in a $5.00 increase in the objective function, and each additional hour of assembly labor results in a $12.50 increase. The marginal value of manufacturing and assembly labor is therefore $5.00 and $12.50, respectively.

The example illustrates that the marginal value of an additional unit of resource is determined from the $c_j - z_j$ value in the final tableau. The absolute value of $c_j - z_j$ for the slack variable for the resource gives the marginal value of that resource. If the slack variable for a resource is in the basis (i.e., unused resource is available), the marginal value of an additional unit of the resource as shown by $c_j - z_j$ is zero. If, however, the slack variable is not in the basis (i.e., all resource has been allocated to production), the marginal value of an additional unit of resource is positive.

Example. Determine the marginal value of an additional unit of resources 1, 2, and 3 for the following linear programming problem:

$$\text{Maximize:} \quad P = \$30x_1 + \$20x_2$$

$$\text{Subject to:} \quad
\begin{array}{llll}
2x_1 + & x_2 \leq 280 & \text{(resource 1)} \\
3x_1 + & 2x_2 \leq 500 & \text{(resource 2)} \\
x_1 + & 3x_2 \leq 420 & \text{(resource 3)} \\
x_1, x_2 \geq 0
\end{array}$$

This problem is solved by adding slack variables x_3, x_4, and x_5 to convert the inequalities to equations and then applying the simplex algorithm. The final tableau for this problem is shown in Table 12.11.

Table 12.11.

c_b	c_j	30	20	0	0	0	
	Basis	x_1	x_2	x_3	x_4	x_5	Solution
30	x_1	1	0	$\frac{3}{5}$	0	$-\frac{1}{5}$	84
0	x_4	0	0	$-\frac{7}{5}$	1	$-\frac{1}{5}$	24
20	x_2	0	1	$-\frac{1}{5}$	0	$\frac{2}{5}$	112
	z_j	30	20	14	0	2	4760
	$c_j - z_j$	0	0	-14	0	-2	

Since x_3 represents slack for resource 1 and the $c_j - z_j$ value for x_3 is -14, the marginal value of resource 1 is $14. The marginal value of resource 2 is given by the $c_j - z_j$ entry for x_4. Since 24 units of resource 2 are in the optimal tableau as slack, resource 2 is currently in excess supply and the marginal value of resource 2 is zero. Similarly, the marginal value of resource 3 is $2.

The value of the resources to the firm comes from the fact that they are used to produce a product that is sold for a profit. Since the products are made from the resources, it follows that the value of the resources in generating profit should be equivalent to the profit made from their utilization. This is, in fact, the case. To illustrate, the marginal values of resources 1, 2, and 3 in Table 12.11 are $14, $0, and $2, respectively. The sum of the marginal values of the resources multiplied by the quantity of resource gives the total value of the resources, i.e.,

$$\$14(280) + \$0(500) + \$2(420) = \$4760$$

The value of the resources is equal to the value of the objective function. This result is true at all iterations.

Example. Show that the value of the resources is equal to the value of the objective function at each iteration for the allocation problem given on p. 395.

The tableaus for this problem are given by Table 12.10. The value of the resources in the initial tableau is

$$\$0(120) + \$0(260) = \$0$$

The value of the resources in the second tableau is

$$\$0(120) + \$12.50(260) = \$3250$$

The value of the resources in the final tableau is

$$\$5(120) + \$12.50(260) = \$3850$$

These values equal the objective function at each iteration.

The marginal value of the resource has a number of uses in decision making. One obvious use is in evaluating the employment of additional resources. In the example, an additional hour of assembly labor is worth $12.50 to the firm in terms of increased profits. If assembly labor can be employed for less than $12.50 per hour, additional laborers should be hired.

Another important use of marginal value occurs in establishing transfer prices for resources between divisions in a firm. Resources in the form of labor, materials, and intermediate products are often transferred from one profit center to another. Knowledge of the marginal value of the resource is, therefore, helpful in establishing the transfer price.

12.3   Minimization

The simplex algorithm applies to both maximization and minimization problems. The only difference in the algorithm involves the selection of the variable to enter the basis. In the maximization problem, the variable with the largest $c_j - z_j$ value is included in the basis. Conversely, the variable with the smallest (i.e., most negative) $c_j - z_j$ value is selected to enter the basis in the minimization problem. The selection of the variable to be removed from the basis and the pivoting calculations are the same for both maximization and minimization problems. The solution is optimal in the minimization problem when all $c_j - z_j$ values are zero or positive.

To illustrate minimization, consider the problem

$$\text{Minimize:} \quad C = 20x_1 + 10x_2$$

$$\text{Subject to:} \quad \begin{aligned} x_1 + 2x_2 &\leq 40 \\ 3x_1 + x_2 &= 30 \\ 4x_1 + 3x_2 &\geq 60 \\ x_1, x_2 &\geq 0 \end{aligned}$$

To apply the simplex algorithm, the inequalities must be converted to equalities. Adding a slack variable to the first inequality and subtracting a surplus variable from the third inequality, we obtain

$$\text{Minimize:} \quad C = 20x_1 + 10x_2 + 0x_3 + 0x_4$$

$$\text{Subject to:} \quad \begin{aligned} 1x_1 + 2x_2 + 1x_3 + 0x_4 &= 40 \\ 3x_1 + 1x_2 + 0x_3 + 0x_4 &= 30 \\ 4x_1 + 3x_2 + 0x_3 - 1x_4 &= 60 \\ x_1, x_2, x_3, x_4 &\geq 0 \end{aligned}$$

The simplex algorithm begins with an initial basic feasible solution. In the examples in the preceding sections, the initial feasible solution was given by including the slack variables in the basis. This procedure does not give a feasible solution for this particular problem. To illustrate, if x_1 and x_2 are equated to zero, the system of equations reduces to

$$1x_3 + 0x_4 = 40$$

$$0x_3 + 0x_4 = 30$$

$$0x_3 - 1x_4 = 60$$

The second equation obviously is not true, i.e., zero is not equal to 30. Consequently, the solution is not feasible.

There is another problem in obtaining the initial basic feasible solution that is not as obvious as that of the equality. For the moment, assume that the equality $3x_1 + x_2 = 30$ is omitted from the original problem. On the basis of this assumption, the problem reduces to

$$\text{Minimize:} \quad C = 20x_1 + 10x_2$$

$$\text{Subject to:} \qquad x_1 + 2x_2 \leq 40$$
$$4x_1 + 3x_2 \geq 60$$
$$x_1, x_2 \geq 0$$

Adding the slack variable and subtracting the surplus variable, we now have the system of equations

$$1x_1 + 2x_2 + 1x_3 + 0x_4 = 40$$

$$4x_1 + 3x_2 + 0x_3 - 1x_4 = 60$$

$$x_1, x_2, x_3, x_4 \geq 0$$

Designating x_3 and x_4 as basic variables in the initial tableau gives the solution $x_1 = 0$, $x_2 = 0$, $x_3 = 40$, and $x_4 = -60$. This solution violates the requirement in the simplex algorithm of $x_j \geq 0$. Thus, even after the original equality is omitted, the solution obtained by including the slack and surplus variables in the basis is not feasible.

12.3.1   ARTIFICIAL VARIABLES

The simplex algorithm requires an initial basic feasible solution. As illustrated by the preceding example, it is not always possible to obtain this initial basic feasible solution by merely adding a slack variable to each "less than or equal to" inequality and subtracting a surplus variable from each "greater than or equal to" inequality. In these cases the problem must be modified by adding *artificial variables*. The initial basic feasible solution to this modified problem is then used as a starting point for applying the simplex algorithm to the original problem.

An artificial variable has no physical interpretation. It is merely a "dummy" variable that is added to constraining equations or inequalities for the purpose of generating an initial basic feasible solution.

The minimization problem discussed above provides an example of the use of artificial variables. The slack and surplus variables in this problem do not provide an initial basic feasible solution. To apply the simplex algorithm, the problem must be modified by adding artificial variables to the second and third constraints. After adding artificial variables, the system of equations is

$$1x_1 + 2x_2 + 1x_3 + 0x_4 + 0A_1 + 0A_2 = 40$$

$$3x_1 + 1x_2 + 0x_3 + 0x_4 + 1A_1 + 0A_2 = 30$$

$$4x_1 + 3x_2 + 0x_3 - 1x_4 + 0A_1 + 1A_2 = 60$$

$$x_1, x_2, x_3, x_4, A_1, A_2 \geq 0$$

This system of equations differs from the original system in that it includes the artificial variables A_1 and A_2. Solutions that include A_1 and A_2 at positive values have no meaning in the linear programming problem. Consequently, the artificial variables cannot have positive values in the final tableau.

The artificial variables merely provide a convenient vehicle for generating an initial basic feasible solution. This solution, obtained by including x_3, A_1, and A_2 in the initial basis, is $x_1 = 0$, $x_2 = 0$, $x_3 = 40$, $x_4 = 0$, $A_1 = 30$, and $A_2 = 60$.

12.3.2 THE "BIG M" METHOD

The basic feasible solution that includes artificial variables is used as a starting point for applying the simplex algorithm. The problem is to minimize the objective function subject to the three constraints. This is accomplished provided the artificial variables have values of zero in the final tableau. If those variables have values of zero, they will have provided a starting point for the simplex algorithm while not affecting the optimal solution.

A simple method is available for assuring that the artificial variables have values of zero in the final tableau. The method is to make the coefficients of the artificial variables in the objective function in a minimization problem extremely large and, conversely, to make the coefficients of the artificial variables in the objective function in a maximization problem extremely small (i.e., large negative numbers). This is analogous in a problem involving minimizing cost to making the artificial variable extremely expensive to produce. Alternatively, in a problem involving maximizing profit, the large negative coefficient has the effect of making the artificial variable extremely costly to produce. Since there are no constraints requiring production of the artificial variable, the simplex algorithm will assure that the artificial variable is not in the basis in the final tableau.

Instead of assigning some arbitrarily large positive or negative numbers as coefficients of the artificial variables in the objective function, it is customary to use the capital letter M. If we adopt this convention, the linear minimization problem becomes

Minimize: $C = 20x_1 + 10x_2 + 0x_3 + 0x_4 + MA_1 + MA_2$

Subject to: $1x_1 + 2x_2 + 1x_3 + 0x_4 + 0A_1 + 0A_2 = 40$
$3x_1 + 1x_2 + 0x_3 + 0x_4 + 1A_1 + 0A_2 = 30$
$4x_1 + 3x_2 + 0x_3 - 1x_4 + 0A_1 + 1A_2 = 60$
$x_1, x_2, x_3, x_4, A_1, A_2 \geq 0$

The tableaus for this problem are given in Table 12.12. Since this is a minimization problem, the variable with the smallest $c_j - z_j$ value is included in the basis. Notice in the initial tableau that $20 - 7M$ is smaller than either

Table 12.12.

Tableau	c_b	Basis	c_j 20 x_1	10 x_2	0 x_3	0 x_4	M A_1	M A_2	Solution
Initial	0	x_3	1	2	1	0	0	0	40
	M	A_1	③	1	0	0	1	0	30 ←
	M	A_2	4	3	0	−1	0	1	60
		z_j	$7M$	$4M$	0	$−M$	M	M	$90M$
		$c_j − z_j$	$20 − 7M$	$10 − 4M$	0	M	0	0	
Second	0	x_3	0	$\frac{5}{3}$	1	0	$-\frac{1}{3}$	0	30
	20	x_1	1	$\frac{1}{3}$	0	0	$\frac{1}{3}$	0	10
	M·	A_2	0	$⑤/③$	0	−1	$-\frac{4}{3}$	1	20 ←
		z_j	20	$\frac{20}{3} + \frac{5}{3}M$	0	$−M$	$\frac{20}{3} - \frac{4}{3}M$	M	$200 + 20M$
		$c_j − z_j$	0	$\frac{10}{3} - \frac{5}{3}M$	0	M	$\frac{7}{3}M - \frac{20}{3}$	0	
Third	0	x_3	0	0	1	1	1	−1	10
	20	x_1	1	0	0	$\frac{1}{5}$	$\frac{3}{5}$	$-\frac{1}{5}$	6
	10	x_2	0	1	0	$-\frac{3}{5}$	$-\frac{4}{5}$	$\frac{3}{5}$	12
		z_j	20	10	0	−2	4	2	240
		$c_j − z_j$	0	0	0	2	$M − 4$	$M − 2$	

$10 − 4M$ or M. Thus, variable x_1 enters the basis. From Simplex Rule 2, variable A_1 leaves the basis. Row operations are then performed to determine the solution values of the variables in the second tableau. The iteration process continues until entries in the $c_j − z_j$ row are zero or positive. The optimal solution is shown in the final tableau as $x_1 = 6$, $x_2 = 12$, $x_3 = 10$, $x_4 = 0$, and $C = 240$.

The linear minimization problem illustrates the types of constraints for which artificial variables are used. These constraints are of the forms $ax_1 + bx_2 \geq c$ and $ax_1 + bx_2 = c$. One surplus and one artificial variable are needed for each greater than or equal to constraint, whereas the equalities require only the artificial variable.

Artificial variables are used to obtain an initial basic feasible solution in both maximization and minimization problems. With the exception that large negative rather than positive numbers are used as the coefficients of the artificial variables in the objective function of a maximization problem, the solution procedure for both maximization and minimization problems is the same. This is illustrated by the following example:

Example.

$$\text{Maximize:} \quad P = x_1 + 2x_2 + 4x_3$$

$$\text{Subject to:} \quad x_1 + x_2 + x_3 \leq 12$$
$$2x_1 - x_2 + x_3 \geq 8$$
$$x_1, x_2, x_3 \geq 0$$

A slack variable is required for the first inequality, while both a surplus and an artificial variable are required for the second inequality. Adding the slack, surplus, and artificial variables gives

$$\text{Maximize:} \quad P = 1x_1 + 2x_2 + 4x_3 + 0x_4 + 0x_5 - MA_1$$

$$\text{Subject to:} \quad 1x_1 + 1x_2 + 1x_3 + 1x_4 + 0x_5 + 0A_1 = 12$$
$$2x_1 - 1x_2 + 1x_3 + 0x_4 - 1x_5 + 1A_1 = 8$$
$$x_1, x_2, x_3, x_4, x_5, A_1 \geq 0$$

The tableaus for the problem are shown in Table 12.13.

An interesting feature of this example is that a variable enters the basis in one of the intermediate tableaus but is not in the basis in the optimal solution. Variable x_1 enters the basis in the second tableau and leaves in the third. Similarly, x_2 enters the basis in the fourth tableau and is replaced by x_5 in the fifth. Although not illustrated by this example, it is also possible for a variable to enter and leave the basis and then reenter in a later iteration.

12.4   Special Cases[8]

The graphical solution procedure was used in Chap. 11 to define and illustrate the cases of no feasible solution, multiple optimal solutions, and unbounded solutions. The relationship between these cases and the simplex algorithm is shown in this section.

12.4.1   NO FEASIBLE SOLUTION

Realistic linear programming applications often have more than 100 variables and constraints. In problems of this size, it is impossible to tell by graphing

---

[8] This section can be omitted without loss of continuity.

Table 12.13.

Tableau		c_j	1	2	4	0	0	$-M$	
	c_b	Basis	x_1	x_2	x_3	x_4	x_5	A_1	Solution
Initial	0	x_4	1	1	1	1	0	0	12
	$-M$	A_1	②	-1	1	0	-1	1	8 ←
		z_j	$-2M$	$+M$	$-M$	0	$+M$	$-M$	$-8M$
		$c_j - z_j$	$1 + 2M$	$2 - M$	$4 + M$	0	$-M$	0	
			↑						
Second	0	x_4	0	$\frac{3}{2}$	$\frac{1}{2}$	1	$\frac{1}{2}$	$-\frac{1}{2}$	8
	1	x_1	1	$-\frac{1}{2}$	②	0	$-\frac{1}{2}$	$\frac{1}{2}$	4 ←
		z_j	1	$-\frac{1}{2}$	$\frac{1}{2}$	0	$-\frac{1}{2}$	$\frac{1}{2}$	4
		$c_j - z_j$	0	$\frac{5}{2}$	$\frac{7}{2}$	0	$\frac{1}{2}$	$-M - \frac{1}{2}$	
					↑				
Third	0	x_4	-1	②	0	1	1	-1	4 ←
	4	x_3	2	-1	1	0	-1	1	8
		z_j	8	-4	4	0	-4	4	32
		$c_j - z_j$	-7	6	0	0	4	$-M - 4$	
				↑					
Fourth	2	x_2	$-\frac{1}{2}$	1	0	$\frac{1}{2}$	①	$-\frac{1}{2}$	2 ←
	4	x_3	$\frac{3}{2}$	0	1	$\frac{1}{2}$	$-\frac{1}{2}$	$\frac{1}{2}$	10
		z_j	5	2	4	3	-1	1	44
		$c_j - z_j$	-4	0	0	-3	1	$-M - 1$	
							↑		
Fifth	0	x_5	-1	2	0	1	1	-1	4
	4	x_3	1	1	1	1	0	0	12
		z_j	4	4	4	4	0	0	48
		$c_j - z_j$	-3	-2	0	-4	0	$-M$	

if the problem has feasible solutions. Instead, feasibility must be determined from the simplex tableau.

To illustrate, consider the problem

$$\text{Maximize:} \quad P = x_1 + 2x_2$$

$$\text{Subject to:} \quad \begin{aligned} x_1 + x_2 &\le 4 \\ x_1 + x_2 &\ge 6 \\ x_1, x_2 &\ge 0 \end{aligned}$$

It can be seen that the constraints are mutually exclusive and, consequently, there can be no feasible solution. This conclusion can also be made from the simplex tableaus for this problem. These are shown in Table 12.14. The simplex algorithm is applied to obtain the tableaus in Table 12.14. The elements in the $c_j - z_j$ row of the second tableau are all zero or negative, indicating for the maximization problem that the solution is optimal. The solution, however, contains an artificial variable. This indicates that the solution shown in the final tableau is not a feasible solution to the original linear programming problem.

The example illustrates the characteristic form of linear programming problems that have no feasible solution. This characteristic is that one or more artificial variables are in the basis at a nonzero level in the final tableau. In the example the variable A_1 was in the basis and had the value $A_1 = 2$.

Table 12.14.

Tableau		c_j	*1*	*2*	*0*	*0*	$-\text{M}$	
	c_b	*Basis*	x_1	x_2	x_3	x_4	A_1	*Solution*
Initial	0	x_3	1	$\textcircled{1}$	1	0	0	4	←
	$-M$	A_1	1	1	0	-1	1	6
		z_j	$-M$	$-M$	0	M	$-M$	$-6M$
		$c_j - z_j$	$1 + M$	$2 + M$	0	$-M$	0	
Second	2	x_2	1	1	1	0	0	4
	$-M$	A_1	0	0	-1	-1	1	2
		z_j	2	2	$2 + M$	M	$-M$	$8 - 2M$
		$c_j - z_j$	-1	0	$-2 - M$	$-M$	0	

Since the $c_j - z_j$ values are all zero or negative, the solution is optimal but not feasible.

12.4.2 MULTIPLE OPTIMAL SOLUTIONS

The existence of multiple optimal solutions to a linear programming problem is determined from the $c_j - z_j$ row of the final tableau. The values of $c_j - z_j$ give the net change in the objective function from including one unit of variable x_j in the basis. As shown by previous examples, the $c_j - z_j$ values for the basic variables in the final tableau are zero.

Multiple optimal solutions to the linear programming problem *exist if the $c_j - z_j$ value for one of the nonbasic variables is zero.* A $c_j - z_j$ value of zero for a nonbasic variable means that the variable can be included in the basis without changing the value of the objective function. If a nonbasic variable can be entered in the basis without changing the value of the objective function, the solution given by including the new variable in the basis is also optimal.

This is illustrated by the problem

$$\text{Maximize:} \quad P = \quad 3x_1 + \quad 5x_2 + \quad 5x_3$$

$$\text{Subject to:} \quad 0.10x_1 + 0.25x_2 + \quad 0x_3 \leq 120$$

$$0.20x_1 + 0.30x_2 + 0.40x_3 \leq 260$$

$$x_1, x_2, x_3 \geq 0$$

The solution is given in Table 12.15. The $c_j - z_j$ row of the final tableau shows that variable x_1 can enter the basis without changing the value of the objective function. To demonstrate, the tableau that includes variable x_1 in the basis is given by Table 12.16. The value of the objective function is unchanged from the tableau in Table 12.15.

The graphical analysis of multiple optimal solutions showed that if more than one optimal solution exists, then an infinite number of optimal solutions exist. These solutions are given by forming a linear combination of the basic solutions.[9] In the example the basic solutions were $x_1 = 0$, $x_2 = 480$, $x_3 = 290$, $x_4 = 0$, $x_5 = 0$ and $x_1 = 1200$, $x_2 = 0$, $x_3 = 50$, $x_4 = 0$, $x_5 = 0$. The linear combination of these solutions is

$$x_1 = b(0) + (1 - b)1200$$

$$x_2 = b(480) + (1 - b)0$$

$$x_3 = b(290) + (1 - b)50$$

$$x_4 = b(0) + (1 - b)0$$

$$x_5 = b(0) + (1 - b)0$$

---

[9] In large-scale applications it is possible to have more than two optimal basic solutions.

Table 12.15.

Tableau		c_j	3	5	5	0	0	
	c_b	Basis	x_1	x_2	x_3	x_4	x_5	Solution
Initial	0	x_4	0.10	0.25	0	1	0	120
	0	x_5	0.20	0.30	0.40	0	1	260 ←
		z_j	0	0	0	0	0	0
		$c_j - z_j$	3	5	5	0	0	
					↑			
Second	0	x_4	0.10	0.25	0	1	0	120 ←
	5	x_3	0.50	0.75	1	0	2.50	650
		z_j	2.50	3.75	5	0	12.50	3250
		$c_j - z_j$	0.50	1.25	0	0	-12.50	
				↑				
Third	5	x_2	0.40	1	0	4	0	480
	5	x_3	0.20	0	1	-3	2.50	290
		z_j	3	5	5	5	12.50	3850
		$c_j - z_j$	0	0	0	-5	-12.50	

Table 12.16.

		c_j	3	5	5	0	0	
c_b		Basis	x_1	x_2	x_3	x_4	x_5	Solution
3		x_1	1	2.50	0	10	0	1200
5		x_3	0	-0.50	1	-5	2.50	50
		z_j	3	5	5	5	12.50	3850
		$c_j - z_j$	0	0	0	-5	-12.50	

where b is a weighting factor with domain $0 \leq b \leq 1$. If, for instance, $b = 0.4$, the solution is $x_1 = 720$, $x_2 = 192$, $x_3 = 146$, $x_4 = 0$, and $x_5 = 0$. This solution satisfies the original constraints and has the optimal objective function value of $P = 3850$.

12.4.3 UNBOUNDED SOLUTIONS

The concept of an unbounded solution was introduced in Chap. 11. The problem used to illustrate the concept was

$$\text{Maximize:} \quad P = \quad x_1 + 2x_2$$
$$\text{Subject to:} \quad -x_1 + x_2 \leq 2$$
$$x_1 + x_2 \geq 4$$
$$x_1, x_2 \geq 0$$

Table 12.17.

c_b	c_j Basis	1 x_1	2 x_2	0 x_3	0 x_4	$-M$ A_1	Solution
0 $-M$	x_3 A_1	-1 1	① 1	1 0	0 -1	0 1	2 ← 4
	z_j	$-M$	$-M$	0	M	$-M$	$-4M$
	$c_j - z_j$	$1 + M$	$2 + M$	0	$-M$	0	
		↑					
2 $-M$	x_2 A_1	-1 ②	1 0	1 -1	0 -1	0 1	2 2 ←
	z_j	$-2 - 2M$	2	$2 + M$	M	$-M$	$4 - 2M$
	$c_j - z_j$	$3 + 2M$	0	$-2 - M$	$-M$	0	
		↑					
2 1	x_2 x_1	0 1	1 0	$\frac{1}{2}$ $-\frac{1}{2}$	$-\frac{1}{2}$ $-\frac{1}{2}$	$\frac{1}{2}$ $\frac{1}{2}$	3 1
	z_j	1	2	$\frac{1}{2}$	$-\frac{3}{2}$	$\frac{3}{2}$	7
	$c_j - z_j$	0	0	$-\frac{1}{2}$	$\frac{3}{2}$	$-M - \frac{3}{2}$	
					↑		

The effect of an unbounded solution on the simplex algorithm is shown in Table 12.17. The final tableau of the table shows that variable x_4 should enter the basis. The elements in the x_4 column of this tableau are, however, negative. According to Simplex Rule 2, negative elements are ignored in forming the ratios used in selecting the variable to leave the basis. Since neither of the basic variables in the final tableau can leave the basis, the change of basis required by Simplex Rule 1 cannot be made.

The condition illustrated by this example occurs for linear programming problems with unbounded solutions. On the basis of Simplex Rule 1, the final solution in Table 12.17 is not optimal and a change of basis is required. The change of basis cannot be made, however, because all the elements in the column headed by the entering variable are zero or negative. If there are no positive elements in this column, the optimal solution is unbounded.

PROBLEMS

1. Solve the following linear programming problems by determining the value of the objective function for all basic solutions to the problem:

(a)   Maximize: $Z = 3x_1 + 1x_2$

  Subject to: $3x_1 + 2x_2 \leq 24$
  $6x_1 + 12x_2 \leq 60$
  $x_1, x_2 \geq 0$

(b)   Maximize: $Z = 10x_1 + 14x_2$

  Subject to: $2x_1 + 3x_2 \leq 12$
  $x_1 + 3x_2 \leq 10$
  $x_1, x_2 \geq 0$

(c)   Maximize: $Z = 7x_1 + 8x_2$

  Subject to: $2x_1 + x_2 \leq 10$
  $x_1 + 3x_2 \leq 16$
  $x_1, x_2 \geq 0$

(d)   Minimize: $Z = 4x_1 + 3x_2$

  Subject to: $x_1 \geq 4$
  $3x_1 + 2x_2 \geq 18$
  $x_2 \geq 0$

2. Solve the following linear programming problems by using the simplex algorithm:

   (a)   Maximize: $Z = 6x_1 + 2x_2$

               Subject to: $4x_1 + x_2 \leq 24$

$$5x_1 + 2x_2 \leq 36$$
$$x_1, x_2 \geq 0$$

   (b)   Maximize: $Z = 8x_1 + 6x_2$

               Subject to: $x_1 \leq 5$

$$x_2 \leq 4$$
$$2x_1 + 5x_2 \leq 20$$
$$x_1, x_2 \geq 0$$

   (c)   Maximize: $Z = 3x_1 + 1x_2 + 2x_3$

               Subject to: $4x_1 + 2x_2 + 2x_3 \leq 36$

$$x_2 + 3x_3 \leq 12$$
$$x_1, x_2, x_3 \geq 0$$

   (d)   Maximize: $Z = 2x_1 + 4x_2 + 3x_3$

               Subject to: $x_1 + 2x_2 \leq 80$

$$x_1 + 4x_2 + 2x_3 \leq 120$$
$$x_1, x_2, x_3 \geq 0$$

3. Solve the following linear programming problems by using the simplex algorithm:

   (a)   Maximize: $Z = 20x_1 + 6x_2 + 8x_3$

               Subject to: $8x_1 + 2x_2 + 3x_3 \leq 200$

$$4x_1 + 3x_2 \leq 100$$
$$x_3 \leq 20$$
$$x_1, x_2, x_3 \geq 0$$

   (b)   Maximize: $Z = 10x_1 + 20x_2 + 16x_3$

               Subject to: $4x_1 + 6x_2 + 6x_3 \leq 6000$

$$x_1 + x_2 + 3x_3 \leq 4000$$
$$x_1, x_2, x_3 \geq 0$$

4. Solve the following linear programming problems by using the simplex algorithm:

   (a)   Minimize: $Z = 2x_1 + 3x_2$

               Subject to: $x_1 + 2x_2 \geq 18$

$$2x_1 + x_2 \geq 24$$
$$x_1, x_2 \geq 0$$

(b)   Minimize: $Z = 20x_1 + 16x_2$

Subject to: $3x_1 + \quad x_2 \geq 6$

$x_1 + \quad x_2 \geq 4$

$x_1, x_2 \geq 0$

5. Solve the following linear programming problems by using the simplex algorithm:

(a)   Minimize: $Z = \quad x_1 + 3x_2$

Subject to: $6x_1 + 2x_2 \geq 12$

$2x_1 + 2x_2 = 8$

$x_1, x_2 \geq 0$

(b)   Minimize: $Z = 10x_1 + 8x_2$

Subject to: $x_1 + 5x_2 \leq 4000$

$2x_1 + 6x_2 \geq 3000$

$x_1 + \quad x_2 = 1000$

$x_1, x_2 \geq 0$

6. The following linear programming problem has multiple optimal solutions. Use the simplex algorithm to find the basic solutions to the problem. Give the linear combinations of the basic solutions that define the multiple optimal solutions.

Maximize: $Z = 3x_1 + 2x_2$

Subject to: $3x_1 + 2x_2 \leq 24$

$x_1 + 2x_2 \leq 12$

$x_1, x_2 \geq 0$

7. The following linear programming problems illustrate the special cases of no feasible solution and unbounded solution. Use the simplex algorithm to specify which problem has no feasible solution and which has an unbounded solution.

(a)   Minimize: $Z = \quad x_1 + \quad 3x_2$

Subject to: $-10x_1 + \quad 6x_2 \geq \quad 60$

$10x_1 + 15x_2 \leq 120$

$x_1, x_2 \geq 0$

(b)   Maximize: $Z = \quad 3x_1 + \quad x_2$

Subject to: $7x_1 + \quad 5x_2 \geq 140$

$-4x_1 + \quad 8x_2 \leq \quad 32$

$x_1, x_2 \geq 0$

8. Two products are manufactured by the Acme Company. Product 1 contributes \$5 to profit and product 2 contributes \$4. Each product requires both assembly labor and finishing labor. Product 1 requires 6 hours of assembly labor and 2 hours of finishing labor. Product 2 requires 4 hours of assembly labor and 8 hours of finishing labor. There

are 120 hours of assembly labor and 160 hours of finishing labor available for scheduling. Use the simplex algorithm to determine the product mix that gives maximum profit.

9. A manufacturer makes three types of decorative tensor lamps; model 1200, model 1201, and model 1202. The cost of raw materials for each lamp is the same; however, the cost of production differs. Each model 1200 lamp requires 0.1 hour of assembly time, 0.2 hour of wiring time, and 0.1 hour of packaging time. The model 1201 requires 0.2 hour of assembly time, 0.1 hour of wiring time, and 0.1 hour of packaging time. The model 1202 requires 0.2 hour of assembly time, 0.3 hour of wiring time, and 0.1 hour of packaging time. The manufacturer makes a profit of $1.20 on each model 1200 lamp, $1.90 on each model 1201 lamp, and $2.40 on each model 1202 lamp. The manufacturer can schedule up to 90 hours of assembly labor, 120 hours of wiring labor, and 100 hours of packaging labor. Assuming that all lamps can be sold, determine the optimal quantities of each model and the marginal values of each resource.

10. A clothing manufacturer is scheduling work for the next week. There are three possible products that can be made: sportcoats, topcoats, and raincoats. The following table gives the profit for each product and the time required in each process:

	Profit	Hours Required per Unit in		
Product	per Unit	Cutting	Sewing	Detailing
Sportcoat	$ 5	1.0	1.0	0.5
Topcoat	8	2.0	1.5	1.0
Raincoat	12	2.0	2.0	1.5

The maximum number of hours that can be scheduled for each process are: cutting, 80 hours; sewing, 60 hours; and detailing, 50 hours. Assuming that all garments produced can be sold, determine the optimal quantities of each item and the marginal values of the three resources.

11. Use the simplex algorithm to determine the optimal quantities of Toots, Wheets, and Honks referred to in Problem 10, Chap. 11. What is the value of an additional unit of each resource used in the manufacture of the noisemakers?

Additional linear programming problems are given in Chap. 11.

SUGGESTED REFERENCES

The references for this chapter are listed in Chap. 11.

Chapter Thirteen

The Transportation and Assignment Problems

By now it should be obvious that linear programming is an important tool for the quantitative analysis of business decisions. The allocation of resources among competing products or activities so that profits are maximized or, alternatively, costs are minimized is a necessary function in a firm. Linear programming provides the manager the information necessary to make an optimal decision in these complex allocation problems.

This chapter considers problems that might be termed derivatives of the general linear programming problem. By this we mean that while certain problems can be formulated and solved as a linear programming problem by using the simplex algorithm, their special mathematical structure allows solution by much more efficient algorithms. Two kinds of problems for which such special algorithms exist are the transportation and the assignment problems. Examples of these important problems, along with the solution algorithms for the problems, are introduced in this chapter.

13.1 The Transportation Problem

The *transportation problem* derives its name from the problem of transporting homogeneous products from various sources of supply to several points of demand. The allocation of the products from the sources of supply to the points of demand is made with the objective of minimizing the total transportation costs or, alternatively, maximizing the profit from the sale of the products. To illustrate, consider a firm that has three factories (i.e., sources of supply) and four warehouses (i.e., points of demand). The cost of

Table 13.1.

Factory (F_i)	Warehouse (W_j)				Factory Capacity
	W_1	W_2	W_3	W_4	
F_1	0.30	0.25	0.40	0.20	100
F_2	0.29	0.26	0.35	0.40	250
F_3	0.31	0.33	0.37	0.30	150
Warehouse Requirement	100	150	200	50	500

shipping from each factory to each warehouse depends on the distance the product must be shipped, freight rates, etc. These costs are shown in Table 13.1.

The cost of shipping from factory i to warehouse j is represented by c_{ij}. The alternative values of c_{ij} are shown in the table. For instance, the cost of transporting 1 unit of product from factory 1 to warehouse 2 is $c_{12} = \$0.25$. The factory capacities and warehouse requirements are also given in the table. In this simplified example, the factory capacities equal the warehouse requirements, i.e., they both sum to 500 units. This represents an example of a *balanced* transportation problem. The more realistic case of the *unbalanced* transportation problem in which factory capacities and warehouse requirements are not equal is introduced later in this section. At this point it is necessary only to observe that an unbalanced transportation problem can be converted to a balanced transportation problem through the addition of an appropriate slack variable.

The objective in this problem is to develop a shipping schedule that minimizes the total transportation cost. The problem can be formulated by letting x_{ij} represent the number of units shipped from factory i to warehouse j. The objective is to minimize the total transportation costs subject to the constraints on factory capacities and warehouse requirements. For the example problem, this can be written as

$$\text{Minimize } C = \quad 0.30x_{11} + 0.25x_{12} + 0.40x_{13} + 0.20x_{14} + 0.29x_{21}$$
$$+ 0.26x_{22} + 0.35x_{23} + 0.40x_{24} + 0.31x_{31} + 0.33x_{32}$$
$$+ 0.37x_{33} + 0.30x_{34}$$

The warehouse constraints are

$$x_{11} + x_{21} + x_{31} = 100$$

$$x_{12} + x_{22} + x_{32} = 150$$

$$x_{13} + x_{23} + x_{33} = 200$$

$$x_{14} + x_{24} + x_{34} = 50$$

The factory constraints are

$$x_{11} + x_{12} + x_{13} + x_{14} = 100$$

$$x_{21} + x_{22} + x_{23} + x_{24} = 250$$

$$x_{31} + x_{32} + x_{33} + x_{34} = 150$$

and the non-negativity constraints are

$$x_{ij} \geq 0, \quad \text{for } i = 1, 2, 3 \quad \text{and} \quad j = 1, 2, 3, 4$$

This transportation problem can be solved by using the simplex algorithm. Because of the special structure of the constraints, however, alternative algorithms that reduce the computational burden are available. Before introducing these algorithms, it will be worthwhile to formally state the problem mathematically and to provide additional examples of the transportation problem.

The standard transportation problem is to optimize an objective function

$$Z = \sum_{i=1}^{m} \sum_{j=1}^{n} c_{ij} x_{ij}, \quad \text{for } i = 1, 2, \ldots, m; \, j = 1, 2, \ldots, n \quad (13.1)$$

subject to the constraints that

$$\sum_{i=1}^{m} x_{ij} = D_j, \quad \text{for } j = 1, 2, \ldots, n \quad\quad\quad (13.2)$$

$$\sum_{j=1}^{n} x_{ij} = S_i, \quad \text{for } i = 1, 2, \ldots, m \quad\quad\quad (13.3)$$

and

$$x_{ij} \geq 0 \quad\quad \text{for all } i \text{ and } j$$

D_j and S_i are non-negative integers that represent, respectively, the demand at the jth facility (warehouse, retail store, etc.) and the supply at the ith source (factory, warehouse, supplier, etc.). An additional requirement is that the sum of the demands equal the sum of the supplies, i.e.,

$$\sum_{i=1}^{m} S_i = \sum_{j=1}^{n} D_j \quad\quad\quad\quad (13.4)$$

For the unbalanced transportation problem, the requirement that the sum of the demands equal the sum of the supplies can be satisfied by creating a fictitious demand facility or supply facility. For instance, if supply exceeds demand, a fictitious demand column is established with zero transportation cost to absorb the excess supply. Similarly, if demand exceeds supply, a fictitious supply row is established with zero transportation cost to absorb the unsatisfied demand. The addition of a fictitious supply or demand to convert an unbalanced transportation problem to the required balanced transportation format is illustrated by the following three examples.

Example. A national firm has three factories and four warehouses. The cost of manufacturing a given product is the same at each of the factories. Because of the locations of the warehouses and factories, the cost of shipping the product from the different factories to the warehouses varies. The capacities of factories 1, 2, and 3 are 5000 units per month, 4000 units per month, and 7000 units per month. The requirements of warehouses 1, 2, 3, and 4 are, respectively, 3000 units per month, 2500 units per month, 3500 units per month, and 4000 units per month. The transportation costs, factory capacities, and warehouse requirements are given in Table 13.2.

Table 13.2.

Factories	Warehouses W_1	W_2	W_3	W_4	Factory Capacity
F_1	15	24	11	12	5000
F_2	25	20	14	16	4000
F_3	12	16	22	13	7000
Warehouse Requirements	3000	2500	3500	4000	13,000 ╲ 16,000

Since the total factory capacity exceeds the total warehouse requirement, the transportation problem is unbalanced. It can be converted to a balanced transportation problem by adding a fictitious warehouse W_5 with requirements of 3000 units. The cost of shipping from each factory to this fictitious warehouse is $0. The balanced tableau is shown in Table 13.3.

The procedure for converting an unbalanced problem in which the demand exceeds the supply to a balanced problem is similar to that shown in Table 13.3. The only difference is that a fictitious row rather than column must be added to the table. The costs associated with this row are again $0. The

Table 13.3.

Factory	W_1	W_2	W_3	W_4	W_5	Factory Capacity
F_1	15	24	11	12	0	5000
F_2	25	20	14	16	0	4000
F_3	12	16	22	13	0	7000
Warehouse Requirements	3000	2500	3500	4000	3000	16,000

supply available from this fictitious source equals the number of units required to balance the table.

   Example. A firm has decided to expand its product line by producing one or more of five possible products. Three of the firm's current plants have the excess capacity to produce these products. The profit margin on each of the proposed products is the same; consequently, the firm wants to minimize the cost of production. The excess capacity of each plant, the potential sales of each product, and the estimated cost of producing the proposed products at the three plants are given in Table 13.4.

Table 13.4.

Plant	Product 1	2	3	4	5	Excess Capacity (Standard Units)
1	22	20	16	23	17	70
2	16	19	14	19	18	80
3	19	16	X	20	21	100
Potential Sales (Standard Units)	60	70	90	65	85	370 250

This problem could be treated as a standard linear programming problem and solved by using the simplex algorithm. If, however, the plants are considered as supply facilities and potential sales of the five products are treated as demands, the problem can also be viewed as a transportation problem.

   To convert the table into a balanced transportation table, a fictitious plant capable of supplying 120 units of capacity must be added. This involves adding a fourth row with a capacity of 120 units and production costs of $0. In addition to this modification, the table shows that product 3 cannot be manufactured at plant 3. To ensure that the optimal solution does not include the use of plant 3 to produce product 3, we can assign an arbitrarily large cost

Table 13.5.

Plant	Product 1	2	3	4	5	Excess Capacity (Standard Units)
1	22	20	16	23	17	70
2	16	19	14	19	18	80
3	19	16	M	20	21	100
4	0	0	0	0	0	120
Potential Sales (Standard Units)	60	70	90	65	85	370

of production for this plant–product combination. Although any large cost would ensure that production of product 3 is not scheduled for plant 3, we shall follow the "Big M" convention introduced in Chap. 12 and assign the letter M as the cost of production. The balanced transportation table is given in Table 13.5.

Example. Bevitz Furniture Company has requested bids from four furniture manufacturers on five different styles of furniture. The quantities of the five styles of furniture required by Bevitz are shown below.

Style	A	B	C	D	E
Quantity	125	75	50	200	175

The four manufacturers have limited production capacities. The total quantities that can be supplied in the time span available are shown below.

Manufacturer	1	2	3	4
Quantity	275	225	175	200

On the basis of the quotes from each manufacturer, Bevitz estimates that the profit per unit for each item sold will vary as shown in Table 13.6. Determine the allocation that maximizes profit.

Table 13.6.

		Style			
	A	B	C	D	E
1	28	35	42	23	15
Manufacturer 2	30	33	45	18	10
3	25	35	48	20	13
4	33	28	40	26	18

This problem can be formulated as a transportation problem. Instead of minimizing transportation cost, however, the objective is to maximize profit.

To apply the transportation algorithm, the problem must be converted to a balanced transportation problem. Since the capacity of the manufacturers exceeds the quantity demanded by Bevitz Furniture, the balance is achieved by adding a fictitious style. The demand for the fictitious style is equal to the unused capacity of the manufacturers, i.e., 250 units. The profit from the fictitious demand is $0. The balanced transportation table is shown in Table 13.7.

Table 13.7.

Manufacturer	*A*	*B*	Style *C*	*D*	*E*	*F*	Quantity Supplied
1	28	35	42	23	15	0	275
2	30	33	45	18	10	0	225
3	25	35	48	20	13	0	175
4	33	28	40	26	18	0	200
Demand	125	75	50	200	175	250	875

13.2 The Initial Basic Feasible Solution

The first step in solving the transportation problem is to develop an initial basic feasible solution. Two methods for obtaining this solution are illustrated in this section. The first method, termed the *northwest corner rule*, provides a straightforward technique for obtaining the initial solution. It suffers, however, when compared to the *minimum (maximum) cell method* in that more iterations are normally required to obtain the optimal solution. The minimum cell method is introduced later in this section.

13.2.1 THE NORTHWEST CORNER RULE

To illustrate the northwest corner rule, consider the transportation problem summarized in Table 13.1. The factory capacities and warehouse requirements for the problem are shown in Table 13.8. The transportation costs (i.e., the c_{ij}'s) are not included in this table. These costs are not relevant when

Table 13.8.

Factory	W_1	Warehouse W_2	W_3	W_4	Factory Capacity
F_1					100
F_2					250
F_3					150
Requirements	100	150	200	50	500

one is using the northwest corner rule to determine the initial solution.

The initial solution is found by beginning in the upper left-hand (i.e., northwest) corner of the tableau and allocating the resource of the first row to the cells in the first row until the resource is exhausted. If a resource is exhausted and a requirement is satisfied by a single allocation, a zero is placed in a neighboring cell. We then move to the second row and continue the allocation until the resources of that row are fully allocated. Again, if a single allocation exhausts both the supply of resource and satisfies the requirement of a column, a zero must be placed in a neighboring cell. This process is continued until all resources are exhausted and all requirements are satisfied.

In this example, the 100 units of capacity of factory 1 are assigned to warehouse 1. This allocation exhausts the capacity of the first row and satisfies the requirements of the first column; consequently, we place a zero in the row 1, column 2 cell and move to the second row. Since the requirements of warehouse 1 have been met, the next allocation is 150 units of factory 2 capacity to warehouse 2. The remaining 100 units of capacity of factory 2 are allocated to warehouse 3. One hundred units of factory 3 capacity complete the requirements of warehouse 3. The final 50 units of factory 3 capacity satisfy the requirements of warehouse 4. The initial solution is shown in Table 13.9.

Table 13.9.

Factory	Warehouse				Factory Capacity
	W_1	W_2	W_3	W_4	
F_1	100	0			100
F_2		150	100		250
F_3			100	50	150
Requirements	100	150	200	50	500

By comparing the solution in Table 13.9 with the constraints for the problem given on p. 415, it can be seen that the solution is feasible. Both the warehouse and the factory constraints are satisfied. The cost of this solution is found by multiplying the cell allocations in Table 13.9 by the transportation costs from Table 13.1 and summing these products. The cost of this initial solution is

$0.30(100) + $0.26(150) + $0.35(100) + $0.37(100) + $0.30(50) = $156.00

Example. Use the northwest corner rule to determine an initial solution for the transportation table shown in Table 13.3.

The factory capacities and the warehouse requirements, including the fictitious warehouse W_5, are given in the transportation table. The allocation is again made by beginning in the upper left-hand corner of the table and assigning capacities to requirements so as to exhaust all capacities and satisfy all requirements. The resulting feasible solution is shown in Table 13.10. The cost of this solution, again found by summing the products of the cell entries and the transportation costs, is $204,000.

Table 13.10.

Factory	Warehouses W₁	W₂	W₃	W₄	W₅	Factory Capacity
F_1	3000	2000				5000
F_2		500	3500	0		4000
F_3				4000	3000	7000
Requirements	3000	2500	3500	4000	3000	16,000

The northwest corner rule provides a systematic, easily understandable method of obtaining an initial basic feasible solution. As mentioned earlier, however, it is inefficient in comparison with alternative methods for obtaining the initial solution. This inefficiency occurs because the costs (profits) are not considered in determining the initial solution. Although the algorithm used to determine the optimal solution is not introduced until the following section, it is reasonable to conclude that the number of iterations required to obtain the optimal solution is dependent on how near the initial solution is to being optimal. This implies that the number of iterations can be reduced by beginning the iterative process from a "near optimal" initial solution. Costs (profits) are used to determine the initial solution in the minimum (maximum) cell method. Consequently, the number of iterations required to obtain the optimal solution is normally reduced by beginning with an initial solution found using the minimum (maximum) cell method.

13.2.2 THE MINIMUM (MAXIMUM) CELL METHOD

To illustrate the minimum (maximum) cell method of obtaining an initial solution, we again consider the transportation problem summarized in Table 13.1. This table, complete with transportation costs, capacities, and requirements, is reproduced as Table 13.11. Notice that the transportation costs from Table 13.1 are placed in the upper left-hand corner of each cell in Table 13.11.

Table 13.11.

Factory	Warehouse				Factory Capacity
	W_1	W_2	W_3	W_4	
F_1	0.30	0.25	0.40	0.20	100
F_2	0.29	0.26	0.35	0.40	250
F_3	0.31	0.33	0.37	0.30	150
Requirements	100	150	200	50	500

The initial solution is obtained by sequentially allocating the resources to the cells with the minimum cost (or alternatively, with the maximum profit). As in the northwest corner rule, if a single allocation exhausts both the capacity of a row and satisfies the requirements of a column, a zero is placed in one of the cells that borders the allocation. Referring to Table 13.11, we see that the minimum transportation cost of $0.20 per unit is between factory 1 and warehouse 4. Since the requirements of warehouse 4 are 50 units, this allocation is entered in cell (1, 4). The allocation is shown in Table 13.12 in the lower right-hand corner of cell (1, 4).

Referring again to Table 13.11, we see that the next least costly allocation is the $0.25 per unit transportation cost from factory 1 to warehouse 2. The 50 units of unallocated capacity from factory 1 are, therefore, allocated to warehouse 2. This allocation is shown in cell (1, 2) in Table 13.12.

Table 13.12.

Factory	Warehouse				Factory Capacity
	W_1	W_2	W_3	W_4	
F_1	0.30	0.25 50	0.40	0.20 50	100
F_2	0.29 100	0.26 100	0.35 50	0.40	250
F_3	0.31	0.33	0.37 150	0.30	150
Requirements	100	150	200	50	500

Again with reference to the table, the next least costly allocation is the $0.26 per unit transportation cost from factory 2 to warehouse 2. The remaining 100 units required by warehouse 2 thus come from factory 2. The allocation is entered in cell (2, 2) in Table 13.12.

The procedure continues by allocating 100 units from factory 2 to warehouse 1, thereby satisfying the demand of warehouse 1. The remaining 50 units of capacity of factory 2 are next allocated to warehouse 3. The initial solution is completed by allocating the 150 units of factory 3 capacity to warehouse 3. The total cost of this initial solution is $150.50. This cost is $5.50 less than the cost of the initial solution obtained by using the northwest corner method.

Example. Use the minimum (maximum) cell method to obtain an initial solution for the transportation problem described in Table 13.3. Compare this solution with that found by using the northwest corner method.

The transportation table for this problem is reproduced in Table 13.13. The transportation costs are entered in the upper left-hand corner of each cell. The initial allocation of factory capacity to warehouse requirements will be entered in the lower right-hand corner of the cell.

Table 13.13.

Factory	Warehouse					Factory Capacity
	W_1	W_2	W_3	W_4	W_5	
F_1	15	24	11	12	0	5000
F_2	25	20	14	16	0	4000
F_3	12	16	22	13	0	7000
Requirements	3000	2500	3500	4000	3000	16,000

The transportation costs from each source to the fictitious warehouse W_5 are $0. Using the minimum cell method, we can allocate units from any of the three factories to W_5. We arbitrarily allocate 3000 units from F_3 to W_5.[1]

The next least costly transportation charge is from factory 1 to warehouse 3. Allocating 3500 units of F_1 capacity to W_3 satisfies the requirements of W_3.

---

[1] An alternative and more efficient procedure is to first allocate units to those cells with positive costs. The allocation to the fictitious row or column with zero cost is then made last.

The process of allocating the factory capacities to the warehouses with the smallest transportation cost is continued until all factory capacities are exhausted and warehouse requirements are satisfied. The initial solution is shown in Table 13.14.

The total transportation cost of this initial solution is $179,500. This represents a reduction of $24,500 from the initial solution of $204,000 obtained by using the northwest corner rule.

Table 13.14.

Factory	Warehouse W_1	W_2	W_3	W_4	W_5	Factory Capacity
F_1	15	24	11 3500	12 1500	0	5000
F_2	25	20 2500	14	16 1500	0	4000
F_3	12 3000	16	22	13 1000	0 3000	7000
Requirements	3000	2500	3500	4000	3000	16,000

Example. Use the minimum (maximum) cell method to determine the initial solution for the Bevitz Furniture Company problem.

The balanced transportation table for the Bevitz Furniture Company was given as Table 13.7. Since the Bevitz problem involves maximizing profit rather

Table 13.15.

Manufacturer	Style A	B	C	D	E	F	Quantity Supplied
1	28	35 75	42	23 125	15 75	0	275
2	30	33	45	18	10 100	0 125	225
3	25	35	48 50	20	13	0 125	175
4	33 125	28	40	26 75	18	0	200
Demand	125	75	50	200	175	250	875

than minimizing transportation cost, the allocations are made to the cells with the largest c_{ij} values rather than to those with the smallest. The initial solution using the maximum cell method is given in Table 13.15. The total profit of this initial solution is $16,100.

13.3 The Stepping-stone Algorithm

After an initial solution to the transportation problem is obtained, alternative solutions must be evaluated. A straightforward method of calculating the effect of alternative allocations is provided by the *stepping-stone algorithm*.

To apply the stepping-stone algorithm, we must first verify that the initial solution is a basic solution. The reader will remember that in the linear programming problem a basic solution is found for a problem of n variables and m constraints by equating $n - m$ of the variables to zero and solving the resulting system of m equations and m variables simultaneously. The procedure differs slightly in a transportation problem.

The transportation problem is constructed so that there are $m + n$ equations, where n represents the number of requirements and m represents the number of resources. It can be shown that one of these $m + n$ equations is redundant and that there are only $m + n - 1$ independent equations. For a solution to be basic, therefore, $m + n - 1$ cells must be occupied. This means that $m + n - 1$ cells in the initial solution must contain either an allocation or a zero.

If the initial solution contains less than $n + m - 1$ occupied cells, the solution is termed *degenerate*. Degeneracy occurs only when the resources of a row are exhausted and the requirements of a column are satisfied by a single allocation. It is eliminated by placing a zero in a cell that borders the allocation. The zero-valued cell is then considered occupied when applying the stepping-stone algorithm.

The stepping-stone algorithm involves transferring 1 unit from an occupied to an unoccupied cell and calculating the change in the objective function. The transfer must be made so as to retain the column and row equalities of the problem. After all unoccupied cells have been evaluated, a reallocation is made to the cell that provides the greatest per unit change in the objective function. Any degeneracies caused by the transfer of units must be removed by placement of zeros in the appropriate cells. The process of evaluating the empty cells and reallocating the units is continued until no further improvement in the objective function is possible. This final allocation is the optimal solution.

To illustrate the stepping-stone algorithm, consider the transportation problem summarized by Table 13.1. The initial solution to this problem, found by using the northwest corner rule, was given in Table 13.9. This solution is repeated in Table 13.16. The transportation costs, i.e., the c_{ij}'s, are included in the table.

Table 13.16.

Factory	Warehouse W_1	W_2	W_3	W_4	Factory Capacity
F_1	0.30 100	0.25 0	0.40	0.20	100
F_2	0.29	0.26 150	0.35 100	0.40	250
F_3	0.31	0.33	0.37 100	0.30 50	150
Requirements	100	150	200	50	500

To determine the effect on the objective function of transferring 1 unit to an unoccupied cell, we must find a *closed path* between the unoccupied cell and occupied cells. The path consists of a series of steps leading from the unoccupied cell to occupied cells and back to the unoccupied cell. In the case of cell (2, 1), for instance, a closed path consists of the series of steps from this cell to cell (1, 1), from cell (1, 1) to cell (1, 2), from cell (1, 2) to cell (2, 2), and from cell (2, 2) back to cell (2, 1). By following this path, we can determine the effect on the objective function of allocating a unit to cell (2, 1).

To illustrate, assume that 1 unit is allocated to cell (2, 1). In order to maintain the column and row equalities in the problem, a unit must be subtracted from cell (1, 1), added to cell (2, 1), and subtracted from cell (2, 2). Notice that this reallocation of units follows the closed path for cell (2, 1).

The net change in the objective function from the reallocation of 1 unit to cell (2, 1) can be found by adding and subtracting the appropriate transportation costs. Again, the closed path is followed; adding one unit to cell (2, 1) increases the objective function by $0.29, subtracting the unit from cell (1, 1) reduces the objective function by $0.30, adding the unit to cell (1, 2) increases the objective function by $0.25, and subtracting the unit from cell (2, 2) reduces the objective function by $0.26. The net decrease in the objective function is, therefore, $0.02. This decrease can be represented in equation form by

$$F_2 W_1 = +F_2 W_1 - F_1 W_1 + F_1 W_2 - F_2 W_2$$

or

$$F_2 W_1 = +0.29 - 0.30 + 0.25 - 0.26 = -\$0.02$$

The net decrease of $-\$0.02$ is entered in the lower right-hand corner of cell (2, 1) in Table 13.17.

The effect of reallocating 1 unit to each of the other unoccupied cells is determined in the same manner. The computations are shown below. It is important to remember that the closed paths are established in order to maintain both column and row equalities.

$$F_3W_2 = +F_3W_2 - F_3W_3 + F_2W_3 - F_2W_2$$

$$F_3W_2 = +0.33 - 0.37 + 0.35 - 0.26 = +\$0.05$$

$$F_1W_3 = +F_1W_3 - F_2W_3 + F_2W_2 - F_1W_2$$

$$F_1W_3 = +0.40 - 0.35 + 0.26 - 0.25 = +\$0.06$$

$$F_2W_4 = +F_2W_4 - F_3W_4 + F_3W_3 - F_2W_3$$

$$F_2W_4 = +0.40 - 0.30 + 0.37 - 0.35 = +\$0.12$$

$$F_3W_1 = +F_3W_1 - F_3W_3 + F_2W_3 - F_2W_2 + F_1W_2 - F_1W_1$$

$$F_3W_1 = +0.31 - 0.37 + 0.35 - 0.26 + 0.25 - 0.30 = -\$0.02$$

$$F_1W_4 = +F_1W_4 - F_3W_4 + F_3W_3 - F_2W_3 + F_2W_2 - F_1W_2$$

$$F_1W_4 = +0.20 - 0.30 + 0.37 - 0.35 + 0.26 - 0.25 = -\$0.07$$

The net change in the objective function caused by reallocating 1 unit to each unoccupied cell is entered in Table 13.17.

The dollar entries in the lower right-hand corner of each unoccupied cell in Table 13.17 represent the net change in the objective function from reallocating 1 unit to the cell. These are, in effect, equivalent to the $c_j - z_j$

Table 13.17.

Factory	Warehouse				Factory Capacity
	W_1	W_2	W_3	W_4	
F_1	0.30 100	0.25 0	0.40 $+\$0.06$	0.20 $-\$0.07$	100
F_2	0.29 $-\$0.02$	0.26 150	0.35 100	0.40 $+\$0.12$	250
F_3	0.31 $-\$0.02$	0.33 $+\$0.05$	0.37 100	0.30 50	150
Requirements	100	150	200	50	500

values in the simplex tableau. The reader will remember that in the simplex algorithm for a minimization problem the variable with the most negative $c_j - z_j$ entry is introduced into the basis. The same procedure is followed in the stepping-stone algorithm, namely, a reallocation is made to the most favorable evaluation, i.e., the cell that provides the largest per unit decrease in the objective function for a minimization problem and the largest per unit increase in the objective function for a maximization problem. The reallocation follows the closed path used to calculate the change in the objective function. As in the simplex algorithm, as many units as possible are reallocated to the cell.

Referring to Table 13.17, note that the largest per unit decrease in the objective function comes from reallocating units to cell (1, 4). The closed path used to evaluate cell (1, 4) was

$$F_1 W_4 = +F_1 W_4 - F_3 W_4 + F_3 W_3 - F_2 W_3 + F_2 W_2 - F_1 W_2$$

The limit on the number of units that can be reallocated to cell (1, 4) is equal to the minimum of the current allocations to cells (3, 4), (2, 3), and (1, 2). The table shows that 50 units can be subtracted from cell (3, 4), 100 units can be subtracted from cell (2, 3), and 0 units can be subtracted from cell (1, 2). Unfortunately, the closed path used to evaluate cell (1, 4) involved subtracting units from cell (1, 2). Since cell (1, 2) has a zero allocation in the initial solution, allocating units to cell (1, 4) would be equivalent to transferring the 0 entry in cell (1, 2) to cell (1, 4). This, of course, would not decrease the value of the objective function.

Rather than merely transferring the 0 entry from cell (1, 2) to cell (1, 4), units can be reallocated to a cell that decreases the value of the objective function. An allocation to cell (2, 1) or cell (3, 1) would decrease the objective function by $0.02 per unit. We arbitrarily select cell (2, 1) for reallocation.

The closed path used to evaluate cell (2, 1) was

$$F_2 W_1 = +F_2 W_1 - F_1 W_1 + F_1 W_2 - F_2 W_2$$

The limit on the number of units that can be added to cell (2, 1) is the 100 units initially assigned to cell (1, 1). This is due to the fact that units must be subtracted from cell (1, 1), and cell (1, 1) contains only 100 units. These units are reallocated to cell (2, 1). In order to maintain the column and row equalities, 100 of the 150 units in cell (2, 2) are reallocated to cell (1, 2). This leaves 50 units in cell (2, 2). The new transportation table is shown in Table 13.18.

The solution shown in Table 13.18 contains six occupied cells. Since there are three rows and four columns in the problem and the number of occupied cells is equal to $m + n - 1$, the solution is not degenerate. Therefore, we need not consider any of the blank cells as occupied at a zero level.

Table 13.18.

Factory	Warehouse				Factory Capacity
	W_1	W_2	W_3	W_4	
F_1	0.30	0.25 100	0.40	0.20	100
F_2	0.29 100	0.26 50	0.35 100	0.40	250
F_3	0.31	0.33	0.37 100	0.30 50	150
Requirements	100	150	200	50	500

The stepping-stone algorithm is used to evaluate the unoccupied cells in Table 13.18. The calculations are shown below.

$$F_1W_1 = +F_1W_1 - F_1W_2 + F_2W_2 - F_2W_1 = +\$0.02$$

$$F_3W_1 = +F_3W_1 - F_3W_3 + F_2W_3 - F_2W_1 = \quad \$0.00$$

$$F_3W_2 = +F_3W_2 - F_3W_3 + F_2W_3 - F_2W_2 = +\$0.05$$

$$F_1W_3 = +F_1W_3 - F_2W_3 + F_2W_2 - F_1W_2 = +\$0.06$$

$$F_1W_4 = +F_1W_4 - F_3W_4 + F_3W_3 - F_2W_3 + F_2W_2 - F_1W_2 = -\$0.07$$

$$F_2W_4 = +F_2W_4 - F_3W_4 + F_3W_3 - F_2W_3 = +\$0.12$$

The evaluations are entered in Table 13.19.

Table 13.19.

Factory	Warehouse				Factory Capacity
	W_1	W_2	W_3	W_4	
F_1	0.30 + $0.02	0.25 100	0.40 + $0.06	0.20 − $0.07	100
F_2	0.29 100	0.26 50	0.35 100	0.40 + $0.12	250
F_3	0.31 $0.00	0.33 + $0.05	0.37 100	0.30 50	150
Requirements	100	150	200	50	500

The table shows that units should be reallocated to cell (1, 4). The reallocation is made by following the closed path that gave the $0.07 per unit decrease in the objective function for the cell. The reallocation is shown in Table 13.20.

Table 13.20.

Factory	Warehouse				Factory Capacity
	W_1	W_2	W_3	W_4	
F_1	0.30	0.25 50	0.40	0.20 50	100
F_2	0.29 100	0.26 100	0.35 50	0.40	250
F_3	0.31	0.33	0.37 150	0.30	150
Requirements	100	150	200	50	500

The stepping-stone is again applied to evaluate the unoccupied cells in Table 13.20. The calculations are shown below.

$$F_1W_1 = +F_1W_1 - F_2W_1 + F_2W_2 - F_1W_2 = +\$0.02$$

$$F_3W_1 = +F_3W_1 - F_3W_3 + F_2W_3 - F_2W_1 = \quad \$0.00$$

$$F_3W_2 = +F_3W_2 - F_3W_3 + F_2W_3 - F_2W_2 = +\$0.05$$

$$F_1W_3 = +F_1W_3 - F_2W_3 + F_2W_2 - F_1W_2 = +\$0.06$$

$$F_2W_4 = +F_2W_4 - F_2W_2 + F_1W_2 - F_1W_4 = +\$0.19$$

$$F_3W_4 = +F_3W_4 - F_3W_3 + F_2W_3 - F_2W_2 + F_1W_2 - F_1W_4 = +\$0.07$$

Since all the evaluations are non-negative, the solution is optimal. This is not, however, a unique solution. The calculations show that the objective function does not change as units are allocated to cell (3, 1). Thus, multiple optimal solutions exist for this problem. The value of the objective function, found by calculating the total transportation cost, is $150.50.

Example. Use the stepping-stone algorithm to determine the optimal allocation for the transportation problem introduced on p. 417.

The initial solution for this problem, found by using the northwest corner rule, was determined on p. 422 and is reproduced in Table 13.21.

Table 13.21.

Factory	W_1		W_2	Warehouse W_3		W_4		W_5		Factory Capacity
F_1	15	3000	24	2000	11		12		0		5000
F_2	25		20	500	14	3500	16	0	0		4000
F_3	12		16		22		13	4000	0	3000	7000
Requirements	3000		2500		3500		4000		3000		16,000

The unoccupied cells are evaluated by establishing closed paths between each unoccupied cell and the occupied cells. The evaluations are

$$F_1W_3 = +F_1W_3 - F_2W_3 + F_2W_2 - F_1W_2 = -\$7$$

$$F_1W_4 = +F_1W_4 - F_2W_4 + F_2W_2 - F_1W_2 = -\$8$$

$$F_1W_5 = +F_1W_5 - F_3W_5 + F_3W_4 - F_2W_4 + F_2W_2 - F_1W_2 = -\$7$$

$$F_2W_1 = +F_2W_1 - F_1W_1 + F_1W_2 - F_2W_2 = +\$14$$

$$F_2W_5 = +F_2W_5 - F_3W_5 + F_3W_4 - F_2W_4 = -\$3$$

$$F_3W_1 = +F_3W_1 - F_3W_4 + F_2W_4 - F_2W_2 + F_1W_2 - F_1W_1 = +\$4$$

$$F_3W_2 = +F_3W_2 - F_3W_4 + F_2W_4 - F_2W_2 = -\$1$$

$$F_3W_3 = +F_3W_3 - F_3W_4 + F_2W_4 - F_2W_3 = +\$11$$

The most favorable evaluation is the $-\$8$ for cell $(1, 4)$. It is impossible to reallocate units to this cell, however, because of the zero allocation in cell $(2, 4)$.

The next most favorable evaluations are for cells $(1, 3)$ and $(1, 5)$. Reallocating units to cell $(1, 5)$ also requires subtracting units from cell $(2, 4)$. Since this would only mean transferring the zero from cell $(2, 4)$ to cell $(1, 5)$, cell $(1, 3)$ rather than cell $(1, 5)$ is selected to receive the allocation. The result is shown in Table 13.22. The evaluations of the unoccupied cells in the new solution are also given. The reader should verify these evaluations by determining the closed paths for each unoccupied cell and calculating the effect on the objective function of reallocating 1 unit to the unoccupied cell.

Table 13.22.

Factory	Warehouse					Factory Capacity
	W_1	W_2	W_3	W_4	W_5	
F_1	15 3000	24 + $7	11 2000	12 − $1	0 $0	5000
F_2	25 + $7	20 2500	14 1500	16 0	0 − $3	4000
F_3	12 − $3	16 − $1	22 + $11	13 4000	0 3000	7000
Requirements	3000	2500	3500	4000	3000	16,000

Total transportation cost can be reduced by reallocating units to cell (3, 1). The reallocation, together with the evaluation of the empty cells, is given in Table 13.23.

Table 13.23.

Factory	Warehouse					Factory Capacity
	W_1	W_2	W_3	W_4	W_5	
F_1	15 1500	24 + $4	11 3500	12 − $4	0 − $3	5000
F_2	25 + $10	20 2500	14 + $3	16 1500	0 − $3	4000
F_3	12 1500	16 − $1	22 + $14	13 2500	0 3000	7000
Requirements	3000	2500	3500	4000	3000	16,000

Transportation cost can be further reduced by reallocating units to cell (1, 4). This allocation is shown in Table 13.24.

An additional reduction in transportation cost is possible by allocating units to cell (2, 5), as shown in Table 13.25.

Allocating units to cell (3, 2) gives Table 13.26.

The transportation cost is again reduced by reallocating units to cell (2, 3). The final reallocation, shown in Table 13.27, gives the optimal solution. The total transportation cost is $167,000.

Table 13.24.

| Factory | Warehouse | | | | | Factory Capacity |
	W_1	W_2	W_3	W_4	W_5	
F_1	15 + \$4	24 + \$8	11 3500	12 1500	0 + \$1	5000
F_2	25 + \$10	20 2500	14 − \$1	16 1500	0 − \$3	4000
F_3	12 3000	16 − \$1	22 + \$10	13 1000	0 3000	7000
Requirements	3000	2500	3500	4000	3000	16,000

Table 13.25.

| Factory | Warehouse | | | | | Factory Capacity |
	W_1	W_2	W_3	W_4	W_5	
F_1	15 + \$4	24 + \$5	11 3500	12 1500	0 + \$1	5000
F_2	25 + \$13	20 2500	14 + \$2	16 + \$3	0 1500	4000
F_3	12 3000	16 − \$4	22 + \$10	13 2500	0 1500	7000
Requirements	3000	2500	3500	4000	3000	16,000

Table 13.26.

| Factory | Warehouse | | | | | Factory Capacity |
	W_1	W_2	W_3	W_4	W_5	
F_1	15 + \$4	24 + \$9	11 3500	12 1500	0 + \$5	5000
F_2	25 + \$9	20 1000	14 − \$2	16 − \$1	0 3000	4000
F_3	12 3000	16 1500	22 + \$10	13 2500	0 + \$4	7000
Requirements	3000	2500	3500	4000	3000	16,000

Table 13.27.

| Factory | Warehouse | | | | | Factory Capacity |
	W_1	W_2	W_3	W_4	W_5	
F_1	15 + $4	24 + $9	11 2500	12 2500	0 + $3	5000
F_2	25 + $11	20 + $2	14 1000	16 + $1	0 3000	4000
F_3	12 3000	16 2500	22 + $10	13 1500	0 + $2	7000
Requirements	3000	2500	3500	4000	3000	16,000

We stated earlier that the number of iterations required to obtain the optimal solution is dependent on how near the initial solution is to being optimal. This is demonstrated by the preceding example. Notice in this example that the fourth tableau, Table 13.24, contains the same allocations as the initial tableau found by using the minimum cell method, Table 13.14. The iterations in the example leading to the fourth tableau would not have been necessary had we begun with the initial solution obtained by using the minimum cell method instead of the initial solution from the northwest corner rule.

13.4　The Assignment Problem

The *assignment problem*, like the transportation problem, is a special case of the linear programming problem. The assignment problem occurs when *n* jobs must be assigned to *n* facilities on a one-to-one basis. The assignment is made with the objective of minimizing the overall cost of completing the jobs, or, alternatively, of maximizing the overall profit from the jobs.

To illustrate the assignment problem, consider the following example. A firm has five jobs that must be assigned to five work crews. Because of varying experience of the work crews, each work crew is not able to complete each job with the same effectiveness. The cost of each work crew to do each job is given by the cost matrix shown in Table 13.28. The objective is to assign the jobs to the work crews in order to minimize the total cost of completing all jobs.

Table 13.28.

Job

	j / i	1	2	3	4	5
	1	41	72	39	52	25
Work	2	22	29	49	65	81
Crew	3	27	39	60	51	40
	4	45	50	48	52	37
	5	29	40	45	26	30

13.4.1 ENUMERATING ALL POSSIBLE ASSIGNMENTS

One way to approach this problem would be to enumerate all possible assignments of work crew i to job j. Although this approach is possible for small problems, it rapidly becomes unmanageable as the size of the problem increases.

The number of possible assignments of n facilities to n jobs on a one-to-one basis is equal to $n(n-1)(n-2)\ldots 1$, or equivalently, by $n!$. Thus, the number of possible assignments in Table 13.28 is $5! = 120$. If only one additional work crew and job were added to the table, the possible number of assignments would increase to $6! = 720$. Obviously, enumerating all possible assignments is feasible for only very small problems. Consequently, it is necessary to develop an alternative solution technique.

13.4.2 LINEAR PROGRAMMING FORMULATION

One alternative to complete enumeration is to formulate the assignment problem as a linear programming problem. If c_{ij} is defined as the cost of assigning facility i to job j and x_{ij} is defined as the proportion of time that facility i is assigned to job j, the linear programming problem is

$$\text{Minimize:} \quad Z = \sum_{i=1}^{n} \sum_{j=1}^{n} c_{ij} x_{ij} \tag{13.5}$$

$$\text{Subject to:} \quad \sum_{i=1}^{n} x_{ij} = 1 \quad \text{for } j = 1, 2, \ldots, n \tag{13.6}$$

$$\sum_{j=1}^{n} x_{ij} = 1 \quad \text{for } i = 1, 2, \ldots, n \tag{13.7}$$

$$x_{ij} \geq 0$$

The first set of constraints is necessary to assure that èach of the j jobs is assigned. The second set of constraints is required to make certain that exactly 100 percent of the time available to a facility is accounted for. For instance, the second set of constraints eliminates the possibility of assigning all jobs to the most efficient facility and no jobs to the other facilities. The final constraint eliminates the possibility of negative-valued variables.

An interesting result of the linear programming solution of the assignment problem is that the variables x_{ij} will have values of either zero or one. This occurs because both the coefficients in the constraining equations and the right-hand-side values are equal to one. The implication of this result is that the optimal assignment always involves a one-to-one matching of facility to job. For instance, if $x_{12} = 1$, then facility 1 is assigned to job 2. On the other hand, if $x_{12} = 0$, then facility 1 is not assigned to job 2. Since the optimal solution can never include variables that have a fractional value, one facility will always be assigned to one and only one job. Conversely, one job will always be assigned to one and only one facility.

13.5   The Assignment Algorithm

The assignment problem can be solved by writing the problem as a linear programming problem and using the simplex algorithm. Fortunately, however, an algorithm that eliminates much of the computational burden required by the simplex algorithm has been developed for the assignment problem.

The algorithm is based on two facts. First, each facility must be assigned to one of the jobs. Second, the relative cost of assigning facility i to job j is not changed by the subtraction of a constant from either a column or a row of the cost matrix. To illustrate, consider the first row in Table 13.28. Since work crew 1 must be assigned to one of the five jobs, and since the relative costs of the assignment are not changed by the subtraction of a constant from each element in the row, we can subtract the minimum element in the row from all other elements in the row. Similarly, the relative costs of assigning work crew 2 to each of the five jobs is not changed by subtracting the minimum element in the second row from all other elements in the second row. The same principle is true for all rows of the assignment table. Table 13.29 gives the *reduced cost matrix* obtained by subtracting the minimum element in each row from all other elements in the row.

In some instances an optimal assignment is possible from the first reduced matrix. *An optimal assignment occurs when an assignment can be made such that the total reduced cost of the assignment is zero.* Since this is not the case in Table 13.29, we must further reduce the cost matrix.

A second reduced cost matrix can be obtained by subtracting the minimum element in each column from all other elements in the column. The rationale

Table 13.29.

Job

i \ j	1	2	3	4	5	*Number Subtracted*
1	16	47	14	27	0	25
2	0	7	27	43	59	22
3	0	12	33	24	13	27
4	8	13	11	15	0	37
5	3	14	19	0	4	26

Work Crew (label for rows 1–5)

for this step is analogous to that used to obtain the first reduced matrix. Since each job must be assigned to one of the work crews, the relative cost of the assignment is not changed by the subtraction of a constant from all elements in the job column. The second reduced cost matrix, obtained by subtracting the minimum element in each column of the first reduced matrix from all other elements in the column, is shown in Table 13.30.

Table 13.30.

Job

i \ j	1	2	3	4	5
1	16	40	3	27	0
2	0	0	16	43	59
3	0	5	22	24	13
4	8	6	0	15	0
5	3	7	8	0	4
Number Subtracted	0	7	11	0	0

Work Crew (label for rows 1–5)

An assignment with a total reduced cost of zero is possible from Table 13.30. The assignment is work crew 1 to job 5, 2 to 2, 3 to 1, 4 to 3, and 5 to 4. The cost of this assignment is found by summing the cost of the assignments in the original cost matrix, Table 13.28, and is $155. This assignment is optimal.

It is not always possible to obtain an optimal assignment from the second reduced cost matrix. To illustrate this case, consider the assignment problem given by Table 13.31. This problem is identical to the original problem with the exception that the final two elements in the fourth column have been interchanged. The reduced cost matrices are obtained in the manner described and are shown in Tables 13.32 and 13.33.

Table 13.31.

Job

j i	1	2	3	4	5
1	41	72	39	52	25
2	22	29	49	65	81
3	27	39	60	51	40
4	45	50	48	26	37
5	29	40	45	52	30

Work Crew corresponds to rows 1–5.

Table 13.32.

Job

j i	1	2	3	4	5	*Number Subtracted*
1	16	47	14	27	0	25
2	0	7	27	43	59	22
3	0	12	33	24	13	27
4	19	24	22	0	11	26
5	0	11	16	23	1	29

Work Crew corresponds to rows 1–5.

Table 13.33.

Job

j i	1	2	3	4	5
1	16	40	0	27	0
2	0	0	13	43	59
3	0	5	19	24	13
4	19	17	8	0	11
5	0	4	2	23	1
Number Subtracted	0	7	14	0	0

Work Crew corresponds to rows 1–5.

An optimal assignment exists if the total reduced cost of the assignment is zero. Since the total reduced cost is not zero, an optimal assignment is not present. Consequently, the cost matrix must be reduced even further.

Table 13.33 can be reduced by use of a very simple technique. The technique involves drawing straight lines, either horizontally or vertically, that cover all zeros in the reduced cost matrix. The zeros must be covered with as few lines as possible. If the minimum number of lines necessary to cover all zeros equals the number of assignments that must be made, an optimal solution has already been found. If the number of lines is less than the number of assignments, an additional reduction is necessary.

Four lines are required to cover the zeros in the reduced cost matrix of Table 13.33. Since five assignments must be made (i.e., the matrix has dimensions of 5 by 5), an additional reduction is necessary. The lines used to cover the zeros are shown in Table 13.34.

Table 13.34.

Job

i \ j	1	2	3	4	5
1	16	40	0	27	0
2	0	0	13	43	59
3	0	5	19	24	13
4	19	17	8	0	11
5	0	4	2	23	1

Work Crew (rows 1–5)

The reduction of Table 13.34 is made by first determining the smallest element not covered by a line. This number is subtracted from each uncovered element in the matrix and is added to those elements covered by two lines. Since the smallest uncovered number in Table 13.34 is 1, each uncovered number is reduced by 1 and each number covered by two lines is increased by 1. The resulting reduced cost matrix is given in Table 13.35.

An assignment with a total reduced cost of zero is possible in Table 13.35. The assignment is work crew 1 to job 3, 2 to 2, 3 to 1, 4 to 4, and 5 to 5. The cost of this assignment is found from the original cost matrix, Table 13.31, and is $151.

The logic underlying the reduction of Table 13.34 to obtain Table 13.35 can be explained as follows. An optimal assignment is not possible in Table

Table 13.35.

Job

i \ j	1	2	3	4	5
1	17	40	0	27	0
2	1	0	13	43	59
3	0	4	18	23	12
4	20	17	8	0	11
5	0	3	1	22	0

Work Crew (rows 1–5)

13.34. Therefore, an additional reduction is necessary. This reduction is made by subtracting the smallest nonzero element from all elements in the matrix. Subtracting 1 from each number in the reduced cost matrix, Table 13.34, gives the cost matrix shown by Table 13.36.

Table 13.36.

Job

		1	*2*	*3*	*4*	*5*
	1	15	39	−1	26	−1
	2	−1	−1	12	42	58
Work	3	−1	4	18	23	12
Crew	4	18	16	7	−1	10
	5	−1	3	1	22	0

Table 13.36 contains negative values. Since the objective is to obtain an assignment with reduced cost of zero, the negative numbers must be eliminated. This can be done by adding a constant to the appropriate rows and/or columns. Adding 1 to each element in the first column gives the matrix shown in Table 13.37.

Table 13.37.

Job

		1	*2*	*3*	*4*	*5*
	1	16	39	−1	26	−1
	2	0	−1	12	42	58
Work	3	0	4	18	23	12
Crew	4	19	16	7	−1	10
	5	0	3	1	22	0

The negative numbers in the first, second, and fourth rows of Table 13.37 can be eliminated by adding 1 to each element in those rows. The resulting reduced cost matrix is identical to the matrix shown by Table 13.35.

The assignment algorithm can be summarized by the following series of steps:

1. Subtract the minimum element of each row in the assignment from all elements of the row. This gives the first reduced cost matrix.

2. Subtract the minimum element in each column of the first reduced matrix from all elements in the column. This gives the second reduced cost matrix.
3. Determine if an assignment with a total reduced cost of zero is possible from the second reduced matrix. If so, this assignment is optimal. If not, proceed to step 4.
4. Draw horizontal and vertical lines to cover all zeros. Use as small a number of lines as possible. Subtract the smallest element not covered by a line from all the elements not covered and add this element to all elements lying at the intersection of two lines.
5. Determine if an optimal assignment is possible. If not, repeat steps 4 and 5 until an optimal assignment is found.

The assignment algorithm, as described, applies only to minimization problems. In order to solve a maximization problem, the assignment matrix must be converted to a new matrix whose elements have *reversed magnitudes*. The easiest way of reversing the magnitudes of the elements in the matrix is to subtract each element in the matrix from the largest element of the matrix. This has the effect of converting the maximization problem to a minimization problem. The algorithm can then be applied to the matrix with reversed magnitudes.

We have indicated that the assignment matrix must be square, i.e., the number of facilities must equal the number of jobs. There are many "real-world" problems, however, in which the number of facilities is greater than the number of jobs or vice versa. These problems can be solved by adding a "dummy" row or column and applying the assignment algorithm to the modified square matrix. For instance, if there had been five jobs and six work crews in the problem discussed earlier in this section, we could have modified the problem to make the matrix square by adding a sixth column representing "idleness" (i.e., job 6). The c_{ij} entries in the column would be either zeros or some measure of the cost of the idle work crew. Conversely, had there been six jobs and five work crews, an additional row representing an imaginary work crew would have been necessary. In these cases, the optimal solution would include either an imaginary job (idle work) or an imaginary work crew.

Example. A management consulting firm has a backlog of four contracts. Work on these contracts must be started immediately. Three project leaders are available for assignment to the contracts. Because of varying work experience of the project leaders, the profit to the consulting firm will vary based on the assignment as shown in Table 13.38. The unassigned contract can be completed by subcontracting the work to an outside consultant. The profit on the subcontract is zero.

In order to determine the optimal assignment for this maximization problem, the magnitude of the profit matrix must be reversed. This is done by

Table 13.38.

		Contract			
		1	*2*	*3*	*4*
	A	13	10	9	11
Project	B	15	17	13	20
Leader	C	6	8	11	7
	Subcontractor	0	0	0	0

subtracting each element in the profit matrix from the largest element of the matrix. The resulting matrix is shown in Table 13.39.

The optimal solution is found by applying the assignment algorithm to the reversed magnitude matrix. The reader should verify that the optimal solution is project leader A to contract 1, B to 4, C to 3, and contract 2 to the outside consultant.

Table 13.39.

		Contract			
		1	*2*	*3*	*4*
	A	7	10	11	9
Project	B	5	3	7	0
Leader	C	14	12	9	13
	Subcontractor	20	20	20	20

Example. In designing a production facility, it is important to locate the work centers in order to minimize the materials handling cost. In a specific example, three work centers are required to manufacture, assemble, and package a product. Four locations are available within the plant. The materials handling cost at each location for the work centers is given by the cost matrix in Table 13.40. Determine the location of work centers that minimizes total materials handling cost.

Table 13.40.

		Location			
		1	*2*	*3*	*4*
	Manufacture	18	15	16	13
Job	Assembly	16	11	X	15
	Packaging	9	10	12	8

To apply the assignment algorithm, two modifications to the cost matrix are required. First, the matrix shows that assembly cannot be performed in location 3. Therefore, the cost of assembly at location 3 is represented by M. Second, a dummy job must be added to the matrix. The cost of this job at each location is zero. The modified cost matrix is given in Table 13.41.

The optimal assignment is manufacturing at location 4, assembly at location 2, and packaging at location 1. Location 3 is not assigned to any of the three jobs, i.e., location 3 is assigned to the dummy row.

Table 13.41.

Location

		1	2	3	4
	Manufacture	18	15	16	13
Job	Assembly	16	11	M	15
	Packaging	9	10	12	8
	Dummy	0	0	0	0

PROBLEMS

1. A manufacturer of inboard motor boats has three assembly plants where different models of the boats are made. The engines for the boats are purchased from a vendor and shipped to the assembly plants from the vendor's two manufacturing facilities. The cost of the engines, with the exception of shipping charges, is the same at each of the vendor's two manufacturing facilities. The supply of engines along with the number of boats scheduled for assembly during the scheduling period is shown below. The shipping costs from each engine plant to the three assembly plants are also shown. The manufacturer wishes to develop a shipping schedule that minimizes the total shipping cost while meeting the assembly requirements.

Engine Plant	Assembly Plant 1	2	3	Engines Available
1	$40	$30	$20	500
2	15	25	35	500
Boats scheduled for assembly	300	300	400	

(a) Establish the northwest corner solution.
(b) Determine the optimal solution by using the stepping-stone algorithm.

2. A farm implement company has manufacturing facilities in Tulsa, Phoenix, and Portland, Oregon. Each of these facilities has the capability to manufacture the four major products made by the company. However, because of differences in equipment and plant design, the cost of manufacturing the four products differs at the three plants. These costs are shown in the table below.

The capacity of the plants also differs. The capacity of each plant and the requirements for each product are shown below. Both capacity and requirements have been expressed in terms of standard units. (This means that capacity and requirements are directly comparable.)

Plant	Product				Capacity (Standard Units)
	A	B	C	D	
Tulsa	$300	$200	$500	$200	75
Phoenix	200	100	200	400	120
Portland	200	300	400	300	105
Requirements (standard units)	65	60	80	95	300

(a) Establish the northwest corner solution.

(b) Determine the optimal solution by using the stepping-stone algorithm.

3. The Baxter Glass Company produces disposable glass containers that are purchased by five soft-drink bottlers. The bottles are sold at a fixed delivered price of $0.25 per case. Orders for the current scheduling period have been received from the five bottlers and are as follows: 30,000 cases from bottler 1; 30,000 cases from bottler 2; 100,000 cases from bottler 3; 50,000 cases from bottler 4; and 40,000 cases from bottler 5.

The bottles are made at three plants. The monthly production capacities of the plants are 75,000 cases at plant 1, 100,000 cases at plant 2, and 125,000 cases at plant 3. The direct costs of production are $0.10 at plant 1, $0.09 at plant 2, and $0.08 at plant 3. The transportation costs of shipping a case from a plant to a bottler are shown below.

Plant	Bottler				
	1	2	3	4	5
1	$0.04	$0.06	$0.10	$0.13	$0.13
2	0.07	0.04	0.09	0.10	0.11
3	0.09	0.08	0.09	0.08	0.12

(a) Formulate the problem as a balanced transportation problem.

(b) Use the minimum cell method to determine the initial solution.

(c) Determine the optimal solution by using the stepping-stone algorithm.

4. The Econo Car Rental Company has a shortage of rental cars in certain cities and an oversupply of cars in other cities. The imbalances are shown in the following table.

City	Cars Short	Cars Excess
Albany	—	15
Boston	—	20
Chicago	28	—
Cleveland	—	15
Dallas	—	20
Detroit	—	30
Kansas City	13	—
Miami	15	—
Philadelphia	24	—

The costs of transporting a car from the cities with an excess to those cities with a shortage are shown below.

From	To:	Transportation Cost ($) per Car		
	Chicago	Kansas City	Miami	Philadelphia
Albany	$140	$210	$240	$ 50
Boston	150	220	230	60
Cleveland	70	190	200	80
Dallas	160	90	180	130
Detroit	60	110	190	90

The company wants to correct the imbalances at the minimum cost.

(a) Formulate the problem as a balanced transportation problem.

(b) Use the minimum cell method to determine the initial solution.

(c) Determine the optimal solution by using the stepping-stone algorithm.

5. The Superior Oil Company has three oil refineries. These refineries are currently operating at less than maximum capacity. Superior has the opportunity to sell certain oil products to a competing oil company. Since the competing firm has alternative sources of supply, the sale will not change Superior's current market position. The potential profit from the sale of the products is shown in the following table:

Plant	Gasoline	Kerosene	Product Diesel	Jet Fuel	Asphalt
1	$0.165	—	$0.140	$0.125	$0.128
2	0.140	$0.146	0.126	—	0.133
3	—	0.139	0.134	0.125	0.130

As indicated in the table, the profit on a product differs at the several plants. In addition, certain products are not available at all plants. This is indicated by the dashes. Superior has excess capacity of 70,000 barrels at plant 1, 90,000 barrels at plant 2, and 40,000 barrels at plant 3. They have been requested to supply part or all of the following products: 60,000 barrels of gasoline, 15,000 barrels of kerosene, 40,000 barrels of diesel fuel, 25,000 barrels of jet fuel, and 20,000 barrels of asphalt. Management wants to schedule production of these products at the three plants in order to maximize profits.

(a) Formulate the problem as a balanced transportation problem.

(b) Use the maximum cell method to obtain the initial solution.

(c) Determine the optimal solution.

6. The Quapaw Company operates three rock quarries in central Oklahoma. Because of the lack of suitable concrete rock in that portion of the state, this company will be the principal supplier of concrete rock for a new turnpike currently under construction.

The concrete rock must be quarried and delivered to six concrete ready-mix plants that are situated along the road site. The quantity of rock that must be delivered to each of these plants is shown below.

	Concrete Plant Site					
	1	*2*	*3*	*4*	*5*	*6*
Rock required (cubic yards)	120,000	80,000	160,000	100,000	80,000	100,000

The cost of the delivered rock is dependent on both the cost of quarrying the rock and the cost of trucking the rock to the ready-mix plant sites. The cost of quarrying the rock is related to factors such as the size of the quarry, the amount of overburden that must be removed, the amount of washing required, etc. These costs, per cubic yard of rock, are $2.40 at quarry 1, $3.20 at quarry 2, and $2.85 at quarry 3. The trucking cost per cubic yard is given in the following table:

From	To:	1	2	3	4	5	6
1		$1.80	$1.60	$0.80	$1.30	$1.40	$2.40
2		1.20	0.90	0.70	1.20	1.30	1.50
3		1.70	1.60	1.30	1.00	1.00	1.50

Because of differences in both equipment and rock, the capacities of the three quarries differ. The capacity of quarry 1 is 300,000 cubic yards, the capacity of quarry 2 is 180,000 cubic yards, and the capacity of quarry 3 is 240,000 cubic yards. The objective is to minimize the cost of the delivered rock.

(a) Formulate the problem as a balanced transportation problem.

(b) Use the minimum cell method to obtain the initial solution.

(c) Determine the optimal solution.

7. The Graham Electronics Company has received a contract from Transcontinental Airlines, Inc., to provide metal detection devices. These devices are used by Transcontinental in the screening of passengers for both domestic and international flights. The contract specifies delivery of the metal detection devices according to the following schedule:

		Quarter		
	First	Second	Third	Fourth
Number	20	30	40	40

Graham has sufficient production capacity to meet the above schedule. However, because of contracts with other customers, the number of metal detection devices that can be manufactured in the next four quarters varies. In addition, Graham expects the cost of manufacture to increase. Maximum production, the unit cost of production, and the unit storage cost are given in the following table:

Quarter	Maximum Production	Unit Cost of Production	Unit Storage Cost per Quarter
1	30	$14,000	$1000
2	45	16,000	1000
3	40	15,000	1000
4	25	17,000	1000

No storage costs are incurred for devices that are delivered in the same quarter as produced. The objective is to schedule production and delivery in order to minimize cost.

(a) Formulate the problem as a balanced transportation problem.

(b) Determine the optimal solution.

8. Lacy's Department Store must place orders for five styles of their very popular hand-carved wooden statues. Lacy's has been able to find only three manufacturers of these particular statues. The number of statues available from each manufacturer is shown below.

Manufacturer	Number of Statues
1	200
2	300
3	450

All statues, regardless of style, are sold at a retail price of $225. The cost of the statues varies, however, from one manufacturer to another. These costs are shown in the following table:

Manufacturer	1	2	Style 3	4	5
1	$200	$190	$140	$210	$160
2	150	200	130	190	160
3	180	150	180	200	250

The potential sales of the statues exceed the total supply. The potential sales are: style 1, 150; style 2, 200; style 3, 350; style 4, 200; style 5, 300. Since the price of all five styles of statues is the same, Lacy's objective is to minimize cost.

(a) Formulate the problem as a balanced transportation problem.

(b) Find the optimal solution.

9. A large oil company has oil reserves in four different locations. This oil must be transported from these locations to five oil refineries. The transportation costs, supplies, and requirements are shown in the following balanced transportation table:

	Transportation Cost (per Barrel) Refinery					Maximum Yearly Supply (Thousands of Barrels)
Reserve	1	2	3	4	5	
1	$6.50	$2.50	$5.00	$3.00	$5.00	1100
2	4.00	0.50	3.00	3.00	3.50	920
3	0.50	6.00	2.00	3.50	3.50	620
4	5.50	8.00	3.50	5.00	1.00	1020
Yearly Demand (Thousands of Barrels)	840	400	1000	600	820	3660

Determine the shipments from the reserves to the refineries that minimize total shipping cost.

10. Five girls in a secretarial pool must be assigned to five different jobs. From past records, the time that each girl takes to do each job is known. These times, in hours, are shown in the following table:

	Job				
Girl	1	2	3	4	5
1	3	7	8	4	6
2	4	8	7	5	6
3	5	7	9	5	7
4	4	6	8	6	8
5	6	9	10	7	9

Assuming that each girl can be assigned to only one job, determine the optimal assignment.

11. A television repair shop employs six technicians. These technicians must be assigned to five different repair jobs. Because of their different specialties and levels of skill, the time required for each technician to complete each job varies. The times required are shown in the following table. All entries are in minutes.
The objective is to minimize the total repair time. The unassigned technician will remain idle. Formulate the problem as an assignment problem and determine the optimal solution.

Technician	1	2	Job 3	4	5
1	180	160	240	300	150
2	190	150	250	320	170
3	150	170	230	340	140
4	170	200	210	310	190
5	220	190	260	330	160
6	190	180	300	310	210

12. A major aerospace company has received a government contract for the production of an advanced fighter/bomber. As is usually the case, a large number of the systems required for the aircraft must be obtained from subcontractors. To illustrate, the avionics have been divided into six different packages and bids have been requested for each package. Eight major avionics manufacturers have submitted bids on the different packages. These bids are shown in the following table. A dash indicates that the avionics manufacturer declined to bid on an individual package. All dollar figures are in millions of dollars.

Manufacturer	Avionics Package 1	2	3	4	5	6
1	$4.2	$6.8	$1.4	$3.6	$9.4	$6.2
2	4.4	—	1.6	3.5	9.5	6.4
3	4.3	7.0	1.5	3.5	9.6	6.1
4	5.0	7.0	1.4	3.7	10.0	6.0
5	5.0	6.9	1.6	3.6	9.5	5.9
6	3.9	6.7	1.7	—	9.6	6.1
7	4.1	6.8	1.5	3.5	9.4	6.1
8	4.2	6.9	1.7	3.5	—	6.0

The scheduling is such that only one avionics package can be subcontracted to any single manufacturer. The aerospace company's objective is, of course, to minimize the cost of procuring the entire avionics system.
(a) Formulate the problem as an assignment problem.
(b) Determine the optimal solution.

13. KECT, the local educational television station, must schedule its weekend television programming. In addition to KECT, there are three network affiliate stations in the local viewing area. The programming for these three network affiliates has already been announced.

On the basis of past experience, KECT realizes that it can attract only a small portion of the total viewing public. Nevertheless, its objective is to maximize its total program exposure.

The station has eight, 1-hour programs that have been selected for the weekend prime-time viewing hours (i.e., 7:00-11:00 p.m. on Saturdays and Sundays). The number of viewers it can expect depends on the popularity of the network shows scheduled during these hours as well as the popularity of KECT's shows. Management's estimates of the number of viewers each show can attract during the prime-time viewing hours are given in the following table. All entries in the table are in thousands of viewers.

					Show				
Day and Time	1	2	3	4	5	6	7	8
Sat. 7:00– 8:00	18	12	25	16	30	5	15	8
8:00– 9:00	14	14	30	18	32	6	18	6
9:00–10:00	12	15	22	18	36	19	17	4
10:00–11:00	10	20	16	14	40	13	16	4
Sun. 7:00– 8:00	20	11	20	15	28	7	14	7
8:00– 9:00	16	12	26	17	34	11	17	5
9:00–10:00	14	14	20	25	40	10	16	3
10:00–11:00	12	16	14	14	30	6	13	2

As an illustration of the table, management expects show 1 to have 18,000 viewers if scheduled on Saturdays from 7:00–8:00 p.m. The number of viewers would drop to 14,000, however, if the show were scheduled from 8:00–9:00 p.m. on the same day.

(a) Formulate the problem as an assignment problem.

(b) Determine the optimal solution.

14. Nash, Smith, and Co., Certified Public Accountants, provides auditing and tax services for a large number of firms in Northern California. Mr. Nash, the managing senior partner of the firm, must assign jobs to the staff (junior) and senior members of the firm.

The accounting firm currently has six staff accountants and three senior accountants available for assignments. The major differences in the two classifications is in experience, the staff accountants normally having less than three years' experience and the senior accountants three to six years' experience.

The firm currently has a backlog of five auditing jobs and three tax jobs that must be assigned. Mr. Nash has estimated the time required by each accountant to complete each job. These entries are shown in the following table in units of days. A dash indicates that the accountant is not qualified for the job.

Staff Accountant*	Audit					Tax		
	1	*2*	*3*	*4*	*5*	*1*	*2*	*3*
Mr. Hill	10	—	5	21	13	—	—	—
Mr. Nutter	12	20	7	18	15	—	15	—
Mr. Hamilton	10	18	6	17	16	18	13	18
Mr. Pryor	8	—	8	19	15	—	19	22
Mr. Pratt	12	22	6	19	13	16	14	20
Mr. Wagner	14	21	7	20	16	—	16	19
Senior Accountant								
Mr. Redman	7	15	5	16	12	10	12	15
Mr. Lochen	8	17	5	15	11	12	14	17
Mr. Savich	8	14	5	15	11	13	14	17

* Entries are in 8-hour days.

The billing rate is $15 per hour for staff accountants and $20 per hour for senior accountants. The objective is to minimize the total billings to the customers.

(a) Formulate the problem as an assignment problem.

(b) Determine the optimal solution.

SUGGESTED REFERENCES

The references for this chapter are listed in Chap. 11.

Chapter Fourteen

Mathematics of Finance

An important application of mathematics in business and economics involves calculating the value of an invested sum of money that earns interest over a period of time. For instance, suppose that one deposits a sum of money in a savings account. The interest paid by the savings institution accrues over a period of time. The total amount of money in the savings account of the individual is determined by using a formula known as the *compound amount formula*. As another example, suppose that an individual borrows a sum of money to purchase a house. The money is normally repaid by monthly installments over a number of years. The monthly amount necessary to repay the loan is calculated by using a formula that gives the periodic payments for an annuity. These, along with other important financial formulas, are discussed in this chapter. These formulas are customarily classified under the heading of *mathematics of finance*.

14.1 Simple and Compound Interest

As the reader is aware, the term *interest* refers to the price that one must pay to borrow a sum of money. This price is normally expressed as a percentage of the sum borrowed. Customarily, this percentage, or, alternatively, the *interest rate*, is based on a time period of one year. To illustrate, an interest rate of 6 percent would, unless specified differently, mean that the price of borrowing a sum of money for one year is equal to 6 percent of the sum of money borrowed. Since the base period for calculating the price of borrowing the sum of money is one year, the interest rate is referred to as the *annual interest rate*.

The term *simple interest* is used when interest is paid only on the original sum of money borrowed. This is contrasted to *compound interest,* in which the interest owed is added to the sum borrowed and subsequent interest is paid on the total of the original sum borrowed and the accumulated interest owed. To illustrate these two forms of interest payment, consider the following two examples.

Example. Bob Short has arranged a loan from a local bank. The sum of money borrowed is $1000 and the simple interest rate is 6 percent. Determine the interest that Bob must pay during the first year.

The interest is calculated by multiplying the interest rate by the sum borrowed. The interest is $1000(0.06), or $60. The total that must be paid the bank at the end of the one-year period is $1060.

Example. Bill Smith also arranged a one-year loan from a local bank. The terms of Bill's loan specify that the interest be calculated at the end of each six-month period, however, rather than at the end of the year. The sum borrowed is again $1000 and the annual interest rate is 6 percent. Assuming Bill repays both the original sum borrowed and the accumulated interest at the end of the one-year period, determine the interest and the total amount that must be paid.

The interest for the first six months is $1000(0.03), or $30. Since Bill does not pay the $30 interest at the end of the first six months, the interest is added to the original sum borrowed. Bill thus owes $1030 during the second six months. The interest charges during the second six months are $1030(0.03) = $30.90. The total owed the bank at the second year is

$$A = \$1000 + \$30 + \$30.90 = \$1060.90$$

These two examples illustrate the effect of compounding: Although the interest rate quoted to both borrowers was 6 percent, Bill Smith found that the loan cost more than 6 percent of the sum borrowed. In both Bob and Bill's loan, the name or *nominal annual interest rate* was 6 percent. Because the interest was compounded on Bill's loan, the *effective annual interest rate* was greater than the nominal rate.

14.2 Compound Amount and Present Value Formulas

One of the more common financial transactions involves the deposit of a certain sum of money in a savings institution. The money deposited earns interest and the interest earned is allowed to accumulate over time. The sum of money deposited is normally referred to as the *principal* and is denoted by

P. The principal plus the interest that has accumulated over the period of time is referred to as the *amount* and is represented by *A*. Since the interest earned during a certain time period is added to the original principal invested rather than withdrawn, the interest is compounded. The amount of money, then, depends on the original principal *P*, the nominal interest rate *i*, the number of years duration of the investment *t*, and the number of compounding periods per year *n*.

The formula for the amount can be most easily developed if we initially assume that the interest is compounded only one time per year. Based on this assumption of annual compounding, the interest earned on the principal *P* during the first year is *iP*. The sum of the principal and interest at the end of the first year is

$$P + iP = P(1 + i)$$

During the second year, interest will be earned on $P(1 + i)$ dollars. The sum of the interest and principal at the end of the second year would thus be

$$P(1 + i) + iP(1 + i) = P(1 + i)^2$$

Continuing in this fashion, we find that the amount that results from the annual compounding of *P* dollars at interest rate *i* is

$$A = P(1 + i)^t \qquad (14.1)$$

This formula is termed the *compound amount* formula.

The compound amount formula is an exponential function with base $(1 + i)$ and parameters *P* and *t*. It is used to determine the amount of money that will accrue if a principal *P* is invested for *t* periods and earns interest at the rate of *i* per period. An identical problem was discussed in Sec. 3.2.3, namely, that of determining the value of an investment *k* that grows at a rate of *r* per period for *x* periods. The formula developed for this problem, Eq. (3.9), and the compound amount formula, Eq. (14.1), are the same. This should come as no surprise since both formulas are used to calculate the time value of an investment.

To illustrate the compound amount formula, assume that an investment of $1000 in securities will grow at an annual rate of 8 percent for ten years. The amount of the investment at the end of the *t*th year is given by $A = \$1000(1.08)^t$. The amount at the end of each of the ten years is shown in Table 14.1.

The values of *A* in Table 14.1 can be calculated from Eq. (14.1) by using logarithms. This approach was demonstrated in Sec. 3.2.3. An alternative approach is to use tables of compound interest factors. A table of compound interest factors gives the value of the expression $(1 + i)^n$ for various values

Table 14.1.

t	1	2	3	4	5	6	7	8	9	10
A	$1080	1166	1260	1360	1469	1587	1714	1851	1999	2159

of i and n. Table A.1 in the Appendix gives these factors. The following examples illustrate the use of the compound amount formula and the compound interest factors.

Example. Table 14.1 gives the amount, A, of an investment of $1000 at the end of each of ten years. The interest rate used in the table was 8 percent. Use Table A.1 in the Appendix to verify the entries in the table.

The compound amount formula

$$A = \$1000(1.08)^t$$

was used to calculate the entries in the table. From Table A.1, the compound amount factor for $i = 0.08$ and $t = 1$ is 1.080000. Multiplying this factor by this initial investment gives

$$A = \$1000(1.080000) = \$1080.00$$

The compound amount factor for $t = 2$ is 1.166400. The amount of the investment at the second year is

$$A = \$1000(1.166400) = \$1166.40$$

Similarly, the amount at the end of the third year is

$$A = \$1000(1.259712) = \$1259.71$$

The remaining entries in Table 14.1 are calculated in the same manner. Notice that the entries in Table 14.1 have been rounded to the nearest dollar.

Example. Determine the interest and amount if $1000 is invested at 6 percent for eight years.

$$A = \$1000(1.06)^8$$
$$A = \$1000(1.593848)$$
$$A = \$1594$$

The interest during the eight-year period totals $594, and the amount at the end of the eight-year period is $1594.

Example. Determine the amount if $5000 is invested at 7 percent for 20 years.

$$A = \$5000(1.07)^{20}$$

$$A = \$5000(3.869684)$$

$$A = \$19,348.42$$

Example. An individual has the option of investing $5000 either in a time deposit account that returns 6 percent interest or in mutual funds. The fee for purchasing the mutual funds is a one-time fee of 8.5 percent paid at the time of purchase of the fund. If the investment has a five-year duration, what approximate interest rate must the fund yield to match that of the time deposit?

The amount at the end of the five-year period from the time deposit is

$$A = \$5000(1.06)^5 = \$6691$$

The principal invested in the mutual fund is

$$P = \$5000 - 0.085(\$5000) = \$4575$$

The interest rate is determined by solving the compound amount formula for i.

$$A = P(1 + i)^t$$

$$\$6691 = \$4575(1 + i)^5$$

$$1.4625 = (1 + i)^5$$

The exact interest rate could be determined by solving the equation for i using logarithms. An approximate solution can be determined from the compound amount factors found in Table A.1 in the Appendix. From Table A.1, we see that the compound amount factor for an interest rate of 8 percent and a period of five years is 1.469328. We thus conclude that the investment in mutual funds must grow at a rate of slightly less than 8 percent to match the investment in time deposits.

The compound amount formula gives the amount of money that will accrue if a sum P is invested for t periods and earns interest at a rate of i per period. Instead of calculating this amount A, suppose that we are given A and asked to determine the principal P. This principal represents the sum of money that must be invested *now* at interest rate i in order to return a sum A at the end of t periods. The sum of money, P, that must be invested now in order to return a sum A in t periods is referred to as the *present value* of A.

To illustrate, assume that one will receive $1000 ten years hence. The present value of this $1000 is equal to the principal that, if currently invested, would amount to $1000 at the end of the ten-year period. The term present value thus refers to the present or current value of an amount of money that will be received at some time in the future.

The present value of an amount of money received t periods in the future with interest of i per period can be determined by solving Eq. (14.1) for P. The present value of an amount of money is

$$P = A(1 + i)^{-t} \tag{14.2}$$

In the present value formula, an amount A received t years from now has a present value of P. The sum P can be invested at rate i and will increase according to Eq. (14.2) to equal A. Consequently, P is termed the present value of A.

Tables of present value factors can be used to evaluate the expression $(1 + i)^{-t}$. Table A.2 in the Appendix is such a table. The following examples illustrate the use of the present value formula and present value factors.

Example. Mr. Smith has an obligation of \$500 due five years from now. If interest is assumed to be 7 percent, what is the present value of the obligation?

$$P = \$500(1.07)^{-5}$$

$$P = \$500(0.712986)$$

$$P = \$356.49$$

Example. Determine the present value of receiving \$1000 ten years hence, assuming an interest rate of 6 percent.

$$P = \$1000(1.06)^{-10}$$

$$P = \$1000(0.558395)$$

$$P = \$558.40$$

Example. Determine the present value of \$10,000 received five years from now, assuming 6 percent annual interest.

$$P = \$10,000(1.06)^{-5}$$

$$P = \$10,000(0.747258)$$

$$P = \$7472.58$$

Example. In a moment of weakness, John Smith promised his wife Mary that they would take an "around the world" vacation trip beginning on their tenth wedding anniversary. John anticipates the trip's costing \$5000. John and Mary were married last month. What amount would John need to deposit now at interest of 5 percent in order to pay for the trip?

$$P = \$5000(1.05)^{-10}$$

$$P = \$5000(0.613913)$$

$$P = \$3069.57$$

Example. An individual has the alternatives of selling an asset for a current cash price of \$6300 or accepting a promissory note for \$10,000 due eight years hence. Determine the approximate interest rate that equates the two alternatives.

$$\$6300 = \$10,000(1 + i)^{-8}$$

$$0.6300 = (1 + i)^{-8}$$

The exact interest rate could be calculated by using logarithm. The approximate interest rate can be determined from Table A.2 by noting that the present value factor for $i = 0.06$ and $t = 8$ years is 0.627412. The approximate interest rate is thus 6 percent.

14.2.1 MULTIPLE COMPOUNDING

Although the interest rate is normally expressed on an annual basis, it is common to compound the interest more than one time per year. For instance, many savings institutions compound interest quarterly or monthly. Quarterly compounding means that the interest earned during a quarter is added to the accumulated investment at the end of each quarter, rather than at the end of the year. Similarly, monthly compounding means that the interest earned during the month is added to the accumulated investment at the end of each month.

In the case of multiple compounding, the nominal annual interest rate does not give the percentage of change in the amount invested during the year. If, for example, \$1 is invested at $i = 4$ percent and compounded quarterly, the percentage of change from the beginning to the end of the year will be 4.06 percent. This rate exceeds the 4 percent nominal rate because the interest is added to the principal four times during the year rather than only at the end of the year. Interest is thus earned on interest during the year. This rate incorporates the effect of compounding f times during the year and is termed the *effective annual interest rate*.

The compound amount formula can be modified to include more than one compounding period per year. If i is the nominal annual interest rate,

t is the number of years, and f is the frequency per year of compounding, the compound interest formula becomes

$$A = P\left(1 + \frac{i}{f}\right)^{ft} \tag{14.3}$$

where i/f is the interest rate per compounding period and ft is the number of periods.

The present value formula, multiple compounding again being assumed, becomes

$$P = A\left(1 + \frac{i}{f}\right)^{-ft} \tag{14.4}$$

Example. Calculate the amount of a \$1000 investment compounded quarterly at 6 percent interest for five years. Compare this amount to the amount if interest is compounded annually.

Compounding quarterly gives

$$A = \$1000\left(1 + \frac{0.06}{4}\right)^{4(5)}$$

$$A = \$1000(1.015)^{20}$$

$$A = \$1346.86$$

Compounding yearly gives

$$A = \$1000(1.06)^5$$

$$A = \$1338.23$$

Example. Determine the present value of \$1000 received in five years if the nominal interest rate is 6 percent and interest is compounded twice per year. Compare this to the present value if interest is compounded yearly.

Compounding twice a year gives

$$P = \$1000\left(1 + \frac{0.06}{2}\right)^{-2(5)}$$

$$P = \$1000(1.03)^{-10}$$

$$P = \$744.09$$

Compounding annually gives

$$P = \$1000(1.06)^{-5}$$

$$P = \$747.26$$

The effective annual interest rate for multiple compounding can easily be determined. If we represent the effective annual interest rate by r and the nominal annual interest rate by i, we can solve for r as follows:

$$P(1 + r)^t = P\left(1 + \frac{i}{f}\right)^{ft}$$

or, by taking the tth root of the expression,

$$(1 + r) = \left(1 + \frac{i}{f}\right)^f$$

Thus,

$$r = \left(1 + \frac{i}{f}\right)^f - 1 \qquad (14.5)$$

Example. Determine the effective annual interest rate if the nominal annual rate of 4 percent is compounded quarterly.

$$r = \left(1 + \frac{0.04}{4}\right)^4 - 1$$

$$r = (1.01)^4 - 1$$

$$r = 0.040604 = 4.0604\%$$

The effective annual rate is 4.0604 percent.

Example. Determine the effective annual interest rate if the nominal annual rate of 6 percent is compounded monthly.

$$r = \left(1 + \frac{0.06}{12}\right)^{12} - 1$$

$$r = (1.005)^{12} - 1$$

$$r = 0.061678 = 6.1678\%$$

14.2.2 CONTINUOUS COMPOUNDING

Certain assets earn on a continuous basis throughout the year. As an example, capital equipment that is used in the manufacture of daily output contributes a continuous flow of earnings to the firm. The rate of return of this equipment is somewhat distorted if it is assumed that the return occurs only once each year. The actual rate of return should be calculated by incorporating the continuous flow of earnings into the formula.

The compound amount formula can be modified to incorporate continuous compounding. The modification requires the limiting process that is customarily presented in differential calculus.[1] The formula for the amount for continuous compounding is given by the exponential function

$$A = Pe^{it} \tag{14.6}$$

where A is the amount, P is the principal, e is the base, i is the nominal annual interest rate, and t is the number of years.

The present value of an amount A received t years hence if the principal is compounded continuously is determined by solving Eq. (14.6) for P. The present value of an amount is

$$P = Ae^{-it} \tag{14.7}$$

Values of the term e^x and e^{-x} for $x = it$ are given in Table A.7 of the Appendix.

The use of Table A.7 in determining the present value and the amount is illustrated by the following examples.

Example. Assume that $1000 is invested for five years at a nominal interest rate of 8 percent. If the interest is compounded continuously, determine the amount at the end of the five-year period.

$$A = \$1000e^{(0.08)5} = \$1000e^{0.40}$$

$$A = \$1000(1.4918) = \$1491.80$$

Example. Determine the present value assuming continuous compounding of $5000 received ten years hence if the nominal interest rate is 10 percent.

$$P = \$500e^{-(0.10)10} = \$5000e^{-1.00}$$

$$P = \$5000(0.367870) = \$1839.35$$

Example. Assume that $10,000 is invested for ten years at a nominal annual interest rate of 6 percent. Compare the amount if the principal is compounded continuously and if it is compounded annually.

For continuous compounding:

$$A = \$10,000e^{(0.06)10} = \$10,000e^{0.6}$$

$$A = \$10,000(1.8221) = \$18,221$$

---

[1] The interested reader may wish to refer to Robert L. Childress, *Calculus for Business and Economics* (Englewood Cliffs, N.J.: Prentice-Hall, Inc., 1972), p. 154.

For annual compounding:

$$A = \$10,000(1.06)^{10}$$

$$A = \$10,000(1.790848) = \$17,908$$

The difference over the ten-year period between continuous compounding and annual compounding is $313.

Example. Determine the present value of a note of $5000 that matures in ten years. Assume continuous compounding and a nominal annual interest rate of 7 percent.

$$P = \$5000e^{-(0.07)10} = \$5000e^{-0.7}$$

$$P = \$5000(0.496585) = \$2482.93$$

Example. Company A has stated that it expects sales to increase from the present level of $1,000,000 by $50,000 per year. Company B expects a continuous growth rate of 6 percent and it reported sales of $900,000 during the past accounting period. Develop predicting formulas for both companies and calculate anticipated sales ten years hence.

For Company A:

$$S = 1,000,000 + 50,000t$$

and

$$S(10) = 1,000,000 + 500,000 = \$1,500,000$$

For Company B:

$$S = \$900,000e^{(0.06)t}$$

and

$$S(10) = \$900,000e^{0.60}$$

$$S(10) = \$1,639,890$$

Example. One share of Miller Growth Fund, a mutual fund, was valued at $10.00 on January 1 and $11.00 on December 31. Determine the rate of return, assuming continuous appreciation throughout the year.

The method of determining i involves using natural logarithms to solve the continuous compounding formula for i. Thus,

$$11.00 = 10.00e^{i(1)}$$

$$\ln (11.00) = \ln (10.00) + i$$

$$i = 2.39790 - 2.30259$$

$$i = 0.09531 = 9.53\%$$

The effective annual interest rate from continuous compounding can be determined by equating Eqs. (14.1) and (14.6). Again representing the effective annual interest rate by r and the nominal annual rate by i, we obtain

$$P(1 + r)^t = Pe^{it}$$

or

$$(1 + r)^t = e^{it}$$

Taking the tth root of the equation gives

$$1 + r = e^i$$

and the effective annual interest rate is

$$r = e^i - 1 \qquad\qquad (14.8)$$

Example. Determine the effective annual interest rate, assuming continuous compounding and a nominal interest rate of $i = 0.10$.

$$r = e^{0.10} - 1$$

$$r = 1.105 - 1$$

$$r = 0.105 = 10.5\%$$

Example. Determine the effective annual interest rate, assuming continuous compounding and a nominal interest rate of $i = 0.05$.

$$r = e^{0.05} - 1$$

$$r = 1.0513 - 1$$

$$r = 0.0513 = 5.13\%$$

Example. Western Pacific Savings and Loan advertises that money deposited in a savings account earns interest of 6 percent compounded daily. Determine the effective annual interest rate paid by Western Pacific.

An exact solution for the effective annual interest rate requires solving Eq. (14.8) for r. That is,

$$r = \left(1 + \frac{0.06}{365}\right)^{365} - 1$$

Solving this equation requires very detailed tables of logarithms. Since these tables are not readily available, we can determine the approximate effective annual interest rate by assuming continuous compounding. The rate is

$$r = e^{0.06} - 1$$

$$r = 1.0618 - 1$$

$$r = 0.0618 = 6.18\%$$

14.3 Annuity

Another common form of a financial agreement is the *annuity*. Annuities consist of a series of equal payments, each payment normally being made at the end of a designated period of time. A common example of an annuity is the mortgage payments on a home loan. Other examples include installment payments on an automobile loan, payments on life insurance, payments for retirement plans, sinking fund payments, and payments to a savings plan.

The *amount of an annuity* is the sum of the periodic payments and the interest earned from these payments. As an example, assume that an individual plans to establish a trust fund for his children's education. The trust fund will be established by annually depositing $1000 in a savings institution. Each payment, or deposit, to the trust fund will be made at the *end* of the year, rather than at the beginning of the year. The payments deposited will earn 6 percent interest compounded annually.

If we let i represent the annual interest rate, t the number of years, p the annuity payment made at the *end of each year*, and A the sum of the annuity payments and interest, we can calculate A as follows:

$$A = p + p(1 + i) + p(1 + i)^2 + \cdots + p(1 + i)^{t-1} \qquad (14.9)$$

The final annuity payment is written first and earns no interest, the next to final payment earns interest for 1 year, and the first payment, which is made at the end of the first year, earns interest for $t - 1$ years. This expression can be solved for A by a simple algebraic maneuver. This consists of multiplying the expression in Eq. (14.9) by $(1 + i)$. This gives

$$A(1 + i) = p(1 + i) + p(1 + i)^2 + p(1 + i)^3 + \cdots + p(1 + i)^t \quad (14.10)$$

Subtracting Eq. (14.9) from Eq. (14.10) gives

$$A(1 + i) - A = p(1 + i)^t - p$$

which when solved for A gives the formula for the amount of an annuity.

$$A = p\left[\frac{(1 + i)^t - 1}{i}\right] \qquad (14.11)$$

We can also determine the periodic payment necessary, when invested at interest rate i for t years, to sum to the amount A. This is found by solving Eq. (14.11) for p.

$$p = A\left[\frac{i}{(1 + i)^t - 1}\right] \qquad (14.12)$$

It is important to remember that these formulas were developed based on the assumption that the payments are made at the end of each year, rather than at the beginning. The formula for the amount of an annuity can be modified to account for payments made at the beginning of the year by replacing p in Eq. (14.11) by $p(1 + i)$. The formula for the periodic payment, Eq. (14.12), is modified by replacing A by $A(1 + i)^{-1}$. The use of these formulas is illustrated by the following examples. In these examples, Table A.3 in the Appendix is used to determine the value of the annuity factor, $[(1 + i)^t - 1]/i$, in Eq. (14.11). The reciprocals of the factors in Table A.3 are used in Eq. (14.12) for periodic payments.

Example. Determine the amount of the trust fund mentioned above if payments are made for ten years. The interest rate is $i = 0.06$, the time period is $t = 10$ years, and the amount is

$$A = p\left[\frac{(1 + i)^t - 1}{i}\right]$$

$$A = \$1000\left[\frac{(1.06)^{10} - 1}{0.06}\right]$$

$$A = \$1000(13.1808)$$

$$A = \$13,180.80$$

Example. Determine the amount in the preceding example if the payments are made at the beginning, rather than the end, of each year.

$$A = p(1 + i)\left[\frac{(1 + i)^t - 1}{i}\right]$$

$$A = \$13,180.80(1.06)$$

$$A = \$13,971.65$$

Example. Mr. Clark plans on investing $1000 per year in a savings plan that earns 5 percent interest compounded annually. Determine the sum of the annuity payments and interest at the end of ten years.

$$A = \$1000\left[\frac{(1.05)^{10} - 1}{0.05}\right]$$

$$A = \$1000(12.5779) = \$12,577.90$$

Example. Sioux Falls Steel Company recently placed a $5 million bond issue with a group of private investors. One of the requirements of the investors was that the firm establish a sinking fund that will retire the bonds in 15 years. If Sioux Falls Steel plans to invest the sinking fund payments in government bonds that earn 6 percent interest, what yearly payment is necessary to retire the $5 million bond issue at the end of 15 years?

$$p = \$5,000,000 \left[\frac{0.06}{(1.06)^{15} - 1} \right]$$

$$p = \$5,0\,0\,0 \left[\frac{1}{23.2760} \right]$$

$$p = \$214,813.54$$

Example. An individual aged 45 wants a savings account balance of $100,000 at age 65. Assuming 5 percent interest compounded annually, what yearly payments should be deposited in the account?

$$p = \$100,000 \left[\frac{0.05}{(1.05)^{20} - 1} \right]$$

$$p = \$100,000 \left[\frac{1}{33.0660} \right] = \$3024.25$$

Eqs. (14.11) and (14.12) were developed based on the assumption of annual annuity payments. In many cases, however, annuity payments are made more than one time per year. For example, most installment loans require monthly payments. Similarly, some savings plans are based on quarterly payments. These cases can be taken into account by replacing i by i/f and t by ft in Eqs. (14.11) and (14.12), where f represents the frequency or number of annuity payments per year. The term i/f is the interest rate per period, and ft is the number of periods or number of annuity payments. We again assume that payments are made at the end of each period and that interest is compounded at the time of the payment.

Example. Determine the amount of an annuity if payments of $100 are deposited quarterly at 4 percent nominal annual interest for five years. Assume that the payments are deposited at the end of each quarter and interest is compounded quarterly.

$$A = p \left[\frac{\left(1 + \dfrac{i}{f} \right)^{ft} - 1}{i/f} \right]$$

$$A = \$100 \left[\frac{(1.01)^{20} - 1}{0.01} \right]$$

$$A = \$100(22.0190)$$

$$A = \$2201.90$$

It is often necessary to calculate the current value of receiving $p per year for t years, assuming an interest rate of i. This sum is termed the *present value of an annuity*. As an example, one might be required to determine the present value of receiving $1000 per year for 10 years if interest is 6 percent per year. Since money has a value over time, the present value of $1000 per year for 10 years will be less than the value of receiving the $10,000 during the first year. The present value of this series of annuity payments will, of course, depend on the interest rate.

The formula for the present value of an annuity is derived by summing the present values of each of the individual annuity payments. The present value of the first payment, made 1 year from now, is $p(1 + i)^{-1}$. The present value of the second payment, made 2 years from the present, is $p(1 + i)^{-2}$. The present value of all payments is the geometric series

$$PV = p(1 + i)^{-1} + p(1 + i)^{-2} + \cdots + p(1 + i)^{-t} \qquad (14.13)$$

If all terms in Eq. (14.13) are multiplied by $(1 + i)$, we obtain

$$PV(1 + i) = p + p(1 + i)^{-1} + \cdots + p(1 + i)^{-(t-1)} \qquad (14.14)$$

Subtracting Eq. (14.13) from Eq. (14.14) gives

$$PV(1 + i) - PV = p - p(1 + i)^{-t}$$

which, when solved for PV, gives

$$PV = p\left[\frac{1 - (1 + i)^{-t}}{i}\right] \qquad (14.15)$$

The periodic payment that has a present value of PV is given by solving Eq. (14.15) for p:

$$p = PV\left[\frac{i}{1 - (1 + i)^{-t}}\right] \qquad (14.16)$$

Present value of an annuity factor used in Eq. (14.15) is given in Table A.4 of the Appendix. The periodic payment factors used in Eq. (14.16) are given by the reciprocal of the entries in Table A.4. The use of the two formulas is illustrated by the following examples.

Example. Determine the present value of receiving $1000 per year for ten years if interest is 6 percent per annum.

$$PV = \$1000\left[\frac{1 - (1.06)^{-10}}{0.06}\right]$$

$$PV = \$1000\,(7.3601)$$

$$PV = \$7360.10$$

Example. Mr. Miller is considering the purchase of a home. The loan balance is $20,000 and the interest rate is 7 percent. Determine the yearly payments for a 25-year loan.

$$p = \$20,000 \left[\frac{0.07}{1 - (1.07)^{-25}} \right]$$

$$p = \$20,000 \left[\frac{1}{11.6536} \right]$$

$$p = \$1716.21$$

PROBLEMS

1. An individual deposits $1000 in a savings institution that pays 6 percent interest compounded annually. Determine the amount at the end of 5 years.

2. An individual purchases a new home for $40,000. If the home increases in value at an annual rate of 5 percent, determine the value of the home at the end of 20 years.

3. Assume that the population of the United States is 220 million. If the population increases at an annual rate of 3 percent, determine the population 50 years hence.

4. The yearly cost of tuition and fees at a certain private university is $2500. If these costs increase at an annual rate of 8 percent, what would the cost be 10 years hence?

5. Determine the present value of $1000 received 10 years hence, assuming interest of 6 percent compounded annually.

6. Assuming interest of 8 percent compounded annually, would a rational investor prefer $5000 at present or $10,000 10 years hence?

7. A college graduate received a job offer with an annual salary of $12,000. His father began working in a similar job 25 years ago with a salary of $6000. If the cost of living has increased at an annual rate of 3 percent, did the father or the son earn more real income on his initial job?

8. Suppose that the rate of increase in the cost of living averages 5 percent. Based on this assumption, a salary of $100,000 25 years hence is equivalent to what salary at the present time?

9. An investor has the opportunity to purchase a second mortgage for $3400. The mortgage matures in 5 years and returns $5000. Assuming annual compounding, determine the approximate rate of interest on the investment.

10. An individual began working in 1960 at an annual salary of $8000. Fifteen years later, the individuals salary had increased to $14,400. Determine the approximate yearly rate of salary increase.

11. The domestic demand for petroleum products was roughly 5 million barrels per day in 1946 and 18 million barrels a day in 1974. Determine the approximate annual growth rate in demand during this 28-year period.

12. The domestic supply of petroleum products was roughly 5 million barrels a day in 1946 and 14 million barrels a day in 1974. Determine the approximate annual growth rate in supply during this 28-year period.

13. Assume in Problem 1 that the savings institution pays 6 percent interest compounded quarterly (i.e., 1.5 percent per quarter). Determine the amount the individual would have in his savings account at the end of 5 years.

14. A certain savings institution pays 6 percent compounded monthly. If $1000 is deposited in this institution, determine the amount at the end of 5 years. Compare this with the amount assuming annual compounding.

15. A nominal annual interest rate of 8 percent compounded quarterly is equivalent to what effective annual interest rate?

16. Would a rational investor prefer 7 percent compounded annually or 6 percent compounded monthly? Why?

17. Answer Problem 16 assuming that the 6 percent is compounded continuously.

18. An investment of $1000 yields a nominal interest rate of 10 percent. Assuming continuous compounding, determine the value of the investment 10 years hence.

19. Determine the present value of $10,000 received 5 years hence. Assume continuous compounding and an interest rate of 12 percent.

20. A nominal annual interest rate of 8 percent compounded continuously is equivalent to what effective annual interest rate?

21. An individual deposits $1000 at the end of each of the next 10 years. If the investment earns 6 percent compounded annually, determine the amount of the investment at the end of the 10-year period.

22. Assume that an individual and his employer are both required to contribute $750 per year to the social-security fund. If this money could instead be invested and earn 6 percent per year, determine the amount in the individual's account at the end of 40 years.

23. An individual plans to deposit $250 at the end of each quarter-year for the next 10 years in a savings institution. If the savings institution pays 6 percent per year, compounded quarterly, determine the amount in the individual's account at the end of the 10-year period.

24. Beginning at age 20 an individual deposits $1742.86 at the end of each year in a savings account that pays 8 percent compounded annually. At what age would the individual have an account balance of $1,000,000?

25. John Smith, a young college graduate, plans to work in a certain job for a period of 10 years and then to start his own company. John estimates that $50,000 will be necessary to start the company. What yearly sum must John save in order to accumulate the $50,000? Assume annual compounding and an interest rate of 6 percent.

26. In order to retain key personnel, a certain company grants bonuses of $50,000 at the end of 10 years of employment. Disregarding any income tax considerations and assuming an interest rate of 7 percent compounded annually, determine the annual salary that is equivalent to this bonus.

27. Determine the present value of $1000 per year for 20 years. Assume annual compounding and an interest rate of 6 percent.

28. Would a rational investor prefer $6000 at present or $1000 per year for the next 10 years? Assume 8 percent interest compounded annually.

29. The winner of a national lottery is offered $100,000 cash or $10,000 per year for the next 20 years. If the winner has an investment opportunity that returns 8 percent compounded annually, which alternative would be preferred?

30. Determine the yearly payments for a $30,000 mortgage loan amortized over 30 years at 8 percent interest compounded annually.

31. The interest rate at a certain bank on new car loans is 1 percent per month. If $3600 is financed, determine the monthly payments on a 3-year loan.

32. Upon reaching retirement age, an individual wants to be able to supplement his retirement income by withdrawing $5000 per year from his savings for 20 years. At the end of the 20 years, the savings account will be depleted. If the individual reaches retirement age in 30 years, determine the yearly savings necessary to achieve this objective. Assume 6 percent interest compounded annually for all calculations.

33. Assume in Problem 32 that the individual wishes to leave his savings intact and withdraw $5000 per year interest from the savings account. Calculate the yearly savings necessary to achieve this objective.

SUGGESTED REFERENCES

ALWAN, A. J., and D. G. PARISI, *Quantitative Methods for Decision Making* (Morristown, N.J.: General Learning Press, 1974).

AYRES, FRANK, JR., *Mathematics of Finance*, Schaum's Outline Series (New York, N.Y.: McGraw-Hill Book Company, Inc., 1967).

BUSH, GRACE A., and JOHN E. YOUNG, *Foundations of Mathematics* (New York, N.Y.: McGraw-Hill Book Company, Inc., 1968).

KEMENY, JOHN G., et al., *Finite Mathematics with Business Applications* (Englewood Cliffs, N.J.: Prentice-Hall, Inc., 1972).

SELBY, SAMUEL M., *Standard Mathematical Tables*, 21st ed. (Cleveland, Ohio: The Chemical Rubber Co., 1973).

VAZSONYI, ANDREW, and RICHARD BRUNELL, *Business Mathematics for Colleges* (Homewood, Ill.: Richard D. Irwin, Inc., 1974).

Appendix

Logarithms:
Laws of Exponents

1. Logarithms as Exponents

Consider the exponential function $y = a^x$, where y represents the value of the base number a raised to the exponent x. The definition of the logarithm of y is as follows: The logarithm of y to the base a is x. This is written as $\log_a(y) = x$. A *logarithm* is *an exponent*. To understand logarithms, it is important to remember the simple fact that a logarithm is an exponent. In fact, it might be helpful for the student to consider the terms logarithm and exponent as synonymous. To illustrate the relationship between a number and its logarithm, consider the following examples.

Example. Determine $\log_{10}(100)$. The logarithm of 100 to the base 10 is the exponent to which 10 must be raised to give 100.
Since $100 = 10^2$, we know that $\log_{10}(100) = 2$.

Example. Determine $\log_{10}(10,000)$.
Since $10,000 = 10^4$, we conclude that $\log_{10}(10,000) = 4$.

Example. Determine $\log_2(8)$.
Since $8 = 2^3$, we conclude that $\log_2(8) = 3$.

Example. Determine $\log_6(36)$.
Since $36 = 6^2$, we conclude that $\log_6(36) = 2$.

Logarithms are used for two purposes in this text. These are (1) in the calculation of the product of two numbers, the quotient of two numbers, or a number raised to a power, and (2) the algebraic manipulation of exponential functions for the purpose of expressing the function in a linear form. To understand the use of logarithms, it will be useful to review the laws of exponents.

2. Laws of Exponents

Let a represent the base and m and n represent exponents. The three laws of exponents are

(i) $a^m \cdot a^n = a^{m+n}$

(ii) $\dfrac{a^m}{a^n} = a^{m-n} = \begin{cases} a^{m-n} & \text{if } m > n \\ 1 & \text{if } m = n \\ \dfrac{1}{a^{n-m}} & \text{if } n > m \end{cases}$

(iii) $(a^m)^n = a^{mn}$

The usefulness of logarithms in calculations of products, quotients, and powers is based upon the laws of exponents. Numbers can be expressed with a common base. The product of two numbers is then equal to the sum of the exponents, the quotient of two numbers is equal to the difference between the exponents, and a number raised to a power is equal to the product of the exponents. These properties of exponents are illustrated by the following examples.

Example. Determine the product of 100 times 1000, using the laws of exponents.

Since $100 = 10^2$ and $1000 = 10^3$, we know that

$$(100)(1000) = (10^2)(10^3) = 10^{2+3} = 10^5 = 100,000$$

Example. Determine the product of 16 times 64, using the laws of exponents.

Since $16 = 2^4$ and $64 = 2^6$, the product of 16 and 64 is

$$(2^4)(2^6) = 2^{4+6} = 2^{10} = 1024$$

Example. Determine the quotient of 1,000,000 divided by 100.

Since $1,000,000 = 10^6$ and $100 = 10^2$, $1,000,000/100 = 10^6/10^2 = 10^{6-2} = 10^4 = 10,000$.

Example. Determine the value of 100 raised to the third power.

Since $100 = 10^2$, we use the third law of exponents to show that $(10^2)^3 = 10^6 = 1,000,000$.

3.   Laws of Logarithms

The laws of exponents can be rewritten in terms of logarithms. If we represent any two nonzero, non-negative numbers as x and y, then the laws of logarithms are

$$\text{(i)} \quad \log_a (xy) = \log_a (x) + \log_a (y)$$

$$\text{(ii)} \quad \log_a (x/y) = \log_a (x) - \log_a (y)$$

$$\text{(iii)} \quad \log_a (x^y) = y \log_a (x)$$

Logarithms are exponents of the base. To determine the value of the product, quotient, or power we must raise the base to the logarithm (or exponent). The process of raising the base to the logarithm is termed finding the antilog of the logarithm. In symbolic form we write

$$(x)(y) = \text{antilog}_a \left[\log_a (x) + \log_a (y)\right] = a^{\log_a(x) + \log_a(y)}$$

The laws of logarithms are illustrated for the examples given in Sec. 2.

Example. Determine the product of 100 times 1000, using logarithms.

$$\log_{10} \left[(100)(1000)\right] = \log_{10} (100) + \log_{10} (1000) = 2 + 3 = 5$$

$$\text{antilog}_{10} (5) = 10^5 = 100{,}000$$

Example. Determine the product of 16 times 64, using logarithms.

$$\log_2 \left[(16)(64)\right] = \log_2 (16) + \log_2 (64) = 4 + 6 = 10$$

$$\text{antilog}_2 (10) = 2^{10} = 1024$$

Example. Determine the quotient of 1,000,000 divided by 100, using logarithms.

$$\log_{10} (1{,}000{,}000/100) = \log_{10} (1{,}000{,}000) - \log_{10} (100) = 6 - 2 = 4$$

$$\text{antilog}_{10} (4) = 10^4 = 10{,}000$$

Example. Determine the value of 100 raised to the third power, using logarithms.

$$\log_{10} (100)^3 = 3 \log_{10} (100) = 3(2) = 6$$

$$\text{antilog}_{10} (6) = 10^6 = 1{,}000{,}000$$

4.   Determining Logarithms

To determine the logarithm of a number, it is necessary to have a table that gives the logarithms of numbers using a specific base. Two bases are commonly used, the base 10 and the base e. Logarithms using the base 10 are termed *common logarithms* and are written as log (x). The absence of a subscript indicates that the base is 10. Logarithms using the base e are termed *Naperian* or *natural logarithms* and are written as ln (x). The symbol "ln" refers to the natural or Naperian logarithm with base e. Any base other than 10 or e is indicated by $\log_a (x)$. We shall first illustrate the common or base 10 logarithms.

A logarithm is made up of two parts, the characteristic and the mantissa. The method of determining the characteristic and the mantissa requires that the number be written in the form: x.xxxx · 10^n. That is, the number for which the logarithm is to be obtained must be written in a form such that a simple integer is to the left of the decimal point with the remaining significant digits to the right of the decimal point. The number x.xxxx is termed the *argument*. The magnitude of the number is expressed as the argument multiplied by a power of 10. The characteristic of the logarithm is n, the power to which 10 is raised. We can illustrate the determination of the characteristic by the following examples.

Example. Determine the characteristic of 121.
Since $121 = 1.21 \times 10^2$, the characteristic is 2. Note also that the argument is 1.21.

Example. Determine the characteristic of 1640.
Since $1640 = 1.640 \times 10^3$, the characteristic is 3.

Example. Determine the characteristic of 0.0653.
Since $0.0653 = 6.53 \times 10^{-2}$, the characteristic is -2.

Example. Determine the characteristic of 7.54.
Since $7.54 = 7.54 \times 10^0$, the characteristic is 0.

The mantissa of the number is the exponent to which the base 10 must be raised to obtain the argument x.xxxx. Since the argument has a value between 1.0000 and 9.9999, the *mantissa must have a value between 0 and 1*. The relationship, again, between the base, the mantissa, and the argument is

$$(\text{base})^{\text{mantissa}} = \text{argument}$$

If we include the characteristic, we see that

$$(\text{base})^{\text{mantissa}} \cdot (\text{base})^{\text{characteristic}} = \text{number}$$

The logarithm of the number is thus the sum of the mantissa and the characteristic. Mantissas for common logarithms are given in Table A.5. As an example, we shall determine the logarithm of 297. We first note that 297 = 2.97×10^2. The characteristic of the logarithm is thus 2. The mantissa of the argument 2.97 is 0.4728. In terms of exponential notation, 297 = $10^{0.4728} \times 10^2$. The logarithm of the number is the exponent to which 10 must be raised to obtain 297. The logarithm of 297 is, therefore, 2.4728, i.e., log (297) = 2.4728. The method of determining logarithms is further illustrated by the following examples.

Example. Log (121) = 2.0828, since 121 = $10^{0.0828} \times 10^2$.

Example. Log (1640) = 3.2148, since 1640 = $10^{0.2148} \times 10^3$.

Example. Log (0.0653) = -1.1851, since 0.0653 = $10^{0.8149} \times 10^{-2}$.

Example. Log (7.54) = 0.8774, since 7.54 = $10^{0.8874} \times 10^0$.

5.   Using Logarithms in Calculations

The first law of logarithms specifies that the product of two numbers is given by the sum of their logarithms.

Example. Multiply 78 by 194.

$$\log 78 = 1.8921$$
$$\log 194 = 2.2878$$
$$\text{sum of logarithms} = 4.1799$$

$$\text{antilog } (4.1799) = 15{,}130$$

The fourth digit from the left, 3, was obtained by linear interpolation.[1]

The second law of logarithms states that the quotient of two numbers is given by the difference between two logarithms.

Example. Divide 657 by 132.

$$\log 657 = 2.8041$$
$$\log 132 = 2.1206$$
$$\text{difference of logarithms} = 0.6835$$

$$\text{antilog } (0.6835) = 4.83$$

---

[1] Linear interpolation is explained in Sec. 7.

Example. Divide 456 by 3648.

$$\log 456 = \quad 2.6590$$
$$\log 3648^2 = \quad 3.5620$$
$$\text{difference of logarithms} = -0.9030$$

This problem requires a slight modification. Negative exponents such as -0.9030 are not included in the table of mantissas. We can, however, rewrite the logarithm as

$$-0.9030 = +0.0970 - 1.0000$$

Since logarithms are exponents, we actually have converted $10^{-0.9030}$ to $10^{0.0970} \cdot 10^{-1.0000}$. Thus, antilog $(0.0970) = 1.25$, and the quotient is $1.25 \cdot 10^{-1} = 0.125$.

The third law of logarithms states that to raise a number to a given power, the logarithm of the number is multiplied by the power.

Example. Raise 6.52 to the 3.25 power.

$$\log (6.52) = 0.8142$$
$$\times 3.25$$
$$\text{product} = 2.6462$$
$$\text{antilog} (2.6462) = 442.8$$

The third law of logarithms also applies in determining roots of a number.

Example. Evaluate $(174)^{0.21}$.

$$\log (174) = 2.2405$$
$$\times 0.21$$
$$\text{product} = 0.4705$$
$$\text{antilog} (0.4705) = 2.953$$

Table A.6 contains natural logarithms, i.e., logarithms with base e. The natural logarithms for numbers from 0 to 999 are given in this table. Both the characteristic and the mantissa are included in the table. Logarithms of numbers that exceed 999 can be determined by writing the number as a power of 10. Thus, the natural logarithm of 1230 is found by expressing 1230 as 123×10^1. The natural logarithm of 1230 is

$$\ln (1230) = \ln (123) + \ln (10)$$
$$\ln (1230) = 4.81218 + 2.30259$$
$$\ln (1230) = 7.11477$$

The laws of logarithms directly apply to natural logarithms.

---

[2] By interpolation.

6. Manipulation of Logarithmic and Exponential Functions

An exponential function may be expressed in terms of logarithms by expressing the terms of both sides of the equal sign in terms of logarithms. Thus,

$$y = kb^{f(x)}$$

can be expressed as

$$\log_a y = \log_a k + f(x) \log_a b$$

Similarly, given the function

$$y = k \log_a f(x)$$

the function can be expressed in its exponential form as

$$a^y = f(x)^k$$

A logarithm may be expressed with a different base by using the following relationship:

$$\log_a x = (\log_b x)(\log_a b)$$

Example. Convert $\log_{10} (25)$ to $\log_e (25)$.

$$\log_e (25) = \log_{10} (25) \log_e (10)$$
$$\log_e (25) = (1.3979)(2.30259) = 3.21888$$

It is also helpful to recognize that

$$\log_a (a) = 1$$

and

$$\log_a (1) = 0$$

7. Linear Interpolation

Linear interpolation provides a method for obtaining the logarithms of numbers containing significant digits not shown in the table of logarithms. To illustrate, suppose that we use linear interpolation to obtain the common logarithm of 2648. Table A.5 gives arguments for numbers with three significant digits. Thus, the logarithm of 2640 is 3.4216 and the logarithm of 2650 is 3.4232. To estimate by linear interpolation the logarithm of 2648, we note that the fourth digit in the number is 8. The logarithm of 2648 should thus equal the logarithm of 2640 plus 0.8 of the difference between the logarithms of 2650 and 2640. Expressed numerically,

$$\log 2648 = \log 2640 + 0.8(\log 2650 - \log 2640)$$
$$\log 2648 = 3.4216 + 0.8(3.4232 - 3.4216)$$
$$\log 2648 = 3.4216 + 0.0013 = 3.4229$$

Using linear interpolation, the logarithm of 2648 is 3.4229.

Example. Use linear interpolation to determine the logarithm of 6324.

$$\log 6324 = \log 6320 + 0.4(\log 6330 - \log 6320)$$

$$\log 6324 = 3.8007 + 0.4(3.8014 - 3.8007)$$

$$\log 6324 = 3.8007 + 0.0003 = 3.8010$$

Linear interpolation is used in the same manner to obtain antilogarithms. To illustrate, we shall determine the number whose logarithm is 4.1799. From Table A.5, antilog (0.1790) is 1.51 and antilog (0.1818) is 1.52. Using linear interpolation, we see that the fourth digit of the number is equal to the proportion $(0.1799 - 0.1790)/(0.1818 - 0.1790)$ or approximately 3. The antilogarithm of 4.1799 is thus 1.513×10^4 or 15,130.

Example. Determine the number whose logarithm is 3.4819.

$$\text{antilog } (0.4814) = 3.03 \qquad \text{antilog } (0.4829) = 3.04$$

$$\frac{0.4819 - 0.4814}{0.4829 - 0.4814} = 0.3$$

$$\text{antilog } (3.4819) = 3.033 \times 10^3 = 3033$$

Tables

Table A.1. Values of $(1 + i)^n$

n	$i = 0.005$	$i = 0.0075$	$i = 0.01$	$i = 0.0125$	$i = 0.015$	$i = 0.02$	$i = 0.03$	$i = 0.04$	$i = 0.05$	$i = 0.06$	$i = 0.07$	$i = 0.08$
1	1.005000	1.007500	1.010000	1.012500	1.015000	1.020000	1.030000	1.040000	1.050000	1.060000	1.070000	1.080000
2	1.010025	1.015056	1.020100	1.025156	1.030225	1.040400	1.060900	1.081600	1.102500	1.123600	1.144900	1.166400
3	1.015075	1.022669	1.030301	1.037971	1.045678	1.061208	1.092727	1.124864	1.157625	1.191016	1.225043	1.259712
4	1.020150	1.030339	1.040604	1.050945	1.061364	1.082432	1.125509	1.169859	1.215506	1.262477	1.310796	1.360489
5	1.025251	1.038067	1.050110	1.064082	1.077284	1.104081	1.159274	1.216653	1.276282	1.338226	1.402552	1.469328
6	1.030378	1.045852	1.061520	1.077383	1.093443	1.126162	1.194052	1.265319	1.340096	1.418519	1.500730	1.586874
7	1.035529	1.053696	1.072135	1.090850	1.109845	1.148686	1.229874	1.315932	1.407100	1.503630	1.605781	1.713824
8	1.040707	1.061599	1.082857	1.104486	1.126493	1.171659	1.266770	1.368569	1.477455	1.593848	1.718186	1.850930
9	1.045911	1.069561	1.093685	1.118292	1.143390	1.195093	1.304773	1.423312	1.551328	1.689479	1.838459	1.999004
10	1.051140	1.077583	1.104622	1.132271	1.160541	1.218994	1.343916	1.480244	1.628895	1.790848	1.967151	2.158925
11	1.056396	1.085664	1.115668	1.146424	1.177949	1.243374	1.384234	1.539454	1.710339	1.898299	2.104852	2.331629
12	1.061678	1.093807	1.126825	1.160755	1.195618	1.268242	1.425761	1.601032	1.795856	2.012196	2.252192	2.518170
13	1.066986	1.102010	1.138093	1.175264	1.213552	1.293607	1.468534	1.665074	1.885649	2.132928	2.404845	2.719624
14	1.072321	1.110276	1.149474	1.189955	1.231756	1.319479	1.512590	1.731676	1.979932	2.260904	2.578534	2.937194
15	1.077683	1.118603	1.160969	1.204829	1.250232	1.345868	1.557967	1.800944	2.078928	2.396558	2.759032	3.172169
16	1.083071	1.126992	1.172579	1.219890	1.268986	1.372786	1.604706	1.872981	2.182875	2.540352	2.952164	3.425943
17	1.088487	1.135445	1.184304	1.235138	1.288020	1.400241	1.652848	1.947900	2.292018	2.692773	3.158815	3.700018
18	1.093929	1.143960	1.196147	1.250577	1.307341	1.428246	1.702433	2.025817	2.406619	2.854339	3.379932	3.996019
19	1.099399	1.152540	1.208109	1.266210	1.326951	1.456811	1.753506	2.106849	2.526950	3.025600	3.616527	4.315701
20	1.104896	1.161184	1.220190	1.282037	1.346855	1.485947	1.806111	2.191123	2.653298	3.207135	3.869684	4.660957
21	1.110420	1.169893	1.232392	1.298063	1.367058	1.515666	1.860295	2.278768	2.785963	3.399564	4.140562	5.033834
22	1.115972	1.178667	1.244716	1.314288	1.387564	1.545980	1.916103	2.369919	2.925261	3.603537	4.430402	5.436540
23	1.121552	1.187507	1.257163	1.330717	1.408377	1.576899	1.973587	2.464716	3.071524	3.819750	4.740530	5.871464
24	1.127160	1.196414	1.269735	1.347351	1.429503	1.608437	2.032794	2.563304	3.225100	4.048935	5.072367	6.341181
25	1.132796	1.205387	1.282432	1.364193	1.450945	1.640606	2.093778	2.665836	3.386355	4.291871	5.427433	6.848475
26	1.138460	1.214427	1.295256	1.381245	1.472710	1.673418	2.156591	2.772470	3.555673	4.549383	5.807353	7.396353
27	1.144152	1.223535	1.308209	1.398511	1.494800	1.706886	2.221289	2.883369	3.733456	4.822346	6.213868	7.988061
28	1.149873	1.232712	1.321291	1.415992	1.517222	1.741024	2.287928	2.998703	3.920129	5.111687	6.648838	8.627106
29	1.155622	1.241957	1.334504	1.433692	1.539981	1.775845	2.356566	3.118651	4.116136	5.418388	7.114257	9.317275
30	1.161400	1.251272	1.347849	1.451613	1.563080	1.811362	2.427263	3.243398	4.321942	5.743491	7.612255	10.062657

484

n	$i=0.005$	$i=0.0075$	$i=0.01$	$i=0.0125$	$i=0.015$	$i=0.02$	$i=0.03$	$i=0.04$	$i=0.05$	$i=0.06$	$i=0.07$	$i=0.08$
31	1.167207	1.260656	1.361327	1.469759	1.586526	1.847589	2.500080	3.373133	4.538039	6.088101	8.145113	10.867669
32	1.173043	1.270111	1.374941	1.488131	1.610324	1.884541	2.575083	3.508059	4.764941	6.453387	8.715271	11.737083
33	1.178908	1.279637	1.388690	1.506732	1.634479	1.922231	2.652335	3.648381	5.003186	6.840590	9.325340	12.676050
34	1.184803	1.289234	1.402577	1.525566	1.658996	1.960676	2.731905	3.794316	5.253348	7.251025	9.978114	13.690134
35	1.190727	1.298904	1.416603	1.544636	1.683881	1.999890	2.813862	3.946089	5.516015	7.686087	10.676581	14.785344
36	1.196681	1.308645	1.430769	1.563944	1.709140	2.039887	2.898278	4.103933	5.791816	8.147252	11.423942	15.968172
37	1.202664	1.318460	1.445076	1.583493	1.734777	2.080685	2.985227	4.268090	6.081407	8.636078	12.223618	17.245626
38	1.208677	1.328349	1.459527	1.603287	1.760798	2.122299	3.074783	4.438813	6.385477	9.154252	13.079271	18.625276
39	1.214721	1.338311	1.474123	1.623328	1.787210	2.164745	3.167027	4.616366	6.704751	9.703507	13.994820	20.115298
40	1.220794	1.348349	1.488864	1.643619	1.814018	2.208040	3.262038	4.801021	7.039989	10.285718	14.974458	21.724522
41	1.226898	1.358461	1.503752	1.664165	1.841229	2.252200	3.359899	4.993061	7.391988	10.902861	16.022670	23.462483
42	1.233033	1.368650	1.518790	1.684967	1.868847	2.297244	2.460696	5.192784	7.761588	11.577033	17.144257	25.339482
43	1.239198	1.378915	1.533978	1.706029	1.896880	2.343189	3.564517	5.400495	8.149667	12.250455	18.344355	27.366640
44	1.245394	1.389256	1.549318	1.727354	1.925333	2.390053	3.671452	5.616515	8.557150	12.985482	19.628460	29.555972
45	1.251621	1.399676	1.564811	1.748946	1.954213	2.437854	3.781596	5.841176	8.985008	13.764611	21.012452	31.920449
46	1.257879	1.410173	1.580459	1.770808	1.983526	2.486611	3.895044	6.074823	9.434258	14.590487	22.472623	34.474085
47	1.264168	1.420750	1.596263	1.792943	2.013279	2.536344	4.011895	6.317816	9.905971	15.465917	24.045707	37.232012
48	1.270489	1.431405	1.612226	1.815355	2.043478	2.587070	4.132252	6.570528	10.401270	16.393872	25.728907	40.210573
49	1.276842	1.442141	1.628348	1.838047	2.074130	2.638812	4.256219	6.833349	10.921333	17.377504	27.529930	43.427420
50	1.283226	1.452957	1.644632	1.861022	2.105242	2.691588	4.383906	7.106683	11.467400	18.420154	29.457025	46.901613
51	1.289642	1.463854	1.661078	1.884285	2.136821	2.745420	4.515423	7.390951	12.040770	19.525364	31.519017	50.653742
52	1.296090	1.474833	1.677689	1.907839	2.168873	2.800328	4.650886	7.686589	12.642808	20.696885	33.725348	54.706041
53	1.302571	1.485894	1.694466	1.931687	2.201406	2.856335	4.790412	7.994052	13.274949	21.938699	36.086122	59.082524
54	1.309083	1.497038	1.711410	1.955833	2.234428	2.913461	4.934125	8.313814	13.938696	23.255020	38.612151	63.809126
55	1.315629	1.508266	1.728525	1.980281	2.267944	2.971731	5.082149	8.646367	14.635631	24.650322	41.315002	68.913856
56	1.322207	1.519578	1.745810	2.005034	2.301963	3.031165	5.234613	8.992222	15.367413	26.129341	44.207052	74.426965
57	1.328818	1.530975	1.763268	2.030097	2.336493	3.091789	5.391651	9.351911	16.133783	27.697101	47.301545	80.381122
58	1.335462	1.542457	1.780901	2.055473	2.371540	3.153624	5.553401	9.725987	16.942572	29.358927	50.612653	86.811612
59	1.342139	1.554026	1.798710	2.081167	2.407113	3.216697	5.720003	10.115026	17.789701	31.120463	54.155539	93.756540
60	1.348850	1.565681	1.816697	2.107181	2.443220	3.281031	5.891603	10.519627	18.679186	32.987691	57.946427	101.257064

Table A.2. Values of $(1 + i)^{-n}$

n	$i = 0.005$	$i = 0.0075$	$i = 0.01$	$i = 0.0125$	$i = 0.015$	$i = 0.02$	$i = 0.03$	$i = 0.04$	$i = 0.05$	$i = 0.06$	$i = 0.07$	$i = 0.08$
1	0.995025	0.992556	0.990099	0.987654	0.985222	0.980392	0.970874	0.961538	0.952381	0.943396	0.934779	0.925926
2	0.990074	0.985167	0.980296	0.975461	0.970662	0.961169	0.942596	0.924556	0.907029	0.889996	0.873439	0.857339
3	0.985149	0.977833	0.970590	0.963418	0.956317	0.942322	0.915142	0.888996	0.863838	0.839619	0.816298	0.793832
4	0.980248	0.970554	0.960980	0.951524	0.942184	0.923845	0.888487	0.854804	0.822702	0.792094	0.762895	0.735030
5	0.975371	0.963329	0.951466	0.939777	0.928260	0.905731	0.862609	0.821927	0.783526	0.747258	0.712986	0.680583
6	0.970518	0.956158	0.942045	0.928175	0.914542	0.887971	0.837484	0.790315	0.746215	0.704961	0.666342	0.630170
7	0.965690	0.949040	0.932718	0.916716	0.901027	0.870560	0.813092	0.759918	0.710681	0.665057	0.622750	0.583490
8	0.960885	0.941975	0.923483	0.905398	0.887711	0.853490	0.789409	0.730690	0.676839	0.627412	0.582009	0.540269
9	0.956105	0.934963	0.914340	0.894221	0.874592	0.836755	0.766417	0.702587	0.644609	0.591898	0.543934	0.500249
10	0.951348	0.928003	0.905287	0.883181	0.861667	0.820348	0.744094	0.675564	0.613913	0.558395	0.508349	0.463193
11	0.946615	0.921095	0.896324	0.872277	0.848933	0.804263	0.722421	0.649581	0.584679	0.526788	0.475093	0.428883
12	0.941905	0.914238	0.887449	0.861509	0.836387	0.778493	0.701380	0.624597	0.556837	0.496969	0.444012	0.397114
13	0.937219	0.907432	0.878663	0.850873	0.824027	0.773033	0.680951	0.600574	0.530321	0.468839	0.414964	0.367698
14	0.932556	0.900677	0.869963	0.840368	0.811849	0.757875	0.661118	0.577475	0.505068	0.442301	0.387817	0.340461
15	0.927917	0.893973	0.861349	0.829993	0.799852	0.743015	0.641862	0.555264	0.481017	0.417265	0.362246	0.315243
16	0.923300	0.887318	0.852821	0.819746	0.788031	0.728446	0.623167	0.533908	0.458112	0.393646	0.338735	0.291890
17	0.918707	0.880712	0.844377	0.809626	0.776385	0.714163	0.605016	0.513373	0.436297	0.371364	0.316574	0.270269
18	0.914136	0.874156	0.836017	0.799631	0.764912	0.700159	0.587395	0.493628	0.415521	0.350344	0.295864	0.250249
19	0.909588	0.867649	0.827740	0.789759	0.753607	0.686431	0.570286	0.474642	0.395734	0.330513	0.276508	0.231712
20	0.905063	0.861190	0.819544	0.780009	0.742470	0.672971	0.553676	0.456387	0.376889	0.311805	0.258419	0.214548
21	0.900560	0.854779	0.811430	0.770379	0.731498	0.659776	0.537549	0.438834	0.358942	0.294155	0.241513	0.198656
22	0.896080	0.848416	0.803396	0.760868	0.720688	0.646839	0.521892	0.421955	0.341850	0.277505	0.225713	0.183941
23	0.891622	0.842100	0.795442	0.751475	0.710037	0.634156	0.506692	0.405726	0.325571	0.261797	0.210947	0.170315
24	0.887186	0.835831	0.787566	0.742197	0.699544	0.621721	0.491934	0.390121	0.310068	0.246979	0.197147	0.157699
25	0.882772	0.829609	0.779768	0.733034	0.689206	0.609531	0.477606	0.375117	0.295303	0.232999	0.184249	0.146018
26	0.878380	0.823434	0.772048	0.723984	0.679021	0.597579	0.463695	0.360689	0.281241	0.219810	0.172195	0.135202
27	0.874010	0.817304	0.764404	0.715046	0.668986	0.585862	0.450189	0.346817	0.267848	0.207368	0.160930	0.125187
28	0.869662	0.811220	0.756836	0.706219	0.659099	0.574375	0.437077	0.333477	0.255094	0.195630	0.150402	0.115914
29	0.865335	0.805181	0.749342	0.697500	0.649359	0.563112	0.424346	0.320651	0.242946	0.184557	0.140563	0.107328
30	0.861030	0.799187	0.741923	0.688889	0.639762	0.552071	0.411987	0.308319	0.231377	0.174110	0.131367	0.099377

n	$i = 0.005$	$i = 0.0075$	$i = 0.01$	$i = 0.0125$	$i = 0.015$	$i = 0.02$	$i = 0.03$	$i = 0.04$	$i = 0.05$	$i = 0.06$	$i = 0.07$	$i = 0.08$
31	0.856746	0.793238	0.734577	0.680384	0.630308	0.541246	0.399987	0.296460	0.220359	0.164255	0.122773	0.092016
32	0.852484	0.787333	0.727304	0.671984	0.620993	0.530633	0.388337	0.285058	0.209866	0.154957	0.114741	0.085200
33	0.848242	0.781472	0.720103	0.663688	0.611816	0.520229	0.377026	0.274094	0.199873	0.146186	0.107235	0.078889
34	0.844022	0.775654	0.712973	0.655494	0.602774	0.510028	0.366045	0.263552	0.190355	0.137912	0.100219	0.073045
35	0.839823	0.769880	0.705914	0.647402	0.593866	0.500028	0.355383	0.253415	0.181290	0.130105	0.093663	0.067635
36	0.835645	0.764149	0.698925	0.639409	0.585090	0.490223	0.345032	0.243669	0.172657	0.122741	0.087535	0.062635
37	0.831487	0.758461	0.692005	0.631515	0.576443	0.480611	0.334983	0.234297	0.164436	0.115793	0.081809	0.057986
38	0.827351	0.752814	0.685153	0.623719	0.567924	0.471187	0.325226	0.225285	0.156605	0.109239	0.076457	0.053690
39	0.823235	0.747210	0.678370	0.616018	0.559531	0.461948	0.315754	0.216621	0.149148	0.103056	0.071455	0.049713
40	0.819139	0.741648	0.671653	0.608413	0.551262	0.452890	0.306557	0.208289	0.142046	0.097222	0.066780	0.046031
41	0.815064	0.736127	0.665003	0.600902	0.543116	0.444010	0.297628	0.200278	0.135282	0.091719	0.062412	0.042621
42	0.811008	0.730647	0.658419	0.593484	0.535089	0.435304	0.288959	0.192575	0.128840	0.086527	0.058329	0.039464
43	0.806974	0.725208	0.651900	0.586157	0.527182	0.426769	0.280543	0.185168	0.122704	0.081630	0.054513	0.036541
44	0.802959	0.719810	0.645445	0.578920	0.519391	0.418401	0.272372	0.178046	0.116861	0.077009	0.050946	0.033834
45	0.798964	0.714451	0.639055	0.571773	0.511715	0.410197	0.264439	0.171198	0.111297	0.072650	0.047613	0.031329
46	0.794989	0.709133	0.632728	0.564714	0.504153	0.402152	0.256737	0.164614	0.105997	0.068538	0.044499	0.029007
47	0.791034	0.703854	0.626463	0.557742	0.496702	0.394268	0.249259	0.158283	0.100949	0.064658	0.041587	0.026859
48	0.787098	0.698614	0.620260	0.550855	0.489362	0.386538	0.241999	0.152195	0.096142	0.060998	0.038867	0.024869
49	0.783182	0.693414	0.614119	0.544056	0.482130	0.378958	0.234950	0.146341	0.091564	0.057546	0.036324	0.023027
50	0.779286	0.688252	0.608039	0.537339	0.475005	0.371528	0.228107	0.140713	0.087204	0.054288	0.033948	0.021321
51	0.775409	0.683128	0.602019	0.530705	0.467985	0.364243	0.221463	0.135301	0.083051	0.051215	0.031727	0.019742
52	0.771551	0.678043	0.596058	0.524153	0.461069	0.357101	0.215013	0.130097	0.079096	0.048316	0.029651	0.018280
53	0.767713	0.672995	0.590156	0.517682	0.454255	0.350100	0.208750	0.125093	0.075330	0.045582	0.027712	0.016926
54	0.763893	0.667986	0.584313	0.511291	0.447542	0.343234	0.202670	0.120282	0.071743	0.043002	0.025899	0.015672
55	0.760093	0.663013	0.578528	0.504979	0.440928	0.336504	0.196767	0.115656	0.068326	0.040567	0.024204	0.014511
56	0.756311	0.658077	0.572800	0.498745	0.434412	0.329906	0.191036	0.111207	0.065073	0.038271	0.022621	0.013436
57	0.752548	0.653178	0.567129	0.492587	0.427992	0.323437	0.185472	0.106930	0.061974	0.036105	0.021141	0.012441
58	0.744804	0.648316	0.561514	0.486506	0.421667	0.317095	0.180070	0.102817	0.059023	0.034061	0.019758	0.011519
59	0.745079	0.643490	0.555954	0.480500	0.415435	0.310878	0.174826	0.098863	0.056213	0.032133	0.018465	0.010666
60	0.741372	0.638700	0.550450	0.474568	0.409296	0.304782	0.169733	0.095060	0.053536	0.030314	0.017257	0.009876

Table A.3. Values of $[(1 + i)^n - 1]/i$

n	i = 0.005	i = 0.0075	i = 0.01	i = 0.0125	i = 0.015	i = 0.02	i = 0.03	i = 0.04	i = 0.05	i = 0.06	i = 0.07	i = 0.08
1	1.0000	1.0000	1.0000	1.0000	1.0000	1.0000	1.0000	1.0000	1.0000	1.0000	1.0000	1.0000
2	2.0050	2.0075	2.0100	2.0125	2.0150	2.0200	2.0300	2.0400	2.0500	2.0600	2.0700	2.0800
3	3.0150	3.0225	3.0301	3.0377	3.0452	3.0604	3.0909	3.1216	3.1525	3.1836	3.2149	3.2464
4	4.0301	4.0452	4.0604	4.0756	4.0909	4.1216	4.1836	4.2464	4.3101	4.3746	4.4399	4.5061
5	5.0503	5.0755	5.1010	5.1266	5.1523	5.2040	5.3091	5.4163	5.5256	5.6370	5.7507	5.8666
6	6.0755	6.1136	6.1520	6.1907	6.2296	6.3081	6.4684	6.6329	6.8019	6.9753	7.1532	7.3359
7	7.1059	7.1594	7.2135	7.2680	7.3230	7.4343	7.6624	7.8982	8.1420	8.3938	8.6540	8.9228
8	8.1414	8.2131	8.2857	8.3589	8.4328	8.5829	8.8923	9.2142	9.5491	9.8974	10.2598	10.6366
9	9.1821	9.2747	9.3685	9.4634	9.5593	9.7546	10.1591	10.5827	11.0265	11.4913	11.9779	12.4875
10	10.2280	10.3443	10.4622	10.5817	10.7027	10.9497	11.4638	12.0061	12.5778	13.1807	13.8164	14.4865
11	11.2792	11.4219	11.5668	11.7139	11.8633	12.1687	12.8077	13.4863	14.2067	14.9716	15.7835	16.6454
12	12.3357	12.5075	12.6825	12.8604	13.0412	13.4121	14.1920	15.0258	15.9171	16.8699	17.8884	18.9771
13	13.3972	13.6013	13.8093	14.0211	14.2368	14.6803	15.6177	16.6268	17.7129	18.8821	20.1406	21.4952
14	14.4642	14.7034	14.9474	15.1964	15.4504	15.9739	17.0863	18.2919	19.5986	21.0150	22.5504	24.2149
15	15.5365	15.8136	16.0969	16.3863	16.6821	17.2934	18.5989	20.0235	21.5785	23.2759	25.1290	27.1521
16	16.6142	16.9322	17.2579	17.5912	17.9324	18.6393	20.1568	21.8245	23.6574	25.6725	27.8880	30.3242
17	17.6973	18.0592	18.4304	18.8111	19.2014	20.0120	21.7615	23.6975	25.8403	28.2128	30.8402	33.7502
18	18.7858	19.1947	19.6147	20.0462	20.4894	21.4123	23.4144	25.6454	28.1323	30.9056	33.9990	37.4502
19	19.8797	20.3386	20.8109	21.2968	21.7967	22.8406	25.1168	27.6712	30.5390	33.7599	37.3789	41.4462
20	20.9791	21.4912	22.0190	22.5629	23.1237	24.2974	26.8703	29.7780	33.0659	36.7855	40.9954	45.7619
21	22.0840	22.6524	23.2392	23.8450	24.4705	25.7833	28.6764	31.9692	35.7192	39.9927	44.8651	50.4229
22	23.1944	23.8222	24.4716	25.1431	25.8376	27.2990	30.5367	34.2479	38.5052	43.3922	49.0057	55.4567
23	24.3104	25.0009	25.7163	26.4574	27.2251	28.8449	32.4528	36.6178	41.4304	46.9958	53.4361	60.8932
24	25.4319	26.1884	26.9735	27.7881	28.6335	30.4219	34.4264	39.0826	44.5019	50.8155	58.1766	66.7647
25	26.5591	27.3848	28.2432	29.1354	30.0630	32.0303	36.4592	41.6459	47.7270	54.8645	63.2490	73.1059
26	27.6919	28.5902	29.5256	30.4996	31.5139	33.6709	38.5530	44.3117	51.1134	59.1563	68.6764	79.9544
27	28.8304	29.8046	30.8209	31.8809	32.9867	35.3443	40.7096	47.0842	54.6691	63.7057	74.4838	87.3507
28	29.9745	31.0282	32.1291	33.2794	34.4815	37.0512	42.9309	49.9675	58.4025	68.5281	80.6976	95.3388
29	31.1244	32.2609	33.4504	34.6954	35.9987	38.7922	45.2188	52.9662	62.3227	73.6397	87.3465	103.9659
30	32.2800	33.5029	34.7849	36.1291	37.5387	40.5681	47.5754	56.0849	66.4388	79.0581	94.4607	113.2832

n	$i=0.005$	$i=0.0075$	$i=0.01$	$i=0.0125$	$i=0.015$	$i=0.02$	$i=0.03$	$i=0.04$	$i=0.05$	$i=0.06$	$i=0.07$	$i=0.08$
31	33.4414	34.7541	36.1327	37.5807	39.1018	42.3794	50.0026	59.3283	70.7607	84.8016	102.0730	123.3458
32	34.6086	36.0148	37.4941	39.0504	40.6883	44.2270	52.5027	62.7014	75.2988	90.8897	110.2181	134.2135
33	35.7817	37.2849	38.8690	40.5386	42.2986	46.1116	55.0778	66.2095	80.0637	97.3431	118.9334	145.9506
34	36.9606	38.5645	40.2577	42.0453	43.9331	48.0338	57.7301	69.8579	85.0669	104.1837	128.2587	158.6266
35	38.1454	39.8538	41.6603	43.5709	45.5921	49.9945	60.4620	73.6522	90.3203	111.4347	138.2368	172.3168
36	39.3361	41.1527	43.0769	45.1155	47.2759	51.9944	63.2759	77.5983	95.8363	119.1208	148.9134	187.1021
37	40.5328	42.4613	44.5076	46.6794	48.9851	54.0343	66.1742	81.7022	101.6281	127.2681	160.3374	203.0703
38	41.7354	43.7798	45.9527	48.2629	50.7199	56.1149	69.1594	85.9703	107.7095	135.9042	172.5610	220.3159
39	42.9441	45.1081	47.4123	49.8862	52.4807	58.2372	72.2342	90.4091	114.0950	145.0584	185.6402	238.9412
40	44.1588	46.4464	48.8864	51.4896	54.2679	60.4020	75.4012	95.0255	120.7997	154.7619	199.6351	259.0565
41	45.3796	47.7948	50.3752	53.1332	56.0819	62.6100	78.6632	99.8265	127.8397	165.0476	214.6095	280.7810
42	46.6065	49.1532	51.8789	54.7973	57.9231	64.8622	82.0231	104.8195	135.2317	175.9505	230.6322	304.2435
43	47.8396	50.5219	53.3978	56.4823	59.7920	67.1595	85.4838	110.0123	142.9933	187.5075	247.7764	329.5830
44	49.0788	51.9008	54.9318	58.1883	61.6889	69.5027	89.0484	115.4128	151.1430	199.7580	266.1208	356.9496
45	50.3242	53.2901	56.4811	59.9157	63.6142	71.8927	92.7198	121.0293	159.7001	212.7435	285.7493	386.5056
46	51.5758	54.6897	58.0459	61.6646	65.5684	74.3306	96.5014	126.8705	168.6851	226.5081	306.7517	418.4260
47	52.8337	56.0999	59.6263	63.4354	67.5519	76.3172	100.3965	132.9453	178.1194	241.0986	329.2243	452.9001
48	54.0978	57.5207	61.2226	65.2284	69.5652	79.3535	104.4083	139.2632	188.0253	256.5645	353.2700	490.1321
49	55.3683	58.9521	62.8348	67.0437	71.6087	81.9406	108.5406	145.8337	198.4266	272.9584	378.9989	530.3427
50	56.6452	60.3942	64.4632	68.8818	73.6828	84.5794	112.7968	152.6670	209.3479	290.3359	406.5289	573.7701
60	69.7700	75.4241	81.6697	88.5745	96.2147	114.0515	163.0534	237.9906	353.5837	533.1281	813.5203	1253.2132
70	83.5661	91.6200	100.6763	110.8720	122.3638	149.9779	230.5940	364.2904	588.5285	967.9321	1614.1341	2720.0800
80	98.0677	109.0725	121.6715	136.1188	152.7109	193.7719	321.3630	551.2449	971.2288	1746.5998	3189.0626	5886.9354
90	113.3109	127.8789	144.8633	164.7050	187.9290	247.1567	443.3489	827.9833	1594.6073	3141.0751	6287.1854	12723.9386
100	129.3337	148.1445	170.4814	197.0723	228.8030	312.2323	607.2877	1237.6237	2610.0251	5638.3680	12381.6617	27484.5157

Table A.4. Values of $[1 - (1 + i)^{-n}]/i$

n	i = 0.005	i = 0.0075	i = 0.01	i = 0.0125	i = 0.015	i = 0.02	i = 0.03	i = 0.04	i = 0.05	i = 0.06	i = 0.07	i = 0.08
1	0.9950	0.9926	0.9900	0.9876	0.9852	0.9803	0.9708	0.9615	0.9523	0.9433	0.9345	0.9259
2	1.9850	1.9777	1.9703	1.9631	1.9558	1.9415	1.9134	1.8860	1.8594	1.8333	1.8080	1.7832
3	2.9702	2.9556	2.9409	2.9265	2.9122	2.8838	2.8286	2.7750	2.7232	2.6730	2.6243	2.5770
4	3.9504	3.9261	3.9019	3.8780	3.8543	3.8077	3.7170	3.6298	3.5459	3.4651	3.3872	3.3121
5	4.9258	4.8894	4.8534	4.8178	4.7826	4.7134	4.5797	4.4518	4.3294	4.2123	4.1001	3.9927
6	5.8963	5.8456	5.7954	5.7460	5.6971	5.6014	5.4171	5.2421	5.0756	4.9173	4.7665	4.6228
7	6.8620	6.7946	6.7281	6.6627	6.5982	6.4719	6.2302	6.0020	5.7863	5.5823	5.3892	5.2063
8	7.8229	7.7366	7.6516	7.5681	7.4859	7.3254	7.0196	6.7327	6.4632	6.2097	5.9712	5.7466
9	8.7790	8.6716	8.5660	8.4623	8.3605	8.1622	7.7861	7.4353	7.1078	6.8016	6.5152	6.2468
10	9.7304	9.5996	9.4713	9.3455	9.2221	8.9825	8.5302	8.1108	7.7217	7.3600	7.0235	6.7100
11	10.6770	10.5207	10.3676	10.2178	10.0711	9.7868	9.2526	8.7604	8.3064	7.8868	7.4986	7.1389
12	11.6189	11.4349	11.2550	11.0793	10.9075	10.5753	9.9540	9.3850	8.8632	8.3838	7.9426	7.5360
13	12.5561	12.3423	12.1337	11.9301	11.7315	11.3483	10.6349	9.9856	9.3935	8.8526	8.3576	7.9037
14	13.4887	13.2430	13.0037	12.7705	12.5433	12.1062	11.2960	10.5631	9.8986	9.2949	8.7454	8.2442
15	14.4166	14.1370	13.8650	13.6005	13.3432	12.8492	11.9379	11.1183	10.3796	9.7122	9.1079	8.5594
16	15.3399	15.0243	14.7178	14.4202	14.1312	13.5777	12.5611	11.6522	10.8377	10.1058	9.4466	8.8513
17	16.2586	15.9050	15.5622	15.2299	14.9076	14.2918	13.1661	12.1656	11.2740	10.4772	9.7632	9.1216
18	17.1727	16.7792	16.3982	16.0295	15.6725	14.9920	13.7535	12.6592	11.6895	10.8276	10.0590	9.3718
19	18.0823	17.6468	17.2260	16.8193	16.4261	15.6784	14.3237	13.1339	12.0853	11.1581	10.3355	9.6035
20	18.9874	18.5080	18.0455	17.5993	17.1686	16.3514	14.8774	13.5903	12.4622	11.4699	10.5940	9.8181
21	19.8879	19.3628	18.8569	18.3696	17.9001	17.0112	15.4150	14.0291	12.8211	11.7640	10.8355	10.0168
22	20.7840	20.2112	19.6603	19.1305	18.6208	17.6580	15.9369	14.4511	13.1630	12.0415	11.0612	10.2007
23	21.6756	21.0533	20.4558	19.8820	19.3308	18.2922	16.4436	14.8568	13.4885	12.3033	11.2721	10.3710
24	22.5628	21.8891	21.2433	20.6242	20.0304	18.9139	16.9355	15.2469	13.7986	12.5503	11.4693	10.5287
25	23.4456	22.7188	22.0231	21.3572	20.7196	19.5234	17.4131	15.6220	14.0939	12.7833	11.6535	10.6747
26	24.3240	23.5422	22.7952	22.0812	21.3986	20.1210	17.8768	15.9827	14.3751	13.0031	11.8257	10.8099
27	25.1980	24.3595	23.5596	22.7962	22.0676	20.7068	18.3270	16.3295	14.6430	13.2105	11.9867	10.9351
28	26.0676	25.1707	24.3164	23.5025	22.7267	21.2812	18.7641	16.6630	14.8981	13.4061	12.1371	11.0510
29	26.9330	25.9759	25.0657	24.2000	23.3760	21.8443	19.1884	16.9837	15.1410	13.5907	12.2776	11.1584
30	27.7940	26.7751	25.8077	24.8889	24.0158	22.3964	19.6004	17.2920	15.3724	13.7648	12.4090	11.2577

n	$i=0.005$	$i=0.0075$	$i=0.01$	$i=0.0125$	$i=0.015$	$i=0.02$	$i=0.03$	$i=0.04$	$i=0.05$	$i=0.06$	$i=0.07$	$i=0.08$
31	28.6508	27.5683	26.5422	25.5692	24.6461	22.9377	20.0004	17.5884	15.5928	13.9290	12.5318	11.3497
32	29.5032	28.3557	27.2695	26.2412	25.2671	23.4683	20.3887	17.8735	15.8026	14.0840	12.6465	11.4349
33	30.3515	29.1371	27.9896	26.9049	25.8789	23.9885	20.7657	18.1476	16.0025	14.2302	12.7537	11.5138
34	31.1955	29.9128	28.7026	27.5604	26.4817	24.4985	21.1318	18.4111	16.1929	14.3681	12.8540	11.5869
35	32.0353	30.6827	29.4085	28.2078	27.0755	24.9986	21.4872	18.6646	16.3741	14.4982	12.9476	11.6545
36	32.8710	31.4468	30.1075	28.8472	27.6606	25.4888	21.8322	18.9082	16.5468	14.6209	13.0352	11.7171
37	33.7025	32.2053	30.7995	29.4787	28.2371	25.9694	22.1672	19.1425	16.7112	14.7367	13.1170	11.7751
38	34.5298	32.9581	31.4846	30.1025	28.8050	26.4406	22.4924	19.3678	16.8678	14.8460	13.1934	11.8288
39	35.3530	33.7053	32.1630	30.7185	29.3645	26.9025	22.8082	19.5844	17.0170	14.9490	13.2649	11.8785
40	36.1722	34.4469	32.8346	31.3269	29.9158	27.3554	23.1147	19.7927	17.1590	15.0462	13.3317	11.9246
41	36.9872	35.1831	33.4996	31.9278	30.4589	27.7994	23.4124	19.9930	17.2943	15.1380	13.3941	11.9672
42	37.7982	35.9137	34.1581	32.5213	30.9940	28.2347	23.7013	20.1856	17.4232	15.2245	13.4524	12.0066
43	38.6052	36.6389	34.8100	33.1074	31.5212	28.6615	23.9819	20.3707	17.5459	15.3061	13.5069	12.0432
44	39.4082	37.3587	35.4554	33.6863	32.0406	29.0799	24.2542	20.5488	17.6627	15.3831	13.5579	12.0770
45	40.2071	38.0732	36.0945	34.2581	32.5523	29.4901	24.5187	20.7200	17.7740	15.4558	13.6055	12.1084
46	41.0021	38.7823	36.7272	34.8228	33.0564	29.8923	24.7754	20.8846	17.8800	15.5243	13.6500	12.1374
47	41.7932	39.4862	37.3536	35.3806	33.5531	30.2865	25.0247	21.0429	17.9810	15.5890	13.6916	12.1642
48	42.5803	40.1848	37.9739	35.9314	34.0425	30.6731	25.2667	21.1951	18.0771	15.6500	13.7304	12.1891
49	43.3635	40.8782	38.5880	36.4755	34.5246	31.0520	25.5016	21.3414	18.1687	15.7075	13.7667	12.2121
50	44.1427	41.5664	39.1961	37.0128	34.9996	31.4236	25.7297	21.4821	18.2559	15.7618	13.8007	12.2334
60	51.7255	48.1734	44.9550	42.0345	39.3802	34.7608	27.6755	22.6234	18.9292	16.1614	14.0391	12.3765
70	58.9394	54.3046	50.1685	46.4696	43.1548	37.4986	29.1234	23.3945	19.3426	16.3845	14.1603	12.4428
80	65.8023	59.9944	54.8882	50.3866	46.4073	39.7445	30.2007	23.9153	19.5964	16.5091	14.2220	12.4735
90	72.3312	65.2746	59.1608	53.8460	49.2098	41.5869	31.0024	24.2672	19.7522	16.5786	14.2533	12.4877
100	78.5426	70.1746	63.0288	56.9013	51.6247	43.0983	31.5989	24.5049	19.8479	16.6175	14.2692	12.4943

Table A.5. Common Logarithms

N	0	1	2	3	4	5	6	7	8	9
10	0000	0043	0086	0128	0170	0212	0253	0294	0334	0374
11	0414	0453	0492	0531	0569	0607	0645	0682	0719	0755
12	0792	0828	0864	0899	0934	0969	1004	1038	1072	1106
13	1139	1173	1206	1239	1271	1303	1335	1367	1399	1430
14	1461	1492	1523	1553	1584	1614	1644	1673	1703	1732
15	1761	1790	1818	1847	1875	1903	1931	1959	1987	2014
16	2041	2068	2095	2122	2148	2175	2201	2227	2253	2279
17	2304	2330	2355	2380	2405	2430	2455	2480	2504	2529
18	2553	2577	2601	2625	2648	2672	2695	2718	2742	2765
19	2788	2810	2833	2856	2878	2900	2923	2945	2967	2989
20	3010	3032	3054	3075	3096	3118	3139	3160	3181	3201
21	3222	3243	3263	3284	3304	3324	3345	3365	3385	3404
22	3424	3444	3464	3483	3502	3522	3541	3560	3579	3598
23	3617	3636	3655	3674	3692	3711	3729	3747	3766	3784
24	3802	3820	3838	3856	3874	3892	3909	3927	3945	3962
25	3979	3997	4014	4031	4048	4065	4082	4099	4116	4133
26	4150	4166	4183	4200	4216	4232	4249	4265	4281	4298
27	4314	4330	4346	4362	4378	4393	4409	4425	4440	4456
28	4472	4487	4502	4518	4533	4548	4564	4579	4594	4609
29	4624	4639	4654	4669	4683	4698	4713	4728	4742	4757
30	4771	4786	4800	4814	4829	4843	4857	4871	4886	4900
31	4914	4928	4942	4955	4969	4983	4997	5011	5024	5038
32	5051	5065	5079	5092	5105	5119	5132	5145	5159	5172
33	5185	5198	5211	5224	5237	5250	5263	5276	5289	5302
34	5315	5328	5340	5353	5366	5378	5391	5403	5416	5428
35	5441	5453	5465	5478	5490	5502	5514	5527	5539	5551
36	5563	5575	5587	5599	5611	5623	5635	5647	5658	5670
37	5682	5694	5705	5717	5729	5740	5752	5763	5775	5786
38	5798	5809	5821	5832	5843	5855	5866	5877	5888	5899
39	5911	5922	5933	5944	5955	5966	5977	5988	5999	6010
40	6021	6031	6042	6053	6064	6075	6085	6096	6107	6117
41	6128	6138	6149	6160	6170	6180	6191	6201	6212	6222
42	6232	6243	6253	6263	6274	6284	6294	6304	6314	6325
43	6335	6345	6355	6365	6375	6385	6395	6405	6415	6425
44	6435	6444	6454	6464	6474	6484	6493	6503	6513	6522
45	6532	6542	6551	6561	6571	6580	6590	6599	6609	6618
46	6628	6637	6646	6656	6665	6675	6684	6693	6702	6712
47	6721	6730	6739	6749	6758	6767	6776	6785	6794	6803
48	6812	6821	6830	6839	6848	6857	6866	6875	6884	6893
49	6902	6911	6920	6928	6937	6946	6955	6964	6972	6981
50	6990	6998	7007	7016	7024	7033	7042	7050	7059	7067
51	7076	7084	7093	7101	7110	7118	7126	7135	7143	7152
52	7160	7168	7177	7185	7193	7202	7210	7218	7226	7235
53	7243	7251	7259	7267	7275	7284	7292	7300	7308	7316
54	7324	7332	7340	7348	7356	7364	7372	7380	7388	7396
55	7404	7412	7419	7427	7435	7443	7451	7459	7466	7474

Table A.5. Common Logarithms (Continued)

N	0	1	2	3	4	5	6	7	8	9
56	7482	7490	7497	7505	7513	7520	7528	7536	7543	7551
57	7559	7566	7574	7582	7589	7597	7604	7612	7619	7627
58	7634	7642	7649	7657	7664	7672	7679	7686	7694	7701
59	7709	7716	7723	7731	7738	7745	7752	7760	7767	7774
60	7782	7789	7796	7803	7810	7818	7825	7832	7839	7846
61	7853	7860	7868	7875	7882	7889	7896	7903	7910	7917
62	7924	7931	7938	7945	7952	7959	7966	7973	7980	7987
63	7993	8000	8007	8014	8021	8028	8035	8041	8048	8055
64	8062	8069	8075	8082	8089	8096	8102	8109	8116	8122
65	8129	8136	8142	8149	8156	8162	8169	8176	8182	8189
66	8195	8202	8209	8215	8222	8228	8235	8241	8248	8254
67	8261	8267	8274	8280	8287	8293	8299	8306	8312	8319
68	8325	8331	8338	8344	8351	8357	8363	8370	8376	8382
69	8388	8395	8401	8407	8414	8420	8426	8432	8439	8445
70	8451	8457	8463	8470	8476	8482	8488	8494	8500	8506
71	8513	8519	8525	8531	8537	8543	8549	8555	8561	8567
72	8573	8579	8585	8591	8597	8603	8609	8615	8621	8627
73	8633	8639	8645	8651	8657	8663	8669	8675	8681	8686
74	8692	8698	8704	8710	8716	8722	8727	8733	8739	8745
75	8751	8756	8762	8768	8774	8779	8785	8791	8797	8802
76	8808	8814	8820	8825	8831	8837	8842	8848	8854	8859
77	8865	8871	8876	8882	8887	8893	8899	8904	8910	8915
78	8921	8927	8932	8938	8943	8949	8954	8960	8965	8971
79	8976	8982	8987	8993	8998	9004	9009	9015	9020	9025
80	9031	9036	9042	9047	9053	9058	9063	9069	9074	9079
81	9085	9090	9096	9101	9106	9112	9117	9122	9128	9133
82	9138	9143	9149	9154	9159	9165	9170	9175	9180	9186
83	9191	9196	9201	9206	9212	9217	9222	9227	9232	9238
84	9243	9248	9253	9258	9263	9269	9274	9279	9284	9289
85	9294	9299	9304	9309	9315	9320	9325	9330	9335	9340
86	9345	9350	9355	9360	9365	9370	9375	9380	9385	9390
87	9395	9400	9405	9410	9415	9420	9425	9430	9435	9440
88	9445	9450	9455	9460	9465	9469	9474	9479	9484	9489
89	9494	9499	9504	9509	9513	9518	9523	9528	9533	9538
90	9542	9547	9552	9557	9562	9566	9571	9576	9581	9586
91	9590	9595	9600	9605	9609	9614	9619	9624	9628	9633
92	9638	9643	9647	9652	9657	9661	9666	9671	9675	9680
93	9685	9689	9694	9699	9703	9708	9713	9717	9722	9727
94	9731	9736	9741	9745	9750	9754	9759	9763	9768	9773
95	9777	9782	9786	9791	9795	9800	9805	9809	9814	9818
96	9823	9827	9832	9836	9841	9845	9850	9854	9859	9863
97	9868	9872	9877	9881	9886	9890	9894	9899	9903	9908
98	9912	9917	9921	9926	9930	9934	9939	9943	9948	9952
99	9956	9961	9965	9969	9974	9978	9983	9987	9991	9996

Table A.6. Natural or Naperian Logarithms
0.010–0.499

N	0	1	2	3	4	5	6	7	8	9
.01	−4.60517	.50986	.42285	.34281	.26870	.19971	.13517	.07454	.01738	*.96332
.02	−3.91202	.86323	.81671	.77226	.72970	.68888	.64966	.61192	.57555	.54046
.03	.50656	.47377	.44202	.41125	.38139	.35241	.32424	.29684	.27017	.24419
.04	.21888	.19418	.17009	.14656	.12357	.10109	.07911	.05761	.03655	.01593
.05	−2.99573	.97593	.95651	.93746	.91877	.90042	.88240	.86470	.84731	.83022
.06	.81341	.79688	.78062	.76462	.74887	.73337	.71810	.70306	.68825	.67365
.07	.65926	.64508	.63109	.61730	.60369	.59027	.57702	.56395	.55105	.53831
.08	.52573	.51331	.50104	.48891	.47694	.46510	.45341	.44185	.43042	.41912
.09	.40795	.39690	.38597	.37516	.36446	.35388	.34341	.33304	.32279	.31264
0.10	−2.30259	.29263	.28278	.27303	.26336	.25379	.24432	.23493	.22562	.21641
.11	.20727	.19823	.18926	.18037	.17156	.16282	.15417	.14558	.13707	.12863
.12	.12026	.11196	.10373	.09557	.08747	.07944	.07147	.06357	.05573	.04794
.13	.04022	.03256	.02495	.01741	.00992	.00248	*.99510	*.98777	*.98050	*.97328
.14	−1.96611	.95900	.95193	.94491	.93794	.93102	.92415	.91732	.91054	.90381
.15	.89712	.89048	.88387	.87732	.87080	.86433	.85790	.85151	.84516	.83885
.16	.83258	.82635	.82016	.81401	.80789	.80181	.79577	.78976	.78379	.77786
.17	.77196	.76609	.76026	.75446	.74870	.74297	.73727	.73161	.72597	.72037
.18	.71480	.70926	.70375	.69827	.69282	.68740	.68201	.67665	.67131	.66601
.19	.66073	.65548	.65026	.64507	.63990	.63476	.62964	.62455	.61949	.61445
0.20	−1.60944	.60445	.59949	.59455	.58964	.58475	.57988	.57504	.57022	.56542
.21	.56065	.55590	.55117	.54646	.54178	.53712	.53248	.52786	.52326	.51868
.22	.51413	.50959	.50508	.50058	.49611	.49165	.48722	.48281	.47841	.47403
.23	.46968	.46534	.46102	.45672	.45243	.44817	.44392	.43970	.43548	.43129
.24	.42712	.42296	.41882	.41469	.41059	.40650	.40242	.39837	.39433	.39030
.25	.38629	.38230	.37833	.37437	.37042	.36649	.36258	.35868	.35480	.35093
.26	.34707	.34323	.33941	.33560	.33181	.32803	.32426	.32051	.31677	.31304
.27	.30933	.30564	.30195	.29828	.29463	.29098	.28735	.28374	.28013	.27654
.28	.27297	.26940	.26585	.26231	.25878	.25527	.25176	.24827	.24479	.24133
.29	.23787	.23443	.23100	.22758	.22418	.22078	.21740	.21402	.21066	.20731
0.30	−1.20397	.20065	.19733	.19402	.19073	.18744	.18417	.18091	.17766	.17441
.31	.17118	.16796	.16475	.16155	.15836	.15518	.15201	.14885	.14570	.14256
.32	.13943	.13631	.13320	.13010	.12701	.12393	.12086	.11780	.11474	.11170
.33	.10866	.10564	.10262	.09961	.09661	.09362	.09064	.08767	.08471	.08176
.34	.07881	.07587	.07294	.07002	.06711	.06421	.06132	.05843	.05555	.05268
.35	−1.04982	.04697	.04412	.04129	.03846	.03564	.03282	.03002	.02722	.02443
.36	.02165	.01888	.01611	.01335	.01060	.00786	.00512	.00239	*.99967	*.99696
.37	−0.99425	.99155	.98886	.98618	.98350	.98083	.97817	.97551	.97286	.97022
.38	.96758	.96496	.96233	.95972	.95711	.95451	.95192	.94933	.94675	.94418
.39	.94161	.93905	.93649	.93395	.93140	.92887	.92634	.92382	.92130	.91879
0.40	−0.91629	.91379	.91130	.90882	.90634	.90387	.90140	.89894	.89649	.89404
.41	.89160	.88916	.88673	.88431	.88189	.87948	.87707	.87467	.87227	.86988
.42	.86750	.86512	.86275	.86038	.85802	.85567	.85332	.85097	.84863	.84630
.43	.84397	.84165	.83933	.83702	.83471	.83241	.83011	.82782	.82554	.82326
.44	.82098	.81871	.81645	.81419	.81193	.80968	.80744	.80520	.80296	.80073
.45	.79851	.79629	.79407	.79186	.78966	.78746	.78526	.78307	.78089	.77871
.46	.77653	.77436	.77219	.77003	.76787	.76572	.76357	.76143	.75929	.75715
.47	.75502	.75290	.75078	.74866	.74655	.74444	.74234	.74024	.73814	.73605
.48	.73397	.73189	.72981	.72774	.72567	.72361	.72155	.71949	.71744	.71539
.49	.71335	.71131	.70928	.70725	.70522	.70320	.70118	.69917	.69716	.69515

From *Standard Mathematical Tables*, 23rd ed., Samuel Selby, © The CRC Press, Inc., 1975. Used by permission of The CRC Press, Inc.
* Characteristic is given on the following row.

Table A.6. Natural or Naperian Logarithms (Continued)
0.500–0.999

N	0	1	2	3	4	5	6	7	8	9
0.50	−0.69315	.69115	.68916	.68717	.68518	.68320	.68122	.67924	.67727	.67531
.51	.67334	.67139	.66934	.66748	.66553	.66359	.66165	.65971	.65778	.65585
.52	.65393	.65201	.65009	.64817	.64626	.64436	.64245	.64055	.63866	.63677
.53	.63488	.63299	.63111	.62923	.62736	.62549	.62362	.62176	.61990	.61804
.54	.61619	.61434	.61249	.61065	.60881	.60697	.60514	.60331	.60148	.59966
.55	.59784	.59602	.59421	.59240	.59059	.58879	.58699	.58519	.58340	.58161
.56	.57982	.57803	.57625	.57448	.57270	.57093	.56916	.56740	.56563	.56387
.57	.56212	.56037	.55862	.55687	.55513	.55339	.55165	.54991	.54818	.54645
.58	.54473	.54300	.54128	.53957	.53785	.53614	.53444	.53273	.53101	.52933
.59	.52763	.52594	.52425	.52256	.52088	.51919	.51751	.51584	.51416	.51249
0.60	−0.51083	.50916	.50750	.50584	.50418	.50253	.50088	.49923	.49758	.49594
.61	.49430	.49266	.49102	.48939	.48776	.48613	.48451	.48289	.48127	.47965
.62	.47804	.47642	.47482	.47321	.47160	.47000	.46840	.46681	.46522	.46362
.63	.46204	.46045	.45887	.45728	.45571	.45413	.45256	.45099	.44942	.44785
.64	.44629	.44473	.44317	.44161	.44006	.43850	.43696	.43541	.43386	.43232
.65	.43078	.42925	.42771	.42618	.42465	.42312	.42159	.42007	.41855	.41703
.66	.41552	.41400	.41249	.41098	.40947	.40797	.40647	.40497	.40347	.40197
.67	.40048	.39899	.39750	.39601	.39453	.39304	.39156	.39008	.38861	.38713
.68	.38566	.38419	.38273	.38126	.37980	.37834	.37688	.37542	.37397	.37251
.69	.37106	.36962	.36817	.36673	.36528	.36384	.36241	.36097	.35954	.35810
0.70	−0.35667	.35525	.35382	.35240	.35098	.34956	.34814	.34672	.34531	.34390
.71	.34249	.34108	.33968	.33827	.33687	.33547	.33408	.33268	.33129	.32989
.72	.32850	.32712	.32573	.32435	.32296	.32158	.32021	.31883	.31745	.31608
.73	.31471	.31334	.31197	.31061	.30925	.30788	.30653	.30517	.30381	.30246
.74	.30111	.29975	.29841	.29706	.29571	.29437	.29303	.29169	.29035	.28902
.75	.28768	.28635	.28502	.28369	.28236	.28104	.27971	.27839	.27707	.27575
.76	.27444	.27312	.27181	.27050	.26919	.26788	.26657	.26527	.26397	.26266
.77	.26136	.26007	.25877	.25748	.25618	.25489	.25360	.25231	.25103	.24974
.78	.24846	.24718	.24590	.24462	.24335	.24207	.24080	.23953	.23826	.23699
.79	.23572	.23446	.23319	.23193	.23067	.22941	.22816	.22690	.22565	.22439
0.80	−0.22314	.22189	.22065	.21940	.21816	.21691	.21567	.21433	.21319	.21196
.81	.21072	.20949	.20825	.20702	.20579	.20457	.20334	.20212	.20089	.19967
.82	.19845	.19723	.19601	.19480	.19358	.19237	.19116	.18995	.18874	.18754
.83	.18633	.18513	.18392	.18272	.18152	.18032	.17913	.17793	.17674	.17554
.84	.17435	.17316	.17198	.17079	.16960	.16842	.16724	.16605	.16487	.16370
.85	−0.16252	.16134	.16017	.15900	.15782	.15665	.15548	.15432	.15315	.15199
.86	.15032	.14966	.14850	.14734	.14618	.14503	.14387	.14272	.14156	.14041
.87	.13926	.13811	.13697	.13582	.13467	.13353	.13239	.13125	.13011	.12897
.88	.12783	.12670	.12556	.12443	.12330	.12217	.12104	.11991	.11878	.11766
.89	.11653	.11541	.11429	.11317	.11205	.11093	.10981	.10870	.10759	.10647
0.90	−0.10536	.10425	.10314	.10203	.10093	.09982	.09872	.09761	.09651	.09541
.91	.09431	.09321	.09212	.09102	.08992	.08883	.08744	.08665	.08556	.08447
.92	.08338	.08230	.08121	.08013	.07904	.07796	.07688	.07580	.07472	.07365
.93	.07257	.07150	.07042	.06935	.06828	.06721	.06614	.06507	.06401	.06294
.94	.06188	.06081	.05975	.05869	.05763	.05657	.05551	.05446	.05340	.05235
.95	.05129	.05024	.04919	.04814	.04709	.04604	.04500	.04395	.04291	.04186
.96	.04082	.03978	.03874	.03770	.03666	.03563	.03459	.03356	.03252	.03149
.97	.03046	.02943	.02840	.02737	.02634	.02532	.02429	.02327	.02225	.02122
.98	.02020	.01918	.01816	.01715	.01613	.01511	.01410	.01309	.01207	.01106
.99	.01005	.00904	.00803	.00702	.00602	.00501	.00401	.00300	.00200	.00100

Table A.6. Natural or Naperian Logarithms (Continued)
1.00–5.49

N	0	1	2	3	4	5	6	7	8	9
1.0	0.00000	.00995	.01980	.02956	.03922	.04879	.05827	.06766	.07696	.08618
.1	.09531	.10436	.11333	.12222	.13103	.13976	.14842	.15700	.16551	.17395
.2	.18232	.19062	.19885	.20701	.21511	.22314	.23111	.23902	.24686	.25464
.3	.26236	.27003	.27763	.28518	.29267	.30010	.30748	.31481	.32208	.32930
.4	.33647	.34359	.35066	.35767	.36464	.37156	.37844	.38526	.39204	.39878
.5	.40547	.41211	.41871	.42527	.43178	.43825	.44469	.45108	.45742	.46373
.6	.47000	.47623	.48243	.48858	.49470	.50078	.50682	.51282	.51879	.52473
.7	.53063	.53649	.54232	.54812	.55389	.55962	.56531	.57098	.57661	.58222
.8	.58779	.59333	.59884	.60432	.60977	.61519	.62058	.62594	.63127	.63658
.9	.64185	.64710	.65233	.65752	.66269	.66783	.67294	.67803	.68310	.68813
2.0	0.69315	.69813	.70310	.70804	.71295	.71784	.72271	.72755	.73237	.73716
.1	.74194	.74669	.75142	.75612	.76081	.76547	.77011	.77473	.77932	.78390
.2	.78846	.79299	.79751	.80200	.80648	.81093	.81536	.81978	.82418	.82855
.3	.83291	.83725	.84157	.84587	.85015	.85442	.85866	.86289	.86710	.87129
.4	.87547	.87963	.88377	.88789	.89200	.89609	.90016	.90422	.90826	.91228
.5	.91629	.92028	.92426	.92822	.93216	.93609	.94001	.94391	.94779	.95166
.6	.95551	.95935	.96317	.96698	.97078	.97456	.97833	.98208	.98582	.98954
.7	.99325	.99695	*.00063	*.00430	*.00796	*.01160	*.01523	*.01885	*.02245	*.02604
.8	1.02962	.03318	.03674	.04028	.04380	.04732	.05082	.05431	.05779	.06126
.9	.06471	.06815	.07158	.07500	.07841	.08181	.08519	.08856	.09192	.09527
3.0	1.09861	.10194	.10526	.10856	.11186	.11514	.11841	.12168	.12493	.12817
.1	.13140	.13462	.13783	.14103	.14422	.14740	.15057	.15373	.15688	.16002
.2	.16315	.16627	.16938	.17248	.17557	.17865	.18173	.18479	.18784	.19089
.3	.19392	.19695	.19996	.20297	.20597	.20896	.21194	.21491	.21788	.22083
.4	.22378	.22671	.22964	.23256	.23547	.23837	.24127	.24415	.24703	.24990
.5	.25276	.25562	.25846	.26130	.26413	.26695	.26976	.27257	.27536	.27815
.6	.28093	.28371	.28647	.28923	.29198	.29473	.29746	.30019	.30291	.30563
.7	.30833	.31103	.31372	.31641	.31909	.32176	.32442	.32708	.32972	.33237
.8	.33500	.33763	.34025	.34286	.34547	.34807	.35067	.35325	.35584	.35841
.9	.36098	.36354	.36609	.36864	.37118	.37372	.37624	.37877	.38128	.38379
4.0	1.38629	.38879	.39128	.39377	.39624	.39872	.40118	.40364	.40610	.40854
.1	.41099	.41342	.41585	.41828	.42070	.42311	.42552	.42792	.43031	.43270
.2	.43508	.43756	.43984	.44220	.44456	.44692	.44927	.45161	.45395	.45629
.3	.45862	.46094	.46326	.46557	.46787	.47018	.47247	.47476	.47705	.47933
.4	.48160	.48387	.48614	.48840	.49065	.49290	.49515	.49739	.49962	.50185
.5	.50408	.50630	.50851	.51072	.51293	.51513	.51732	.51951	.52170	.52388
.6	.52606	.52823	.53039	.53256	.53471	.53687	.53902	.54116	.54330	.54543
.7	.54756	.54969	.55181	.55393	.55604	.55814	.56025	.56235	.56444	.56653
.8	.56862	.57070	.57277	.57485	.57691	.57898	.58104	.58309	.58515	.58719
.9	.58924	.59127	.59331	.59534	.59737	.59939	.60141	.60342	.60543	.60744
5.0	1.60944	.61144	.61343	.61542	.61741	.61939	.62137	.62334	.62531	.62728
.1	.62924	.63120	.63315	.63511	.63705	.63900	.64094	.64287	.64481	.64673
.2	.64866	.65058	.65250	.65441	.65632	.65823	.66013	.66203	.66393	.66582
.3	.66771	.66959	.67147	.67335	.67523	.67710	.67896	.68083	.68269	.68455
.4	.68640	.68825	.69010	.69194	.69378	.69562	.69745	.69928	.70111	.70293

Table A.6. Natural or Naperian Logarithms (Continued)
5.50–9.99

N	0	1	2	3	4	5	6	7	8	9
.5	.70475	.70656	.70838	.71019	.71199	.71380	.71560	.71740	.71919	.72098
.6	.72277	.72455	.72633	.72811	.72988	.73166	.73342	.73519	.73695	.73871
.7	.74047	.74222	.74397	.74572	.74746	.74920	.75094	.75267	.75440	.75613
.8	.75786	.75958	.76130	.76302	.76473	.76644	.76815	.76985	.77156	.77326
.9	.77495	.77665	.77834	.78002	.78171	.78339	.78507	.78675	.78842	.79009
6.0	i.79176	.79342	.79509	.79675	.79840	.80006	.80171	.80336	.80500	.80665
.1	.80829	.80993	.81156	.81319	.81482	.81645	.81808	.81970	.82132	.82294
.2	.82455	.82616	.82777	.82938	.83098	.83258	.83418	.83578	.83737	.83896
.3	.84055	.84214	.84372	.84530	.84688	.84845	.85003	.85160	.85317	.85473
.4	.85630	.85786	.85942	.86097	.86253	.86408	.86563	.86718	.86872	.87026
.5	.87180	.87334	.87487	.87641	.87794	.87947	.88099	.88251	.88403	.88555
.6	.88707	.88858	.89010	.89160	.89311	.89462	.89612	.89762	.89912	.90061
.7	.90211	.90360	.90509	.90658	.90806	.90954	.91102	.91250	.91398	.91545
.8	.91692	.91839	.91986	.92132	.92279	.92425	.92571	.92716	.92862	.93007
.9	.93152	.93297	.93442	.93586	.93730	.93874	.94018	.94162	.94305	.94448
7.0	1.94591	.94734	.94876	.95019	.95161	.95303	.95445	.95586	.95727	.95869
.1	.96009	.96150	.96291	.96431	.96571	.96711	.96851	.96991	.97130	.97269
.2	.97408	.97547	.97685	.97824	.97962	.98100	.98238	.98376	.98513	.98650
.3	.98787	.98924	.99061	.99198	.99334	.99470	.99606	.99742	.99877	*.00013
.4	2.00148	.00283	.00418	.00553	.00687	.00821	.00956	.01089	.01223	.01357
.5	.01490	.01624	.01757	.01890	.02022	.02155	.02287	.02419	.02551	.02683
.6	.02815	.02946	.03078	.03209	.03340	.03471	.03601	.03732	.03862	.03992
.7	.04122	.04252	.04381	.04511	.04640	.04769	.04898	.05027	.05156	.05284
.8	.05412	.05540	.05668	.05796	.05924	.06051	.06179	.06306	.06433	.06560
.9	.06686	.06813	.06939	.07065	.07191	.07317	.07443	.07568	.07694	.07819
8.0	2.07944	.08069	.08194	.08318	.08443	.08567	.08691	.08815	.08939	.09063
.1	.09186	.09310	.09433	.09556	.09679	.09802	.09924	.10047	.10169	.10291
.2	.10413	.10535	.10657	.10779	.10900	.11021	.11142	.11263	.11384	.11505
.3	.11626	.11746	.11866	.11986	.12106	.12226	.12346	.12465	.12585	.12704
.4	.12823	.12942	.13061	.13180	.13298	.13417	.13535	.13653	.13771	.13889
.5	.14007	.14124	.14242	.14359	.14476	.14593	.14710	.14827	.14943	.15060
.6	.15176	.15292	.15409	.15524	.15640	.15756	.15871	.15987	.16102	.16217
.7	.16332	.16447	.16562	.16677	.16791	.16905	.17020	.17134	.17248	.17361
.8	.17475	.17589	.17702	.17816	.17929	.18042	.18155	.18267	.18380	.18493
.9	.18605	.18717	.18830	.18942	.19054	.19165	.19277	.19389	.19500	.19611
9.0	2.19722	.19834	.19944	.20055	.20166	.20276	.20387	.20497	.20607	.20717
.1	.20827	.20937	.21047	.21157	.21266	.21375	.21485	.21594	.21703	.21812
.2	.21920	.22029	.22138	.22246	.22354	.22462	.22570	.22678	.22786	.22894
.3	.23001	.23109	.23216	.23324	.23431	.23538	.23645	.23751	.23858	.23965
.4	.24071	.24177	.24284	.24390	.24496	.24601	.24707	.24813	.24918	.25024
.5	.25129	.25234	.25339	.25444	.25549	.25654	.25759	.25863	.25968	.26072
.6	.26176	.26280	.26384	.26488	.26592	.26696	.26799	.26903	.27006	.27109
.7	.27213	.27316	.27419	.27521	.27624	.27727	.27829	.27932	.28034	.28136
.8	.28238	.28340	.28442	.28544	.28646	.28747	.28849	.28950	.29051	.29152
.9	.29253	.29354	.29455	.29556	.29657	.29757	.29858	.29958	.30058	.30158

Table A.6.   Natural or Naperian Logarithms (Continued)
0–499

N	0	1	2	3	4	5	6	7	8	9
0	− ∞	0.00000	0.69315	1.09861	.38629	.60944	.79176	.94591	*.07944	*.19722
1	2.30259	.39790	.48491	.56495	.63906	.70805	.77259	.83321	.89037	.94444
2	.99573	*.04452	*.09104	*.13549	*.17805	*.21888	*.25810	*.29584	*.33220	*.36730
3	3.40120	.43399	.46574	.49651	.52636	.55535	.58352	.61092	.63759	.66356
4	.68888	.71357	.73767	.76120	.78419	.80666	.82864	.85015	.87120	.89182
5	.91202	.93183	.95124	.97029	.98898	*.00733	*.02535	*.04305	*.06044	*.07754
6	4.09434	.11087	.12713	.14313	.15888	.17439	.18965	.20469	.21951	.23411
7	.24850	.26268	.27667	.29046	.30407	.31749	.33073	.34381	.35671	.36945
8	.38203	.39445	.40672	.41884	.43082	.44265	.45435	.46591	.47734	.48864
9	.49981	.51086	.52179	.53260	.54329	.55388	.56435	.57471	.58497	.59512
10	4.60517	.61512	.62497	.63473	.64439	.65396	.66344	.67283	.68213	.69135
11	.70048	.70953	.71850	.72739	.73620	.74493	.75359	.76217	.77068	.77912
12	.78749	.79579	.80402	.81218	.82028	.82831	.83628	.84419	.85203	.85981
13	.86753	.87520	.88280	.89035	.89784	.90527	.91265	.91998	.92725	.93447
14	.94164	.94876	.95583	.96284	.96981	.97673	.98361	.99043	.99721	*.00395
15	5.01064	.01728	.02388	.03044	.03695	.04343	.04986	.05625	.06260	.06890
16	.07517	.08140	.08760	.09375	.09987	.10595	.11199	.11799	.12396	.12990
17	.13580	.14166	.14749	.15329	.15906	.16479	.17048	.17615	.18178	.18739
18	.19296	.19850	.20401	.20949	.21494	.22036	.22575	.23111	.23644	.24175
19	.24702	.25227	.25750	.26269	.26786	.27300	.27811	.28320	.28837	.29330
20	5.29832	.30330	.30827	.31321	.31812	.32301	.32788	.33272	.33754	.34233
21	.34711	.35186	.35659	.36129	.36598	.37064	.37528	.37990	.38450	.38907
22	.39363	.39816	.40268	.40717	.41165	.41610	.42053	.42495	.42935	.43372
23	.43808	.44242	.44674	.45104	.45532	.45959	.46383	.46806	.47227	.47646
24	.48064	.48480	.48894	.49306	.49717	.50126	.50533	.50939	.51343	.51745
25	.52146	.52545	.52943	.53339	.53733	.54126	.54518	.54908	.55296	.55683
26	.56068	.56452	.56834	.57215	.57595	.57973	.58350	.58725	.59099	.59471
27	.59842	.60212	.60580	.60947	.61313	.61677	.62040	.62402	.62762	.63121
28	.63479	.63835	.64191	.64545	.64897	.65249	.65599	.65948	.66296	.66643
29	.66988	.67332	.67675	.68017	.68358	.68698	.69036	.69373	.69709	.70044
30	5.70378	.70711	.71043	.71373	.71703	.72031	.72359	.72685	.73010	.73334
31	.73657	.73979	.74300	.74620	.74939	.75257	.75574	.75890	.76205	.76519
32	.76832	.77144	.77455	.77765	.78074	.78383	.78690	.78996	.79301	.79606
33	.79909	.80212	.80513	.80814	.81114	.81413	.81711	.82008	.82305	.82600
34	.82895	.83188	.83481	.83773	.84064	.84354	.84644	.84932	.85220	.85507
35	.85793	.86079	.86363	.86647	.86930	.87212	.87493	.87774	.88053	.88332
36	.88610	.88888	.89164	.89440	.89715	.89990	.90263	.90536	.90808	.91080
37	.91350	.91620	.91889	.92158	.92426	.92693	.92959	.93225	.93489	.93754
38	.94017	.94280	.94542	.94803	.95064	.95324	.95584	.95842	.96101	.96358
39	.96615	.96871	.97126	.97381	.97635	.97889	.98141	.98394	.98645	.98896
40	5.99146	.99396	.99645	.99894	*.00141	*.00389	*.00635	*.00881	*.01127	*.01372
41	6.01616	.01859	.02102	.02345	.02587	.02828	.03069	.03309	.03548	.03787
42	.04025	.04263	.04501	.04747	.04973	.05209	.05444	.05678	.05912	.06146
43	.06379	.06611	.06843	.07074	.07304	.07535	.07764	.07993	.08222	.08450
44	.08677	.08904	.09131	.09357	.09582	.09807	.10032	.10256	.10479	.10702
45	.10925	.11147	.11368	.11589	.11810	.12030	.12249	.12468	.12687	.12905
46	.13123	.13340	.13556	.13773	.13988	.14204	.14419	.14633	.14847	.15060
47	.15273	.15486	.15698	.15910	.16121	.16331	.16542	.16752	.16961	.17170
48	.17379	.17587	.17794	.18002	.18208	.18415	.18621	.18826	.19032	.19236
49	.19441	.19644	.19848	.20051	.20254	.20456	.20658	.20859	.21060	.21261

Table A.6. Natural or Naperian Logarithms (Continued)
500–999

N	0	1	2	3	4	5	6	7	8	9
50	6.21461	.21661	.21860	.22059	.22258	.22456	.22654	.22851	.23048	.23245
51	.23441	.23637	.23832	.24028	.24222	.24417	.24611	.24804	.24998	.25190
52	.25383	.25575	.25767	.25958	.26149	.26340	.26530	.26720	.26910	.27099
53	.27288	.27476	.27664	.27852	.28040	.28227	.28413	.28600	.28786	.28972
54	.29157	.29342	.29527	.29711	.29895	.30079	.30262	.30445	.30628	.30810
55	.30992	.31173	.31355	.31536	.31716	.31897	.32077	.32257	.32436	.32615
56	.32794	.32972	.33150	.33328	.33505	.33683	.33859	.34036	.34212	.34388
57	.34564	.34739	.34914	.35089	.35263	.35437	.35611	.35784	.35957	.36130
58	.36303	.36475	.36647	.36819	.36990	.37161	.37332	.37502	.37673	.37843
59	.38012	.38182	.38351	.38519	.38688	.38856	.39024	.39192	.39359	.39526
60	6.30693	.39859	.40026	.40192	.40357	.40523	.40688	.40853	.41017	.41182
61	.41346	.41510	.41673	.41836	.41999	.42162	.42325	.42487	.42649	.42811
62	.42972	.43133	.43294	.43455	.43615	.43775	.43935	.44095	.44254	.44413
63	.44572	.44731	.44889	.45047	.45205	.45362	.45520	.45677	.45834	.45990
64	.46147	.46303	.46459	.46614	.46770	.46925	.47080	.47235	.47389	.47543
65	.47697	.47851	.48004	.48158	.48311	.48464	.48616	.48768	.48920	.49072
66	.49224	.49375	.49527	.49677	.49828	.49979	.50129	.50279	.50429	.50578
67	.50728	.50877	.51026	.51175	.51323	.51471	.51619	.51767	.51915	.52062
68	.52209	.52356	.52503	.52649	.52796	.52942	.53088	.53233	.53379	.53524
69	.53669	.53814	.53959	.54103	.54247	.54391	.54535	.54679	.54822	.54965
70	6.55108	.55251	.55393	.55536	.55678	.55820	.55962	.56103	.56244	.56386
71	.56526	.56667	.56808	.56948	.57088	.57228	.57368	.57508	.57647	.57786
72	.57925	.58064	.58203	.58341	.58479	.58617	.58755	.58893	.59030	.59167
73	.59304	.59441	.59578	.59715	.59851	.59987	.60123	.60259	.60394	.60530
74	.60665	.60800	.60935	.61070	.61204	.61338	.61473	.61607	.61740	.61874
75	.62007	.62141	.62274	.62407	.62539	.62672	.62804	.62936	.63068	.63200
76	.63332	.63463	.63595	.63726	.63857	.63988	.64118	.64249	.64379	.64509
77	.64639	.64769	.64898	.65028	.65157	.65286	.65415	.65544	.65673	.65801
78	.65929	.66058	.66185	.66313	.66441	.66568	.66696	.66823	.66950	.67077
79	.67203	.67330	.67456	.67582	.67708	.67834	.67960	.68085	.68211	.68336
80	6.68461	.68586	.68711	.68835	.68960	.69084	.69208	.69332	.69456	.69580
81	.69703	.69827	.69950	.70073	.70196	.70319	.70441	.70564	.70686	.70808
82	.70930	.71052	.71174	.71296	.71417	.71538	.71659	.71780	.71901	.72022
83	.72143	.72263	.72383	.72503	.72623	.72743	.72863	.72982	.73102	.73221
84	.73340	.73459	.73578	.73697	.73815	.73934	.74052	.74170	.74288	.74406
85	.74524	.74641	.74759	.74876	.74993	.75110	.75227	.75344	.75460	.75577
86	.75693	.75809	.75926	.76041	.76157	.76273	.76388	.76504	.76619	.76734
87	.76849	.76964	.77079	.77194	.77308	.77422	.77537	.77651	.77765	.77878
88	.77992	.78106	.78219	.78333	.78446	.78559	.78672	.78784	.78897	.79010
89	.79122	.79234	.79347	.79459	.79571	.79682	.79794	.79906	.80017	.80128
90	6.80239	.80351	.80461	.80572	.80683	.80793	.80904	.81014	.81124	.81235
91	.81344	.81454	.81564	.81674	.81783	.81892	.82002	.82111	.82220	.82329
92	.82437	.82546	.82655	.82763	.82871	.82979	.83087	.83195	.83303	.83411
93	.83518	.83626	.83733	.83841	.83948	.84055	.84162	.84268	.84375	.84482
94	.84588	.84694	.84801	.84907	.85013	.85118	.85224	.85330	.85435	.85541
95	.85646	.85751	.85857	.85961	.86066	.86171	.86276	.86380	.86485	.86589
96	.86693	.86797	.86901	.87005	.87109	.87213	.87316	.87420	.87523	.87626
97	.87730	.87833	.87936	.88038	.88141	.88244	.88346	.88449	.88551	.88653
98	.88755	.88857	.88959	.89061	.89163	.89264	.89366	.89467	.89568	.89669
99	.89770	.89871	.89972	.90073	.90174	.90274	.90375	.90475	.90575	.90675

Table A.7. Exponential Functions

x	e^x	e^{-x}	x	e^x	e^{-x}
0.00	1.0000	1.000000	0.50	1.6487	0.606531
0.01	1.0101	0.990050	0.51	1.6653	.600496
0.02	1.0202	.980199	0.52	1.6820	.594521
0.03	1.0305	.970446	0.53	1.6989	.588605
0.04	1.0408	.960789	0.54	1.7160	.582748
0.05	1.0513	0.951229	0.55	1.7333	0.576950
0.06	1.0618	.941765	0.56	1.7507	.571209
0.07	1.0725	.932394	0.57	1.7683	.565525
0.08	1.0833	.923116	0.58	1.7860	.559898
0.09	1.0942	.913931	0.59	1.8040	.554327
0.10	1.1052	0.904837	0.60	1.8221	0.548812
0.11	1.1163	.895834	0.61	1.8404	.543351
0.12	1.1275	.886920	0.62	1.8589	.537944
0.13	1.1388	.878095	0.63	1.8776	.532592
0.14	1.1503	.869358	0.64	1.8965	.527292
0.15	1.1618	0.860708	0.65	1.9155	0.522046
0.16	1.1735	.852114	0.66	1.9348	.516851
0.17	1.1853	.843665	0.67	1.9542	.511709
0.18	1.1972	.835270	0.68	1.9739	.506617
0.19	1.2092	.826959	0.69	1.9937	.501576
0.20	1.2214	0.818731	0.70	2.0138	0.496585
0.21	1.2337	.810584	0.71	2.0340	.491644
0.22	1.2461	.802519	0.72	2.0544	.486752
0.23	1.2586	.794534	0.73	2.0751	.481909
0.24	1.2712	.786628	0.74	2.0959	.477114
0.25	1.2840	0.778801	0.75	2.1170	0.472367
0.26	1.2969	.771052	0.76	2.1383	.467666
0.27	1.3100	.763379	0.77	2.1598	.463013
0.28	1.3231	.755784	0.78	2.1815	.458406
0.29	1.3364	.748264	0.79	2.2034	.453845
0.30	1.3499	0.740818	0.80	2.2255	0.449329
0.31	1.3634	.733447	0.81	2.2479	.444858
0.32	1.3771	.726149	0.82	2.2705	.440432
0.33	1.3910	.718924	0.83	2.2933	.436049
0.34	1.4049	.711770	0.84	2.3164	.431711
0.35	1.4191	0.704688	0.85	2.3396	0.427415
0.36	1.4333	.697676	0.86	2.3632	.423162
0.37	1.4477	.690734	0.87	2.3869	.418952
0.38	1.4623	.683861	0.88	2.4109	.414783
0.39	1.4770	.677057	0.89	2.4351	.410656
0.40	1.4918	0.670320	0.90	2.4596	0.406570
0.41	1.5068	.663650	0.91	2.4843	.402524
0.42	1.5220	.657047	0.92	2.5093	.398519
0.43	1.5373	.650509	0.93	2.5345	.394554
0.44	1.5527	.644036	0.94	2.5600	.390628
0.45	1.5683	0.637628	0.95	2.5857	0.386741
0.46	1.5841	.631284	0.96	2.6117	.382893
0.47	1.6000	.625002	0.97	2.6379	.379083
0.48	1.6161	.618783	0.98	2.6645	.375311
0.49	1.6323	.612626	0.99	2.6912	.371577
0.50	1.6487	0.606531	1.00	2.7183	0.367870

From *Standard Mathematical Tables*, 23rd ed., Samuel Selby, © The CRC Press, Inc., 1975. Used by permission of The CRC Press, Inc.

Table A.7. Exponential Functions (Continued)

x	e^x	e^{-x}	x	e^x	e^{-x}
1.0	2.718	0.368	5.5	244.7	0.0041
1.1	3.004	0.333	5.6	270.4	0.0037
1.2	3.320	0.301	5.7	298.9	0.0033
1.3	3.669	0.273	5.8	330.3	0.0030
1.4	4.055	0.247	5.9	365.0	0.0027
1.5	4.482	0.223	6.0	403.4	0.0025
1.6	4.953	0.202	6.1	445.9	0.0022
1.7	5.474	0.183	6.2	492.8	0.0020
1.8	6.050	0.165	6.3	544.6	0.0018
1.9	6.686	0.150	6.4	601.8	0.0017
2.0	7.389	0.135	6.5	665.1	0.0015
2.1	8.166	0.122	6.6	735.1	0.0014
2.2	9.025	0.111	6.7	812.4	0.0012
2.3	9.974	0.100	6.8	897.8	0.0011
2.4	11.023	0.091	6.9	992.3	0.0010
2.5	12.18	0.082	7.0	1,096.6	0.0009
2.6	13.46	0.074	7.1	1,212.0	0.0008
2.7	14.88	0.067	7.2	1,339.4	0.0007
2.8	16.44	0.061	7.3	1,480.3	0.0007
2.9	18.17	0.055	7.4	1,636.0	0.0006
3.0	20.09	0.050	7.5	1,808.0	0.00055
3.1	22.20	0.045	7.6	1,998.2	0.00050
3.2	24.53	0.041	7.7	2,208.3	0.00045
3.3	27.11	0.037	7.8	2,440.6	0.00041
3.4	29.96	0.033	7.9	2,697.3	0.00037
3.5	33.12	0.030	8.0	2,981.0	0.00034
3.6	36.60	0.027	8.1	3,294.5	0.00030
3.7	40.45	0.025	8.2	3,641.0	0.00027
3.8	44.70	0.022	8.3	4,023.9	0.00025
3.9	49.40	0.020	8.4	4,447.1	0.00022
4.0	54.60	0.018	8.5	4,914.8	0.00020
4.1	60.34	0.017	8.6	5,431.7	0.00018
4.2	66.69	0.015	8.7	6,002.9	0.00017
4.3	73.70	0.014	8.8	6,634.2	0.00015
4.4	81.45	0.012	8.9	7,332.0	0.00014
4.5	90.02	0.011	9.0	8,103.1	0.00012
4.6	99.48	0.010	9.1	8,955.3	0.00011
4.7	109.95	0.009	9.2	9,897.1	0.00010
4.8	121.51	0.008	9.3	10,938	0.00009
4.9	134.29	0.007	9.4	12,088	0.00008
5.0	148.4	0.0067	9.5	13,360	0.00007
5.1	164.0	0.0061	9.6	14,765	0.00007
5.2	181.3	0.0055	9.7	16,318	0.00006
5.3	200.3	0.0050	9.8	18,034	0.00006
5.4	221.4	0.0045	9.9	19,930	0.00005

Table A.8.   Squares and Square Roots
1–1000

N	N^2	$\sqrt{N}$	N	N^2	$\sqrt{N}$	N	N^2	$\sqrt{N}$
			50	2 500	7.071 068	100	10 000	10.00000
1	1	1.000 000	51	2 601	7.141 428	101	10 201	10.04988
2	4	1.414 214	52	2 704	7.211 103	102	10 404	10.09950
3	9	1.732 051	53	2 809	7.280 110	103	10 609	10.14889
4	16	2.000 000	54	2 916	7.348 469	104	10 816	10.19804
5	25	2.236 068	55	3 025	7.416 198	105	11 025	10.24695
6	36	2.449 490	56	3 136	7.483 315	106	11 236	10.29563
7	49	2.645 751	57	3 249	7.549 834	107	11 499	10.34408
8	64	2.828 427	58	3 364	7.615 773	108	11 664	10.39230
9	81	3.000 000	59	3 481	7.681 146	109	11 881	10.44031
10	100	3.162 278	60	3 600	7.745 967	110	12 100	10.48809
11	121	3.316 625	61	3 721	7.810 250	111	12 321	10.53565
12	144	3.464 102	62	3 844	7.874 008	112	12 544	10.58301
13	169	3.605 551	63	3 969	7.937 254	113	12 769	10.63015
14	196	3.741 657	64	4 096	8.000 000	114	12 906	10.67708
15	225	3.872 983	65	4 225	8.062 258	115	13 225	10.72381
16	256	4.000 000	66	4 356	8.124 038	116	13 456	10.77033
17	289	4.123 106	67	4 489	8.185 353	117	13 689	10.81665
18	324	4.242 641	68	4 624	8.246 211	118	13 924	10.86278
19	361	4.358 899	69	4 761	8.306 624	119	14 161	10.90871
20	400	4.472 136	70	4 900	8.366 600	120	14 400	10.95445
21	441	4.582 576	71	5 041	8.426 150	121	14 641	11.00000
22	484	4.690 416	72	5 184	8.485 281	122	14 884	11.04536
23	529	4.795 832	73	5 329	8.544 004	123	15 129	11.09054
24	576	4.898 979	74	5 476	8.602 325	124	15 376	11.13553
25	625	5.000 000	75	5 625	8.660 254	125	15 625	11.18034
26	676	5.099 020	76	5 776	8.717 798	126	15 876	11.22497
27	729	5.196 152	77	5 929	8.774 964	127	16 129	11.26943
28	784	5.291 503	78	6 084	8.831 761	128	16 384	11.31371
29	841	5.385 165	79	6 241	8.888 194	129	16 641	11.35782
30	900	5.477 226	80	6 400	8.944 272	130	16 900	11.40175
31	961	5.567 764	81	6 561	9.000 000	131	17 161	11.44552
32	1 024	5.656 854	82	6 724	9.055 385	132	17 424	11.48913
33	1 089	5.744 563	83	6 889	9.110 434	133	17 689	11.53256
34	1 156	5.830 952	84	7 056	9.165 151	134	17 956	11.57584
35	1 225	5.916 080	85	7 225	9.219 544	135	18 225	11.61895
36	1 296	6.000 000	86	7 396	9.273 618	136	18 496	11.66190
37	1 369	6.082 763	87	7 569	9.327 379	137	18 769	11.70470
38	1 444	6.164 414	88	7 744	9.380 832	138	19 044	11.74734
39	1 521	6.244 998	89	7 921	9.433 981	139	19 321	11.78983
40	1 600	6.324 555	90	8 100	9.486 833	140	19 600	11.83216
41	1 681	6.403 124	91	8 281	9.539 392	141	19 881	11.87434
42	1 764	6.480 741	92	8 464	9.591 663	142	20 164	11.91638
43	1 849	6.557 439	93	8 649	9.643 651	143	20 449	11.95826
44	1 936	6.633 250	94	8 836	9.695 360	144	20 736	12.00000
45	2 025	6.708 204	95	9 025	9.746 794	145	21 025	12.04159
46	2 116	6.782 330	96	9 216	9.797 959	146	21 316	12.08305
47	2 209	6.855 655	97	9 409	9.848 858	147	21 609	12.12436
48	2 304	6.928 203	98	9 604	9.899 405	148	21 904	12.16553
49	2 401	7.000 000	99	9 801	9.949 874	149	22 201	12.20656

Table A.8. Squares and Square Roots (Continued)
1–1000

N	N^2	$\sqrt{N}$	N	N^2	$\sqrt{N}$	N	N^2	$\sqrt{N}$
150	22 500	12.24745	200	40 000	14.14214	250	62 500	15.81139
151	22 801	12.28821	201	40 401	14.17745	251	63 001	15.84298
152	23 104	12.32883	202	40 804	14.21267	252	63 504	15.87451
153	23 409	12.36932	203	41 209	14.24781	253	64 009	15.90597
154	23 716	12.40967	204	41 616	14.28286	254	64 516	15.93738
155	24 025	12.44990	205	42 025	14.31782	255	65 025	15.96872
156	24 336	12.49000	206	42 436	14.35270	256	65 536	16.00000
157	24 649	12.52996	207	42 849	14.38749	257	66 049	16.03122
158	24 964	12.56981	208	43 264	14.42221	258	66 564	16.06238
159	25 281	12.60952	209	43 681	14.45683	259	67 081	16.09348
160	25 600	12.64911	210	44 100	14.49138	260	67 600	16.12452
161	25 921	12.68858	211	44 521	14.52584	261	68 121	16.15549
162	26 244	12.72792	212	44 944	14.56022	262	68 644	16.18641
163	26 569	12.76715	213	45 369	14.59452	263	69 169	16.21727
164	26 896	12.80625	214	45 796	14.62874	264	69 696	16.24808
165	27 225	12.84523	215	46 225	14.66288	265	70 225	16.27882
166	27 556	12.88410	216	46 656	14.69694	266	70 756	16.30951
167	27 889	12.92285	217	47 089	14.73092	267	71 289	16.34013
168	28 224	12.96148	218	47 524	14.76482	268	71 824	16.37071
169	28 561	13.00000	219	47 961	14.79865	269	72 361	16.40122
170	28 900	13.03840	220	48 400	14.83240	270	72 900	16.43168
171	29 241	13.07670	221	48 841	14.86607	271	73 441	16.46208
172	29 584	13.11488	222	49 284	14.89966	272	73 984	16.49242
173	29 929	13.15295	223	49 729	14.93318	273	74 529	16.52271
174	30 276	13.19091	224	50 176	14.96663	274	75 076	16.55295
175	30 625	13.22876	225	50 625	15.00000	275	75 625	16.58312
176	30 976	13.26650	226	51 076	15.03330	276	76 176	16.61325
177	31 329	13.30413	227	51 529	15.06652	277	76 729	16.64332
178	31 684	13.34166	228	51 984	15.09967	278	77 284	16.67333
179	32 041	13.37909	229	52 441	15.13275	279	77 841	16.70329
180	32 400	13.41641	230	52 900	15.16576	280	78 400	16.73320
181	32 761	13.45362	231	53 361	15.19868	281	78 961	16.76305
182	33 124	13.49074	232	53 824	15.23155	282	79 524	16.79286
183	33 489	13.52775	233	54 289	15.26434	283	80 089	16.82260
184	33 856	13.56466	234	54 756	15.29706	284	80 656	16.85230
185	34 225	13.60147	235	55 225	15.32971	285	81 225	16.88194
186	34 596	13.63818	236	55 696	15.36229	286	81 796	16.91153
187	34 969	13.67479	237	56 169	15.39480	287	82 369	16.94107
188	35 344	13.71131	238	56 644	15.42725	288	82 944	16.97056
189	35 721	13.74773	239	57 121	15.45962	289	83 521	17.00000
190	36 100	13.78405	240	57 600	15.49193	290	84 100	17.02939
191	36 481	13.82027	241	58 081	15.52417	291	84 681	17.05872
192	36 864	13.85641	242	58 564	15.55635	292	85 264	17.08801
193	37 249	13.89244	243	59 049	15.58846	293	85 849	17.11724
194	37 636	13.92839	244	59 536	15.62050	294	86 436	17.14643
195	38 025	13.96424	245	60 025	15.65248	295	87 025	17.17556
196	38 416	14.00000	246	60 516	15.68439	296	87 616	17.20465
197	38 809	14.03567	247	61 009	15.71623	297	88 209	17.23369
198	39 204	14.07125	248	61 504	15.74802	298	88 804	17.26268
199	39 601	14.10674	249	62 001	15.77973	299	89 401	17.29162

Table A.8.   Squares and Square Roots (Continued)
1–1000

N	N^2	$\sqrt{N}$	N	N^2	$\sqrt{N}$	N	N^2	$\sqrt{N}$
300	90 100	17.32051	350	122 500	18.70829	400	160 000	20.00000
301	90 601	17.34935	351	123 201	18.73499	401	160 801	20.02498
302	91 204	17.37815	352	123 904	18.76166	402	161 604	20.04994
303	91 809	17.40690	353	124 609	18.78829	403	162 409	20.07486
304	92 416	17.43560	354	125 316	18.81489	404	163 216	20.09975
305	93 025	17.46425	355	126 025	18.84144	405	164 025	20.12461
306	93 636	17.49286	356	126 736	18.86796	406	164 836	20.14944
307	94 249	17.52142	357	127 449	18.89444	407	165 649	20.17424
308	94 864	17.54993	358	128 164	18.92089	408	166 464	20.19901
309	95 481	17.57840	359	128 881	18.94730	409	167 231	20.22375
310	96 100	17.60682	360	129 600	18.97367	410	168 100	20.24846
311	96 721	17.63519	361	130 321	19.00000	411	168 921	20.27313
312	97 344	17.66352	362	131 044	19.02630	412	169 744	20.29778
313	97 969	17.69181	363	131 769	19.05256	413	170 569	20.32249
314	98 596	17.72005	364	132 496	19.07878	414	171 396	20.34699
315	99 225	17.74824	365	133 225	19.10497	415	172 225	20.37155
316	99 856	17.77639	366	133 956	19.13113	416	173 056	20.39608
317	100 489	17.80449˙	367	134 689	19.15724	417	173 889	20.42058
318	101 124	17.83255	368	135 424	19.18333	418	174 724	20.44505
319	101 761	17.86057	369	136 161	19.20937	419	175 561	20.46949
320	102 400	17.88854	370	136 900	19.23538	420	176 400	20.49390
321	103 041	17.91647	371	137 641	19.26136	421	177 241	20.51828
322	103 684	17.94436	372	138 384	19.28730	422	178 084	20.54264
323	104 329	17.97220	373	139 129	19.31321	423	178 929	20.56696
324	104 976	18.00000	374	139 876	19.33908	424	179 776	20.59126
325	105 625	18.02776	375	140 625	19.36492	425	180 625	20.61553
326	106 276	18.05547	376	141 376	19.39072	426	181 476	20.63977
327	106 929	18.08314	377	142 129	19.41649	427	182 329	20.66398
328	107 584	18.11077	378	142 884	19.44222	428	183 184	20.68816
329	108 241	18.13836	379	143 641	19.46792	429	184 041	20.71232
330	108 900	18.16590	380	144 400	19.49359	430	184 900	20.73644
331	109 561	18.19341	381	145 161	19.51922	431	185 761	20.76054
332	110 224	18.22087	382	145 924	19.54483	432	186 624	20.78461
333	110 889	18.24829	383	146 689	19.57039	433	187 489	20.80865
334	111 556	18.27567	384	147 456	19.59592	434	188 356	20.83267
335	112 225	18.30301	385	148 225	19.62142	435	189 225	20.85665
336	112 896	18.33030	386	148 996	19.64688	436	190 096	20.88061
337	113 569	18.35756	387	149 769	19.67232	437	190 969	20.90454
338	114 244	18.38478	388	150 544	19.69772	438	191 844	20.92845
339	114 921	18.41195	389	151 321	19.72308	439	192 721	20.95233
340	115 600	18.43909	390	152 100	19.74842	440	193 600	20.97618
341	116 281	18.46619	391	152 881	19.77372	441	194 481	21.00000
342	116 964	18.49324	392	153 664	19.79899	442	195 364	21.02380
343	117 649	18.52026	393	154 449	19.82423	443	196 249	21.04757
344	118 336	18.54724	394	155 236	19.84943	444	197 136	21.07131
345	119 025	18.57418	395	156 025	19.87461	445	198 025	21.09502
346	119 716	18.60108	396	156 816	19.89975	446	198 916	21.11871
347	120 409	18.62794	397	157 609	19.92486	447	199 809	21.14237
348	121 104	18.65476	398	158 404	19.94994	448	200 704	21.16601
349	121 801	18.68154	399	159 201	19.97498	449	201 601	21.18962

Table A.8. Squares and Square Roots (Continued)
1–1000

N	N^2	$\sqrt{N}$	N	N^2	$\sqrt{N}$	N	N^2	$\sqrt{N}$
450	202 500	21.21320	500	250 000	22.36068	550	302 500	23.45208
451	203 401	21.23676	501	251 001	22.38303	551	303 601	23.47339
452	204 304	21.26029	502	252 004	22.40536	552	304 704	23.49468
453	205 209	21.28380	503	253 009	22.42766	553	305 809	23.51595
454	206 116	21.30728	504	254 016	22.44994	554	306 916	23.53720
455	207 025	21.33073	505	255 025	22.47221	555	308 025	23.55844
456	207 936	21.35416	506	256 036	22.49444	556	309 136	23.57965
457	208 849	21.37756	507	257 049	22.51666	557	310 249	23.60085
458	209 764	21.40093	508	258 064	22.53886	558	311 364	23.62202
459	210 681	21.42429	509	259 081	22.56103	559	312 481	23.64318
460	211 600	21.44761	510	260 100	22.58318	560	313 600	23.66432
461	212 521	21.47091	511	261 121	22.60531	561	314 721	23.68544
462	213 444	21.49419	512	262 144	22.62742	562	315 844	23.70654
463	214 369	21.51743	513	263 169	22.64950	563	316 969	23.72762
464	215 296	21.54066	514	264 196	22.67157	564	318 096	23.74868
465	216 225	21.56386	515	265 225	22.69361	565	319 225	23.76973
466	217 156	21.58703	516	266 256	22.71563	566	320 356	23.79075
467	218 089	21.61018	517	267 289	22.73763	567	321 489	23.81176
468	219 024	21.63331	518	268 324	22.75961	568	322 624	23.83275
469	219 961	21.65641	519	269 361	22.78157	569	323 761	23.85372
470	220 900	21.67948	520	270 400	22.80351	570	324 900	23.87467
471	221 841	21.70253	521	271 441	22.82542	571	326 041	23.89561
472	222 781	21.72556	522	272 484	22.84732	572	327 184	23.91652
473	223 729	21.74856	523	273 529	22.86919	573	328 329	23.93742
474	224 676	21.77154	524	274 576	22.89105	574	329 476	23.95830
475	225 625	21.79449	525	275 625	22.91288	575	330 625	23.97916
476	226 576	21.81742	526	276 676	22.93469	576	331 776	24.00000
477	227 529	21.84033	527	277 729	22.95648	577	332 929	24.02082
478	228 484	21.86321	528	278 784	22.97825	578	334 084	24.04163
479	229 441	21.88607	529	279 841	23.00000	579	335 241	24.06242
480	230 400	21.90890	530	280 900	23.02173	580	336 400	24.08319
481	231 361	21.93171	531	281 961	23.04344	581	337 561	24.10394
482	232 324	21.95450	532	283 024	23.06513	582	338 724	24.12468
483	233 289	21.97726	533	284 089	23.08679	583	339 889	24.14539
484	234 256	22.00000	534	285 156	23.10844	584	341 056	24.16609
485	235 225	22.02272	535	286 225	23.13007	585	342 225	24.18677
486	236 196	22.04541	536	287 296	23.15167	586	343 396	24.20744
487	237 169	22.06808	537	288 369	23.17326	587	344 569	24.22808
488	238 144	22.09072	538	289 444	23.19483	588	345 744	24.24871
489	239 121	22.11334	539	290 521	23.21637	589	346 921	24.26932
490	240 100	22.13594	540	291 600	23.23790	590	348 100	24.28992
491	241 081	22.15852	541	292 681	23.25941	591	349 281	24.31049
492	242 064	22.18107	542	293 764	23.28089	592	350 464	24.33105
493	243 049	22.20360	543	294 849	23.30236	593	351 649	24.35159
494	244 036	22.22611	544	295 936	23.32381	594	352 836	24.37212
495	245 025	22.24860	545	297 025	23.34524	595	354 025	24.39262
496	246 016	22.27106	546	298 116	23.36664	596	355 216	24.41311
497	247 009	22.29350	547	299 209	23.38803	597	356 409	24.43358
498	248 004	22.31591	548	300 304	23.40940	598	357 604	24.45404
499	249 001	22.33831	549	301 401	23.43075	599	358 801	24.47448

Table A.8. Squares and Square Roots (Continued)
1–1000

N	N^2	$\sqrt{N}$	N	N^2	$\sqrt{N}$	N	N^2	$\sqrt{N}$
600	360 000	24.49490	650	422 500	25.49510	700	490 000	26.45751
601	361 201	24.51530	651	423 801	25.51470	701	491 401	26.47640
602	362 404	24.53569	652	425 104	25.53429	702	492 804	26.49528
603	363 609	24.55606	653	426 409	25.56386	703	494 209	26.51415
604	364 816	24.57641	654	427 716	25.57342	704	495 616	26.53200
605	366 025	24.59675	655	429 025	25.59297	705	497 025	26.55184
606	367 236	24.61707	656	430 336	25.61250	706	498 436	26.57066
607	368 449	24.63737	657	431 649	25.63201	707	499 849	26.58947
608	369 664	24.65766	658	432 964	25.65151	708	501 264	26.60827
609	370 881	24.67793	659	434 281	25.67100	709	502 681	26.62705
610	372 100	24.69818	660	435 600	25.69047	710	504 100	26.64583
611	373 321	24.71841	661	436 921	25.70992	711	505 521	26.66458
612	374 544	24.73863	662	438 244	25.72936	712	506 944	26.68383
613	375 769	24.75884	663	439 569	25.74829	713	508 369	26.70206
614	376 996	24.77902	664	440 896	25.76820	714	509 796	26.72078
615	378 225	24.79919	665	442 225	25.78759	715	511 225	26.73948
616	379 456	24.81935	666	443 556	25.80698	716	512 656	26.75818
617	380 689	24.83948	667	444 889	25.82634	717	514 089	26.77686
618	381 924	24.85961	668	446 224	25.84570	718	515 524	26.79552
619	383 161	24.87971	669	447 561	25.86503	719	516 961	26.81418
620	384 400	24.89980	670	448 900	25.88436	720	518 400	26.83282
621	385 641	24.91987	671	450 241	25.90367	721	519 841	26.85144
622	386 884	24.93993	672	451 584	25.92296	722	521 284	26.87006
623	388 129	24.95997	673	452 929	25.94224	723	522 729	26.88966
624	389 376	24.97999	674	454 276	25.96151	724	524 176	26.90725
625	390 625	25.00000	675	455 625	25.98076	725	525 625	26.92582
626	391 876	25.01999	676	456 976	26.00000	726	527 076	26.94439
627	393 129	25.03997	677	458 329	26.01922	727	528 529	26.96294
628	394 384	25.05993	678	459 684	26.03843	728	529 984	26.98148
629	395 641	25.07987	679	461 041	26.05763	729	531 441	27.00000
630	396 900	25.09980	680	462 400	26.07681	730	532 900	27.01851
631	398 161	25.11971	681	463 761	26.09598	731	534 361	27.03701
632	399 424	25.13961	682	465 124	26.11513	732	535 824	27.03550
633	400 689	25.15949	683	466 489	26.13427	733	537 289	27.02397
634	401 956	25.17936	684	467 856	26.15339	734	538 756	27.09243
635	403 225	25.19921	685	469 225	26.17250	735	540 225	27.11088
636	404 496	25.21904	686	470 596	26.19160	736	541 696	27.12932
637	405 769	25.23886	687	471 969	26.21068	737	543 169	27.14774
638	407 044	25.25866	688	473 344	26.22975	738	544 644	27.16616
639	408 321	25.27845	689	474 721	26.24881	739	546 121	27.18455
640	409 600	25.29822	690	476 100	26.26785	740	547 600	27.20294
641	410 881	25.31798	691	477 481	26.28688	741	549 081	27.22132
642	412 164	25.33772	692	478 864	26.30589	742	550 564	27.23968
643	413 449	25.35744	693	480 249	26.32489	743	552 049	27.25803
644	414 736	25.37716	694	481 636	26.34388	744	553 536	27.27636
645	416 025	25.39685	695	483 025	26.36285	745	555 025	27.29469
646	417 319	25.41653	696	484 416	26.38181	746	556 516	27.31300
647	418 609	25.43619	697	485 809	26.40076	747	558 009	27.33130
648	419 904	25.45584	698	487 204	26.41969	748	559 504	27.34959
649	421 201	25.47510	699	488 601	26.43861	749	561 001	27.36786

Table A.8. Squares and Square Roots (Continued)
1–1000

N	N²	$\sqrt{N}$	N	N²	$\sqrt{N}$	N	N²	$\sqrt{N}$
750	562 500	27.38613	800	640 000	28.28427	850	722 500	29.15476
751	564 001	27.40438	801	641 601	28.30194	851	724 201	29.17190
752	565 504	27.42262	802	643 204	28.31960	852	725 904	29.18904
753	567 009	27.44085	803	644 809	28.33725	853	727 609	29.20616
754	568 516	27.45906	804	646 416	28.35489	854	729 316	29.22328
755	570 025	27.47726	805	648 025	28.37252	855	731 025	29.24038
756	571 536	27.49545	806	649 636	28.39014	856	732 736	29.25748
757	573 049	27.51363	807	651 249	28.40775	857	734 449	29.27456
758	574 564	27.53180	808	652 864	28.42534	858	736 164	29.29164
759	576 081	27.54995	809	654 481	28.44293	859	737 881	29.30870
760	577 600	27.56810	810	656 100	28.46050	860	739 600	29.32576
761	579 121	27.58623	811	657 721	28.47806	861	741 321	29.34280
762	580 644	27.60435	812	659 344	28.49561	862	743 044	29.35984
763	582 169	27.62245	813	660 969	28.51315	863	744 769	29.37686
764	583 696	27.64055	814	662 596	28.53069	864	746 496	29.39388
765	585 225	27.65863	815	664 225	28.54820	865	748 225	29.41088
766	586 756	27.67671	816	665 856	28.56571	866	749 956	29.42788
767	588 289	27.69476	817	667 489	28.58321	867	751 689	29.44486
768	589 824	27.71281	818	669 124	28.60070	868	753 424	29.46184
769	591 361	27.73085	819	670 761	28.61818	869	755 161	29.47881
770	592 900	27.74887	820	672 400	28.63564	870	756 900	29.49576
771	594 441	27.76689	821	674 041	28.65310	871	758 641	29.51271
772	595 984	27.78489	822	675 684	28.67054	872	760 384	29.52965
773	597 529	27.80288	823	677 329	28.68798	873	762 129	29.54657
744	599 076	27.82086	824	678 976	28.70540	874	763 876	29.56349
775	600 625	27.83882	825	680 625	28.72281	875	765 625	29.58040
776	602 176	27.85678	826	682 276	28.74022	876	767 376	29.59730
777	603 729	27.87472	827	683 929	28.75761	877	769 129	29.61419
778	605 284	27.89265	828	685 584	28.77499	878	770 884	29.63106
779	606 341	27.91057	829	687 241	28.79236	879	772 641	29.64793
780	608 400	27.92848	830	688 900	28.80972	880	774 400	29.66479
781	609 961	27.94638	831	690 561	28.82707	881	776 161	29.68164
782	611 524	27.96426	832	692 224	28.84441	882	777 924	29.69848
783	613 089	27.98214	833	693 889	28.86174	883	779 689	29.71532
784	614 656	28.00000	834	695 556	28.87906	884	781 456	29.73214
785	616 225	28.01785	835	697 225	28.89637	885	783 225	29.74895
786	617 796	28.03569	836	698 896	28.91366	886	784 996	29.76575
787	619 369	28.05352	837	700 569	28.93095	887	786 769	29.78255
788	620 944	28.07134	838	702 244	28.94823	888	788 544	29.79933
789	622 521	28.08914	839	703 921	28.96550	889	790 321	29.81610
790	624 100	28.10694	840	705 600	28.98275	890	792 100	29.83287
791	625 681	28.12472	841	707 281	29.00000	891	793 881	29.84962
792	627 264	28.14249	842	708 964	29.01724	892	795 664	29.86637
793	628 849	28.16026	843	710 649	29.03446	893	797 449	29.88341
794	630 436	28.17801	844	712 336	29.05168	894	799 236	29.89983
795	632 025	28.19574	845	714 025	29.06888	895	801 025	29.91655
796	633 616	28.21347	846	715 716	29.08608	896	802 816	29.93326
797	635 209	28.23119	847	717 409	29.10326	897	804 609	29.94996
798	636 804	28.24889	848	719 104	29.12044	898	806 404	29.96665
799	638 401	28.26659	849	720 801	29.13760	899	808 201	29.98333

Table A.8.   Squares and Square Roots (Continued)
1–1000

N	N^2	$\sqrt{N}$	N	N^2	$\sqrt{N}$
900	810 000	30.00000	950	902 500	30.82207
901	811 801	30.01666	951	904 401	30.83829
902	813 604	30.03331	952	906 304	30.85450
903	815 409	30.04996	953	908 209	30.87070
904	817 216	30.06659	954	910 116	30.88689
905	819 025	30.08322	955	912 025	30.90307
906	820 836	30.09983	956	913 936	30.91925
907	822 649	30.11644	957	915 849	30.93542
908	824 464	30.13304	958	917 764	30.95158
909	826 281	30.14963	959	919 681	30.96773
910	828 100	30.16621	960	921 600	30.98387
911	829 921	30.18278	961	923 521	31.00000
912	831 744	30.19934	962	925 444	31.01612
913	833 569	30.21589	963	927 369	31.03224
914	835 396	30.23243	964	929 296	31.04835
915	837 225	30.24897	965	931 225	31.06445
916	839 056	30.26549	966	933 156	31.08054
917	840 889	30.28201	967	935 089	31.09662
918	842 724	30.29851	968	937 024	31.11270
919	844 561	30.31501	969	938 961	31.12876
920	846 400	30.33150	970	940 900	31.14482
921	848 241	30.34798	971	942 841	31.16087
922	850 084	30.36445	972	944 784	31.17691
923	851 929	30.38092	973	946 729	31.19295
924	853 776	30.39737	974	948 676	31.20897
925	855 625	30.41381	975	950 625	31.22499
926	857 476	30.43025	976	952 576	31.24100
927	859 329	30.44667	977	954 529	31.25700
928	861 184	30.46309	978	956 484	31.27299
929	863 041	30.47950	979	958 441	31.28898
930	864 900	30.49590	980	960 400	31.30495
931	866 761	30.51229	981	962 361	31.32092
932	868 624	30.52868	982	964 324	31.33688
933	870 489	30.54505	983	966 289	31.35283
934	872 356	30.56141	984	968 256	31.36877
935	874 225	30.57777	985	970 225	31.38471
936	876 096	30.59412	986	972 196	31.40064
937	877 969	30.61046	987	974 169	31.41656
938	879 844	30.62679	988	976 144	31.43247
939	881 721	30.64311	989	978 121	31.44837
940	883 600	30.65942	990	980 100	31.46427
941	885 481	30.67572	991	982 081	31.48015
942	887 364	30.69202	992	984 064	31.49603
943	889 249	30.70831	993	986 049	31.51190
944	891 136	30.72458	994	988 036	31.52777
945	893 025	30.74085	995	990 025	31.54362
946	894 916	30.75711	996	992 016	31.55947
947	896 809	30.77337	997	994 009	31.57531
948	898 704	30.78961	998	996 004	31.59114
949	900 601	30.80584	999	998 001	31.60696
			1000	1 000 000	31.62278

Table A.9. Binomial Distribution

$$b(x \mid n, p) = \binom{n}{x} p^x(1 - p)^{n-x}$$

N	X	.01	.02	.03	.04	.05	.10	.15	.20	.25	.30	.40	.50
							P						
2	0	.9801	.9404	.9409	.9216	.9025	.8100	.7225	.6400	.5625	.4900	.3600	.2500
	1	.0198	.0392	.0582	.0768	.0950	.1800	.2550	.3200	.3750	.4200	.4800	.5000
	2	.0001	.0004	.0009	.0016	.0025	.0100	.0255	.0400	.0625	.0900	.1600	.2500
3	0	.9703	.9412	.9127	.8847	.8574	.7290	.6141	.5120	.4219	.3430	.2160	.1250
	1	.0294	.0576	.0847	.1106	.1354	.2430	.3251	.3840	.4219	.4410	.4320	.3750
	2	.0003	.0012	.0026	.0046	.0071	.0270	.0574	.0960	.1406	.1890	.2880	.3750
	3	.0000	.0000	.0000	.0001	.0001	.0010	.0034	.0080	.0156	.0270	.0640	.1250
4	0	.9606	.9224	.8853	.8493	.8145	.6561	.5220	.4096	.3164	.2401	.1296	.0625
	1	.0388	.0753	.1095	.1416	.1715	.2916	.3685	.4096	.4019	.4116	.3456	.2500
	2	.0006	.0023	.0051	.0088	.0135	.0486	.0975	.1536	.2109	.2646	.3456	.3750
	3	.0000	.0000	.0001	.0002	.0005	.0036	.0115	.0256	.0469	.0756	.1536	.2500
	4	.0000	.0000	.0000	.0000	.0000	.0001	.0005	.0016	.0039	.0081	.0256	.0625
5	0	.9510	.9039	.8587	.8154	.7738	.5905	.4437	.3277	.2373	.1681	.0778	.0313
	1	.0480	.0922	.1328	.1699	.2036	.3280	.3915	.4096	.3955	.3602	.2592	.1563
	2	.0010	.0038	.0082	.0142	.0214	.0729	.1382	.2048	.2637	.3087	.3456	.3125
	3	.0000	.0001	.0003	.0006	.0011	.0081	.0244	.0512	.0879	.1323	.2304	.3125
	4	.0000	.0000	.0000	.0000	.0000	.0004	.0022	.0064	.0146	.0283	.0768	.1563
	5	.0000	.0000	.0000	.0000	.0000	.0000	.0001	.0003	.0010	.0024	.0102	.0313
6	0	.9415	.8858	.8330	.7828	.7351	.5314	.3771	.2621	.1780	.1176	.0467	.0156
	1	.0571	.1085	.1546	.1957	.2321	.3543	.3993	.3932	.3560	.3025	.1866	.0938
	2	.0014	.0055	.0120	.0204	.0305	.0984	.1762	.2458	.2966	.3241	.3110	.2344
	3	.0000	.0002	.0005	.0011	.0021	.0146	.0415	.0819	.1318	.1852	.2765	.3125
	4	.0000	.0000	.0000	.0000	.0001	.0012	.0055	.0154	.0330	.0595	.1382	.2344
	5	.0000	.0000	.0000	.0000	.0000	.0001	.0004	.0015	.0044	.0102	.0369	.0938
	6	.0000	.0000	.0000	.0000	.0000	.0000	.0000	.0001	.0002	.0007	.0041	.0156
7	0	.9321	.8681	.8080	.7514	.6983	.4783	.3206	.2097	.1335	.0824	.0280	.0078
	1	.0659	.1240	.1749	.2192	.2573	.3720	.3960	.3670	.3115	.2471	.1306	.0547
	2	.0020	.0076	.0162	.0274	.0406	.1240	.2097	.2753	.3115	.3177	.2613	.1641
	3	.0000	.0003	.0008	.0019	.0036	.0230	.0617	.1147	.1730	.2269	.2903	.2734
	4	.0000	.0000	.0000	.0001	.0002	.0026	.0109	.0287	.0577	.0972	.1935	.2734
	5	.0000	.0000	.0000	.0000	.0000	.0002	.0012	.0043	.0115	.0250	.0774	.1641
	6	.0000	.0000	.0000	.0000	.0000	.0000	.0001	.0004	.0013	.0036	.0172	.0547
	7	.0000	.0000	.0000	.0000	.0000	.0000	.0000	.0000	.0001	.0002	.0016	.0078
8	0	.9227	.8508	.7837	.7214	.6634	.4305	.2725	.1678	.1001	.0576	.0168	.0039
	1	.0746	.1389	.1939	.2405	.2793	.3826	.3847	.3355	.2670	.1977	.0896	.0313
	2	.0026	.0099	.0210	.0351	.0515	.1488	.2376	.2936	.3115	.2965	.2090	.1096
	3	.0001	.0004	.0013	.0029	.0054	.0331	.0839	.1468	.2076	.2541	.2787	.2188

Table A.9. Binomial Distribution (Continued)

N	X	.01	.02	.03	.04	.05	.10	.15	.20	.25	.30	.40	.50
8	4	.0000	.0000	.0001	.0002	.0004	.0046	.0185	.0459	.0865	.1361	.2322	.2734
	5	.0000	.0000	.0000	.0000	.0000	.0004	.0026	.0092	.0231	.0467	.1239	.2188
	6	.0000	.0000	.0000	.0000	.0000	.0000	.0002	.0011	.0038	.0100	.0413	.1094
	7	.0000	.0000	.0000	.0000	.0000	.0000	.0000	.0001	.0004	.0012	.0079	.0313
	8	.0000	.0000	.0000	.0000	.0000	.0000	.0000	.0000	.0000	.0001	.0007	.0039
9	0	.9135	.8337	.7602	.6925	.6302	.3874	.2316	.1342	.0751	.0404	.0101	.0020
	1	.0830	.1531	.2116	.2597	.2985	.3874	.3679	.3020	.2253	.1556	.0605	.0176
	2	.0034	.0125	.0262	.0433	.0629	.1722	.2597	.3020	.3003	.2668	.1612	.0703
	3	.0001	.0006	.0019	.0042	.0077	.0446	.1069	.1762	.2336	.2668	.2508	.1641
	4	.0000	.0000	.0001	.0003	.0006	.0074	.0283	.0661	.1168	.1715	.2508	.2461
	5	.0000	.0000	.0000	.0000	.0000	.0008	.0050	.0165	.0389	.0735	.1672	.2461
	6	.0000	.0000	.0000	.0000	.0000	.0001	.0006	.0028	.0087	.0210	.0743	.1641
	7	.0000	.0000	.0000	.0000	.0000	.0000	.0000	.0003	.0012	.0039	.0212	.0703
	8	.0000	.0000	.0000	.0000	.0000	.0000	.0000	.0000	.0001	.0004	.0035	.0176
	9	.0000	.0000	.0000	.0000	.0000	.0000	.0000	.0000	.0000	.0000	.0003	.0020
10	0	.9044	.8171	.7374	.6648	.5987	.3487	.1969	.1074	.0563	.0282	.0060	.0010
	1	.0914	.1667	.2281	.2770	.3151	.3874	.3474	.2684	.1877	.1211	.0403	.0098
	2	.0042	.0153	.0317	.0519	.0746	.1937	.2759	.3020	.2816	.2335	.1209	.0439
	3	.0001	.0008	.0026	.0058	.0105	.0574	.1298	.2013	.2503	.2668	.2150	.1172
	4	.0000	.0000	.0001	.0004	.0010	.0112	.0401	.0881	.1460	.2001	.2508	.2051
	5	.0000	.0000	.0000	.0000	.0001	.0015	.0085	.0264	.0584	.1029	.2007	.2461
	5	.0000	.0000	.0000	.0000	.0000	.0001	.0012	.0055	.0162	.0368	.1115	.2051
	7	.0000	.0000	.0000	.0000	.0000	.0000	.0001	.0008	.0031	.0090	.0425	.1172
	8	.0000	.0000	.0000	.0000	.0000	.0000	.0000	.0001	.0004	.0014	.0106	.0439
	9	.0000	.0000	.0000	.0000	.0000	.0000	.0000	.0000	.0000	.0001	.0016	.0098
	10	.0000	.0000	.0000	.0000	.0000	.0000	.0000	.0000	.0000	.0000	.0001	.0010
11	0	.8953	.8007	.7153	.6382	.5688	.3138	.1673	.0859	.0422	.0198	.0036	.0005
	1	.0995	.1798	.2433	.2925	.3293	.3835	.3248	.2362	.1549	.0932	.0266	.0054
	2	.0050	.0183	.0376	.0609	.0867	.2131	.2866	.2953	.2581	.1998	.0887	.0269
	3	.0002	.0011	.0035	.0076	.0137	.0710	.1517	.2215	.2581	.2568	.1774	.0806
	4	.0000	.0000	.0002	.0006	.0014	.0158	.0536	.1107	.1721	.2201	.2365	.1611
	5	.0000	.0000	.0000	.0000	.0001	.0025	.0132	.0388	.0803	.1321	.2207	.2256
	6	.0000	.0000	.0000	.0000	.0000	.0003	.0023	.0097	.0268	.0566	.1471	.2256
	7	.0000	.0000	.0000	.0000	.0000	.0000	.0003	.0017	.0064	.0173	.0701	.1611
	8	.0000	.0000	.0000	.0000	.0000	.0000	.0000	.0002	.0011	.0037	.0234	.0806
	9	.0000	.0000	.0000	.0000	.0000	.0000	.0000	.0000	.0001	.0005	.0052	.0269
	10	.0000	.0000	.0000	.0000	.0000	.0000	.0000	.0000	.0000	.0000	.0007	.0054
	11	.0000	.0000	.0000	.0000	.0000	.0000	.0000	.0000	.0000	.0000	.0000	.0005
12	0	.8864	.7847	.6938	.6127	.5404	.2824	.1422	.0687	.0317	.0138	.0022	.0002
	1	.1074	.1922	.2575	.3064	.3413	.3766	.3012	.2062	.1267	.0712	.0174	.0029
	2	.0060	.0216	.0438	.0702	.0988	.2301	.2924	.2835	.2323	.1678	.0639	.0161
	3	.0002	.0015	.0045	.0098	.0173	.0852	.1720	.2362	.2581	.2397	.1419	.0537

Table A.9.   Binomial Distribution (Continued)

N	X	.01	.02	.03	.04	.05	.10	.15	.20	.25	.30	.40	.50
12	4	.0000	.0001	.0003	.0009	.0021	.0213	.0683	.1329	1936	2311	.2128	.1208
	5	.0000	.0000	.0000	.0001	.0002	.0038	.0193	.0532	.1032	.1585	.2270	.1934
	6	.0000	.0000	.0000	.0000	.0000	.0005	.0040	.0155	.0401	.0792	.1766	.2256
	7	.0000	.0000	.0000	.0000	.0000	.0000	.0006	.0033	.0115	.0291	.1009	.1934
	8	.0000	.0000	.0000	.0000	.0000	.0000	.0001	.0005	.0024	.0078	.0420	.1208
	9	.0000	.0000	.0000	.0000	.0000	.0000	.0000	.0001	.0004	.0015	.0125	.0537
	10	.0000	.0000	.0000	.0000	.0000	.0000	.0000	.0000	.0000	.0002	.0025	.0161
	11	.0000	.0000	.0000	.0000	.0000	.0000	.0000	.0000	.0000	.0000	.0003	.0029
	12	.0000	.0000	.0000	.0000	.0000	.0000	.0000	.0000	.0000	.0000	.0000	.0002
13	0	.8775	.7690	.6730	.5882	.5133	.2542	.1209	.0550	.0238	.0097	.0013	.0001
	1	.1152	.2040	.2706	.3186	.3512	.3672	.2774	.1787	.1029	.0540	.0113	.0016
	2	.0070	.0250	.0502	.0797	.1109	.2448	.2937	.2680	.2059	.1388	.0453	.0095
	3	.0003	.0019	.0057	.0122	.0214	.0997	.1900	.2457	.2517	.2181	.1107	.0349
	4	.0000	.0001	.0004	.0013	.0028	.0277	.0838	.1535	.2097	.2337	.1845	.0873
	5	.0000	.0000	.0000	.0001	.0003	.0055	.0266	.0691	.1258	.1803	.2214	.1571
	6	.0000	.0000	.0000	.0000	.0000	.0008	.0063	.0230	.0559	.1030	.1968	.2095
	7	.0000	.0000	.0000	.0000	.0000	.0001	.0011	.0058	.0186	.0442	.1312	.2095
	8	.0000	.0000	.0000	.0000	.0000	.0000	.0001	.0011	.0047	.0142	.0656	.1571
	9	.0000	.0000	.0000	.0000	.0000	.0000	.0000	.0001	.0009	.0034	.0243	.0873
	10	.0000	.0000	.0000	.0000	.0000	.0000	.0000	.0000	.0001	.0006	.0065	.0349
	11	.0000	.0000	.0000	.0000	.0000	.0000	.0000	.0000	.0000	.0001	.0012	.0095
	12	.0000	.0000	.0000	.0000	.0000	.0000	.0000	.0000	.0000	.0000	.0001	.0016
	13	.0000	.0000	.0000	.0000	.0000	.0000	.0000	.0000	.0000	.0000	.0000	.0001
14	0	.8687	.7536	.6528	.5647	.4877	.2288	.1028	.0440	.0178	.0068	.0008	.0001
	1	.1229	.2153	.2827	.3294	.3593	.3559	.2539	.1539	.0832	.0407	.0073	.0009
	2	.0081	.0286	.0568	.0892	.1229	.2570	.2912	.2501	.1802	.1134	.0317	.0056
	3	.0003	.0023	.0070	.0149	.0259	.1142	.2056	.2501	.2402	.1943	.0845	.0222
	4	.0000	.0001	.0006	.0017	.0037	.0349	.0998	.1720	.2202	.2290	.1549	.0611
	5	.0000	.0000	.0000	.0001	.0004	.0078	.0352	.0860	.1468	.1963	.2066	.1222
	6	.0000	.0000	.0000	.0000	.0000	.0013	.0093	.0322	.0734	.1262	.2066	.1833
	7	.0000	.0000	.0000	.0000	.0000	.0002	.0019	.0092	.0280	.0618	.1574	.2095
	8	.0000	.0000	.0000	.0000	.0000	.0000	.0003	.0020	.0082	.0232	.0918	.1833
	9	.0000	.0000	.0000	.0000	.0000	.0000	.0000	.0003	.0018	.0066	.0408	.1222
	10	.0000	.0000	.0000	.0000	.0000	.0000	.0000	.0000	.0003	.0014	.0136	.0611
	11	.0000	.0000	.0000	.0000	.0000	.0000	.0000	.0000	.0000	.0002	.0033	.0222
	12	.0000	.0000	.0000	.0000	.0000	.0000	.0000	.0000	.0000	.0000	.0005	.0056
	13	.0000	.0000	.0000	.0000	.0000	.0000	.0000	.0000	.0000	.0000	.0001	.0009
	14	.0000	.0000	.0000	.0000	.0000	.0000	.0000	.0000	.0000	.0000	.0000	.0001
15	0	.8601	.7386	.6333	.5421	.4633	.2059	.0874	.0352	.0134	.0047	.0005	.0000
	1	.1303	.2261	.2938	.3388	.3658	.3432	.2312	.1319	.0668	.0305	.0047	.0005
	2	.0092	.0323	.0636	.0988	.1348	.2669	.2856	.2309	.1559	.0916	.0219	.0032
	3	.0004	.0029	.0085	.0178	.0307	.1285	.2184	.2501	.2252	.1700	.0634	.0139

Table A.9.   Binomial Distribution (Continued)

N	X	.01	.02	.03	.04	.05	.10	.15	.20	.25	.30	.40	.50
15	4	.0000	.0002	.0008	.0022	.0049	.0428	.1156	.1876	.2252	.2186	.1268	.0417
	5	.0000	.0000	.0001	.0002	.0006	.0105	.0449	.1032	.1651	.2061	.1859	.0916
	6	.0000	.0000	.0000	.0000	.0000	.0019	.0132	.0430	.0917	.1472	.2066	.1527
	7	.0000	.0000	.0000	.0000	.0000	.0003	.0030	.0138	.0393	.0811	.1771	.1964
	8	.0000	.0000	.0000	.0000	.0000	.0000	.0005	.0035	.0131	.0348	.1181	.1964
	9	.0000	.0000	.0000	.0000	.0000	.0000	.0001	.0007	.0034	.0116	.0612	.1527
	10	.0000	.0000	.0000	.0000	.0000	.0000	.0000	.0001	.0007	.0030	.0245	.0916
	11	.0000	.0000	.0000	.0000	.0000	.0000	.0000	.0000	.0001	.0006	.0074	.0417
	12	.0000	.0000	.0000	.0000	.0000	.0000	.0000	.0000	.0000	.0001	.0016	.0139
	13	.0000	.0000	.0000	.0000	.0000	.0000	.0000	.0000	.0000	.0000	.0003	.0032
	14	.0000	.0000	.0000	.0000	.0000	.0000	.0000	.0000	.0000	.0000	.0000	.0005
16	0	.8515	.7238	.6143	.5204	.4401	.1853	.0743	.0281	.0100	.0033	.0003	.0000
	1	.1376	.2363	.3040	.3469	.3706	.3294	.2096	.1126	.0535	.0228	.0030	.0002
	2	.0104	.0362	.0705	.1084	.1463	.2745	.2775	.2111	.1336	.0732	.0150	.0018
	3	.0005	.0034	.0102	.0211	.0359	.1423	.2285	.2463	.2079	.1465	.0468	.0085
	4	.0000	.0002	.0010	.0029	.0061	.0514	.1311	.2001	.2252	.2040	.1014	.0278
	5	.0000	.0000	.0001	.0003	.0008	.0137	.0555	.1201	.1802	.2099	.1623	.0667
	6	.0000	.0000	.0000	.0000	.0001	.0028	.0180	.0550	.1101	.1649	.1983	.1222
	7	.0000	.0000	.0000	.0000	.0000	.0004	.0045	.0197	.0524	.1010	.1889	.1746
	8	.0000	.0000	.0000	.0000	.0000	.0001	.0009	.0055	.0197	.0487	.1417	.1964
	9	.0000	.0000	.0000	.0000	.0000	.0000	.0001	.0012	.0058	.0185	.0840	.1746
	10	.0000	.0000	.0000	.0000	.0000	.0000	.0000	.0002	.0014	.0056	.0392	.1222
	11	.0000	.0000	.0000	.0000	.0000	.0000	.0000	.0000	.0002	.0013	.0142	.0667
	12	.0000	.0000	.0000	.0000	.0000	.0000	.0000	.0000	.0000	.0002	.0040	.0278
	13	.0000	.0000	.0000	.0000	.0000	.0000	.0000	.0000	.0000	.0000	.0008	.0085
	14	.0000	.0000	.0000	.0000	.0000	.0000	.0000	.0000	.0000	.0000	.0001	.0018
	15	.0000	.0000	.0000	.0000	.0000	.0000	.0000	.0000	.0000	.0000	.0000	.0002
17	0	.8429	.7093	.5958	.4996	.4181	.1668	.0631	.0225	.0075	.0023	.0002	.0000
	1	.1447	.2461	.3133	.3539	.3741	.3150	.1893	.0957	.0426	.0169	.0019	.0001
	2	.0117	.0402	.0775	.1180	.1575	.2800	.2673	.1914	.1136	.0581	.0102	.0010
	3	.0006	.0041	.0120	.0246	.0415	.1556	.2359	.2393	.1893	.1245	.0341	.0052
	4	.0000	.0003	.0013	.0036	.0076	.0605	.1457	.2093	.2209	.1868	.0796	.0182
	5	.0000	.0000	.0001	.0004	.0010	.0175	.0668	.1361	.1914	.2081	.1379	.0472
	6	.0000	.0000	.0000	.0000	.0001	.0039	.0236	.0680	.1276	.1784	.1839	.0944
	7	.0000	.0000	.0000	.0000	.0000	.0007	.0065	.0267	.0668	.1201	.1927	.1484
	8	.0000	.0000	.0000	.0000	.0000	.0001	.0014	.0084	.0279	.0644	.1606	.1855
	9	.0000	.0000	.0000	.0000	.0000	.0000	.0003	.0021	.0093	.0276	.1070	.1855
	10	.0000	.0000	.0000	.0000	.0000	.0000	.0000	.0004	.0025	.0095	.0571	.1484
	11	.0000	.0000	.0000	.0000	.0000	.0000	.0000	.0001	.0005	.0026	.0242	.0944
	12	.0000	.0000	.0000	.0000	.0000	.0000	.0000	.0000	.0001	.0006	.0081	.0472
	13	.0000	.0000	.0000	.0000	.0000	.0000	.0000	.0000	.0000	.0001	.0021	.0182

Table A.9. Binomial Distribution (Continued)

N	X	.01	.02	.03	.04	.05	.10	.15	.20	.25	.30	.40	.50
17	14	.0000	.0000	.0000	.0000	.0000	.0000	.0000	.0000	.0000	.0000	.0004	.0052
	15	.0000	.0000	.0000	.0000	.0000	.0000	.0000	.0000	.0000	.0000	.0001	.0010
	16	.0000	.0000	.0000	.0000	.0000	.0000	.0000	.0000	.0000	.0000	.0000	.0001
18	0	.8345	.6951	.5780	.4796	.3972	.1501	.0536	.0180	.0056	.0016	.0001	.0000
	1	.1517	.2554	.3217	.3597	.3763	.3002	.1704	.0811	.0338	.0126	.0012	.0001
	2	.0130	.0443	.0846	.1274	.1683	.2835	.2556	.1723	.0958	.0458	.0069	.0006
	3	.0007	.0048	.0140	.0283	.0473	.1680	.2406	.2297	.1704	.1046	.0246	.0031
	4	.0000	.0004	.0016	.0044	.0093	.0700	.1592	.2153	.2130	.1681	.0614	.0117
	5	.0000	.0000	.0001	.0005	.0014	.0218	.0787	.1507	.1988	.2017	.1146	.0327
	6	.0000	.0000	.0000	.0000	.0002	.0052	.0301	.0816	.1436	.1873	.1655	.0708
	7	.0000	.0000	.0000	.0000	.0000	.0010	.0091	.0350	.0820	.1376	.1892	.1214
	8	.0000	.0000	.0000	.0000	.0000	.0002	.0022	.0120	.0376	.0811	.1734	.1669
	9	.0000	.0000	.0000	.0000	.0000	.0000	.0004	.0033	.0139	.0386	.1284	.1855
	10	.0000	.0000	.0000	.0000	.0000	.0000	.0001	.0008	.0042	.0149	.0771	.1669
	11	.0000	.0000	.0000	.0000	.0000	.0000	.0000	.0001	.0010	.0046	.0374	.1214
	12	.0000	.0000	.0000	.0000	.0000	.0000	.0000	.0000	.0002	.0012	.0145	.0708
	13	.0000	.0000	.0000	.0000	.0000	.0000	.0000	.0000	.0000	.0002	.0045	.0327
	14	.0000	.0000	.0000	.0000	.0000	.0000	.0000	.0000	.0000	.0000	.0011	.0117
	15	.0000	.0000	.0000	.0000	.0000	.0000	.0000	.0000	.0000	.0000	.0002	.0031
	16	.0000	.0000	.0000	.0000	.0000	.0000	.0000	.0000	.0000	.0000	.0000	.0006
	17	.0000	.0000	.0000	.0000	.0000	.0000	.0000	.0000	.0000	.0000	.0000	.0001
19	0	.8262	.6812	.5606	.4604	.3774	.1351	.0456	.0144	.0042	.0011	.0001	.0000
	1	.1586	.2642	.3294	.3645	.3774	.2852	.1529	.0685	.0268	.0093	.0008	.0000
	2	.0144	.0485	.0917	.1367	.1787	.2852	.2428	.1540	.0803	.0358	.0046	.0003
	3	.0008	.0056	.0161	.0323	.0533	.1796	.2428	.2182	.1517	.0869	.0175	.0018
	4	.0000	.0005	.0020	.0054	.0112	.0798	.1714	.2182	.2023	.1491	.0467	.0074
	5	.0000	.0000	.0002	.0007	.0018	.0266	.0907	.1636	.2023	.1916	.0933	.0222
	6	.0000	.0000	.0000	.0001	.0002	.0069	.0374	.0955	.1574	.1916	.1451	.0518
	7	.0000	.0000	.0000	.0000	.0000	.0014	.0122	.0443	.0974	.1525	.1797	.0961
	8	.0000	.0000	.0000	.0000	.0000	.0002	.0032	.0166	.0487	.0981	.1797	.1442
	9	.0000	.0000	.0000	.0000	.0000	.0000	.0007	.0051	.0198	.0514	.1464	.1762
	10	.0000	.0000	.0000	.0000	.0000	.0000	.0001	.0013	.0066	.0220	.0976	.1762
	11	.0000	.0000	.0000	.0000	.0000	.0000	.0000	.0003	.0018	.0077	.0532	.1442
	12	.0000	.0000	.0000	.0000	.0000	.0000	.0000	.0000	.0004	.0022	.0237	.0961
	13	.0000	.0000	.0000	.0000	.0000	.0000	.0000	.0000	.0001	.0005	.0085	.0518
	14	.0000	.0000	.0000	.0000	.0000	.0000	.0000	.0000	.0000	.0001	.0024	.0222
	15	.0000	.0000	.0000	.0000	.0000	.0000	.0000	.0000	.0000	.0000	.0005	.0074
	16	.0000	.0000	.0000	.0000	.0000	.0000	.0000	.0000	.0000	.0000	.0001	.0018
	17	.0000	.0000	.0000	.0000	.0000	.0000	.0000	.0000	.0000	.0000	.0000	.0003
20	0	.8179	.6676	.5438	.4420	.3585	.1216	.0388	.0115	.0032	.0008	.0000	.0000
	1	.1652	.2725	.3364	.3683	.3774	.2702	.1368	.0576	.0211	.0068	.0005	.0000
	2	.0159	.0528	.0988	.1458	.1887	.2852	.2293	.1369	.0669	.0278	.0031	.0002
	3	.0010	.0065	.0183	.0364	.0596	.1901	.2428	.2054	.1339	.0716	.0123	.0011

Table A.9. Binomial Distribution (Continued)

N	X	.01	.02	.03	.04	.05	.10	.15	.20	.25	.30	.40	.50
20	4	.0000	.0006	.0024	.0065	.0133	.0898	.1821	.2182	.1897	.1304	.0350	.0046
	5	.0000	.0000	.0002	.0009	.0022	.0319	.1028	.1746	.2023	.1789	.0746	.0148
	6	.0000	.0000	.0000	.0001	.0003	.0089	.0454	.1091	.1686	.1916	.1244	.0370
	7	.0000	.0000	.0000	.0000	.0000	.0020	.0160	.0545	.1124	.1643	.1659	.0739
	8	.0000	.0000	.0000	.0000	.0000	.0004	.0046	.0222	.0609	.1144	.1797	.1201
	9	.0000	.0000	.0000	.0000	.0000	.0001	.0011	.0074	.0271	.0654	.1597	.1602
	10	.0000	.0000	.0000	.0000	.0000	.0000	.0002	.0020	.0099	.0308	.1171	.1762
	11	.0000	.0000	.0000	.0000	.0000	.0000	.0000	.0005	.0030	.0120	.0710	.1602
	12	.0000	.0000	.0000	.0000	.0000	.0000	.0000	.0001	.0008	.0039	.0355	.1201
	13	.0000	.0000	.0000	.0000	.0000	.0000	.0000	.0000	.0002	.0010	.0146	.0739
	14	.0000	.0000	.0000	.0000	.0000	.0000	.0000	.0000	.0000	.0002	.0049	.0370
	15	.0000	.0000	.0000	.0000	.0000	.0000	.0000	.0000	.0000	.0000	.0013	.0148
	16	.0000	.0000	.0000	.0000	.0000	.0000	.0000	.0000	.0000	.0000	.0003	.0046
	17	.0000	.0000	.0000	.0000	.0000	.0000	.0000	.0000	.0000	.0000	.0000	.0011
	18	.0000	.0000	.0000	.0000	.0000	.0000	.0000	.0000	.0000	.0000	.0000	.0002
25	0	.7778	.6035	.4670	.3604	.2774	.0718	.0172	.0038	.0008	.0001	.0000	.0000
	1	.1964	.3079	.3611	.3754	.3650	.1994	.0759	.0236	.0063	.0014	.0000	.0000
	2	.0238	.0754	.1340	.1877	.2305	.2659	.1607	.0708	.0251	.0074	.0004	.0000
	3	.0018	.0118	.0318	.0600	.0930	.2265	.2174	.1358	.0641	.0243	.0019	.0001
	4	.0001	.0013	.0054	.0137	.0269	.1384	.2110	.1867	.1175	.0572	.0071	.0004
	5	.0000	.0001	.0007	.0024	.0060	.0646	.1564	.1960	.1645	.1030	.0199	.0016
	6	.0000	.0000	.0001	.0003	.0010	.0239	.0920	.1633	.1828	.1472	.0442	.0053
	7	.0000	.0000	.0000	.0000	.0001	.0072	.0441	.1108	.1654	.1712	.0800	.0143
	8	.0000	.0000	.0000	.0000	.0000	.0018	.0175	.0623	.1241	.1651	.1200	.0322
	9	.0000	.0000	.0000	.0000	.0000	.0004	.0058	.0294	.0781	.1336	.1511	.0609
	10	.0000	.0000	.0000	.0000	.0000	.0001	.0016	.0118	.0417	.0916	.1612	.0974
	11	.0000	.0000	.0000	.0000	.0000	.0000	.0004	.0040	.0189	.0536	.1465	.1328
	12	.0000	.0000	.0000	.0000	.0000	.0000	.0001	.0012	.0074	.0268	.1139	.1550
	13	.0000	.0000	.0000	.0000	.0000	.0000	.0000	.0003	.0025	.0115	.0760	.1550
	14	.0000	.0000	.0000	.0000	.0000	.0000	.0000	.0001	.0007	.0042	.0434	.1328
	15	.0000	.0000	.0000	.0000	.0000	.0000	.0000	.0000	.0002	.0013	.0212	.0974
	16	.0000	.0000	.0000	.0000	.0000	.0000	.0000	.0000	.0000	.0004	.0088	.0609
	17	.0000	.0000	.0000	.0000	.0000	.0000	.0000	.0000	.0000	.0001	.0031	.0322
	18	.0000	.0000	.0000	.0000	.0000	.0000	.0000	.0000	.0000	.0000	.0009	.0143
	19	.0000	.0000	.0000	.0000	.0000	.0000	.0000	.0000	.0000	.0000	.0002	.0053
	20	.0000	.0000	.0000	.0000	.0000	.0000	.0000	.0000	.0000	.0000	.0000	.0016
	21	.0000	.0000	.0000	.0000	.0000	.0000	.0000	.0000	.0000	.0000	.0000	.0004
	22	.0000	.0000	.0000	.0000	.0000	.0000	.0000	.0000	.0000	.0000	.0000	.0001
50	0	.6050	.3642	.2181	.1299	.0769	.0052	.0003	.0000	.0000	.0000	.0000	.0000
	1	.3056	.3716	.3372	.2706	.2025	.0286	.0026	.0002	.0000	.0000	.0000	.0000
	2	.0756	.1858	.2555	.2762	.2611	.0779	.0113	.0011	.0001	.0000	.0000	.0000
	3	.0122	.0607	.1264	.1842	.2199	.1386	.0319	.0044	.0004	.0000	.0000	.0000

Table A.9. Binomial Distribution (Continued)

N	X	.01	.02	.03	.04	.05	.10	P .15	.20	.25	.30	.40	.50
50	4	.0015	.0145	.0459	.0902	.1360	.1809	.0661	.0128	.0016	.0001	.0000	.0000
	5	.0001	.0027	.0131	.0346	.0658	.1849	.1072	.0295	.0049	.0006	.0000	.0000
	6	.0000	.0004	.0030	.0108	.0260	.1541	.1419	.0554	.0123	.0018	.0000	.0000
	7	.0000	.0001	.0006	.0028	.0086	.1076	.1575	.0870	.0259	.0048	.0000	.0000
	8	.0000	.0000	.0001	.0006	.0024	.0643	.1493	.1169	.0463	.0110	.0002	.0000
	9	.0000	.0000	.0000	.0001	.0006	.0333	.1230	.1364	.0721	.0220	.0005	.0000
	10	.0000	.0000	.0000	.0000	.0001	.0152	.0890	.1398	.0985	.0386	.0014	.0000
	11	.0000	.0000	.0000	.0000	.0000	.0061	.0571	.1271	.1194	.0602	.0035	.0000
	12	.0000	.0000	.0000	.0000	.0000	.0022	.0328	.1033	.1294	.0838	.0076	.0001
	13	.0000	.0000	.0000	.0000	.0000	.0007	.0169	.0755	.1261	.1050	.0147	.0003
	14	.0000	.0000	.0000	.0000	.0000	.0002	.0079	.0499	.1110	.1189	.0260	.0008
	15	.0000	.0000	.0000	.0000	.0000	.0001	.0033	.0299	.0888	.1223	.0415	.0020
	16	.0000	.0000	.0000	.0000	.0000	.0000	.0013	.0164	.0648	.1147	0606	.0044
	17	.0000	.0000	.0000	.0000	.0000	.0000	.0005	.0082	.0432	.0983	.0808	.0087
	18	.0000	.0000	.0000	.0000	.0000	.0000	.0001	.0037	.0264	.0772	.0987	.0160
	19	.0000	.0000	.0000	.0000	.0000	.0000	.0000	.0016	.0148	.0558	.1109	.0270
	20	.0000	.0000	.0000	.0000	.0000	.0000	.0000	.0006	.0077	.0370	.1146	.0419
	21	.0000	.0000	.0000	.0000	.0000	.0000	.0000	.0002	.0036	.0227	.1091	.0598
	22	.0000	.0000	.0000	.0000	.0000	.0000	.0000	.0001	.0016	.0128	.0959	.0788
	23	.0000	.0000	.0000	.0000	.0000	.0000	.0000	.0000	.0006	.0067	.0778	.0960
	24	.0000	.0000	.0000	.0000	.0000	.0000	.0000	.0000	.0002	.0032	.0584	.1080
	25	.0000	.0000	.0000	.0000	.0000	.0000	.0000	.0000	.0001	.0014	.0405	.1123
	26	.0000	.0000	.0000	.0000	.0000	.0000	.0000	.0000	.0000	.0006	.0259	.1080
	27	.0000	.0000	.0000	.0000	.0000	.0000	.0000	.0000	.0000	.0002	.0154	.0960
	28	.0000	.0000	.0000	.0000	.0000	.0000	.0000	.0000	.0000	.0001	.0084	.0788
	29	.0000	.0000	.0000	.0000	.0000	.0000	.0000	.0000	.0000	.0000	.0043	.0598
	30	.0000	.0000	.0000	.0000	.0000	.0000	.0000	.0000	.0000	.0000	.0020	.0419
	31	.0000	.0000	.0000	.0000	.0000	.0000	.0000	.0000	.0000	.0000	.0009	.0270
	32	.0000	.0000	.0000	.0000	.0000	.0000	.0000	.0000	.0000	.0000	.0003	.0160
	33	.0000	.0000	.0000	.0000	.0000	.0000	.0000	.0000	.0000	.0000	.0001	.0087
	34	.0000	.0000	.0000	.0000	.0000	.0000	.0000	.0000	.0000	.0000	.0000	.0044
	35	.0000	.0000	.0000	.0000	.0000	.0000	.0000	.0000	.0000	.0000	.0000	.0020
	36	.0000	.0000	.0000	.0000	.0000	.0000	.0000	.0000	.0000	.0000	.0000	.0008
	37	.0000	.0000	.0000	.0000	.0000	.0000	.0000	.0000	.0000	.0000	.0000	.0003
	38	.0000	.0000	.0000	.0000	.0000	.0000	.0000	.0000	.0000	.0000	.0000	.0001

Table A.10.   Standard Normal Cumulative Distribution

Z	.00	.01	.02	.03	.04	.05	.06	.07	.08	.09
−3.8	.0001	.0001	.0001	.0001	.0001	.0001	.0001	.0001	.0001	.0001
−3.7	.0001	.0001	.0001	.0001	.0001	.0001	.0001	.0001	.0001	.0001
−3.6	.0002	.0002	.0001	.0001	.0001	.0001	.0001	.0001	.0001	.0001
−3.5	.0002	.0002	.0002	.0002	.0002	.0002	.0002	.0002	.0002	.0002
−3.4	.0003	.0003	.0003	.0003	.0003	.0003	.0003	.0003	.0003	.0002
−3.3	.0005	.0005	.0005	.0004	.0004	.0004	.0004	.0004	.0004	.0003
−3.2	.0007	.0007	.0006	.0006	.0006	.0006	.0006	.0005	.0005	.0005
−3.1	.0010	.0009	.0009	.0009	.0008	.0008	.0008	.0008	.0007	.0007
−3.0	.0014	.0013	.0013	.0012	.0012	.0011	.0011	.0011	.0010	.0010
−2.9	.0019	.0018	.0018	.0017	.0016	.0016	.0015	.0015	.0014	.0014
−2.8	.0026	.0025	.0024	.0023	.0023	.0022	.0021	.0021	.0020	.0019
−2.7	.0035	.0034	.0033	.0032	.0031	.0030	.0029	.0028	.0027	.0026
−2.6	.0047	.0045	.0044	.0043	.0041	.0040	.0039	.0038	.0037	.0036
−2.5	.0062	.0060	.0059	.0057	.0055	.0054	.0052	.0051	.0049	.0048
−2.4	.0082	.0080	.0078	.0076	.0073	.0071	.0069	.0068	.0066	.0064
−2.3	.0107	.0104	.0102	.0099	.0096	.0094	.0091	.0089	.0087	.0084
−2.2	.0139	.0136	.0132	.0129	.0125	.0122	.0119	.0116	.0113	.0110
−2.1	.0179	.0174	.0170	.0166	.0162	.0158	.0154	.0150	.0146	.0143
−2.0	.0228	.0222	.0217	.0212	.0207	.0202	.0197	.0192	.0188	.0183
−1.9	.0287	.0281	.0274	.0268	.0262	.0256	.0250	0244	.0239	.0233
−1.8	.0359	.0352	.0344	.0336	.0329	.0322	.0314	.0307	.0301	.0294
−1.7	.0446	.0436	.0427	.0418	.0409	.0401	.0392	.0384	.0375	.0367
−1.6	.0548	.0537	.0526	.0516	.0505	.0495	.0485	.0475	.0465	.0455
−1.5	.0668	.0655	.0643	.0630	.0618	.0606	.0594	.0582	.0571	.0559
−1.4	.0808	.0793	.0778	.0764	.0749	.0735	.0721	.0708	.0694	.0681
−1.3	.0968	.0951	.0934	.0918	.0901	.0885	.0869	.0853	.0838	.0823
−1.2	.1151	.1131	.1112	.1094	.1075	.1057	.1038	.1020	.1003	.0985
−1.1	.1357	.1335	.1314	.1292	.1271	.1251	.1230	.1210	.1190	.1170
−1.0	.1587	.1562	.1539	.1515	.1492	.1469	.1446	.1423	.1401	.1379
−0.9	.1841	.1814	.1788	.1762	.1736	.1711	.1685	.1660	.1635	.1611
−0.8	.2119	.2090	.2061	.2033	.2005	.1977	.1949	.1922	.1894	.1867
−0.7	.2420	.2389	.2358	.2327	.2296	.2266	.2236	.2206	.2177	.2148
−0.6	.2743	.2709	.2676	.2643	.2611	.2578	.2546	.2514	.2483	.2451
−0.5	.3085	.3050	.3015	.2981	.2946	.2912	.2877	.2843	.2810	.2776
−0.4	.3446	.3409	.3372	.3336	.3300	.3264	.3228	.3192	.3156	.3121
−0.3	.3821	.3783	.3745	.3707	.3669	.3632	.3594	.3557	.3520	.3483
−0.2	.4207	.4168	.4129	.4090	.4052	.4013	.3947	.3936	.3897	.3859
−0.1	.4602	.4562	.4522	.4483	.4443	.4404	.4364	.4325	.4286	.4247
−0.0	.5000	.4960	.4920	.4880	.4840	.4801	.4761	.4721	.4681	.4641

Table A.10.   Standard Normal Cumulative Distribution (Continued)

Z	.00	.01	.02	.03	.04	.05	.06	.07	.08	.09
0.0	.5000	.5040	.5080	.5120	.5160	.5199	.5239	.5279	.5319	.5359
0.1	.5398	.5438	.5478	.5517	.5557	.5596	.5636	.5675	.5714	.5753
0.2	.5793	.5832	.5871	.5910	.5948	.5987	.6026	.6064	.6103	.6141
0.3	.6179	.6217	.6255	.6293	.6331	.6368	.6406	.6443	.6480	.6517
0.4	.6554	.6591	.6628	.6664	.6700	.6736	.6772	.6808	.6844	.6879
0.5	.6915	.6950	.6985	.7019	.7054	.7088	.7123	.7157	.7190	.7224
0.6	.7257	.7291	.7324	.7357	.7389	.7422	.7454	.7486	.7517	.7549
0.7	.7580	.7611	.7642	.7673	.7704	.7734	.7764	.7794	.7823	.7852
0.8	.7881	.7910	.7939	.7967	.7995	.8023	.8015	.8078	.8106	.8133
0.9	.8159	.8186	.8212	.8238	.8264	.8289	.8315	.8340	.8365	.8389
1.0	.8413	.8438	.8461	.8485	.8508	.8531	.8554	.8577	.8599	.8621
1.1	.8643	.8665	.8686	.8708	.8729	.8749	.8770	.8790	.8810	.8830
1.2	.8849	.8869	.8888	.8906	.8925	.8943	.8962	.8980	.8997	.9015
1.3	.9032	.9049	.9066	.9082	.9099	.9115	.9131	.9147	.9162	.9177
1.4	.9192	.9207	.9222	.9236	.9251	.9265	.9279	.9292	.9306	.9319
1.5	.9332	.9345	.9357	.9370	.9382	.9394	.9406	.9418	.9429	.9441
1.6	.9452	.9463	.9474	.9484	.9495	.9505	.9515	.9525	.9535	.9545
1.7	.9554	.9564	.9573	.9582	.9591	.9599	.9608	.9616	.9625	.9633
1.8	.9641	.9648	.9656	.9664	.9671	.9678	.9686	.9693	.9699	.9706
1.9	.9713	.9719	.9726	.9732	.9738	.9744	.9750	.9756	.9761	.9767
2.0	.9772	.9778	.9783	.9788	.9793	.9798	.9803	.9808	.9812	.9817
2.1	.9821	.9826	.9830	.9834	.9838	.9842	.9846	.9850	.9854	.9857
2.2	.9861	.9864	.9868	.9871	.9875	.9878	.9881	.9884	.9887	.9890
2.3	.9893	.9896	.9898	.9901	.9904	.9906	.9909	.9911	.9913	.9916
2.4	.9918	.9920	.9922	.9924	.9927	.9929	.9931	.9932	.9934	.9936
2.5	.9938	.9940	.9941	.9943	.9945	.9946	.9948	.9949	.9951	.9952
2.6	.9953	.9955	.9956	.9957	.9959	.9960	.9961	.9962	.9963	.9964
2.7	.9965	.9966	.9967	.9968	.9969	.9970	.9971	.9972	.9973	.9974
2.8	.9974	.9975	.9976	.9977	.9977	.9978	.9979	.9979	.9980	.9981
2.9	.9981	.9982	.9982	.9983	.9984	.9984	.9985	.9985	.9986	.9986
3.0	.9986	.9987	.9987	.9988	.9988	.9989	.9989	.9989	.9990	.9990
3.1	.9990	.9991	.9991	.9991	.9992	.9992	.9992	.9992	.9993	.9993
3.2	.9993	.9993	.9994	.9994	.9994	.9994	.9994	.9995	.9995	.9995
3.3	.9995	.9995	.9995	.9996	.9996	.9996	.9996	.9996	.9996	.9997
3.4	.9997	.9997	.9997	.9997	.9997	.9997	.9997	.9997	.9997	.9998
3.5	.9998	.9998	.9998	.9998	.9998	.9998	.9998	.9998	.9998	.9998
3.6	.9998	.9998	.9999	.9999	.9999	.9999	.9999	.9999	.9999	.9999
3.7	.9999	.9999	.9999	.9999	.9999	.9999	.9999	.9999	.9999	.9999
3.8	.9999	.9999	.9999	.9999	.9999	.9999	.9999	.9999	.9999	.9999

Selected Answers to Odd Numbered Questions

CHAPTER 1

1. No. The individuals that will vote in an upcoming election are not known with certainty.
3. Yes. It is not necessary to have a listing of all family units with incomes of less than $6000 for the group to be classified as a set.
5. $\{a, b, c\}, \{a, b\}, \{a, c\}, (b, c), \{a\}, \{b\}, \{c\}, \{\ \}$
7. 64
9. (a) Q' (b) ϕ (c) U
11. 60
13. (a) 20   (b) 280
15. (a) 1100   (b) 250   (c) 50   (d) 1400
17. 12
19. 630
21. Cities where revenue is less than $5000 per month.

CHAPTER 2

1. (a) 0   (c) 2   (e) -0.5 (g) -3
3. (a) $f(x) = 3000 + \frac{2}{3}x$ (c) $f(x) = 3500 + 10x$
   (e) $p = 11.50 - 0.001q$ (g) $S = 10,000 + 500t$
5. (a) 2.236   (c) 2.828   (e) 11.180   (g) 6.403

7. $d_{AB} = 70.71$, $d_{BC} = 55.90$, $d_{CA} = 103.08$

9. Yes

11. (a) relation   (c) function

13.

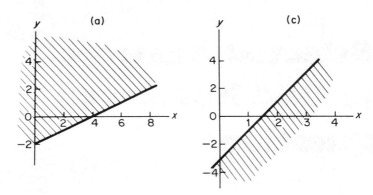

15. (a) $2x_1 + 3x_2 = 80$ for $0 \le x_1 \le 40$ and $0 \le x_2 \le \frac{80}{3}$
    (c) $20x_1 + 60x_2 + 80x_3 = 1200$ for $0 \le x_1 \le 60, 0 \le x_2 \le 20$,
    and $0 \le x_3 \le 15$

17. $A = 6, B = 3$

19. Consistent with an infinite number of solutions.

21. (a) Consistent, unique solution $x = 7, y = 5$
    (c) Inconsistent

23. (a) 1600 units   (b) $10,000

25. 10,000 units

27. (a) $66,666.67   (b) $80,000   (c) $100,000

CHAPTER 3

1.

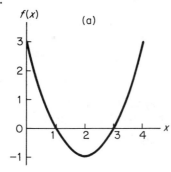

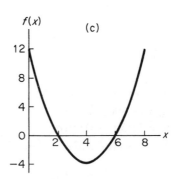

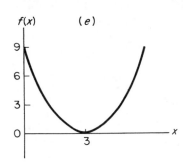

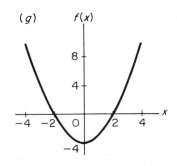

3. (a) $f(x) = 3 - 4x + x^2$ (c) $f(x) = -11 - 2.5x + 1.5x^2$
     (e) $f(x) = -4 + x^2$ (g) $f(x) = 16 - 4x - 2x^2$

5. (a) $x = 1, x = 3$ (c) $x = 2, x = 6$
     (e) $x = 3$ (g) $x = -2, x = 2$

7. (a) $x = -1, f(-1) = 18$, maximum
     (c) $x = 3, f(3) = -15.5$, maximum
     (e) $x = 6, f(6) = -28$, minimum

9. $c(x) = 12.50 - 4.0833x + 0.4167x^2$. Minimum average cost of \$2.50 occurs for $x = 5$ hundred units.

11. $f(x) = -133 + 6.483x - 0.0633x^2$. Maximum earnings are \$32,960 at $x = 51.2$ years.

13. $P(x) = -3900 + 105.3x - 0.533x^2$. Maximum profit of \$1301 occurs when $x = 99$ units are rented.

15. (a) $\ln k = 0.28769, c = 0.20273$, and $f(x) = 1.33e^{0.203x}$
     (c) $\ln k = 2.64315, c = -0.17028$, and $f(x) = 14.04e^{-0.170x}$
     (e) $\ln k = 1.61734, c = -0.23105$, and $f(x) = 5.04e^{-0.231x}$
     (g) $\ln k = 4.60517, c = 0.0392$, and $f(x) = 100e^{0.039x}$

17. $f(t) = 10e^{0.05068t}, f(12 \text{ years}) = \$18.4$ million, $r = 5.2\%$

19. $r = 5.9\%$

21. $s = 5x_1 + 8x_2 + 6x_3, \$30,000$

23. (a) Profit $= rR + uU + pP$ (b) $r = \$0.05, u = \$0.03, p = \$0.04$

CHAPTER 4

1. $2 \cdot 3 \cdot 2 = 12$

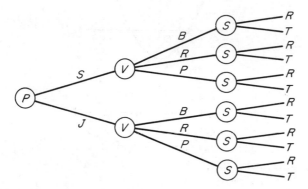

3. 6

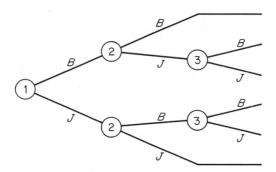

5. $2 \cdot 2 \cdot 1 = 4$

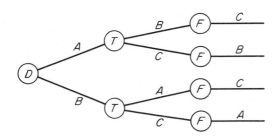

7. $2 \cdot 2 \cdot 3 = 12$

9. $10 \cdot 30 \cdot 7 \cdot 3 = 27,300$

11. $108(4!) = 2592$

13. $6! = 720$

15. $5! = 120$

17. $_5P_3 = 30$

19. $_{24}P_3 = 12,144$

21. $5! = 120$

23. $\dfrac{9!}{2!\,2!\,2!} = 45,360$ in "professor;" $\dfrac{7!}{2!} = 2520$ in "student"

25. $_8P_8(5, 3) = 56$

27. (a) $_{15}P_{15}(5, 4, 6) = 626,220$ (b) $15! = 1,394,852,659,200$

29. $_2P^8 = 256$

31. $_{10}P^7 = 10^7$ and $8 \cdot _{10}P^6 = 8 \cdot 10^6$

33. $_{15}C_5 = 3003$

35. $_{12}C_2 = 66$

37. 630

39. $\dfrac{1}{3!} \cdot _{12}C_{4,4,4} = 5775$

41. $_7C_{2,3,1,1} = 420$

43. $\dbinom{n}{r} = \dbinom{n}{n-r} = \dfrac{n!}{r!\,(n-r)!}$

45. $A^4 - 12A^3B + 54A^2B^2 - 108AB^3 + 81B^4$

CHAPTER 5

1. (a) $\{(mm), (mf), (fm), (ff)\}$
   (b) $\{(aaa), (aad), (ada), (daa), (add), (dad), (dda), (ddd)\}$
   (c) $\{(xyz), (xzy), (yxz), (yzx), (zxy), (zyx)\}$
   (d) $\{(gggg), (gggd), (ggdg), (gdgg), (dggg), (ggdd), (gdgd), (dggd),$
       $(dgdg), (ddgg), (gddg), (dddg), (ddgd), (dgdd), (gddd), (dddd)\}$

3. $\{$(Smith, California), (Smith, Oregon), (Smith, Arizona),
   (Jones, California), (Jones, Oregon), (Jones, Arizona),
   (Clark, California), (Clark, Oregon), (Clark, Arizona)$\}$

5.

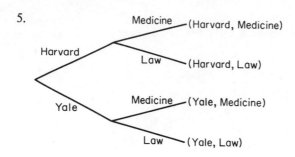

7.

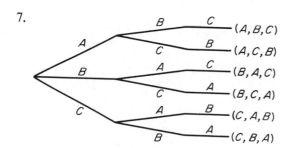

9. $S = \{0, 1, 2, \ldots\}$, Countably infinite

11. $S = \{(ggg), (ggd), (gdg), (dgg), (gdd), (dgd), (ddg), (ddd)\}$
   (a) $\{(ggg)\}$, Simple
   (b) $\{(ggg), (ggd), (gdg), (dgg)\}$, Composite
   (c) $\{(ggd), (gdg), (dgg), (gdd), (dgd), (ddg), (ddd)\}$, Composite
   (d) $\{(ddd)\}$, Simple
   (e) $\{(dgg), (dgd), (ddg), (ddd)\}$, Composite

13. (a) 8   (b) 57   (c) 68   (d) 32
   (e) 100   (f) 0   (g) 17   (h) 40

15. $P(A \cup B) = 0.60$, $P(A' \cap B) = 0.20$, $P(A \cup B)' = 0.40$

17. 0.25

19. (a) $P(\text{Motor Vehicle}) = 54,800/203,000,000 = 0.00024$
   $P(\text{Aircraft}) = 1400/203,000,000 = 0.0000069$
   (b) No. To determine the relative safety, we should use a measure of deaths per passenger mile.

21. $_{13}C_5/_{52}C_5 = 1287/2,598,960 = 0.000495$

23. $(_4C_2 \cdot \, _5C_1)/_9C_3 = 0.357$

25. (a) Objective   (b) 0.32   (c) 0.30   (d) 0.345
    (e) No. The joint probability, $P(N \cap C)$, does not equal the product of the marginal probabilities of N and C.

27. $P(L) = 0.40$, $P(B) = 0.70$, $P(L \cap B) = 0.28$

29.

	B	B'	Total
A	0.42	0.18	0.60
A'	0.04	0.36	0.40
Total	0.46	0.54	1.00

(a) 0.46   (b) 0.36

CHAPTER 6

1. (a) $\{x \mid x = 0, 1, 2, 3, 4\}$, Discrete
   (b) $\{t \mid t \geq 0\}$, Continuous
   (c) $\{d \mid d \geq 0\}$, Continuous
   (d) $\{x \mid x = 0, 1, 2, \ldots, 10\}$, Discrete

3. (a)

x	0	1	2	3
$P(X = x)$	$\frac{1}{8}$	$\frac{3}{8}$	$\frac{3}{8}$	$\frac{1}{8}$

(b)

x	0	1	4	9
$P(X = x)$	$\frac{1}{8}$	$\frac{3}{8}$	$\frac{3}{8}$	$\frac{1}{8}$

(c)

x	3	1	-1	-3
$P(X = x)$	$\frac{1}{8}$	$\frac{3}{8}$	$\frac{3}{8}$	$\frac{1}{8}$

5. (a) $\frac{8}{50}$ (b) $\frac{33}{50}$ (c) $\frac{9}{50}$ (d) $\frac{33}{50}$

7. (b) $\frac{3}{9}$ (c) $\frac{1}{9}$ (d) $\frac{5}{9}$

11. $E(X) = 1\frac{1}{3}$, $\text{Var}(X) = 1\frac{5}{9}$
13. $E(X) = 2.8$, $\text{Var}(X) = 1.36$
15. (a) 3   (b) 3
17. median $= 3$, mode $= 4$
19. 0.764
21. $E(X) = 3$, $P(X = 3) = 0.2668$
23. $n = 6$
25. $P(X \geq 7 \mid n = 10, p = 0.25) = 0.0035$
27. (a) 0.3085   (b) 0.5328   (c) 0.1057   (d) 0.1747
29. 0.7745
31. $P(X \geq 46)$ is 0.7881 for A and 0.8790 for B
     $P(X \leq 40)$ is 0.0228 for A and 0.0475 for B

CHAPTER 7

1. 0.75
3. (a) 0.75   (b) $\frac{1}{9}$
5. 89,000
9. $\bar{x} = 75$, $s^2 = 115$
11. $\bar{x} = \$500$, $s^2 = 166,667$
13. Contest described in pb. 12
17. 0.0228
19. (a) 0.8664   (b) 0.7886
21. $16.82 \geq \mu \geq 15.18$
23. $59,548 \geq \mu \geq 58,452$
25. μ is estimated as 1000 and σ as 300
27. For $H_0: \mu \geq 2000$ and $H_1: \mu < 2000$, accept H_0 if $\bar{x} \geq 1976.7$ and accept H_1 if $\bar{x} < 1976.7$
29. (a) $H_0: \mu = 5.00$, $H_1: \mu \neq 5.00$ (b) 0.0456

CHAPTER 8

1. (a) a_2 (b) a_3

3. (a)

	Route	
Freeway Condition	*Surface Street*	*Freeway*
Clear	40	30
Conjested	40	60

(b) Freeway
(c) Surface street

5. No

7. (a) Strategy 1   (b) Strategy 2,   7 percent

9. (a)

		Number of Copies Stocked				
Demand	Prob.	10	11	12	13	14
10	0.10	$2.00	$1.70	$1.40	$1.10	$0.80
11	0.15	2.00	2.20	1.90	1.60	1.30
12	0.25	2.00	2.20	2.40	2.10	1.80
13	0.30	2.00	2.20	2.40	2.60	2.30
14	0.20	2.00	2.20	2.40	2.60	2.80

(b) 13
(c) 12
(d) 12 copies with EMV = $2.23

11. (a)

		Pairs of Shoes (in thousands)				
Demand	Prob.	1	2	3	4	5
1	0.30	$5	$ 2	−$1	−$4	−$7
2	0.30	5	10	7	4	1
3	0.20	5	10	15	12	9
4	0.10	5	10	15	20	17
5	0.10	5	10	15	20	25

(b) 3000 pairs with EMV = $7800
(c) $4200

13. $p \geq 0.25$

15. (a)

		Number of Copies Stocked			
Demand	Prob.	4	5	6	7
4	0.20	$2.00	$1.50	$1.00	$0.50
5	0.20	2.00	2.50	2.00	1.50
6	0.35	2.00	2.50	3.00	2.50
7	0.25	2.00	2.50	3.00	3.50

(b) 6
(c) $0.425
(d) 6

CHAPTER 9

1. (a) $(-4, -2, 7)$ (b) $(4, 2, -7)$
   (c) The product of two row vectors is not defined.
   (d) $(24, 14, 17)$

3. (a) $\begin{pmatrix} 14 \\ 2 \\ 8 \end{pmatrix}$ (b) $\begin{pmatrix} 4 \\ 10 \\ -2 \end{pmatrix}$ (c) $\begin{pmatrix} 2 \\ 16 \\ -6 \end{pmatrix}$ (d) $\begin{pmatrix} 15 \\ -1 \\ 10 \end{pmatrix}$

5. 192

7. (a) $\begin{pmatrix} 5 & 0 \\ 5 & 12 \\ -4 & 9 \end{pmatrix}$ (c) $\begin{pmatrix} 7 & 20 \\ 3 & 12 \\ 8 & 7 \end{pmatrix}$

9. $A(B + C) = AB + AC = \begin{pmatrix} 72 & 84 \\ -12 & -60 \\ 60 & 96 \end{pmatrix}$

11. $\begin{pmatrix} 10 & 16 & 6 \\ -10 & -16 & -6 \\ 20 & 32 & 12 \end{pmatrix}$

13. $AI = IA = \begin{pmatrix} 6 & 3 & 8 \\ 4 & 2 & 5 \\ 3 & 1 & 7 \end{pmatrix}$

15. (a) $x_1 = 4, x_2 = 6$ (b) $x_1 = 6, x_2 = 4$
    (c) $x_1 = 3, x_2 = 5, x_3 = 4$ (d) $x_1 = 8, x_2 = 4, x_3 = 6$

17. (a) $\begin{pmatrix} \frac{7}{2} & -\frac{3}{2} \\ -2 & 1 \end{pmatrix}$ (b) $\begin{pmatrix} -\frac{2}{11} & -\frac{3}{11} \\ \frac{5}{11} & \frac{2}{11} \end{pmatrix}$

    (c) $\begin{pmatrix} 2 & -3 & 1 \\ 1 & 3 & -2 \\ -1 & -1 & 1 \end{pmatrix}$ (d) $\begin{pmatrix} -2 & -1 & -1 \\ -3 & -2 & -5 \\ -1 & -1 & -5 \end{pmatrix}$

CHAPTER 10

1. (a)
$$\begin{array}{c} \\ a_1 \\ a_2 \\ a_3 \end{array} \begin{array}{ccc} a_1 & a_2 & a_3 \\ \begin{pmatrix} 0 & 1 & 0 \\ 0 & \frac{1}{2} & \frac{1}{2} \\ \frac{1}{3} & 0 & \frac{2}{3} \end{pmatrix} \end{array}$$

   (b)
$$\begin{array}{c} \\ a_1 \\ a_2 \\ a_3 \end{array} \begin{array}{ccc} a_1 & a_2 & a_3 \\ \begin{pmatrix} 0 & \frac{1}{2} & \frac{1}{2} \\ 0 & 0 & 1 \\ \frac{1}{2} & 0 & \frac{1}{2} \end{pmatrix} \end{array}$$

3. (a) $p_{11} = \frac{2}{5}, p_{21} = \frac{2}{5}, p_{31} = 1$

   (b) $p_{13} = \frac{1}{2}, p_{23} = 1, p_{31} = \frac{2}{3}$

5. (a)   (b)

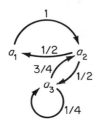 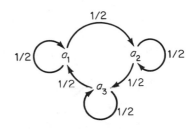

7. (a) $\begin{pmatrix} 0.51 & 0.49 \\ 0.42 & 0.58 \end{pmatrix}$ (b) $\begin{pmatrix} 0.46 & 0.54 \\ 0.30 & 0.70 \end{pmatrix}$

9. $p^{(2)} = \begin{pmatrix} \frac{1}{4} & \frac{1}{4} & \frac{1}{2} \\ \frac{1}{2} & 0 & \frac{1}{2} \\ \frac{1}{2} & \frac{1}{4} & \frac{1}{4} \end{pmatrix}$ $p^{(3)} = \begin{pmatrix} \frac{3}{8} & \frac{1}{8} & \frac{1}{2} \\ \frac{1}{2} & \frac{1}{4} & \frac{1}{4} \\ \frac{3}{8} & \frac{1}{4} & \frac{3}{8} \end{pmatrix}$

11. $t = (0.455, 0.545)$

13. 0.42

15. 0.36

17. 0.6561

19. (a) $p_1 = 0.714, p_2 = 0.286$ (b) $p_1 = 0.526, p_2 = 0.474$

21. (a) $p_1 = \frac{1}{6}, p_2 = \frac{2}{6}, p_3 = \frac{3}{6}$ (b) $p_1 = \frac{2}{7}, p_2 = \frac{1}{7}, p_3 = \frac{4}{7}$

23. 63.6 percent for Station A, 22.8 percent for Station B, and 13.6 percent for Station C.

25. Matrices b and d

27. (a) $\begin{array}{cc} & 1 \quad 3 \\ \begin{array}{c} 2 \\ 4 \end{array} & \begin{pmatrix} \frac{4}{9} & \frac{5}{9} \\ \frac{2}{3} & \frac{1}{3} \end{pmatrix} \end{array}$

   (b) For initial state 2, the expected number of steps before absorbtion is $2\frac{2}{9}$.
       For initial state 4, the expected number of steps before absorbtion is $1\frac{1}{3}$.

   (c) $\begin{array}{cc} & 2 \quad 4 \\ \begin{array}{c} 2 \\ 4 \end{array} & \begin{pmatrix} \frac{4}{3} & \frac{8}{9} \\ 0 & \frac{4}{3} \end{pmatrix} \end{array}$

29. If Mike begins with three quarters and Joe with two quarters, the probability of Mike winning all five quarters is 0.6 and of Joe winning all five quarters is 0.4.

CHAPTER 11

1. (a)

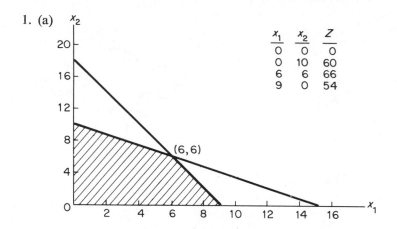

x_1	x_2	Z
0	0	0
0	10	60
6	6	66
9	0	54

Z has a maximum value of 66 when $x_1 = 6$ and $x_2 = 6$.

(c)

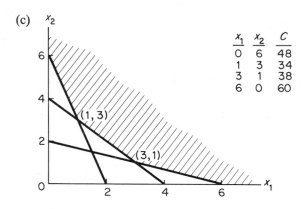

x_1	x_2	C
0	6	48
1	3	34
3	1	38
6	0	60

C has a minimum value of 34 when $x_1 = 1$ and $x_2 = 3$.

3. (a)

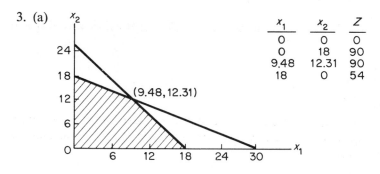

x_1	x_2	Z
0	0	0
0	18	90
9.48	12.31	90
18	0	54

Multiple optimal solutions: $x_1 = 9.48$, $x_2 = 12.31$ and $x_1 = 0$, $x_2 = 18$ are the extreme points. The maximum value of the objective function is $Z = 90$.

(c)

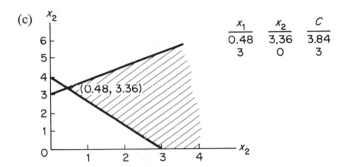

x_1	x_2	C
0.48	3.36	3.84
3	0	3

C has a minimum value of 3 when $x_1 = 3$ and $x_2 = 0$.

5. Let: x_1 = number of units of Model A
   x_2 = number of units of Model B

Maximize: $R = 10x_1 + 20x_2$

Subject to:
$$1.0x_1 + 1.2x_2 \leq 1200$$
$$0.8x_1 + 2.0x_2 \leq 1600$$
$$0.5x_1 \leq 500$$
$$x_1 \geq 200$$
$$x_2 \geq 100$$

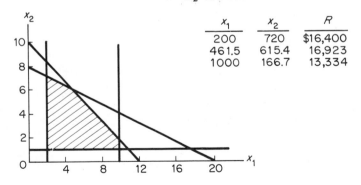

x_1	x_2	R
200	720	$16,400
461.5	615.4	16,923
1000	166.7	13,334

Maximum profit of $16,923 occurs for 461.5 units of A and 615.4 units of B.

7. Let: x_1 = number of Type A wallets
   x_2 = number of Type B wallets

Maximize: $P = 0.80x_1 + 0.60x_2$

Subject to:
$$2x_1 + x_2 \leq 1000$$
$$x_1 + x_2 \leq 800$$
$$x_1 \leq 450$$
$$x_2 \leq 700$$
$$x_1, x_2 \geq 0$$

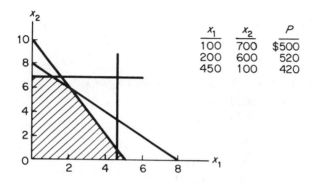

x_1	x_2	P
100	700	$500
200	600	520
450	100	420

Maximum profit of $520 occurs for 200 Type A wallets and 600 Type B wallets.

9. Let: x_1 = number of ounces of Supplement 1
   x_2 = number of ounces of Supplement 2

Minimize: $C = 0.03x_1 + 0.04x_2$

Subject to:
$$6x_1 + 12x_2 \geq 48$$
$$20x_1 + 10x_2 \geq 80$$
$$10x_1 + 10x_2 \geq 60$$
$$x_1, x_2 \geq 0$$

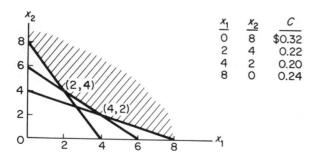

x_1	x_2	C
0	8	$0.32
2	4	0.22
4	2	0.20
8	0	0.24

C has a minimum cost of \$0.20 when 4 ounces of Supplement 1 and 2 ounces of Supplement 2 are used.

11. Let x_j = number of each product manufactured, for $j = 1, 2, 3$.

Maximize: $Z = 0.15x_1 + 0.12x_2 + 0.09x_3$

Subject to:
$$0.03x_1 + 0.02x_2 + 0.03x_3 \leq 1000$$
$$0.11x_1 + 0.14x_2 + 0.20x_3 \leq 4000$$
$$0.30x_1 + 0.20x_2 + 0.26x_3 \leq 9000$$
$$0.08x_1 + 0.07x_2 + 0.08x_3 \leq 3000$$
$$x_1 \geq 9000, x_2 \geq 9000, x_3 \geq 6000.$$

13. Let x_{ijk} represent the amount of grain i in feed j sold at price k, for $i = 1$(Wheat) and 2(Barley), $j = 1$(Fertilex) and 2(Multiplex), and $k = 1$(regular price) and 2(discount price).

Maximize: $Z = 1.10x_{111} + 0.94x_{112} + 1.23x_{211} + 1.07x_{212}$
$$+ 0.86x_{121} + 0.69x_{122} + 0.99x_{221} + 0.82x_{222}$$

Subject to:
$$x_{111} + x_{112} + x_{121} + x_{122} \leq 1000$$
$$x_{211} + x_{212} + x_{221} + x_{222} \leq 1200$$
$$x_{111} + x_{211} \leq 99$$
$$x_{121} + x_{221} \leq 99$$
$$x_{111} \geq 2x_{211}$$
$$x_{112} \geq 2x_{212}$$
$$x_{221} \geq 2x_{121}$$
$$x_{222} \geq 2x_{122}$$
$$x_{ijk} \geq 0.$$

15. Let x_{ij} represent the amount of blending component i in brand j for $i = 1, 2$ and $j = 1, 2, 3$.

Maximize: $Z = 0.23x_{11} + 0.28x_{12} + 0.33x_{13} + 0.13x_{21}$
$$+ 0.18x_{22} + 0.23x_{23}$$

Subject to:
$$x_{11} + x_{12} + x_{13} \leq 4000$$
$$x_{21} + x_{22} + x_{23} \leq 2000$$
$$10x_{11} + 50x_{21} \geq 20(x_{11} + x_{21})$$
$$10x_{12} + 50x_{22} \geq 30(x_{12} + x_{22})$$
$$10x_{13} + 50x_{23} \geq 40(x_{13} + x_{23})$$
$$x_{ij} \geq 0$$

17. Let x_{ij} represent job i scheduled during period j, for $i = 1, 2$ and $j = 1, 2$.

Assumptions: 1) Jobs scheduled during the day can be completed during the day or night.

2) Jobs scheduled during the night are constrained by the night capacity and the overflow of day jobs.

Maximize: $Z = 275(x_{11} + x_{12}) + 125(x_{21} + x_{22}) + 225(x_{31} + x_{32})$

Subject to: $1200(x_{11} + x_{12}) + 1400(x_{21} + x_{22})$
$$+ 800(x_{31} + x_{32}) \leq 13,400$$
$$20(x_{11} + x_{12}) + 15(x_{21} + x_{22})$$
$$+ 35(x_{31} + x_{32}) \leq 400$$
$$100(x_{11} + x_{12}) + 60(x_{21} + x_{22})$$
$$+ 80(x_{31} + x_{32}) \leq 1050$$
$$1200x_{12} + 1400x_{22} + 800x_{32} \leq 9200$$
$$20x_{12} + 15x_{22} + 35x_{32} \leq 250$$
$$100x_{12} + 60x_{22} + 80x_{32} \leq 650$$
$$x_{ij} \geq 0$$

19. Let x_j represent the exposures in media j, for $j = 1, 2, 3$.

Media	Effectiveness Coefficient
1	$0.80(0.4) + 0.70(0.2) + 0.15(0.4) = 0.52$
2	$0.70(0.4) + 0.80(0.2) + 0.20(0.4) = 0.52$
3	$0.20(0.4) + 0.60(0.2) + 0.40(0.4) = 0.36$

Maximize: $Z = 0.52(600,000)x_1 + 0.52(800,000)x_2 + 0.36(300,000)x_3$

Subject to: $600x_1 + 800x_2 + 450x_3 \leq 20,000$
$$x_1 \leq 12, x_2 \leq 24, x_3 \leq 12$$
$$x_1 \geq 3, x_2 \geq 6, x_3 \geq 2$$
$$x_{2j} \geq 1 \quad \text{for } j = 1, 2, 3$$
$$x_{1j} \geq 0 \quad \text{for } j = 1, 2, 3$$

21. A graph of the relationships between scores and study hours shows that a minimum of two hours of study is required in Q.M. and five hours in Mkt. and Acc. to attain the minimum grade of 70. Furthermore, the graph shows that a study hour in Mkt. is worth 2 percentage points, a study hour in Q.M. is worth 5 percentage points, and a study hour in Acc. is worth 6 percentage points.

Taking into account the weight of the exams and the number of units of each course, the linear programming problem is

Maximize: $Z = 2(.5)(2)(x_1) + 5(.25)(4)(x_2) + 6(.20)(4)(x_3)$

Subject to: $x_1 \geq 5, x_2 \geq 2, x_3 \geq 5,$
$$x_2 \leq 8, x_3 \leq 10, x_1 + x_2 + x_3 \leq 20,$$

where x_j = the number of study hours on each of the j subjects, for $j = 1(\text{Mkt.})$, $2(\text{Q.M.})$, and $3(\text{Acc.})$.

CHAPTER 12

1. (a) $x_1 = 8$, $x_2 = 0$, $Z = 24$ (c) $x_1 = 2.8$, $x_2 = 4.4$, $Z = 54.8$
3. (a) $x_1 = 13.75$, $x_2 = 15$, $x_3 = 20$, $Z = 525$
   (b) $x_1 = 0$, $x_2 = 1000$, $x_3 = 0$, $Z = 20{,}000$
5. (a) $x_1 = 4$, $x_2 = 0$, $Z = 4$ (b) $x_1 = 250$, $x_2 = 750$, $Z = 8500$
7. (a) Nonfeasible solution   (b) Unbounded solution
9. Produce no model 1200 lamps, 75 model 1201 lamps, and 375 model 1202 lamps. Total profit is $1042.40. Marginal value of assembly labor is $8.25, of wiring labor is $2.50, and of packaging labor is $0.
11. Produce 70 Toots, 0 Wheets, and 90 Honks. Total profit is $121.50. A marginal hour of assembly labor is worth $0.90 and an additional ounce of sequin powder is worth $0.10.

CHAPTER 13

1. (a)

	A_1	A_2	A_3	Available
E_1	300	200		500
E_2		100	400	500
Scheduled	300	300	400	1000

(b)

	A_1	A_2	A_3	Available
E_1		100	400	500
E_2	300	200		500
Scheduled	300	300	400	1000

The total shipping cost is $20,500.

3. (a)

	1	2	3	4	5	6	Capacity
1	0.14	0.16	0.20	0.23	0.23	0	75
2	0.16	0.13	0.18	0.19	0.20	0	100
3	0.17	0.16	0.17	0.16	0.20	0	125
Req'd.	30	30	100	50	40	50	300

(b)

	1	2	3	4	5	6	Capacity
1	30		5		40		75
2		30	70				100
3			25	50		50	125
Req'd.	30	30	100	50	40	50	300

(c)

	1	2	3	4	5	6	Capacity
1	30					45	75
2		30	25		40	5	100
3			75	50			125
Req'd.	30	30	100	50	40	50	300

The total cost is $41,350.

5. (a)

Plant	gas.	kero.	diesel	jet	asph.	slack	Capacity
1	0.165	$-M$	0.140	0.125	0.138	0	70,000
2	0.140	0.146	0.126	$-M$	0.133	0	90,000
3	$-M$	0.139	0.134	0.134	0.130	0	40,000
Req'd.	60,000	15,000	40,000	25,000	20,000	40,000	200,000

(b)

Plant	gas.	kero.	diesel	jet	asph.	slack	Capacity
1	60,000		10,000				70,000
2		15,000	15,000		20,000	40,000	90,000
3			15,000	25,000			40,000
Req'd.	60,000	15,000	40,000	25,000	20,000	40,000	200,000

The solution shown in (b) is optimal. Profit is $24,400.

7. (a)

	1	2	3	4	Slack	Capacity
1	$14	$15	$16	$17	0	30
2	M	$16	$17	$18	0	45
3	M	M	$15	$16	0	40
4	M	M	M	$17	0	25
Req'd.	20	30	40	40	10	140

(b)

	1	2	3	4	Slack	Capacity
1	20	10				30
2	—	20	0	15	10	45
3	—	—	40			40
4	—	—	—	25		25
Req'd.	20	30	40	40	10	140

The total cost is $1,895,000.

9.

	1	2	3	4	5	Supply
1		240	260	600		1100
2	220	160	540			920
3	620					620
4			200		820	1020
Req'd.	840	400	1000	600	820	3660

The total transportation cost is $8,110,000.

11. The initial reduced cost matrix, shown below, does not permit an optimal assignment.

	1	2	3	4	5	6
1	30	10	30	0	10	0
2	40	0	40	20	30	0
3	0	20	20	40	0	0
4	20	50	0	10	50	0
5	70	40	50	30	20	0
6	40	30	90	10	70	0

The second reduced cost matrix is

	1	2	3	4	5	6
1	30	10	30	0	10	10
2	40	0	40	20	30	10
3	0	20	20	40	0	10
4	20	50	0	10	50	10
5	60	30	40	20	10	0
6	30	20	80	0	60	0

Since an assignment cannot be made, an additional reduction is necessary.

	1	2	3	4	5	6
1	20	0	20	Ⓞ	0	10
2	40	Ⓞ	40	30	30	20
3	Ⓞ	20	20	50	0	20
4	20	50	Ⓞ	20	50	20
5	50	20	30	20	Ⓞ	0
6	20	10	70	0	50	Ⓞ

An optimal assignment is shown by the circled elements. The total time is 970 minutes.

13. (a) Since the problem involves maximizing the number of viewers, it is necessary to reverse the magnitudes of the entries. This reversed magnitude tableau is

	1	2	3	4	5	6	7	8
1	22	28	15	24	10	35	25	32
2	26	26	10	22	8	34	22	34
3	28	25	18	22	4	21	23	36
4	30	20	24	26	0	27	24	36
5	20	29	20	25	12	33	26	33
6	24	28	14	23	16	29	23	35
7	26	26	20	15	0	30	24	37
8	28	24	26	26	10	34	27	38

(b)

	1	2	3	4	5	6	7	8
1	3	2	3	0	1	8	1	Ⓞ
2	9	2	Ⓞ	0	1	9	0	4
3	15	5	12	4	1	Ⓞ	5	10
4	20	3	21	11	Ⓞ	9	9	13
5	Ⓞ	2	7	0	2	5	1	0
6	6	3	3	0	8	3	Ⓞ	4
7	16	9	17	Ⓞ	0	12	9	4
8	11	Ⓞ	16	4	3	9	5	7

An optimal assignment is shown by the circled elements. The total number of viewers is 175 thousand.

CHAPTER 14

1. $1338.23
3. 964 million
5. $558.40
7. Father
9. 8 percent
11. 4.5 percent
13. $1346.86
15. 8.24 percent
17. The investor would prefer 7 percent compounded annually.
19. $5488
21. $13,181
23. $14,067
25. $3793
27. $11,470
29. $PV = \$98,181$. The investor would therefore prefer the $100,000 cash payment.
31. $119.57
33. $1096

Index

fluid blending, 358-62
media selection, 362-65
portfolio selection, 362-65
product mix, 353-56
Logarithms:
base a, 474
base e, 477
common logarithm, 477
as an exponent, 474
natural or Naperian, 477
use in calculations, 477
Logic:
conjunction, 25
connectives, 24
exclusive disjunction, 25
inclusive disjunction, 25
negation, 24
and set operations, 24-28
statement, 24
Logical argument, 26-28

Marginal probability, 160
Markov chain:
absorbing, 325-31
definition, 306, 308
fundamental matrix, 327-28
k-step transition matrix, 310-16
regular transition matrix, 319
steady state probability distribution, 319,
323-24
transition diagram, 308-10
transition matrix, 308-9
transition probability, 307 8
Markov process, 307 fn, 308
Matrix:
augmented, 290
coefficient, 287
definition, 271, 277-78
dimension or order, 278
elements or components, 278
fundamental, 327-28
identity, 285
inverse, 295-97
k-step transition, 310-16
null, 285
probability, 308 fn
transition, 308-9
Matrix algebra:
addition, 279
inversion, 295-301
multiplication, 280
multiplication by a scalar, 280
row operations, 289
subtraction, 279
Maximax criterion, 251-52
Maximin criterion, 252-54
Maximum likelihood criterion, 254-55
Mean, 185-86 (*see also* Expected value)
Media selection problems, 262-65
Median, 187-89
Minimum cell solution, 422-26
Mode, 190-91
Multiplication:
of matrices, 280
of vectors, 274
Multivariate functions, 101-2
Mutually exclusive events, 149-50

$n!$, 113
n-tuple, 20
Natural logarithm, 477
Nominal interest rate, 455
Normal probability distribution:
density function, 203
mean or expected value, 203
standard deviation, 203
table of values, 516
Northwest corner solution, 420-22
Null hypothesis, 233

Null matrix, 295
Null set, 6

Objective function, 341-44
Objective probability, 151-52
Opportunity loss, 256 fn
Outer product, 283

Pascal's triangle, 133-34
Payoff table, 245
Permutations:
of distinguishable objects, 112-15
of indistinguishable objects, 115-18
with repetitions, 118-19
Point estimate, 230
Population, 215-16
Portfolio selection problems, 362-65
Present value:
of an amount, 458-59
of an annuity, 469
Principal, 455-56
Probabilistic sample, 216
Probability:
assignment of, 151-59
axioms, 153-56
conditional, 162-63
empirical, 152
joint, 160
marginal, 160
objective, 151-52
subjective, 151-53
table, 159-62
transition, 307-8
Probability function:
binomial, 203
continuous, 182-85
definition, 174, 178-79
density function, 182-85
discrete, 179-82
hypergeometric, 197-98
mass function, 179-80
normal, 203
Product mix problems, 353-56
Proper subset, 6

Quadratic formula, 81-84
Quadratic function:
establishing, 76-81
functional form, 74-75
graphing, 75-76
maximum value of, 84-87
minimum value of, 84-87

Random variable:
continuous, 176-77
definition, 175-76
discrete, 176-77
expected value or mean, 185-86
median, 187-89
mode, 190-91
standard deviation, 194
variance, 191-94
Range:
of a function, 34, 45
of a relation, 48-49
Regular transition matrix, 319
Relation:
and the Cartesian product, 48-50
definition, 48
domain of, 48-49
range of, 48-49
Risk, 246-48
Row operations, 289
Row vector, 272

Saddle point, 254 fn
Sample:
mean, 217-18
reason for, 215-16
variance, 219-20